PROGRESS IN ACTIVITY-BASED ANALYSIS

Related Books

HENSHER & BUTTON (eds.) Handbooks in Transport 6-Volume Set

AXHAUSEN (ed.) Travel Behaviour Research
DOHERTY & LEE-GOSSELIN (eds.) Integrated Land-Use and Transport Models
GÄRLING et al. (eds.) Theoretical Foundations of Travel Choice Modeling
HENSHER (ed.) Travel Behaviour Research: The Leading Edge
MAHMASSANI (ed.) Travel Behavior Research Opportunities and Applications Challenges

PROGRESS IN ACTIVITY-BASED ANALYSIS

Edited by

Harry Timmermans
Eindhoven University of Technology
The Netherlands

Emerald

United Kingdom – North America – Japan
India – Malaysia – China

Emerald Group Publishing Limited
Howard House, Wagon Lane, Bingley BD16 1WA, UK

First edition 2005

British Library Cataloguing in Publication Data
A catalogue record for this book is available from the British Library

ISBN: 978-0-08-044581-6

INVESTOR IN PEOPLE

CONTENTS

CONTENTS v

LIST OF FIGURES ix

LIST OF TABLES xiii

PREFACE xvii

Activity-Based Approaches: Models, Data and Applications
Harry Timmermans xix

CHAPTER 1
Contributions to Understanding Joint Relationships among Activity and Travel Variables
Ram M. Pendyala and Xin Ye 1

CHAPTER 2
A Dynamic Programming Approach for the Activity Generation and Scheduling Problem
Anders Karlström 25

CHAPTER 3
Simulation of Daily Activity Patterns
Ondrej Pribyl and Konstadinos G. Goulias 43

CHAPTER 4
Simulation Daily Activity Patterns through the Identification of Sequential Dependencies
Davy Janssens, Geert Wets, Tom Brijs and Koen Vanhoof 67

CHAPTER 5
Adjustments of Activity Timing and Duration in an Agent-Based Traffic Flow Simulation
Michael Balmer, Bryan Raney and Kai Nagel 91

CHAPTER 6
The Scheduling Agent -Using Sesam to Implement a Generator of Activity Programs
Guido Rindsfüser and Franziska Klügl 115

CHAPTER 7
Activity-Based Analysis of Travel Demand Using Cognitive Agents
Rosaldo J. F. Rossetti and Ronghui Liu 139

CHAPTER 8
Adaptation of Time Use Patterns to Simulated Travel Times in a Travel Demand Model
Georg Hertkorn and Peter Wagner 161

CHAPTER 9
An Integrated Framework for Modelling Short- and Long-Run Household Decision-Making
Eric Miller 175

CHAPTER 10
Strategies for Resolving Activity Scheduling Conflicts: An Empirical Analysis
Matthew Roorda and Eric Miller 203

CHAPTER 11
Sequential and Simultaneous Choice Structures for Modelling Intra-Household Interactions in Regional Travel Models
Peter Vovsha, John Gliebe, Eric Petersen and Frank Koppelman 223

CHAPTER 12
An Empirical Comparison of Alternative Models of Household Time Allocation
Junyi Zhang, Akimasa Fujiwara, Harry Timmermans and Aloys Borgers 259

CHAPTER 13
Incorporating Latent Determinants in Activity Time Allocation Modelling: Application to Value of Activity Time Estimation
K. P. Nepal, D. Fukuda and T. Yai 285

CHAPTER 14
An Analysis of Activity Type Classification and Issues Related to the *With Whom* and *For Whom* Questions of an Activity Diary
Konstadinos G. Goulias and Tae-Gyu Kim 309

CHAPTER 15
Recent Developments in Activity Diary-Based Surveys and Analysis: Some Japanese Case Studies
Kazuo Nishii, Kuniaki Sasaki, Ruichi Kitamura and Katsunao Kondo 335

CHAPTER 16
A Data Collection Strategy for Perceived and Observed Flexibility in the Spatio-Temporal Organisation of Household Activities and Associated Travel
Martin E.H. Lee-Gosselin 355

CHAPTER 17
The Design and Implementation of an Activity Scheduling Survey Using The Internet
Tomás Ruiz 373

CHAPTER 18
Activity-Based Travel Forecasting Models in The United States: Progress Since 1995 and Prospects for the Future
Peter Vovsha, Mark Bradley and John L. Bowman 389

CHAPTER 19
Two Applications of GIS-Based Activity-Travel Simulators
Nobuaki Ohmori, Noboru Harata and Katsutoshi Ohta 415

CHAPTER 20
The Relation Between Motives and Frequency of Telework: A Qualitative Study from The Oslo Region on Telework and Transport Effects
Randi J. Hjorthol 437

CHAPTER 21
Changing Travel Characteristics and Activity Travel Patterns of Households Through Telecommuting?
Andrea Glogger, Thomas Zängler and Georg Karg 457

CHAPTER 22
Communication and Travel Behaviour: Two Facets of Human Activity Patterns
Claudia Nobis, Barbara Lenz and Colin Vance 471

CHAPTER 23
Choices of Activity- and Travel-Change Options for Reduced Car Use
Peter Loukopoulos, Tommy Gärling, Cecilia Jakobsson, Solveig Meland and Satoshi Fujii 489

LIST OF FIGURES

Figure 2.1 The Probability of Departing from Home Before and After a Change from 8 a.m. *39*
Figure 3.1 An Example of an Activity Pattern *46*
Figure 3.2 Overview of the Model Estimation Phase *48*
Figure 3.3 Overview of the Simulation Phase *49*
Figure 3.4 The Principle of Simulating Joint Activities for Multiple-Adult Households *54*
Figure 3.5 An Observed Activity Profile for One-Adult Household *55*
Figure 3.6 A Simulated Activity Profile for One-Adult Household *56*
Figure 3.7 Comparison of the Average Number of Episodes Between Observed and Simulated Patterns (24 hours) *59*
Figure 3.8 Comparison of Number of Episodes Between Observed and Simulated Patterns (Conducted between 7 a.m. and 8 a.m.) *60*
Figure 3.9 Comparison of Number of Episodes between Observed and Simulated Patterns (Conducted between 8 a.m. and 9 a.m.) *60*
Figure 3.10 Comparison of Number of Episodes Between Observed and Simulated Patterns (Conducted between 5 p.m. and 6 p.m.) *61*
Figure 3.11 Comparison of Number of Episodes Between Observed and Simulated Patterns (Conducted between 6 p.m. and 7 p.m.) *61*
Figure 4.1 Overview of the Simulation Framework *70*
Figure 4.2 The Construction of Activity Bundles with Maximum Likelihood Estimates *73*
Figure 4.3 Description of the Simulation Procedure for Generating a General Skeleton *76*
Figure 4.4 Description of the Simulation Framework for Generating Time Information *77*
Figure 4.5 Description of the Simulation Framework for Generating Location Information *79*
Figure 4.6 Difference in Duration Probability Distributions for Three Discretized Intervals (5 Randomly Chosen Activities) *83*
Figure 4.7 Origin-Destination Bubble Chart for the Training Data *85*
Figure 5.1 Switzerland Network *101*
Figure 5.2 Average Travel Times of Routes-Only-Initial-Times-Extern and Times-Routes-Initial-Times-Extern *105*
Figure 5.3 Average Scores of Routes-Only-Initial-Times-Extern and Times-Routes-Initial-Times-Extern *105*
Figure 5.4 Arrival and Departure Histograms when the Initial Plans have "Plausible" Departure Times *106*
Figure 5.5 Initial Plans with Externally Defined Departure Times: Comparison to Traffic Count Data *107*

Figure 5.6 Average Travel Times of Routes-Only-Initial-Times-All6a.m. and Times-Routes-Initial-Times-All6a.m. *108*
Figure 5.7 Average Scores of Routes-Only-Initial-Times-All6a.m. and Times-Routes-Initial-Times-all6a.m. *109*
Figure 5.8 Arrival and Departure Histograms when in the Initial Plans Everybody Departs at 6 a.m. *109*
Figure 5.9 Departure Time 6 a.m. Plans: Comparison to Traffic Count Data *110*
Figure 6.1 Schematic Representation of the Main Structure and Components *122*
Figure 6.2 Schematic Representation of the Simulation Concept *123*
Figure 6.3 Schematic Representation of Observed Scheduling Behaviour of a Single Person for a Single Activity *124*
Figure 6.4 The Agents' Activity Graph *126*
Figure 6.5 A Situation within a Simulation in SeSAm (Screenshot) *131*
Figure 6.6 Examples of Generated Activity Patterns (Screenshot from Browser) *132*
Figure 7.1 A Two-Layered Architecture for the Driver Agent *144*
Figure 7.2 The Basic BDI Architecture (Wooldridge, 1999) *146*
Figure 7.3 A Basic Transport Analysis Framework *149*
Figure 7.4 MADAM+DRACULA Simulation Framework *151*
Figure 7.5 Activity Journeys with Lateness Thresholds *154*
Figure 7.6 Activity Journeys with both Relative Earliness and Lateness Thresholds *155*
Figure 7.7 Activity Journeys with both Absolute Earliness and Lateness Thresholds *156*
Figure 8.1 Structure of the Model *163*
Figure 8.2 Nodes on the First Two Levels of the CHAID-Decision Tree with Mode Choice Probabilities *164*
Figure 8.3 Adaptation of Starting Times and Durations to Actual Travel Times *167*
Figure 8.4 Population Density in the City of Cologne *170*
Figure 8.5 Left: Roads with Less Traffic; Right: Roads with More Traffic as a Consequence of the Bridge Closure *171*
Figure 9.1 Stress-based Search Process *184*
Figure 9.2 Maslow's Hierarchy of Needs *187*
Figure 9.3 The Decision-Making Unit (DMU) Agent *191*
Figure 9.4 The Decision Management Agent *193*
Figure 9.5 The Household DMU *197*
Figure 10.1 Classes of Activity Conflict *205*
Figure 10.2 The Proportion of Conflict Classes by Original Activity Duration *216*
Figure 10.3 Conflict Resolution Strategies by Original Activity Duration *216*
Figure 10.4 Precedence Rankings Assumed in TASHA vs. those Observed in CHASE *218*
Figure 11.1 Operational Classification of Intra-Household Interactions *224*
Figure 11.2 Classification of Daily Activity Patterns *227*

Figure 11.3 DAP Type Choice Structure *229*
Figure 11.4 Parallel Choice Structure Applied for DAP *231*
Figure 11.5 Example of Model Structure for Joint Non-Mandatory Activity *232*
Figure 11.6 General Choice Structure for Joint Participation *238*
Figure 11.7 General Choice Structure for Ride-Sharing *241*
Figure 11.8 Escorting Choice Tree – General Case *248*
Figure 11.9 Escorting Choice Tree for a Single Chauffeur *249*
Figure 11.10 Escorting Choice Tree for a Single Child Tour *250*
Figure 11.11 General Choice Structure for Maintenance Task Allocation *253*
Figure 12.1 Time Allocation Pattern of the Elderly and Non-Elderly People *273*
Figure 12.2 Modal Shares for the Elderly and Non-Elderly People *273*
Figure 12.3 Influence Patterns of Attributes of Households and Their Members *279*
Figure 13.1 The Activity Time Allocation Model with Latent Determinants *293*
Figure 14.1 First Example of a Portion of Husband-Wife Schedule *330*
Figure 14.2 Second Example of a Portion of Husband-Wife Schedule *330*
Figure 15.1 Share of Total Activity Time *341*
Figure 15.2 Time-Space Prism and Time Allocation by Day of the Week *344*
Figure 15.3 Average Time Use for Out-of-Home Activities *351*
Figure 16.1 Detail of OPFAST Paper Instruments, Completed and Faxed Daily by Respondents *364*
Figure 17.1 Demographic and Socio-Economic Screen *376*
Figure 17.2 Example of a Completed Daily Activity Agenda Screen *377*
Figure 17.3 Activity Calendar Screen *378*
Figure 17.4 Add Dialog Screen *380*
Figure 17.5 Fatigue Effects of Respondents: Survey Days Collected Considering Subsequent Designated Days in the Activity Calendar *381*
Figure 17.6 Fatigue Effects of Respondents: type of Activity Calendar Assigned by the Number of Survey Days Completed *381*
Figure 17.7 Average Number of Observed Daily Activities by Day of the Week and Activity Type *384*
Figure 17.8 Average Number of Observed Daily Trips by Day of the Week and Mode Type *384*
Figure 18.1 Household Activity and Travel Model Hierarchy for Model Systems in Development *401*
Figure 19.1 Concept of GIS Data Use in GIS-Based Activity-Travel Simulator (SMAP) *418*
Figure 19.2 Basic Structure of GIS-Based Activity-Travel Simulator (SMAP) *419*
Figure 19.3 Representation of Activity-Travel Patterns and Available Opportunities Within a Space-Time Prism in SMAP-E *423*
Figure 19.4 Representation of the Feasibility of Engaging in the Target Activity on SMAP-E *424*

Figure 19.5 Operational Variables in the Simulation Exercises in SMAP-E *424*
Figure 19.6 Snapshot of SMAP-L *428*
Figure 19.7 Activity Scheduling in SMAP-L *428*
Figure 19.8 Modification of Activity Schedule *430*
Figure 21.1 Temporal Daily Trip Distribution of Telecommuters by Activity Group *465*
Figure 22.1 Correlation Between Number of Out-Door-Activities and ICT Use *477*
Figure 22.2 Correlation of Communication and Travel Behaviour Depending on Age *477*
Figure 22.3 Correlation of New Media (Internet and Mobile Phone) Use and Travel Behaviour *479*
Figure 22.4 Flexibility in Time for Different Leisure Activities *480*
Figure 22.5 Assessment of ICT Statements Depending on Intensity of ICT Use *481*

List of Tables

Table 1.1 Measures of Fit for Joint Activity Episode Timing-Duration Models *16*
Table 1.2 Measures of Fit for Joint Activity Episode Timing-Mode Choice Models *19*
Table 1.3 Measures of Fit for Joint Non-Work Tour Complexity-Mode Choice Models *21*
Table 1.4 Measures of Fit for Joint Work Tour Complexity-Mode Choice Models *22*
Table 3.1 The Average MSE and P-Values for the Evaluation of Activities Profiles *57*
Table 3.2 Strength of Relationship Between the Observed and Simulated Patterns *58*
Table 3.3 Similarity of the Frequency of the Number of Activities Between Observed and Simulated Patterns *62*
Table 4.1 A First-Order Example of Activity Bundles *74*
Table 4.2 The Final First-Order Transition Probability Matrix *75*
Table 4.3 An Example of a 5 x 5 Second-Order Probability Matrix *81*
Table 4.4 Comparing the Observed and Predicted Mean Number of Tours by Differentiating Between the Order of Activity Bundles *82*
Table 4.5 Comparing Observed and Simulation Probability Distributions of Departure Times for the Training and Test Set *84*
Table 5.1 Bias and Error of Routes-Only-Initial-Times-Extern and Times-Routes-Initial-Times-Extern Compared to Field Data at 7-8 a.m. *107*
Table 5.2 Bias and Error of Routes-Only-Initial-Times-All6a.m. and Times-Routes-Initial-Times-All6a.m. Compared to Field Data at 7-8 a.m. *110*
Table 7.1 Tolerances at Arrival Time *152*
Table 7.2 Base Belief Entries *153*
Table 8.1 Travel Demand by Residence on the Western or Eastern side of Rhiver Rhine for Scenario A: Normal Traffic Network Conditions and Scenario B: Deutzer Brücke Closed *169*
Table 8.2 Trips Crossing the Rhine by Residential Location *171*
Table 10.1 A Typology of Priority and Preference *206*
Table 10.2 Original Activities Ordered by Probability of Modification / Deletion *210*
Table 10.3 Optimal Precedence Ranking for Conflicting Activities *212*
Table 10.4 Optimal Precedence Rankings by Age and Household Income *213*
Table 10.5 Strategies Used to Resolve Conflicts *214*
Table 10.6 Conflict Resolution Strategies by Conflict Class *214*
Table 10.7 Conflict Resolution Strategies by Activity Type *215*
Table 11.1 Coordinated DAP Model – A Matrix View *228*
Table 11.2 Joint Participation Model – A Matrix View *233*

Table 11.3 Ride-Sharing Model – A Matrix View *242*
Table 11.4 Escorting Tour Construction Model – A Matrix View *246*
Table 11.5 Task Allocation Model – A Matrix View *251*
Table 12.1 Estimation Results of Multi-Linear Household Time Allocation Model *275*
Table 12.2 Estimation Results of Iso-Elastic Household Time Allocation Model *276*
Table 12.3 Correlation between the Observed and Estimated Activity Time *277*
Table 12.4 Proportion of Zero Observations of Activity Time *277*
Table 12.5 Relative Influence of Travel Time by Mode *278*
Table 13.1 Summary of Activities and Groups *299*
Table 13.2 Descriptive Statistics of Daily Time Allocations (Minutes) *299*
Table 13.3 Structural Equations Parameters *301*
Table 13.4 Measurement Equations Parameters *302*
Table 13.5 Summary of LISREL Goodness of Fit Statistics *302*
Table 13.6 Parameters of Activity Time Allocation Models without and with Latent Variables *303*
Table 13.7 Values of Activity Time (Yen/minute) *304*
Table 13.8 Values of Activity Time with Respect to Wage Rate *304*
Table 14.1 A Selection of Sample Characteristics *312*
Table 14.2 Codes for With Whom and for Whom Questions in the Activity Diary *313*
Table 14.3 Daily Average Time Allocation for the With Whom Question by Gender *314*
Table 14.4 Daily Average Episode Frequencies "With Whom" by Gender *315*
Table 14.5 Daily Average Number of Trips Among Persons Making at Least One Trip *316*
Table 14.6 Daily Averages for the For Whom Question by Gender *318*
Table 14.7 Daily Average Episode Frequencies "for whom" by Gender *318*
Table 14.8 Daily Average Number of Trips Among Persons Making at Least a Trip (for Whom) *319*
Table 14.9 Solo and Joint Activity Clusters *323*
Table 14.10 Average Membership Probabilities for Solo and Joint Activity Clusters *325*
Table 14.11 Self-Serving and Altruistic Activity Clusters *327*
Table 14.12 Average Membership Probabilities for Self-Serving and Altruistic Activity Clusters *328*
Table 14.13 Within Cluster Membership Probabilities and Composition *329*
Table 15.1 Changes in Socio-Economic Circumstances and the Corresponding Issues in Activity-Based Analysis *337*
Table 15.2 General Purposes of Data Collection and Analytical Issues in AD Surveys *338*
Table 15.3 Types of Questionnaire Sheet in the Pre-AD Survey *343*
Table 15.4 Trip Production Rate on Weekdays by Age and Gender *344*
Table 15.5 Participation Rate and Average Duration by Day of the Week *345*

Table 15.6 AD Surveys on Activity Patterns of Night Express Bus Users *346*
Table 15.7 Average Participation Rate by Activity and Time Interval *348*
Table 15.8 Average Activity Duration by Trip Purpose (in Minutes) *348*
Table 15.9 Question Items and Contents of SCAT *350*
Table 15.10 Frequencies of Telecommunication by Day of the Week *350*
Table 15.11 Distribution of Telecommunications by Content and Mode *350*
Table 16.1 Perceived Spatio-Temporal Flexibility of Activities *363*
Table 17.1 Total Response *377*
Table 17.2 Activities and Trips per Respondent per Day for Planned and Executed Schedules per Type of Activity Calendar Completed *382*
Table 17.3 Scheduling Steps per Respondent per Day in Different Time Horizons for those whose Executed Schedule differs from Previously Planned Schedule *382*
Table 17.4 Modified/Added Attributes in Modify-Decisions per Respondent per Day x 10^{-2} in Different Time Horizons for those whose Executed Schedule differs from the Planned Schedule *383*
Table 18.1 Key Features of Activity-Based Model Systems Developed for Selected U.S. Metropolitan Planning Organizations since 1995 *394*
Table 18.2 Modelled Activity-Travel Purpose and Generation/Participation Categories *399*
Table 19.1 Comparison of the Three Versions of SMAP *420*
Table 19.2 Scheduling Patterns and Elements of Pre-Planned Activity Schedules *430*
Table 20.1 Sample Characteristics and Frequency of Teleworking *444*
Table 20.2 Relation between Motives and Transport Effects *450*
Table 21.1 Comparison of Average Trip Frequency and Travel Distance per Telecommuter and Working Week *461*
Table 21.2 Comparison of Average Trip Frequency and Travel Distance per Household Member and Working Week *461*
Table 21.3 Comparison of Average Trip Frequency and Travel Distance per Household and Working Week *461*
Table 21.4 Comparison of Travel Time of Telecommuters by Activity Group *462*
Table 21.5 Comparison of Travel Time of Households by Activity Group *463*
Table 21.6 Comparison of Travel Time of Telecommuters for Commuting by Mode of Transport *463*
Table 21.7 Comparison of Travel Time of Households for Commuting by Mode of Transport *464*
Table 21.8 Comparison of Average Trip and Tour Frequency of Household Members by Survey Day *466*
Table 21.9 Comparison of Tours of Household Members by Number of Out-of-Home Activities and Survey Day *466*

Table 22.1 Number of Locations and Average Trip Distances Depending on Intensity of ICT Use *482*
Table 22.2 Relation Between Communication and Mobility Types *483*
Table 22.3 Socio-Demographic Characteristics of Communication and Mobility Groups *484*
Table 22.4 Parameter Values of Variables Used in the Linear Regression Analysis *485*
Table 22.5 Results of the Linear Regression Analysis *486*
Table 23.1 Adaptation Alternatives and their Hypothesized Associated Costs *492*
Table 23.2 Sample Descriptives *494*
Table 23.3 Degree of Change by Percentage of Respondents Choosing Each Change Option, by Gender, Mean Age, and Mean Income *495*
Table 23.4 Results of Multinomial Logit Analyses *496*

PREFACE

In 1995, we organised an international workshop on activity-based analysis. After a slow start in the 1980s, the interest in activity-based models seemed to increase rapidly and several individual scholars and groups had just started building their new models. The 1995 conference saw Bowman and Ben Akiva's nested logit model, Pendyala and Kitamura's Amos model, the first micro-simulation models, using representative activity-travel schedules, Kwan's constrained-based GISICAS and several other examples of prototype models and analyses. Results were reported in Ettema and Timmermans (eds.) *Activity-Based Approaches to Travel Analysis*, Pergamon, which was published in 1997.

Now, after a decade, a lot has happened. Some of the models mentioned above have now been used in practice. Conventional nested logit models have been gradually replaced by more flexible structures. Bhat has introduced such an advanced econometric system for travel demand analysis. Several other micro-simulation models have been developed and some of these have already been applied in practice. Whereas in 1995, rule-based systems were only briefly mentioned, Arentze and Timmermans' Albatross system provides evidence that multi-agent, rule-based systems offer a viable alternative to the more commonly used utility-maximizing models. In addition to these and many other modeling efforts, new technology has provided new opportunities to collect the data required for activity-based analysis. In additional to the collection of conventional activity-travel diary data, interest in collecting data about scheduling and re-scheduling behaviour has increased.

In other words, many of the original ideas associated with activity-based analysis have matured, and hence the timing seemed right to organise another conference. This volume contains some selected papers that were originally presented at this conference, which took place in Maastricht, The Netherlands, May 28-31, 2004. All these papers went through a review process before being accepted for this volume.

Theo Arentze (Eindhoven University of Technology) and Geert Wets (LUC) joined me in organising this conference. Kai Axhausen (ETH, IATBR), Chandra Bhat (University of Texas at Austin, TRB Commission A1C102)), Kostas Goulias (UCSB, TRB Taskforce on Moving Activity-Based Models to Practice) and Ram Pendyala (University of South Florida, TRB Commission A1C104) served on the advisory committee.

The papers included in the volume indicate that indeed much progress has been made in the field of activity-based analysis and modelling. The first models are now being applied in practice. In addition, and perhaps more importantly, several of the chapters in this book evidence that new topics are being addressed, holding much promise and excitement for the near future.

Harry Timmermans
Editor

Eindhoven University of Technology
Urban Planning Group
Eindhoven
February 2005

ACTIVITY-BASED APPROACHES: MODELS, DATA AND APPLICATIONS

Harry J.P. Timmermans, Eindhoven University of Technology, Eindhoven, The Netherlands

The rapidly growing interest in activity-based models and analysis in transportation research and planning can be viewed as an attempt of increasing the degree of complexity in our representation of the links between individual and household behaviour, characteristics of the urban environments, the transportation system, and the more general policy context. Several experiences with the application of conventional, simpler models of transport demand stimulated this trend. For example, the prediction of "leave-your-car-at-home-for-the-work-commute" policies turned out to be biased because other household members started to use the car that was left at home for trips they used to made on foot or by bike, leading to an overprediction of policy success. Similarly, while teleworking seemed a good remedy for reducing mobility, sometimes the effects were less positive or even negative because individuals and households were forced or given an opportunity to reschedule their activities and related travel patterns. These and other examples illustrate that predictions of travel demand, based on an analysis of single day travel patterns may be seriously biased because there are various kinds of direct and indirect effects between household members, times of days, days of the week, and activity types.

Over the last decades, transportation research has therefore seen a gradual shift from trip-based, via tour-based to activity-based models. Many operational models are based on the well-know utility-maximising, nested logit (discrete choice) framework, adding additional nests to the specification of the model, and extending the number of variables. Other scholars argued that such algebraic models have some inherent limitations in capturing the kind of complexity that is necessary, and developed computational process and (agent-based) micro-simulation models.

The trend towards increased complexity and detail also led to a need for better and more detailed data. Although some activity-based models have been estimated using conventional travel surveys, activity-travel diaries are required if the model also takes into account the duration of the activity and possible substitution between in-home and out-of-home activities. If, in addition, the focus shifts from long-term, equilibrium forecasts to short-term dynamics, there is also the need for data about behavioural change and activity-travel rescheduling behaviour. To the extent that the models simulate continuous activity-travel patterns, as opposed to broader time-of-day effects, data are required on a much more refined time scale. New technology, such as cellular phones and global positioning systems, provided researchers with alternative ways of collecting such data. The attempt of improving behavioural realism also led to data collection efforts to capture activity scheduling and rescheduling processes.

Although the level of sophistication in data collection, analysis and modelling of travel patterns has certainly increased, there is no a priori guarantee that indeed the accuracy of our forecasts will improve. Moreover, like people, transport planners also show some reluctance to change, sometimes for good reasons. Hence, before activity-based approaches become standard practice in transportation planning, academics will have to show that their new set of toys indeed improve the accuracy of travel forecasts, or allow policy-makers to address issues that cannot be addressed by conventional four-step approaches.

MODELS

The tendency of incorporating increased complexity in activity-based models can be detected in several aspects of these models. In the first chapter of this volume, *Pendyala and Ye* address the issue of jointly modelling several dimensions of activity and travel behaviour in simultaneous equations frameworks in an attempt to overcome the sequential choice mechanism implied in the traditional four-step travel demand modelling process. Based on previous model estimation efforts, the chapter documents the nature of the joint relationship between activity episode scheduling and duration, trip chaining and mode choice, and activity episode scheduling and mode choice. The results of the model estimation suggest that there are distinct relationships among these variables and that these relationships vary by market segment.

An element of increased complexity concerns the incorporation of uncertainty in activity scheduling decisions. Operational activity-based models typically implicitly or explicitly assume that individuals maximise their utility on the basis of perfect and full information, which of course is not a very realistic assumption. In Chapter 2, *Karlström* explores the applicability of a dynamic programming approach based on micro-economic utility to incorporate uncertainty in activity scheduling decisions. The model is formulated and the results of a computational experiment, showing the potential value of the suggested approach, are reported. Questions still open to future research concern the problem whether the approach can also be applied to complex schedules and whether the model can be generalised to household decision-making.

On of the common approaches in activity-based modelling of travel demand has been to extract typical activity profiles from data and use these in a micro-simulation to predict activity patterns in a new area or for some future point in time. The next two chapters in this volume report progress in this research tradition. First, in Chapter 3, *Pribyl and Goulias* propose a model that simulates individual's daily activity-travel patterns, incorporating the interactions among members of particular household. Cluster analysis is used to classify activity patterns. Decision trees, particularly the CHAID algorithm, are used to take into account the personal and household characteristics. The model is evaluated using data from the Centre County, Pennsylvania that were collected during the Fall of 2002 and Spring of 2003.

Whereas the focus in Chapter 3 is on incorporating interactions among household members, in Chapter 4, *Janssens, Wets, Brijs and Vanhoof* develop an approach for capturing sequential information and dependencies that are present in activity patterns. This information is used to generate a skeleton of activities and transport modes that is used for simulating time and location information. The suggested approaches for generating these facets are heuristic in nature and easy to comprehend. It is shown that the simulation results of the different facets are satisfactory and that the heuristic framework developed, when further elaborated and generalized to multiple facets, is a viable candidate for extracting activity profiles that can be used in micro-simulation approaches.

As a particular type of micro-simulation approach, agent-based simulation is rapidly gaining interest in transportation research. The following three chapters provide examples of multi-agent systems, illustrating their potential. In Chapter 5, *Balmer, Raney and Nagel* propose in the context of route choice behaviour to a truly agent-based representation of the traffic system and the assignment process, in which each person remains individually identifiable throughout the whole simulation process. This approach allows a completely consistent modelling of the behavioural processes related to transportation. Their models allows to call arbitrary "strategy generation" modules, which are able to use individualized (or aggregated) performance to update plans of each agent, which can be fed back into a traffic micro-simulation. They demonstrate, besides the use of a routing module, a time allocation module, which makes travellers adjust the timing of all their activities, i.e. not just departure time, throughout the day. Validation and sensitivity tests are discussed.

Rindsfüser and Klügl discuss in Chapter 6 the application of the Sesam multi-agent platform to the activity scheduling problem. The implementation of "The Scheduling Agent" as a single agent, organising an individual activity program in a multi-agent simulation system is based on the concept of time-dependent gathering and evaluation of information about dynamic environmental features and the agents' own attributes followed by the agents' decisions. With every time step during the simulation run the attributes of all agents (including a world with resources – also modelled as classes of single agents) are updated, based on specific model assumptions. During the simulation every new day starts with the day-dependent habitual program chosen to be executed. During a simulated day each agent continuously evaluates his situation (status of the environment, status of his own attributes, etc.). Then, the agent has some options to act (operate), based on several rules and assumptions as well as on a few models of individual behaviour. The realised agents' behaviour results in an observable activity sequence (pattern).

Another interesting application is described in Chapter 7, where *Rossetti and Liu* report on the use of cognitive agents to support the activity-based analysis of demand generation in urban traffic networks. Demand is regarded as the result of the cognitive process carried out by each individual of the population. The approach proposed allows for the diversity of activity parameters that may influence daily journey decision-making. Preliminary experiments aim at demonstrating the potential of the cognitive model in designing and implementing activity-based travel demand. Emphasis is given to analysing scheduled delay at activity destinations accounting for arrival time constraints, while different selection behaviours for departure time are simulated within a framework that combines a multi-agent demand model with the microscopic representation of the movement. Simulation results show that drivers are more likely to meet activity arrival constraints with an absolute earliness-lateness tolerance window than when tolerance thresholds are relative to travel time.

The relationship between departure time and traffic flows is modelled by *Hertkorn and Wagner*. An application of their model is described in Chapter 8. An activity-based microscopic travel demand model is presented in which travel demand is derived from observed diary data. The data are used to generate activity patterns with variable start times and durations for the different episodes. The spatial environment of the travellers is described by the positions of locations where activities can be performed. Based on these data and on travel times for the different modes, the destinations for trips are determined together with the travel modes. The flexibility of start times is used to adapt the schedules to the travel times that result from the actual destination and mode choice. The travel demand model generates trip tables for a traffic flow simulation to compute travel times. Using these in another run of the travel demand simulation, a feedback loop is established. The effect of the feedback between travel demand and traffic network performance is investigated in a case study for the City of Cologne. Travel demand and mode choice predicted by the model are discussed in a scenario in which one of the bridges over the River Rhine in the city centre is closed.

Progress in activity-based modelling is also reflected in recent advances in integrated land use – transportation modelling. In Chapter 9, *Miller* discusses the theoretical underpinning of the ILUTE system. It has very interesting features. Activity agendas are formulated around projects, which keep together a series of related activities, required to accomplish a particular goal. The household is the central decision-making unit, implying that links between activities can be modelled in an integrated fashion. The concept of stress is proposed to model dynamics in activity scheduling. Finally, heuristics are assumed to model activity scheduling behaviour. This is accomplished by a model called TASHA. In Chapter 10, *Roorda and Miller* report the results of empirical analyses, conducted to find rules for activity rescheduling decisions in response to scheduling conflicts that can be used as input to this model. Although further research is needed, the results suggest that such an approach is promising.

Intra-household interactions has long been recognized to constitute an important aspect in modelling activity-travel patterns. Although there has been some quantitative analyses of such interactions, only recently a rapidly growing body of research on various aspects of modelling intra-household interactions and group decision making mechanisms as well as first attempts to incorporate intra-household interactions in regional travel demand models can be observed. In this volume, two chapters address this problem. First, in Chapter 11, *Vosvsha, Gliebe, Peterson and Koppelman* propose a general framework for incorporating intra-household interactions in regional travel demand models. The proposed approach distinguishes between three principal levels of intra-household interactions: Coordinated principal daily pattern types, Episodic joint activity and travel, and Intra-household allocation of maintenance activities. Model structures are discussed, with emphasis on the advantages of the simultaneous approach relative to the sequential method, as well as implications for practical applications. In Chapter 12, *Zhang, Fujiwara, Timmermans and Borgers* compare the results of two of their models (a multilinear model and an iso-elastic model) in a study of household time allocation in small towns in Japan. Both models have a high goodness-of-fit. The interpretation of the estimated parameters between the two models however would suggest differences in relative influence of household members.

Chapter 13 is also concerned with modelling time allocation, but in this case of individuals. *Nepal, Fukuda and Yai* apply the latent class time allocation framework, suggested by Ben Akiva and his co-workers to estimate the value of activity time. The parameters of the latent variables turned out to be highly significant, and the estimates of the values of activity time for different activities were realistic. The findings suggest that this model, which incorporates latent determinants of time allocation in a traditional activity time allocation model, is valuable not only for modelling activity time allocation, but also in calculating the value of activity time.

In Chapter 14, *Goulias and Kim* argue that the formulation and specification of activity analysis models require better understanding of time allocation behaviour that goes beyond the more recent within household analyses. In addition, since very little is known about perceived selfish and altruistic behaviour, we need analyses that explain how these affect travel behaviour and time allocation. They report the results of analyses, using the CentreSIM survey, a two-day time use/activity diary of more than 1400 persons. In particular, they analysed answers to with whom and for whom questions to identify differences within a day and among the different days of a week, accounting for person and household characteristics. Significant differences between solo and joint participation and between self-serving and altruistic behaviour were observed among the persons that work in different ways (part time and full time), among the different school age children, and among persons that may appear to have reasons to stay home.

DATA

The second set of chapters in this volume is concerned with aspects of data collection. Chapter 15 represents the shift from models to data collection. *Nishii, Sasaki, Kitamura and Kondo* discuss three activity-travel diaries collected in Japan, and illustrate the richness of such data for analysis and modelling. Especially, the analysis of time use offers answers to questions that are impossible to addresses with traditional travel surveys, focusing on trip characteristics.

As indicated before, developments in activity-based modelling have led to an increased interest in collection data of decision processes underlying the organisation of activities in time and space. In Chapter 16, *Lee-Gosselin* argues that an appreciation of those decision processes requires observations at the individual and household levels, with due regard to interactions beyond the household, within social networks and with a wide range of other sources of activity opportunity and constraint. He discusses OPFAST, an instrument package developed and implemented in a longitudinal panel survey in Quebec City, Canada. It was designed to be complementary to the use of the method known as CHASE in an overall strategy involving a parallel panel survey in Toronto. Different methods were used for subsequent waves, and these are briefly described. OPFAST places particular emphasis on recording respondents' perceptions of the flexibility available to them in organising their activities in time and space. Initial inferences are made from the implemented panel surveys about advantages, difficulties and costs. It is concluded that there is a case for combined strategies using both CHASE and OPFAST.

Ruiz in Chapter 17 discusses a similar example. In particular, he describes how the Internet can be used to collect activity-travel scheduling data in the short and medium term. A web-based questionnaire was designed, where respondents were asked to record their scheduling decisions for one to four non-consecutive days as they were adopted over time. Respondents were first contacted by e-mail, in which they were asked to participate in the survey. The e-mail provided a hyperlink to the web-based questionnaire. Participants who entered the website were first requested to provide the data about their demographic and socio-economic characteristics. Second, respondents were asked to make a plan based on their activities and associated travels for several non-consecutive days. Finally, respondents corrected their planned agenda through adding, deleting or modifying activities and travels in order to describe the real executed schedule. Overall, the results showed the typical low-response rate for an Internet survey. The analysis of the data suggested a number of interesting results concerning to different characteristics of the scheduling process according to the time horizon.

APPLICATIONS

The final series of chapter discusses applications or specific themes associated with activity-based analysis. First, In Chapter 18, *Vovsha, Bradley and Bowman* review the applications of activity-based models in the United States. Various types of integrity are viewed as the main advantage of activity-based models. Applications in Portland, San Francisco, New York and Columbus are discussed in light of these potential advantages. In addition, micro-simulation and a greater level of detail are identified as strong components of activity-based approaches. Finally, misconceptions and concerns related to the application of activity-based approaches are highlighted.

Whereas the previous chapter is most of all concerned with utility-maximising models, constrained models have traditionally been developed in the time geography tradition. In Chapter 19, *Ohmori, Harata and Ohta* discuss the development and application of GIS-based activity-travel simulators. One application was developed especially for the purpose of instructing students in understanding the theory of space-time prisms/accessibility and travel behaviour under spatio-temporal constraints. A second application involved a decision-support system for activity planning using interactive surveys to collect information about the activity scheduling process of tourists' leisure activities. These two applications illustrate the usefulness of geographic information systems for these types of applications.

Activity-based approaches are especially useful for particular problems. One of these problems is the substitution between travel and in-home activities. The next two chapters of this volume are both concerned with teleworking. *Hjorthol* in Chapter 20 discusses the results of a qualitative study, which sheds some lights on factors influencing teleworking and underlying motives. *Glogger, Zängler and Karg*, in Chapter 21, report the results of a quantitative analyses. Interestingly, they conclude that telecommuting does not reduce distances travelled, casting doubt on the effectiveness of such policies.

Nobis, Lenz and Vance's contribution, described in Chapter 22, is concerned with the larger issue of ICT and travel. Their basic idea is that one of the major achievements of recent technological developments, both in transport and communication, is people's greater spatial and temporal flexibility. Using the data of the first wave of the 'DLR's ICT and Mobility Panel', they explore the relationship between communication and mobility behaviour. The data show that within all age groups a high number of trips correlates with high ICT use. Generally, a distinction of ICT devices should be made as the effect of the particular media can be different. In exploring the questions of what factors influence travel demand and the nature of the role played by ICT use in this context, the expectation is confirmed that young and employed men are more likely to have a high travel demand. However, the influence is considerably less strong than generally assumed. Instead, the use of ICT, and in particular the mobile phone, turned out to be the strongest influencing factor in a regression model presented in the paper.

Finally, Chapter 23 addresses behavioural change in car use. *Loukopoulos, Gärling, Jakobsson, Meland and Fujii* outline a conceptual framework based on self-regulation theory with the purpose of analyzing adaptations of household car use. Changes in car-use options resulting from the implementation of policies designed to reduce car use are assumed to influence long-term, strategic choices of car-use reduction or change goals as well as the day-to-day, operational choices of activity/travel change options in order to attain these goals. Analyses are reported of retrospective survey data on activity/travel changes after the introduction of a toll ring in Trondheim, Norway.

1

CONTRIBUTIONS TO UNDERSTANDING JOINT RELATIONSHIPS AMONG ACTIVITY AND TRAVEL VARIABLES

Ram M. Pendyala, University of South Florida, Tampa, USA

Xin Ye, University of South Florida, Tampa, USA

INTRODUCTION

Understanding joint relationships among multiple endogenous variables has been of much interest to researchers in the field of activity and travel behaviour modelling (Fujii and Kitamura, 2000; Golob, 2003). Structural equations models and econometric simultaneous equations models have been developed and presented in the literature with a view to unravelling the joint causal relationships among activity and travel behaviour variables. Bhat (1997, 1998, 2001) and Bhat and Singh (2000) conducted a series of research that aims at integrating ordered/unordered discrete and continuous endogenous variables into econometric simultaneous equations modelling frameworks thus providing the means of studying relationships among mixed variables.

Based on much of the work in the recent past on joint estimation of simultaneous equations models while accommodating flexible error covariance structures, the authors have further studied the joint relationships among several activity and travel variables involving unordered discrete variables and continuous variables. This chapter presents a series of joint econometric model formulations, in which discrete and continuous endogenous variables are included. The modelling methods presented in this paper directly contribute to a deeper understanding of the joint simultaneous relationships and mutual interactions among several activity and travel variables. In particular, the

Progress in Activity-Based Analysis edited by H. Timmermans

authors use a variety of approaches to infer the nature of the relationships between the following pairs of variables: (i) Maintenance activity timing and activity episode duration; (ii) Non-work activity/trip timing and mode choice, and (iii) Activity sequencing (trip chaining) and mode choice.

The joint nature of maintenance activity timing and episode duration may be considered in two ways. On the one hand, the timing of an activity may dictate its duration. For example, if an activity is being pursued in the peak period, then it may be of a shorter duration. On the other hand, the duration of activity may dictate its timing. For example, activities of longer duration may be undertaken during off-peak periods. There are then two possible causal relationships linking activity timing and duration.

Similarly, there are two possible causal structures linking non-work activity timing and mode choice. On the one hand, the timing of an activity may dictate the mode chosen for travelling to that activity. A non-work trip in the off-peak period may be undertaken by automobile because there is little congestion and transit service may be limited. On the other hand, the mode chosen may dictate the timing of a non-work activity. For example, the choice of the automobile may lead to an activity being undertaken in the off-peak period, say, to avoid congestion.

Finally, one can also postulate two possible causal structures between the complexity of trip chaining (or activity sequencing) and mode choice. On the one hand, the complexity of the trip chain or activity sequence may motivate the selection of the automobile as the mode of transport. On the other hand, the selection of the automobile as the mode of transport may motivate the undertaking of complex tours and trip chains. Thus, there are two possible causal structures between the complexity of trip chaining and mode choice.

The preceding discussion illustrates the need for a deeper understanding of the causal relationships underlying activity and travel characteristics. In the absence of such an understanding, activity-based travel modelling systems cannot accurately portray and reflect joint relationships among endogenous variables. Consequently, forecasts and policy impacts estimated from such models may be highly erroneous.

It must be noted that there is considerable uncertainty with respect to the extent to which causal relationships can be inferred from quantitative travel survey data using econometric and statistical modelling methods. A true understanding of causal relationships and decision processes can be probably be best obtained using qualitative research methods that probe behavioural processes and provide detailed process data. As such, the results documented in this paper are subject to further research and validation using more detailed process-oriented methodologies.

The remainder of this chapter is organized as follows. The next section describes the modelling methodologies employed in this paper. The third section describes the data sets used for model estimation. Model estimation results are presented in the fourth section while concluding remarks are furnished in the final section of the paper.

MODELLING METHODOLOGIES

This section presents the modelling methodologies that can be employed in the context of examining joint relationships among endogenous variables.

Discrete-Continuous Econometric Modelling Framework

An application to the analysis between activity timing and activity episode duration is presented here. The presentation follows that in Pendyala and Bhat (2004).

Let i be an index for time of day of activity participation ($i = 1, 2,\ldots, I$) and let q be an index for observations ($q = 1, 2,\ldots, Q$). Consider the following equation system:

$$\begin{aligned} u_{qi}^* &= \beta_i' z_{qi} + \gamma_i a_q + \varepsilon_{qi} \\ a_q &= \theta' x_q + \delta' D_q + \omega_q \end{aligned} \qquad (1)$$

$\varepsilon_{qi} \sim$ i.i.d. Gumbel(0,1), $\omega_q \sim \mathrm{N}(0,\sigma^2)$.

where u_{qi}^* is the indirect (latent) utility associated with the ith time of day for the qth observation, D_q is a vector of the time of day dummy variables of length I, δ is a column vector of coefficients, i.e. ($\delta_1, \delta_2 \ldots \delta_I$), representing the effects of different times of the day of activity participation on activity duration, ε_{qi} is a standard extreme-value (Gumbel) distributed error term assumed to be independently and identically distributed across times of the day and observations, a_q is the logarithm of activity duration (to address the possibility of negative durations) and γ is its coefficient. The error term ω_q is assumed to be i.i.d. normally distributed across observations with a mean of zero and variance of σ^2. In Equation 1, the time of day alternative i will be chosen (i.e., D_{qi} =1) if the utility of that alternative is the maximum of I alternatives. Defining

$$v_{qi} = \max_{j=1,2,\ldots,I,\ j\neq i} u^{*}_{qj} - \varepsilon_{qi}\,, \tag{2}$$

the utility maximizing condition for the choice of the ith alternative may be written as: $D_{qi} = 1$ if and only if $\beta_i' z_{qi} > v_{qi}$. Let $F_i(v_{qi})$ represent the marginal distribution function of v_{qi} implied by the assumed IID extreme value distribution for the error terms ε_{qi} (i=1,2,…,I). Using the properties that the maximum over identically distributed extreme value random terms is extreme value distributed and the difference of two identically distributed extreme values terms is logistically distributed, the implied distribution for v_{qi} may be derived as:

$$F_i(y) = \Pr(v_{qi} < y) = \frac{\exp(y)}{\exp(y) + \sum_{j\neq i} \exp(\beta_j' z_{qj})} \tag{3}$$

Therefore,

$$\Pr(D_{qi} = 1) = F_i(\beta_i' z_{qi} + \gamma_i a_q) \tag{4}$$

$$\Pr(D_{qi} = 0) = 1 - F_i(\beta_i' z_{qi} + \gamma_i a_q) \tag{5}$$

Both $F_i(y)$ and $\Phi^{-1}(y)$ (inverse of standard normal cumulative distribution function) are monotone increasing functions, so

$$\Pr(D_{qi} = 1) = \Pr[\beta_i' z_{qi} + \gamma_i a_q > v_{qi}] = \Pr\{\Phi^{-1}[F_i(\beta_i' z_{qi} + \gamma_i a_q)] > \Phi^{-1}[F_i(v_{qi})]\}, \tag{6}$$

Let $v^{*}_{qi} = \Phi^{-1}[F_i(v_{qi})]$, then

$$\Pr(D_{qi} = 1) = \Pr\{\Phi^{-1}[F_i(\beta_i' z_{qi} + \gamma_i a_q)] > v^{*}_{qi}\}, \tag{7}$$

It can be easily shown that $v_{qi}{}^{*}$ is standard normally distributed. One can introduce a new latent variable:

$$D^{*}_{qi} = \Phi^{-1}[F_i(\beta_i' z_{qi} + \gamma_i a_q)] - v^{*}_{qi}\,, \tag{8}$$

which is able to indicate binary response of D_{qi} since

$$\Pr(D_{qi}^* > 0) = \Pr(\Phi^{-1}[\mathrm{F_i}(\beta_i' z_{qi} + \gamma_i a_q)] - v_{qi}^* > 0) = \Pr(D_{qi} = 1) \tag{9}$$

$$\Pr(D_{qi}^* < 0) = \Pr(\Phi^{-1}[\mathrm{F_i}(\beta_i' z_{qi} + \gamma_i a_q)] - v_{qi}^* < 0) = \Pr(D_{qi} = 0) \tag{10}$$

Equation system (1) may now be rewritten as:

$$D_{qi}^* = \Phi^{-1}[\mathrm{F_i}(\beta_i' z_{qi} + \gamma_i a_q)] - v_{qi}^*,\ D_q = 0 \ \textit{if}\ D_{qi}^* < 0,\ D_q = 1 \ \textit{if}\ D_{qi}^* > 0$$

$$a_q = \theta' x_q + \delta' D_q + \omega_q \tag{11}$$

A correlation ρ_i between the error terms v_{qi}^* and ω_q is allowed to accommodate common unobserved factors influencing the time of day choice for activity participation and the duration of the participation. Since a_q is partially determined by ω_q and v_{qi}^* is correlated with ω_q if ρ_i is unequal to zero, a_q is apparently correlated with random error term v_{qi}^* in the first equation. Similarly, D_q is also correlated with random error term ω_q in the second equation. The endogenous nature of dependent variables D_q and a_q entails the full-information maximum likelihood method to jointly estimate their corresponding parameters γ and δ. Limited-information maximum likelihood estimation (sequential estimation) does not provide consistent estimators for the coefficients of endogenous variables.

In Equations (4), replacing a_q with the second equation of (1), one obtains:

$$\Pr(D_{qi} = 1) = \mathrm{F_i}(\beta_i' z_{qi} + \gamma_i \theta' x_q + \gamma_i \delta_i + \gamma_i \omega_q) \tag{12}$$

Similarly, it can be shown that

$$\Pr(D_{qi} = 0) = 1 - \mathrm{F_i}(\beta_i' z_{qi} + \gamma_i \theta' x_q + \gamma_i \delta_j + \gamma_i \omega_q) \ \textit{if}\ D_{qj} = 1 \tag{13}$$

$$\Pr(D_{qi} = 1) + \Pr(D_{qi} = 0) = 1, \text{ then } \gamma_i \delta_i = \gamma_i \delta_j \tag{14}$$

Three possible restrictions may be imposed on the modelling coefficients to satisfy Equation (14):

1) $\gamma_i \neq 0$ and $\delta_i = \delta_j \neq 0$, which implies that the continuous variable appears in the right hand side of the equation for the discrete choice and a vector of dummy variables corresponding to the discrete choice also appear in the model for the continuous variable. However, the coefficients on the dummy variables must be mutually identical. The modelling specification constraint by

this condition is practically meaningless, since discrete variables ought to have varied impacts on the continuous variable and thus have unequal coefficients.

2) $\gamma_i \neq 0$ and $\delta_i = \delta_j = 0$, which implies that the continuous variable appears in the utility function of the discrete choice variable but the discrete choice variable does not appear in the model for the continuous variable. This restriction will lead to a recursive structure for the endogenous variables, where the continuous variable is predetermined and then influences the discrete variable.

3) $\gamma_i = 0$, in which case Equation (14) is always satisfied; then δ_i and δ_j can take any unequal values. This restriction will lead to the other recursive structure, where the discrete variable is predetermined and then influences the continuous variable.

Accordingly, the condition of logical consistency only allows two alternative recursive structures. The first is the case where $\gamma \neq 0$ and $\delta = 0$, i.e., the continuous variable affects the discrete variable. In this case, the continuous endogenous variable a_q is predetermined and appears as an explanatory variable in the utility functions u_{qi}^*. The full-information likelihood function for estimating parameters in this case is equal to:

$$L = \prod_{q=1}^{Q} \left\{ \prod_{i=1}^{I} \left[\frac{1}{\sigma} \phi(l_q) \Phi(b_{qi}) \right]^{D_{qi}} \right\}, \tag{15}$$

where $\phi(.)$ is the standard normal density function, and l_q and b_{qi} are defined as follows:

$$l_q = \left(\frac{a_q - \theta' x_q}{\sigma} \right), \quad b_{qi} = \left(\frac{\Phi^{-1} \mathrm{F}_i (\beta_i' z_{qi} + \gamma a_q) - \rho_i l_q}{\sqrt{1 - \rho_i^2}} \right) \tag{16}$$

The second case is when $\gamma = 0$ and $\delta \neq 0$, i.e., the discrete variable affects the continuous variable. In this case, the vector of discrete variables D_q is predetermined and serves as an explanatory variable vector in the linear model for the continuous endogenous variable, a_q.

The full-information likelihood function is the same as equation (15), but here

$$l_q = \left(\frac{a_q - \theta' x_q - \delta' D_q}{\sigma} \right), \quad b_{qi} = \left(\frac{\Phi^{-1} \mathrm{F}_i (\beta_i' z_{qi}) - \rho_i l_q}{\sqrt{1 - \rho_i^2}} \right) \tag{17}$$

Recursive Simultaneous Bivariate Probit Model

This modelling methodology is utilized in the application to the analysis of the relationship between trip timing and mode choice and between trip chain type and mode choice. The discussion here follows that presented in Tringides *et al.* (2004). If the trip's departure time choice (peak vs. off-peak) and mode choice (SOV vs. non-SOV) are treated as two binary choices, the bivariate probit model can be formulated at the trip level to simultaneously analyze their probabilities with accommodation of random error correlation. The general formulation is as follows:

$$\begin{cases} M_q^* = \gamma' z_q + \alpha T_q + \varepsilon_q \\ \\ T_q^* = \beta' x_q + \eta M_q + \omega_q \end{cases} \tag{18}$$

where,

q is an index for observations of trips (q = 1, 2, … , Q);
M_q^* and T_q^* are latent variables representing the mode choice and departure time for trip q, respectively;
M_q = 1 if $M_q^* > 0$, = 0 otherwise (i.e., M_q is a dummy variable indicating whether trip q uses the SOV mode);
T_q = 1, if $T_q^* > 0$, = 0 otherwise (i.e., T_q is a dummy variable indicating whether trip q is made in the peak period);
z_q and x_q are vectors of explanatory variables for M_q^* and T_q^*, respectively;
γ, β are two vectors of model coefficients associated with the explanatory variables z_q and x_q, respectively;
α is a scalar coefficient for T_q to measure the impact of departure time choice on mode choice;
η is a scalar coefficient for M_q to measure the impact of mode choice on departure time choice;
ε_q and ω_q are random error terms, which are standard bivariate normally distributed with zero means, unit variances, and correlation ρ, i.e. $\varepsilon_q, \omega_q \sim \phi_2(0,0,1,1,\rho)$.

Based on this normality assumption, one can derive the probability of each possible combination of binary choices for trip q:

$$\Pr(M = 0, T = 0) = \Phi_2[-\gamma' z, -\beta' x, \rho] \tag{19}$$

$$\Pr(M = 1, T = 0) = \Phi_1[-(\beta' x + \eta)] - \Phi_2[-\gamma' z, -(\beta' x + \eta), \rho] \tag{20}$$

$$\Pr(M=0,T=1)=\Phi_1[-(\gamma' z+\alpha)]-\Phi_2[-(\gamma' z+\alpha),-\beta' x,\rho] \tag{21}$$

$$\Pr(M=1,T=1)=1-\Phi_1[-(\gamma' z+\alpha)]-\Phi_1[-(\beta' x+\eta)] \\ +\Phi_2[-(\gamma' z+\alpha),-(\beta' x+\eta),\rho] \tag{22}$$

where $\Phi_1[\cdot]$ and $\Phi_2[\cdot]$ are cumulative distribution function for standard univariate and bivariate normal distribution, respectively.

The sum of the probabilities for the four combinations of two binary choices should be equal to one, i.e.,

$$\Pr(M=0,T=0)+\Pr(M=1,T=0)+\Pr(M=0,T=1)+\Pr(M=1,T=1)=1 \tag{23}$$

Substituting equations (19) through (22) into equation (23), it can be shown that

$$\Phi_2[-\gamma' z,-\beta' x,\rho]+\Phi_2[-(\gamma' z+\alpha),-(\beta' x+\eta),\rho]$$

$$=\Phi_2[-\gamma' z,-(\beta' x+\eta),\rho]+\Phi_2[-(\gamma' z+\alpha),-\beta' x,\rho] \tag{24}$$

This equation does not hold unless either α or η is equal to zero. This requirement, known as the logical consistency condition, will lead to two different recursive simultaneous modelling structures (Maddala, 1983), suggesting two different causal relationships. The first structure is defined by $\alpha=0$, $\eta\neq 0$ and implies that mode choice influences departure time choice. Thus:

$$\begin{cases} M_q^* = \gamma' z_q + \varepsilon_q \\ T_q^* = \beta' x_q + \eta M_q + \omega_q \end{cases} \tag{25}$$

In this structure, mode choice is predetermined as per the first functional relationship. Then, the choice of mode is specified as a dummy variable in the second functional relationship for departure time choice to directly measure the impact of mode choice on time-of-day choice.

In the second structure $\alpha\neq 0$ and $\eta=0$, implying that departure time choice influences mode choice. Thus,

$$\begin{cases} M_q^* = \gamma' z_q + \alpha T_q + \varepsilon_q \\ T_q^* = \beta' x_q + \omega_q \end{cases} \tag{26}$$

Departure time choice is predetermined as per the second functional relationship. The trip departure time is specified as an explanatory variable influencing mode choice as per the first functional relationship.

Thus, the desirable feature of the bivariate probit model in which the coefficients of two endogenous dummy variables do not coexist in both functional relationships provides an appropriate modelling framework to analyze the causality between trip departure time and mode choice. The endogenous nature of one of the dependent variables in the simultaneous equation system can be ignored in formulating the likelihood function. To facilitate formulating likelihood functions, equations (19) through (22) can be rewritten in a format including only the cumulative distribution function of the standard bivariate normal distribution (Greene, 2003). The corresponding likelihood functions can be summarized by the following general formulations for the two different unidirectional causal structures:

1) $\alpha = 0$, $\eta \neq 0$ (Mode Choice $\rightarrow$ Departure Time Choice)

$$L = \prod_{q=1}^{Q} \{\Phi_2[\mu_q \gamma' z_q, \tau_q(\beta' x_q + \eta M_q), \mu_q \tau_q \rho]\} \tag{27}$$

2) $\alpha \neq 0$, $\eta = 0$ (Departure Time Choice $\rightarrow$ Mode Choice)

$$L = \prod_{q=1}^{Q} \{\Phi_2[\mu_q(\gamma' z_q + \alpha T_q), \tau_q \beta' x_q, \mu_q \tau_q \rho]\} \tag{28}$$

where $\mu_q = 2M_q - 1$ and $\tau_q = 2T_q - 1$.

Simultaneous Logit Model

This modelling methodology is presented in the context of an application to the analysis of the joint relationship between trip chaining type choice and mode choice (i.e., trip chain type choice $\leftrightarrow$ mode choice). The bivariate probit model only allows the unidirectional causal relationship between two binary choices. To accommodate plausible bidirectional causality, the simultaneous logit model can be applied (Schmidt *et al.*, 1975; Ouyang *et al.*, 2002). The simultaneous logit model is introduced in the context of trip chaining type choice and mode choice. This modelling methodology may be considered an extension of the multinomial logit model commonly used in transportation modelling practice. In the simultaneous logit model, the logarithm of the ratio of probabilities for two alternatives to be selected from one choice set is assumed to equal a linear combination of a set of explanatory variables. One dummy variable indicating the choice of tour complexity may be added into the set of explanatory variables for mode choice. Similarly, one dummy variable indicating mode choice may be added into the set of explanatory variables for tour complexity. The model may be formulated as follows:

$$\ln\left[\frac{\Pr(M_q = 1 \mid T_q)}{\Pr(M_q = 0 \mid T_q)}\right] = \gamma' z_q + \alpha T_q \tag{29}$$

$$\ln\left[\frac{\Pr(T_q = 1 \mid M_q)}{\Pr(T_q = 0 \mid M_q)}\right] = \beta' x_q + \eta M_q \tag{30}$$

where

q is an index for observations of tour ($q = 1, 2, \dots, Q$);
M_q is a dummy variable indicating whether tour q uses the auto mode;
T_q is a dummy variable indicating whether tour q is complex;
z_q is a vector of explanatory variables for M_q;
x_q is a vector of explanatory variables for T_q;
γ, β are two vectors of model coefficients associated with the explanatory variables z_q and x_q, respectively;
α is a scalar coefficient for T_q to measure the impact of tour's complexity on mode choice;
η is a scalar coefficient for M_q to measure the impact of mode choice on the choice of tour complexity.

By rewriting equations (29) and (30) across two possible values that T_q and M_q can take, one gets:

$$\ln\left[\frac{\Pr(M_q = 1, T_q = 0)}{\Pr(M_q = 0, T_q = 0)}\right] = \gamma' z_q \tag{31}$$

$$\ln\left[\frac{\Pr(M_q = 1, T_q = 1)}{\Pr(M_q = 0, T_q = 1)}\right] = \gamma' z_q + \alpha \tag{32}$$

$$\ln\left[\frac{\Pr(T_q = 1, M_q = 0)}{\Pr(T_q = 0, M_q = 0)}\right] = \beta' x_q \tag{33}$$

$$\ln\left[\frac{\Pr(T_q = 1, M_q = 1)}{\Pr(T_q = 0, M_q = 1)}\right] = \beta' x_q + \eta \tag{34}$$

The sum of the probabilities for the four combinations of binary choices should be equal to one, i.e.,

$$\begin{aligned}&\Pr(M_q = 0, T_q = 0) + \Pr(M_q = 1, T_q = 0)\\&+ \Pr(M_q = 0, T_q = 1) + \Pr(M_q = 1, T_q = 1) = 1\end{aligned} \tag{35}$$

By converting simultaneous equations (31) through (34), it can be shown that

$$\begin{aligned}&\Pr(M_q = 1, T_q = 1) = \Pr(M_q = 0, T_q = 0)\exp(\gamma' z_q + \beta' x_q + \alpha)\\&= \Pr(M_q = 0, T_q = 0)\exp(\gamma' z_q + \beta' x_q + \eta)\end{aligned} \tag{36}$$

For logical consistency, α must be equal to η. Endogenous dummy variables T_q and M_q are allowed to coexist in the simultaneous equation system. By replacing η with α and solving the simultaneous equations (31) through (35), the probability for each combination is formulated as follows:

$$P_{00q} = \Pr(M_q = 0, T_q = 0) = 1/\Delta_q \tag{37}$$

$$P_{10q} = \Pr(M_q = 1, T_q = 0) = \exp(\gamma' z_q)/\Delta_q \tag{38}$$

$$P_{01q} = \Pr(M_q = 0, T_q = 1) = \exp(\beta' x_q)/\Delta_q \tag{39}$$

$$P_{11q} = \Pr(M = 1, T = 1)_q = \exp(\gamma' z_q + \beta' x_q + \alpha)/\Delta_q \tag{40}$$

where,

$$\Delta_q = 1 + \exp(\gamma' z_q) + \exp(\beta' x_q) + \exp(\gamma' z_q + \beta' x_q + \alpha) \tag{41}$$

Finally, the likelihood function may be formulated as follows:

$$L = \prod_{q=1}^{Q} (P_{00q})^{(1-M_q)(1-T_q)} (P_{10q})^{M_q(1-T_q)} (P_{01q})^{(1-M_q)T_q} (P_{11q})^{M_q T_q} \tag{42}$$

DATA SETS

This section describes the data sets that were used for estimating the joint model systems and inferring the nature of the causal relationships among selected activity and travel variables.

Data Set for Analyzing Relationship between Activity Timing and Episode Duration

The data set for the analysis between activity timing and activity episode duration is derived from a comprehensive household travel survey that was administered in 1996 in the Tampa Bay Region of Florida. The survey was a traditional trip diary survey and was not an activity or time use survey. The survey was a mail-out mail-back survey that collected household and person socio-economic and demographic characteristics together with detailed information about all trips undertaken over a 24 hour period. Households were asked to return one complete diary for every household member (including children). After extensive checking and data integrity screening, a final respondent sample of 5261 households was obtained. From these 5261 households, a total of 9066 persons returned usable trip diaries. The 9066 persons reported information for a total of 31459 trips (through the 24 hour trip diary). The trip file was used to create an out-of-home activity file where individual activity records were created from the trip records. This activity file included information about activity type, activity timing, activity duration, and other variables pertinent to each activity episode. Based on a time of day distribution of all trips in the data set, four distinct time periods were identified. They are:

- AM peak : 7:15 AM – 9:15 AM
- Midday : 9:16 AM – 3:15 PM
- PM peak : 3:16 PM – 6:15 PM
- Off peak : 6:16 PM – 7:14 AM

Maintenance activities included the following three activity (trip) types: Shopping, personal business, and errands; Medical/dental; Serve passenger or child. These activity records were extracted from the original file to create two maintenance activity record files, one for commuters and one for non-commuters. Commuters were defined as driving age individuals who commuted to a work place on the travel diary day, while non-commuters were defined as driving age individuals who did not commute to a work place (made zero work trips) on the travel diary day. Note that a worker (employed person) who did not commute on the travel diary day would still be classified as a non-commuter for the purpose of this analysis. Also, children under the age of 16 were excluded from the analysis completely. Maintenance activity records that had full information (no missing data) were extracted to create commuter and non-commuter data files for the modelling effort in this paper.

Maintenance activities were pursued by 2904 individuals residing in 2386 households. Of these individuals, 1023 were commuters and they reported 1351 maintenance activities. The remaining 1881 individuals were non-commuters and they reported 2899 maintenance activities. The commuter and non-commuter maintenance activity episode data sets included complete socio-economic and activity information for the respective samples.

Data Set for Analyzing Relationship between Non-Work Trip Timing and Mode Choice

The dataset used to analyze the relation between trip timing and mode choice is drawn from the Southeast Florida Regional Household Travel Survey, which was conducted during 1999 in Miami-Dade, Broward, and Palm Beach counties. Households agreeing to participate in the survey were mailed a survey package including a 24-hour travel diary for each member of the household. As with most household travel surveys, this survey collected detailed socio-demographic and trip information for each person in the household. The survey provided a respondent sample of 11,426 persons reporting a total of 33,082 trips. The socio-economic, demographic, and travel characteristics of the respondent sample were generally consistent with those of the population in the region. The analysis focuses on the relationship between time-of-day choice and mode choice for non-work trips made by adults. For this reason, all non-work trips made by persons 18 years of age or older were extracted from the original dataset. In addition, the analysis distinguishes between workers (employed) and non-workers (unemployed) in an attempt to capture the effect of potential differences in temporal and modal choice flexibility between these two groups. For example, workers might link their non-work trips to the commute while non-workers might make use of their travel flexibility to avoid congestion during peak hours. From the original trip data set, all non-work trips that had complete information including household and person socio-economic data, trip attribute data, and modal level of service data were extracted. This subsample of trips included a total of 14,410 non-work trips of which 7,947 were made by 2,710 workers and 6,463 were made by 1,741 non-workers.

Data Set for Analyzing Relationship between Trip Chaining and Mode Choice

The data set used for analysis of relation between trip chaining and mode choice is extracted from the Swiss Travel Microcensus 2000. The survey respondent sample consists of 27,918 households from 26 cantons in Switzerland. The person sample was formed by randomly selecting one person over 6 years old from each household with less than 4 household members and two persons over 6 years old from each household with 4 or more members. As a result of this sampling scheme, the person respondent sample consisted of 29,407 persons. All of the persons in the person sample were asked to report their travel in a one-day trip diary. The resulting trip data set includes 103,376 trips reported by 29,407 interviewed persons (including the possibility of some respondents making zero trips on the survey day).

Data corresponding to respondents from the Canton of Zurich was extracted to reduce the data to a more manageable size and to control for possible area specific effects. The Zurich subsample includes 5,128 households from which 5,241 persons provided travel information. A trip chain is defined in this analysis as a complete home-to-home journey where the origin of the first trip is home and the destination of the last trip is home. No intermediate home stop can exist within a trip chain. Whenever the home location is reached, a chain is formed. A tour-level data set was formed by aggregating the trip data set to the tour level. All person and household characteristics were merged into the tour level data set. In most cases, a single mode was prevalent for the trip chain. In cases where multiple modes were prevalent within the same trip chain or tour, a single mode was assigned based on the whether or not the auto mode was used in the chain. If the auto mode was used for any segment in the trip chain, then the chain was assigned an auto mode. Each tour was classified as a simple or complex tour depending on whether it had one intermediate stop or more than one intermediate stop within the chain. In addition, tours were also classified as work-based tours and non work-based tours. Any tour that included a work stop (regardless of the presence of other types of stops) was classified as a work-based tour while any tour that included only non-work stops was classified as a non work-based tour. It was felt that the causal relationships governing work-based tours may be different from those governing non work-based tours. This is because the presence of a work stop may impose a certain amount of spatial and temporal rigidity on the activity/travel behaviour of the individual in the context of that tour. The constraints associated with the work activity may lead to a different causal structure underlying trip chain formation and mode choice. The Zurich subsample included 4,901 non-work tours and 1,711 work tours.

MODEL ESTIMATION RESULTS

In this section, the major findings from the model estimation results are presented. Detailed model specifications and estimation results may be found for the activity timing-duration model systems in Pendyala and Bhat (2004), activity timing-mode choice model systems in Tringides *et al.* (2004), and trip chaining-mode choice model systems in Ye and Pendyala (2004).

Activity Episode Timing and Duration

Non-commuters: *Activity timing* → *Activity episode duration.* The time of day indicators are all statistically significant in the model. Relative to the off-peak period, all other periods are characterized by shorter maintenance activity episodes as reflected by the negative coefficients. The midday indicator has the most negative coefficient suggesting that the maintenance activity episodes in this period are the shortest. The error correlation between midday activity participation and activity duration is the only statistically significant error correlation.

Non-commuters: *Activity episode duration* → *Activity timing.* The activity duration variable affects time of day choice in the AM peak and PM peak activity participation equations. It does not enter the midday activity participation equation. The coefficients associated with the duration variable are positive indicating that non-commuters are not constrained with respect to the lengths of their activity episodes in these time periods. Also, the error correlation between midday activity participation and activity duration is the only statistically significant error correlation.

An assessment of the causal structures is performed using the goodness-of-fit measures as both model structures offered plausible results. The adjusted likelihood ratio index at zero is computed as:

$$\bar{\rho}_0^2 = 1 - \frac{L(\beta) - k}{L(0)} \tag{43}$$

where, k is the number of parameters as shown in the table. Similarly, the likelihood ratio index at sample shares is computed as:

$$\bar{\rho}_c^2 = 1 - \frac{L(\beta) - k}{L(C)} \tag{44}$$

Table 1.1
Measures of Fit for Joint Activity Episode Timing-Duration Models

	Non-Commuter Model		Commuter Model	
Summary Statistic	**Time of Day → Duration**	**Duration → Time of Day**	**Time of Day → Duration**	**Duration → Time of Day**
Sample size	2899	2899	1351	1351
Number of parameters[a]	22	21	21	22
Log-Likelihood				
At convergence	-5584.14	-5316.82	-2990.86	-2990.60
At market share[b]	-5756.11	-5756.11	-3058.38	-3058.38
At zero[c]	-7072.00	-7072.00	-3216.60	-3216.60
Adjusted Likelihood Ratio				
$\bar{\rho}_0^2$	0.207	0.245	0.064	0.063
$\bar{\rho}_c^2$	0.027	0.073	0.016	0.016

[a] The number of parameters does not include the constant and variance term in the log-linear duration model.

[b] The log-likelihood at sample shares corresponds to the likelihood function value of the joint model with only alternative specific constants in the MNL time of day model, and with only the constant and variance term in the log-linear duration equation. All correlation terms are zero.

[c] The log-likelihood at zero corresponds to the likelihood function value of the joint model with no variables in the MNL time of day model, and with only the constant and variance (standard deviation) term in the log-linear duration equation. All correlation terms are zero.

A comparison of the adjusted likelihood ratio indices, as shown in Table 1.1, provides a mechanism for comparing two non-nested models. Ben-Akiva and Lerman (1985) note that, for estimations involving more than 250 observations, if the adjusted likelihood ratio indices differ by more than 0.01, then the model with the lower index is almost certainly the incorrect model. As the difference in indices for the non-commuter models is substantially greater than 0.01, it may be safely concluded that the model representing the causal structure where duration affects time of day is the more appropriate one. Thus, activity episode duration drives activity timing (scheduling) for maintenance activities of non-commuters.

Commuters. In the log-linear duration model, time-of-day variables are significant in influencing activity duration. As expected, episode duration tends to be longer in the off-peak period as reflected by the positive coefficient associated with the off-peak period indicator. In the causal structure where activity duration affects time of day choice is considered, activity duration is found to be significant in all time period equations. Relative to the off-peak period, the negative coefficients associated with activity duration suggest that there is lower propensity to pursue longer activity episodes in the AM-peak, midday, or PM-peak periods. Within these three periods, the lowest propensity is seen in the AM-peak period and the highest propensity is seen in the midday periods. It is found that none of the error correlation terms are statistically significant regardless of the causal structure considered. Thus, in effect, the joint model reduces to an independent model system where time of day and duration models may be estimated separately in a sequential fashion. This finding suggests that there is only a loose relationship between time of day choice and activity episode duration for commuters.

The comparison between two alternative causal structures shows that the adjusted likelihood ratio indices, as shown in Table 1.1, are very similar and have differences less than the 0.01 value required to help identify the correct model. In addition, the model fit is substantially poorer than the fits obtained in the context of the non-commuter samples. These findings coupled with the finding that none of the error correlation terms are statistically significant suggest that time of day choice and activity episode duration are correlated albeit with only a loose causal relationship between them. It does not appear that one decision precedes or necessarily determines the other. This conclusion may be explained by the fact that work schedules tend to dictate time of day participation and activity durations for commuters. Thus, activity episode duration and activity timing of maintenance activities are only loosely related to one another for commuters and may be modelled independently in a single-equation framework.

Non-Work Trip Timing and Mode Choice

Workers. In the model where departure time choice is assumed to affect mode choice (timing → mode choice), it is found that the dummy variable representing peak period departure time choice significantly affects the choice of SOV as the mode for non-work trips. The coefficient is negative indicating that a departure time choice in the peak period tends to lower the propensity to drive alone for non-work trips. There are two important possible explanations for this. First, it is possible that peak period non-work trips are primarily serve passenger trips where a worker is dropping off or picking up a child at school or day care on the way to and from work. As nearly one-half of the households in the sample have at least one child, this is likely to be a strong explanation for this relationship. Second, it is possible that some workers are choosing to use alternative modes of transportation for their non-work trips to avoid the frustration of driving alone in congested conditions during the peak period. Random error correlation is positive and statistically significant at the 0.05 level of significance.

In the model where mode choice is assumed to affect departure time choice (mode → timing), it is found that that the SOV mode choice contributes negatively to peak period departure time choice as evidenced by a negative coefficient associated with the SOV choice variable in the departure time choice model. In addition, the random error correlation is statistically significant. These indications are consistent with those found in the first causal structure.

The adjusted likelihood ratio index as a goodness-of-fit measure can be used for testing and comparing non-nested relationships in discrete choice models. To choose between two models (say, 1 and 2), Ben-Akiva and Lerman (1985) provide a test where under the null hypothesis that model 1 is the true specification, the following holds asymptotically:

$$\Pr(\bar{\rho}_2^2 - \bar{\rho}_1^2 > z) \leq \Phi\{-[-2z\,L(0) + (K_2 - K_1)]^{1/2}\}, z > 0 \tag{45}$$

where,

$\bar{\rho}_i^2$ is the adjusted likelihood ratio index at zero for model $i = 1, 2$, as in equation (43);

K_i is the number of parameters in model i;

Φ is the standard normal cumulative distribution function;

$L(0)$ is the log-likelihood value at zero; if all N observations in the sample have all J alternatives, $L(0) = N \ln(1/J)$.

Table 1.2 shows that the difference between the adjusted likelihood ratio indices for the two worker models is 0.0002 with the model in which departure time choice precedes mode choice showing the better fit. Applying equation (45) yields a bounding probability of almost zero; therefore, it can be said with a high degree of confidence (99 percent confidence or better) that the model corresponding to the causal structure "departure time choice → mode choice" is statistically dominant in the worker sample (for non-work trips). This may be behaviourally explained by considering the typical work schedule constraints faced by workers. As workers tend to link their non-work trips with the commute to and from work, the departure time choice is predetermined in conjunction with the work schedule that takes precedence over all else. The mode choice is then simply determined by the mode that has been chosen for the commute trip as the non-work trips are part of a larger trip chaining mechanism. Thus, departure time choice (timing) decision precedes the mode choice decision for non-work trips of workers.

Non-Workers. The model in which departure time choice influences mode choice appears to reject the paradigm of simultaneity. The coefficient of the dummy endogenous variable indicating peak-period departure in the mode choice model is negative, but not statistically significant. Moreover, the random error correlation is also not statistically significant at the 0.05 level. Both of these findings indicate that this model specification does not support the notion of simultaneity in departure time and mode choice for non-work trips made by non-workers. As these findings are quite counter-intuitive, the authors feel that this causal structure is not appropriate to describe the behaviour of non-workers.

Table 1.2
Measures of Fit for Joint Activity Episode Timing-Mode Choice Models

	Workers	Workers	Non-Workers	Non-Workers
	Dep Time→Mode	**Mode→Dep Time**	**Dep Time→Mode**	**Mode→Dep Time**
Sample size	7947	7947	6463	6463
No. of parameters	18	20	20	20
Log-Likelihood				
At convergence	-9912.779	-9908.679	-7448.404	-7440.233
At market share	-10417.222	-10417.222	-7964.838	-7964.838
At zero	-11016.881	-11016.881	-8959.620	-8959.620
Likelihood Ratio				
ρ_0^2	0.1002	0.1006	0.1298	0.1308
ρ_c^2	0.0484	0.0488	0.0648	0.0659
$\bar{\rho}_0^2$	0.0986	0.0984	0.1664	0.1674
$\bar{\rho}_c^2$	0.0467	0.0465	0.0623	0.0634

Estimation results in the model where mode choice precedes time-of-day choice supports the hypothesis of simultaneity between departure time choice and mode choice. The coefficient of mode choice (SOV) in the departure time choice model is negative and statistically significant at the 0.05 level of significance. In general, the model indicates that non-workers are likely to avoid travelling in the peak period (negative constant in the departure time choice model) and using the SOV mode further contributes to avoiding the peak period. In addition, the random error correlation is positive and statistically significant at the 0.05 level of significance. The significant error correlation supports the notion of a simultaneous relationship between time of day choice and mode choice and is intuitively more consistent with travel behaviour hypotheses. Thus, from a qualitative and intuitive standpoint, it appears that the causal model in which departure time choice precedes mode choice is more applicable to workers' non-work trips while the opposite causal structure in which mode choice precedes departure time choice is more applicable to the non-worker sample.

The results of the analysis of the relationship between departure time choice and mode choice are quite intuitive and may be explained from a behavioural standpoint. One may conjecture that individuals first make decisions regarding those choices that are more constrained and then proceed to decisions that are less constrained. In the case of workers, the timing of a non-work activity is likely to be more constrained due to the rigidity of the work schedules around which non-work activities must be accommodated. Thus workers first determine when a non-work activity will be pursued and then proceed to the mode choice decision. On the other hand, non-workers are likely to be more mode constrained than time of day constrained as the worker(s) in the household may have taken the automobiles to work and so on. Then, non-workers first determine the mode that will be used for the non-work trip and then proceed to determining when they will make the trip. They are less constrained with respect to the timing decision because of the absence of rigid work schedules. Thus, mode choice decision precedes the departure time choice decision for non-work trips of non-workers.

Complexity of Trip Chaining and Mode Choice

Non-Work Tours (Workers and Non-Workers). In the model in which tour complexity influences mode choice, the coefficient for tour complexity is statistically significant and positive in the mode choice model. This lends credence to the hypothesis that the need to make a complex tour is likely to increase dependency on the auto mode. In addition, the error correlation is found to be statistically significant and this is indicative of the validity of the assumption that non-work tour complexity and mode choice should be modelled in a simultaneous equations framework. Interestingly, in the alternative recursive causal structure where mode choice influences tour type choice, it is found that mode choice significantly affects tour complexity and that the choice of auto is positively associated with the formation of complex tours. Thus it appears from this model that the choice of the automobile mode for a tour contributes positively to the formation of multi-stop trip chains. In addition, the error correlation is significant and negative. Estimation results of simultaneous logit model appear to support the notion that there is a bidirectional causality between mode choice and tour complexity. The significantly positive joint dependence parameter, α, shows the presence of significant positive correlation between auto mode choice and tour complexity.

The test of non-nested models, as described in earlier, is applied to conduct a statistical comparison between the alternative causal structures so as to identify the most appropriate simultaneous structure. As shown in Table 1.3, the difference in adjusted likelihood ratios is approximately 0.004 between the model where "tour complexity → mode choice" and the models where "mode choice → tour complexity" and "mode choice ↔ tour complexity". According to equation (45), the calculated bounding probability on the right hand side of the expression is almost zero. Thus, it may be concluded that the model "tour complexity → mode choice" is more closely capturing the causal structure underlying the relationship between mode choice and tour complexity. The significantly better goodness-of-fit of the model suggests that the causal structure where the complexity of the tour affects mode choice is statistically, and possibly behaviourally, dominant in the population for non-work tours. This result suggests that the complexity of the trip chain drives mode choice decisions for non-work tours. Thus the auto mode is chosen (or not chosen) as a consequence of the complexity of the activity agenda/schedule/pattern that the person needs to undertake.

Table 1.3
Measures of Fit for Joint Non-Work Tour Complexity-Mode Choice Models

	Tour Complexity → Mode Choice	Mode Choice → Tour Complexity	Mode Choice ↔ Tour Complexity
Sample size	4901	4901	4901
Number of parameters	20	18	19
Log-likelihood			
At convergence	-5179.689	-5207.736	-5203.343
At market share	-5734.170	-5734.170	-5734.170
At zero	-6794.229	-6794.229	-6794.229
Likelihood Ratio			
ρ_0^2	0.2376	0.2335	0.2342
ρ_c^2	0.0967	0.0918	0.0926
$\bar{\rho}_0^2$	0.2347	0.2309	0.2314
$\bar{\rho}_c^2$	0.0932	0.0887	0.0893

Work Tours (Workers Only). In the model where "tour complexity → auto mode choice", it is found that tour complexity has a positive impact on auto mode choice. This is consistent with expectations, trends in the data, and the models of non-work tours. The coefficient associated with tour complexity variable in the mode choice model is positive and statistically significant. Thus the model supports the notion that a complex tour or trip chaining pattern contributes to the choice of auto as the mode for the tour. In addition, the error correlation is negative and statistically significant, once again supporting the simultaneous equations formulation of the relationship between tour complexity and mode choice.

In the model where "auto mode choice → tour complexity", the coefficient associated with the auto mode choice variable in the tour complexity equation is statistically significant and positive indicating that the choice of auto mode contributes positively to the formation of complex multi-stop trip chains. However, unlike other models, the error correlation is statistically insignificant. Thus, this model suggests that tour complexity and mode choice can be modelled as two independent equations where mode choice affects tour complexity in a recursive unidirectional causal structure. In the simultaneous logit model for work tours, the joint dependence parameter, α, is found to be statistically significant and positive. This model supports the notion that there is a significant and positive bidirectional causal relationship between tour complexity and auto mode choice. In comparing the models, the seemingly better model "tour complexity → mode choice" has an adjusted likelihood ratio index that is only 0.001 greater than those of the models in the other two causal structures, as shown in Table 1.4.

Table 1.4
Measures of Fit for Joint Work Tour Complexity-Mode Choice Models

	Tour Complexity → Mode Choice	Mode Choice → Tour Complexity	Mode Choice ↔ Tour Complexity
Sample size	1711	1711	1711
Number of parameters	16	16	15
Log-likelihood			
At convergence	-2076.249	-2078.843	-2079.840
At market share	-2354.340	-2354.340	-2354.340
At zero	-2371.950	-2371.950	-2371.950
Likelihood Ratio			
ρ_0^2	0.1247	0.1236	0.1232
ρ_c^2	0.1181	0.1170	0.1166
$\bar{\rho}_0^2$	0.1179	0.1168	0.1168
$\bar{\rho}_c^2$	0.1113	0.1102	0.1102

The bounding probabilities, as per the right hand side of equation (45), are calculated to be 0.036 and 0.067, respectively. The statistical test rejects the causal structure where mode choice drives the complexity of the work tour (i.e., mode choice → tour complexity). However, the test fails to reject the simultaneous logit model, i.e., bidirectional simultaneous causality, at the 0.05 level of significance. In addition to this non-nested test, the insignificance of the random error correlation in the model of mode choice affecting tour complexity suggests that the assumed causal structure in that joint model may not be valid. Thus, for work tours, two possible causal structures can not be rejected from this analysis. Either, the decision to make a complex work tour tends to result in the choice of the auto mode or both of these decisions are made contemporaneously. The results suggest that for work tours, two joint relationships appear equally valid to explain the causality between complexity of trip chain and mode choice. Either the complexity of the trip chain precedes the mode choice decision or both decisions are made contemporaneously.

CONCLUSIONS

This paper summarizes a series of research efforts exploring the joint relationships among activity and travel variables including trip chaining pattern, mode choice, time-of-day choice, and activity episode duration. Joint econometric simultaneous equations modelling methodologies accommodating mixed variables and flexible error covariance structures are used for analyzing the joint relationships. A discrete-continuous model framework is used to model the relationship between a discrete choice variable (activity timing) and a continuous variable (activity duration) while the bivariate probit and simultaneous logit modelling approaches are used to analyze relationships between two binary choice variables. The similarity between discrete-continuous modelling framework and the bivariate probit model is that both endogenous variables cannot coexist in the model specification to ensure logical consistency and model identification in a full-information maximum likelihood estimation framework. This property leads to two alternative recursive structures representing two unidirectional causal relationships between the endogenous variables. The performance comparisons between the alternative recursive causal structures help identify the dominant causal relationship between the endogenous variables. In turn, a knowledge of the dominant causal relationships helps in the development of accurate activity-based travel demand model systems that intend to capture behavioural mechanisms at the level of the individual traveller. In addition, the knowledge of the true causal relationships underlying decision processes will help in the accurate assessment and impact analysis of alternative transportation policies such as variable pricing, parking pricing, and telecommuting.

The analysis and findings presented in this paper have direct implications for the development of activity-based travel demand modelling systems. In the context of individual travel behaviour micro-simulation, the development and application of these model systems calls for the ability to accurately represent causal relationships that are prevalent in the population. The analysis between trip timing and activity duration suggests that for non-commuters, the model for forecasting maintenance activity duration ought to be applied before the model for scheduling the activity. The analysis between trip timing and mode choice suggests that, for workers, the trip scheduling model may precede the mode choice model, but in contrast, for non-workers, the mode choice model may precede the trip scheduling model. The analysis between trip chaining formation and mode choice suggests that the activity agenda or tour formation step may precede the mode choice step for both non-work and work tours, although one may also use a contemporaneous relationship for the latter. Future research efforts should focus on analyzing whether these findings regarding causal relationships between these travel variables hold in other data sets as well. In addition, the modelling framework for causal analysis between two discrete choices can be extended to consider multinomial choice situations.

ACKNOWLEDGEMENTS

The authors thank Dr. Chandra Bhat, University of Texas at Austin, for his insights and help on the initial work that motivated this exploration into joint relationships among activity and travel variables. The authors gratefully acknowledge comments provided by anonymous reviewers that helped enhance the paper.

REFERENCES

Aptech (1996). *GAUSS 3.2.* Aptech Systems, Maple Valley, Washington.

Ben-Akiva, M. and S.R. Lerman (1985). *Discrete Choice Analysis: Theory and Application to Travel Demand.* The MIT Press, Cambridge, MA.

Bhat, C.R. (1997). Work travel mode choice and number of non-work commute stops. *Transpn. Res. B*, **31**, 41-54.

Bhat, C.R. (1998). A model of post-home arrival activity participation behaviour. *Transpn. Res. B*, **32**, 387-400.

Bhat, C.R. (2001). Modelling the commute activity-travel pattern of workers: formulation and empirical analysis. *Transpn. Sc.*, **35**, 61-79.

Bhat, C.R. and S.K. Singh (2000). A comprehensive daily activity-travel generation model system for workers. *Transpn. Res. A*, **34**, 1-22.

Fujii, S. and R. Kitamura (2000). Evaluation of trip-inducing effects of new freeways using a structural equations model system of commuters' time use and travel. *Transpn. Res.B*, **34**, 339-354.

Golob, T.F. (2003). Structural equation modelling for travel behaviour research. *Transpn. Res. B*, **37**, 1-25.

Greene, W.H. (2002). *LIMDEP Version 8.0: Reference Guide.* Econometric Software, Inc., Plainview, NY.

Greene, W.H. (2003). *Econometric Analysis.* Fifth Edition. Pearson Education, Inc., NJ.

Lee, L.F. (1983). Generalized econometric models with selectivity. *Econom.*, **51**, 507-512.

Maddala, G.S. (1983). *Limited-Dependent and Qualitative Variables in Econometrics.* Cambridge University Press, Cambridge, MA.

Ouyang, Y., V. Shankar and T. Yamamoto (2002). Modelling the simultaneity in injury causation in multi-vehicle collisions. *Proc. 81st Ann. Meet. Transpn. Res. B.,* Washington, D.C. (CD-Rom).

Pendyala, R.M. and C.R. Bhat (2004). An exploration of the relationship between timing and duration of maintenance activities. *Transpn.*, **31**, 429-456.

Schmidt, P. and R.P. Strauss (1975). Estimation of models with jointly dependent qualitative variables: a simultaneous logit approach. *Econom.*, **43**, 745-755.

Tringides, C.A., X. Ye and R.M. Pendyala (2004). Alternative formulations of joint model systems of departure time choice and mode choice for non-work trips. *Transpn. Res. Rec.* (in press).

Ye, X., R.M. Pendyala and G. Gottardi (2004). An exploration of the relationship between auto mode choice and complexity of trip chaining patterns. *Proc. 83rd Ann. Meet. Transpn. Res. B.,* Washington, D.C. (CD-Rom).

2

A Dynamic Programming Approach for the Activity Generation and Scheduling Problem

Anders Karlström, Royal Institute of Technology, Stockholm, Sweden

Introduction and Motivation

Travel is derived demand and a trip should not be seen in isolation. In fact, a trip may be a consequence of decisions and behaviour at a much earlier stage. As a consequence, the theoretical superiority of activity-based approaches over trip-based models has been well documented during the past several decades. The comparison between trip-based approaches and activity-based approaches has been discussed at length elsewhere (e.g., Timmermans, 2000). In characterizing the area of activity-based modelling, activity-based approaches can be partitioned into many different dimensions. For the purpose of this chapter, it is useful to distinguish between models in two different dimensions.

First, we discriminate between approaches that are founded in a microeconomic theory and those for which no such claim is made. Transport models has been used extensively in project evaluation during many decades, and it seems clear that a model has to be consistent with microeconomic theory if it should be used for cost benefit assessment and welfare economic analysis. However, travel models well rooted in microeconomic theory have been criticized to the extent that they make unrealistic behavioural assumptions. Therefore, it is useful to discriminate between different approaches also on the basis of behavioural content. That is, whether a model claims to actually model the decision process of individuals (and households) or whether the model is more confined

Progress in Activity-Based Analysis edited by H. Timmermans

to replicate the pattern of travel decisions rather than the decision process. A related distinction is whether it is assumed that individuals behave with high rationality or low rationality. This is logically disjunct from whether individuals are utility maximizers or, for that matter, whether the model is consistent with microeconomic theory.

One inherent difficulty with the econometric approach is the computational complexity of finding, in a utility maximizing framework, the optimal activity pattern, including activity generation, allocation and scheduling. The current state-of-the-art in microeconometric models has been noted to suffer from problems related to combinatorics. For instance, if there are 10 activities that can take place at 100 locations, and there are 5 modes and 100 time periods, there are 10^{13} alternative schedules facing the individual (Bowman and Ben-Akiva, 2001). Seen from this perspective, it is not surprising that current econometric models have to be constrained in some way to be tractable, for instance by not allowing for all sequences of activities. In contrast, we argue that the activity pattern problem is quite tractable without making a priori restrictions with respect to allowed trip patterns. Econometric estimation of models of similar size is regularly made in the literature on labour supply and social security.

Theoretically, first we need to develop a theoretical framework that allows for sequential decision making in an uncertain environment. The suggested approach relies on the framework of Markov decision processes, and we will demonstrate that the dynamic framework in fact makes the activity pattern problem quite manageable. That is, although the combinatorics still look daunting, we can solve the decision problem for one individual without a priori restricting the travel patterns. If we want to estimate the model econometrically, we need a probabilistic formulation of the decision problem. If we choose, for instance, a logit model as the discrete choice model, the model looks a bit similar to the state-of-art activity-based econometric models. In comparison, we will argue that there are a number of advantages asociated with the suggested approach. First, our model is computationally efficient. We do not have to constrain the activity patterns into a priori activity patterns. The dynamic programming approach is efficient in solving these kind of Markov decision processes, and the activity pattern problem is small enough such that we can allow for all possible combinations. As such, it is not combinatorics that is the limiting factor for relevant models.

Second, by formulating a Markov decision problem, we can allow for explicit sequential decision making in an uncertain environment. For instance, the travel time to work in the morning may be stochastic, although with a known distribution. This may be taken into account when deciding on departure time for the work trip in the morning. When arriving at work, this uncertainty is dissolved. If the travel time was unusual long, then a shopping trip may be postponed to the next day. The method also has the potential of modelling information. If information of expected travel conditions is provided to the individual, it may affect the activity pattern throughout the day.

Finally, a shift in departure time for work in the morning may influence the departure time from work in the afternoon.

Third, since the dynamic programming approach is computationally efficient, and we can solve the utility maximizing problem for one individual for one day, we are able to introduce between-day dependencies. That is, if an individual switches to public transportation (as a response to congestion pricing, for instance), a shopping trip on the way home from work may be postponed to the weekend, for instance. The use of different time-scales for individuals' agendas fits well with this approach. It is natural to consider planning within a day, then for several days, and longer (months, and years). In this chapter, we will demonstrate the approach by considering one day and between-days interrelations only. Note that in-between day interdependencies and uncertainty/rescheduling are usually not considered in econometric models. The topic however has attracted attention in proposals for computational process models (e.g., Gärling and Young, 2000; Doherty *et al.*, 2005). Smash (Ettema *et al.*, 2000) and Aurora (Joh *et al.*, 2002, 2003, 2004) are operational models of rescheduling behaviour. The latter model has been extended to rescheduling under uncertainty (Arentze and Timmermans, 2004) and the case of information provision (Sun *et al.*, 2005).

In this chapter, we follow the econometric approach of estimating a model that replicates the actual choices of individuals without making any claim with respect to the decision process individuals actually use. Still, we cannot overlook the serious and valid criticism that it is unrealistic to think that individuals actually are able to solve very complicated utility maximizing problems. In our case, in a household setting where there are interpersonal relations, our activity pattern problem becomes much more difficult to solve. If we have a problem that we cannot solve even if we use the most sophisticated current algorithm using the best available computer power, it is unrealistic to think that individuals are able to find the optimal solution. Therefore, we need suboptimal algorithms to solve such problems. Such algorithms (in the class of state-of-art algorithms for our class of problems) are based on algorithms that are also used in models which are not based on micro-economics, but rather based on heuristic (behavioural) algorithms, such as computational process models and rule-based approaches. Approximation methods, such as function approximation used in the reinforcement learning literature, become necessary as the dimensions of the problem grow. A rule-based approach based on decision trees (e.g., Arentze and Timmermans, 2000) may be viewed as a piecewise linear approximation of the value function in our setting. Algorithms based on reinforcement-learning, neural networks, and decision-trees have indeed proven to be efficient in finding good decision policies in the setting of Markov decision problems. These ideas will only be touched upon in this chapter.

LITERATURE

This is not the place for a detailed review of activity-based models. Excellent reviews can be found elsewhere (e.g., Timmermans, 2000; McNally, 2000; Bowman and Ben-Akiva, 2001). To put the current work into perspective it is useful to follow the argument of Nagel and Marchal (2003, pp. 20–21). First, note that the main difficulty with activity-based modelling is combinatorics. Even for one individual, with only 10 destinations, three modes, and 100 time intervals, the number of activity patterns can easily be 10^{13} (Bowman and Ben-Akiva, 2001). What an activity modeller needs therefore is good search methods. Second, the main purpose of this chapter is to argue that a problem of the above size is quite tractable using efficient search methods. We will argue that such a problem can easily be solved with dynamic programming. Third, solving the deterministic problem is much easier than solving the problem in a random utility framework. However, for activity problems of the size indicated above, for one individual, it is shown that the problem is still tractable even with a random utility formulation.

Fourth, although the one-individual problem is tractable, the household activity model is much larger in terms of computational complexity. To be able to handle such problems, it may be useful to use other search techniques. Nagel and Marchal (2003) suggested search heuristics from computer science. Indeed, one of the approaches we have tested is to use reinforcement learning algorithms with function approximation (state aggregation) to solve the activity scheduling problem. One way to solve very large dynamic programming problems is to use approximation methods. Intuitively, the indirect utility of being in a certain state at time t is not solved exactly, but rather approximated by some function. In the reinforcement learning literature, many such approximations have been used. Reinforcement learning has been also used in the activity based literature, but most often such methods has been confined to the route choice problem, rather than the activity scheduling problem (Charypar *et al.*, 2004). We are not aware of such algorithms for the scheduling problem. On the other hand, rule-based methods using decision trees has been advocated for the activity scheduling problem, which may be seen as a way of solving the DP problem with function approximation (see Bertsekas and Tsitsiklis, 1996).

A DYNAMIC MICRO-ECONOMIC FRAMEWORK

In this section we will develop the micro-economic framework for sequential decision making in an uncertain environment. We will here use the principle of dynamic programming (DP) for solving Markov decision problems (MDP). In the economic literature, dynamic programming has been used to model decision making in a dynamic setting with uncertainties, for instance in modelling labour supply, savings and retirement decision (Rust and Phelan, 1997; French, 2001, and Karlstrom *et al.*

2004), investment decisions and durable consumption (Rust, 1987). Discussions of dynamic programming models in econometrics and economics can be found in Rust (1994, 1996) and Adda and Cooper (2003).

In the following we will formulate the activity pattern problem for one individual as an MDP. Let us first describe the concept of a (discrete time) Markov decision process, in which the decision maker being in state s_t at time $t = 1,\ldots,T$ takes action a_t that will determine the immediate utility $u(s_t, a_t)$ and also determine the distribution of the next period's state s_{t+1}, represented by a Markov transition probability matrix. The individual seeks a decision rule $a_t = d(s_t)$ that solves

$$V(s) = \max_{a_t} E\left\{\sum_{t=0}^{T} \beta^t u(s_t, a_t) \mid s_0 = s\right\} \tag{1}$$

where E denotes expectation with respect to the stochastic process (s_t, a_t) induced by the decision rule, β is the discount rate, and s_0 is state at time t=0.

Let us consider an example. Consider an individual with children and full-time work with flexible working hours. For now, assume that the individual has no interrelations with other individuals in a household. Hence, suppose the individual knows that (s)he will pick up children at school in the afternoon, more specifically in a given time window. In the morning, the individual will have to decide at what time to travel to work. The travel time to the workplace is uncertain. The individual may know the empirical distribution of travel time to work, including extreme travel times due to extreme weather or accidents, etc. When deciding the departure time for travelling to work, the uncertainty is taken into account, as well as the rest of the agenda for the rest of the day. When deciding when to leave work, the uncertainty of the morning travel time has disappeared, and this information is taken into account at that time. All travel times may be stochastic, with different variability depending on time of day, destination, and mode. At each time period, the individual can choose destinations (including the option of not travelling) and mode.

To see how this example of an activity pattern problem can be formulated as a MDP, we first discretize the day into a finite number of time periods $t = 1,\ldots,T$. Assume that there is a finite number of locations $x = \{1, \ldots, J\}$. To travel, we assume that there is a finite set of modes available $m = \{1, \ldots, M\}$. Thus, we have a finite action space, $a_t \in J \times M$. Now, what are the state variables? First, one state state variable is the time t, and another is the location x. To know whether we have a car available, we need a dummy variable δ_{car}. We also need a dummy variable δ_{child} to know the status of the child. For instance, if the child may be brought to day-care and the individual is a single-parent, one state variable is whether the child is at the day-care or not. Furthermore, we will need three stock variables. First, we need a variable $K \in [0, \overline{K}]$ as a proxy variable indicating the need for maintenance shopping. If it is close to zero, we will have to go shopping before we can derive any utility while being at home at dinner or breakfast time (or lunch). It will be useful to think of K as the amount of food in the fridge. Second, the individual may have flexible working hours. The second stock variable is therefore flex $F \in [\underline{F}, \overline{F}]$. Finally, money is needed for certain activities. The last stock variable that we will use in this example is money $Y \in R$.

As the example indicates, for the purpose of this chapter, we will confine our interest in discrete choice decision processes. That is, the action space $A(s)$ is assumed to be finite for each $s \in S$. In the example above, we can choose among, say 10 destinations and three modes. Even with this simplification, the optimization problem is rather daunting. For each sequence of decisions $(a_1,...,a_T)$ we need to evaluate a $T + 1$ dimensional integral to find the objective function. And the number of decision sequences may be large. In the example above, we can take 10 x 3 = 30 decisions at each time period.

Fortunately, the problem can be solved by dynamic programming. First, notice that for the finite horizon problem, where $T < \infty$, the problem can be solved by backward recursion. The value function in the final period T is given by

$$V_T = \max_{a_t \in A(s_T)} u(s_T, a_T) \tag{2}$$

where $A(s_t)$ is the action space available being in state s_t. For each preceding time period, we can solve

$$V_t(s_t) = \max_{a_t \in A(s_t)} u(s_t, a_t) + \beta \int V_{t+1}(s_{t+1}) p(ds_{t+1} \mid s_t, a_t) \tag{3}$$

where β is a discount factor.

Equation (3) is known as the Bellman's equation, and the existence for a value function $V(s)$ follows by the fact that the Bellman operator is a contraction mapping (see, e.g., Rust, 1994). Furthermore, it follows that a decision rule $d(s)$, which is found by doing the argmax in the Bellman operator, is an optimal decision rule for the infinite-horizon MDP problem.

Bellman's optimality principle is a powerful tool. For large problems, it is difficult to solve since the state space grows exponentially with the number of variables. However, it should be clear that it is an inherent characteristic of the problem and not of the solution method. The curse of dimensionality is inherent of the problem. What we can hope for is to find efficent algorithms for the problem at hand. Dynamic programming is more efficient than other methods, such as linear programming, for problem sizes by a factor of hundreds (Bertsekas and Tsitsiklis, 1996).

One Day Activity: The Finite Horizon Problem

We will start by formulating a problem with a finite horizon, i.e. an activity pattern for one day and one individual. Let s_t denote the state vector at time $t = 0, 1, ..., T$. The states are locations x_i, time t, car availability δ_{car}, and children task indicator δ_{child}. In addition, we have three stock variables, that is K which is the proxy variable for the need of mandatory shopping, F which represents the currently available flex hours, and Y which is the available money. In our simple example, we will have seven locations: home, work, day care, local shopping centre, shopping mall, and two other locations (recreational). When the individual wakes up in the morning, she will know that at the end of the day she will have at most visited these seven locations (but possibly more than once, or not at all). The time is continuous, but, depending on the choice of solution method, it may be useful to discretize time. In that case, there will be six minutes intervals during the morning (6 – 9 a.m.) and afternoon (3 – 6 p.m.), and fifteen minutes intervals in the evening (6 – 8 p.m.). To make things simple, if the individual is working during the day, there will only be one long interval during that period of time. However, we allow for flexible working hours, so the departure time from home and work is endogeneous. If the individual checks out from work earlier than the fixed work supply, flex hours variable F will be decreased. δ_{car} is a dummy variable that indicates whether the car is available to the individual. In the morning, this is set exogenously in a household allocation stage. If the individual chooses the car as mode of transportation, when deciding mode for the next movement, car will be available. In our experiment, the individual will always take the car if it is available and the location is not home. δ_{child} is a dummy variable indicating whether the individual has fulfilled a child movement task that has been allocated in the household allocation stage. For instance, if the individual is to pick up the children after work, δ_{child} is set to zero until the child(ren) has been picked up.

For the finite horizon problem, we first need the continuation pay-off at the end of the day. First, assume that this is given by a function $J(F,K,Y)$. This is the expected pay-off for the rest of the life, given that the individual at the end of the day ends up with stock variables (F,K,Y). Hence, these are the only variables that matter in-between days. We will assume that individuals will end the day at home, with the car in the garage, and children not still at the day care. Depending on the decisions and activities taken during the day, there will be consequences for the following day, but these are limited to the state variables associated with what is in the fridge (K), available flex hours (F), and available money (Y). Hence, the value function at the end of the day is given by:

$$V_T(s_T) = J(F,K,Y) \tag{4}$$

Now, at time $t = T$-1 we have

$$V_{T-1}(s_t) = \max_{a_t} u(a_t,s_t) + \beta\sum_{s'} P(s'|\, s_t, a_t)V_T(s') \tag{5}$$

where $u(s_t,a_t)$ is the immediate pay off when being in state s_t and taking action a_t. Since this decision is taken during the day, it is natural to set the discount factor β (close) to one.

Although there are a huge number of combinations of activities, this problem can be solved efficiently by backward induction. To make the computation even faster, there are a number of computational methods that may be considered. First, we may use a non-uniform discretization grid. In fact, in a discrete decision problem such as this randomization can be shown to break the curse of dimensionality (Rust, 1997a, 1997b) in the sense of worst-case complexity. Second, since we use discretization both in time and flex hours, it may be the case that a decision does not give exactly a flex hour that is in the discretization set. In such cases, we will use interpolation to find the intermediate values. Randomization has a "self-approximation" property that is useful and in fact is the reason why the curse of dimensionality is broken in the case of a discrete MDP problem such as ours. Experience with programming this model indicates that it is quite feasible to compute the activity scheduling (and generation) pattern using the suggested approach. How this translates into feasibility of *estimation* will be touched upon below.

Between-Day Interrelations: The Infinite Horizon Problem

In the preceding subsection we assumed that we already knew the in-between days continuation pay-off function $J(F,K,Y)$. We want the continuation pay off to be consistent with the expected value of the value function for the rest of the life, given that the individual behaves according to the decision rule $d(s)$. In the literature on dynamic programming, such a decision rule is also known as a policy, and we will use this term below. Using the decision rule d(s) throughout a day, the following should hold

$$J(z) = E\left\{\sum_{t=0}^{T} u(s_t, d(s))\right\} + \gamma \sum_{z'} Q(z' \mid a, d(s)) J(z') \tag{6}$$

where γ a discount factor, and we denote $z = (K, F, Y)$. Q is the transition probability matrix. Let $r(z,d(s)) = E\{\sum_t u(s_t, d(s))\}$ denote the expected sum of immediate pay-offs during the day, given that the day starts in state z and one follows the policy $d(s)$ during the day. Then (6) can be seen as a linear equation system:

$$J(\mathbf{z}) = (\mathbf{I} - \gamma \mathbf{Q})^{-1} \mathbf{R} \tag{7}$$

where **R** is the vector of $r(z,d(s))$, and **I** is the $|z| \times |z|$ identity matrix. Now, (7) is a linear equation system with size $|z|$. For instance, if K, F, and Y each is discretized into seven values, (7) is a linear equation system with 173 unknowns, which is easily solved.

A less straightforward part is the calculation of the expected value of the utility stream as a result of a given policy $d(s)$. In a stochastic environment, even a deterministic policy may result in stochastic pay off. Similarly, even a deterministic policy can result in different values of z at the end of the day. For instance, if travel times to work are stochastic, then an unusual long travel time to work in the morning may result in a postponed shopping trip, as compared with an unusual short travel time to work in the morning. In our example, the only stochastic component is travel time. Travel time during peak hours by car is stochastic, but with a known distribution. Starting at state z in the morning, following the policy $d(s)$, we want to calculate the expected stream of immediate pay-offs during the day, but also the distribution of states z' we end up with at the end of the day. We have tested different methods, but for now we use simulation. We simulate the path for each individual following policy $d(s)$. Since the stochastic component is very limited, and the state variables are only those associated with shopping, flex time and income, the resulting transition probability matrix Q is very sparse. Note that if travel times were deterministic, there is no need for simulation, and the calculation of $J(z)$ according to equation (7) is straightforward.

Solution Methods and Estimation

The one-day finite horizon problem for one individual with a moderately sized state space is best solved with backward induction, as indicated above. The infinite horizon problem is more difficult to solve. There are many different solution methods (value iteration, Q-learning, etc.), but as indicated above we will use policy iteration. That is, we will fix the one-day policy $d(s)$ and solve equation system (7) to find the between-day continuation pay-offs $J(z)$. With these new between-days continuation payoffs, we solve the one-day activity scheduling problem with backward induction to find a new policy $d(s)$. Policy iteration is known to converge quite fast, and it is to be preferred if the associated equation system can be solved efficiently. In our case, the size of the equation system is quite small, so policy iteration is appropriate.

Estimating the model is typically more difficult than just solving the DP problem for each individual, since we need to do it repeatedly for different parameter vectors, and we may also need gradient information. To date, DP econometric models have been estimated by maximum likelihood. One method is the nested fixed point poly-algorithm (NFXP) due to Rust (1994, 1996). In this algorithm, the DP problem for one individual is solved by policy iteration for a given parameter vector, and a hill-climbing algorithm is used to find the maximum likelihood parameter vector, using Newton-Kantorovich iterations to achieve fast convergence towards the maximum likelihood solution. In a more recent algorithm, Aguirregabiria and Mira (2002) show that the ML solution can be attained by swapping the fixed-point iterations and the hill-climbing maximum likelihood iterations. Under certain assumptions, this method should be computationally faster than the NFXP algorithm.

We have not started to estimate the model yet. We do believe that estimation will be faster with the swapped iteration method, but in our case we are not only interested in estimating the model. We are also interested in computation of the model with given estimated parameter vectors. The choice of estimation method will also have consequences for the computational effort (and programming effort) to compute the model for policy applications. We have not decided on what estimator to use. It should be noted that the computational time on current computers to solve the one-individual scheduling problem is well below one second, although we have made no effort to optimize the code. The same holds true even for the infinite horizon problem below (since it converge quite fast and the associated equation system is small). Therefore, we believe it is feasible to estimate this model. So far, we have shown that we can calibrate the model to give realistic mode choice shares, departure time distributions, and maintenance shopping frequencies.

Household Allocation and Bounded Rationality

We have demonstrated above that the activity scheduling problem for one individual can efficiently be solved in a dynamic mircoeconomic framework by dynamic programming. However, in our examples we have taken the allocation of activities within a household as given, which is a limitation in households with more than one adult. More specifically, we have allocated the pick up or drop off of children, the maintenance shopping and the use of the car to one individual.

Unfortunately, technically, the size of the state space explodes for a household. For instance, if there are five thousand states for one individual, there will be at least 25 million states for a household with two individuals. Hence, it is difficult to see that we can solve the household activity allocation and scheduling problem with straightforward brutal methods, as we were able to do for the one-individual activity scheduling problem. Therefore, we will try two different approaches for the household problem.

The first obvious option is to constrain household interaction into just a few aspects. Different activity-based models focus on different aspects (e.g., Gliebe and Koppelman, 2005; Srinivasan and Bhat, 2005). In any case, the problem is significantly simplified if we were to focus on just a few aspects of household interactions, for instance escorting children and mainenence shopping for out-of-home maintenance activities, and car availability for resource allocation. Thereby we can to a large extent reduce the household allocation problem to a small number of individual activity scheduling problems. That is, suppose there are four possible combinations of escorting children (A drops off children and B picks then up, or the other way around, or A does both, or B does both), two combinations of car availability, and two for maintenance shopping. Then, the household allocation problem is essentially a choice among 16 combinations, each of which can be solved by solving the individual scheduling problem for each of the two individuals. This problem is already of tractable size.

However, if we add more and more decisions to the household allocation problem, the problems again grows at an alarming rate. Therefore, the above approach will be useful to determine the major household allocation decisions, but it is not flexible enough to allow for more general household allocation problems (including in-house maintenance and out-of-home joint activities). One way to overcome this problem would be to use approximations methods. Since the problem has considerable structure, one may be successful in using the efficient methods for solving the one-individual activity scheduling problems to approximate the household activity problem by state aggregation. It should be emphasized, as stated in the introduction, that such an approximation method could be casted into a decision-tree framework, or a reinforcement learning framework. The reconsilidation of these methods is left for further research.

Multiagent Models and Interaction

As noted in the introduction, utility maximization is not a prerequisite for a micro-economic foundation. This line of research has recently been particular active in game theory and the theory of learning games. In our context, things become much more complicated (and interesting) in a multi-agent setting, where agents are interacting with each other. In such a setting, it is not convincing to argue that agents know the preferences and the actions chosen by all other relevant individuals (Fudenberg and Levine, 1998). Instead, it may be argued that individuals use more low-rational strategies to play the game.

The travel departure time is an integral part of our activity model. We let the individual observe a distribution of travel times, and from this information the individual solves the activity scheduling problem that gives the departure time. In the morning peak hour, however, the individuals are interacting with each other. Ideally, one would like the dynamic traffic assignment to be a part of the activity scheduling problem. This raises a number of theoretical questions. In fact, Nagel *et al.* (2000) have pointed out the lack of theoretical results. For instance, what is the outcome of such an interaction? From recent theory of learning in games (e.g., Fudenberg and Levine, 1998) we know that simple adaptive learning rules can sometimes be shown to converge, but global convergence has been shown only under fairly strict assumptions. Even if the learning algorithms do converge, to what do they converge? Traditionally, this problem has been approach with the concept of NE, Nash equilibrium (Nagel and Marchal, 2003). However, there is theoretical evidence (Fudenberg and Levine, 1998) that the characteristics of the equilibrium depend on the detail of the learning algorithm, and the outcome frequently may be a correlated equilibrium (CE), rather than a NE. What solution concept should be used is an open question (Greenwald *et al.*, 2001). Furthermore, there may be multiple equilibria. In summary, there remains much theoretical work before we can say anything definite about which equilibrium will be the outcome (see also Williams, 2001).

As is evident from the discussion in this subsection, the theoretical considerations for activity scheduling and dynamic traffic assignment are overlapping. Likewise, the algorithms for solving route choice and activity scheduling are overlapping. In fact, the problem as described in this chapter relates to the problem of the stochastic shortest path.

Comparison with Activity-Based Nested Logit Models

The dynamic programming model may be compared with state-of-the-art nested logit models, as described by Bowman and Ben-Akiva (2001). The DP model is somewhat similar to such a model, as soon as we allow for idiosyncratic error components, for instance extreme value distributed error terms yielding a logit specification for the choice probabilities. This is in fact the common practice when estimating econometric DP models (see Rust, 1996). It is therefore natural to compare the DP approach with structural logit specifications. In the econometric literature, one alternative to a dynamic programming model that has been proposed is the so-called option value model (Stock and Wise, 1990). In the context of an optimal stopping problem, given an option value model, the individuals are not supposed to solve the complicated dynamic programming model, but rather solve a more simple model where individuals compare the utility of retirement in period t with the maximum expected utility of retirement in the future. Hence, the option value model is model of bounded rationality. Although there is little doubt of the theoretical superiority of the dynamic programming approach, there has been a number of empirical studies to investigate which model best fits empirical data (Butler *et al.*, 2003).

The difference between the DP approach and using nested logit specification is the way in which utilities from the future (the rest of the day) are nested into a value function at the present time. In the DP setting, we take the expected maximum of values, whereas in the nested logit formulation we take the maximum of expectations. Furthermore, the decision tree in DP retains the sequential decision process, whereas in the nested logit formulation the timing of events is decided in one (or a few) steps. For instance, time departure choices to and from work may be one level in the nested logit formulation. Therefore, the nested logit formulation is not dynamically consistent, although the numerical outcome may (or may not) be similar.

It is not difficult to construct examples where a structural form nested logit formulation will be at its disadvantage (one such case is when we have uncertainties that dissolve during the day). However, in the activity modelling setting, more important are computational issues. In the labour supply optimal stopping context, the logit formulation is quite simple, typically a binary logit model, while in the activity modelling setting, the structural logit models are rather complex nested logit models. This means that to estimate such a model, we first have to impose some a priori structure (partition into nests). Second, the estimated models may turn out to be difficult to interpret. A common problem in the estimation is that the dissimilarity parameters (or logsum parameters) exhibit values that are not easily interpreted in a stochastic utility maximization setting.

In contrast, there are no such parameters in the DP model. Instead, we ensure that the indirect utilities are consistent (even dynamically consistent) by solving the scheduling problem backwards (for the finite horizon problem), or solving the equation system (7) (for the infinite horizon problem). Thus, the advantage of the DP approach is that (i) we do not have to impose some a priori structure (nests or restrict activity patterns) and (ii) the estimated model will be dynamically consistent. However, there is a price to be paid: we have to solve for the indirect utilities (value functions). As in the labour supply literature, the theoretical superiority of the dynamical programming approach follows from the explicit modelling of sequential decisions. However, it remains an open question which model will fit empirical data best. The exact trade-off of computational burden and programming effort is also not clear (and probably subjective and idiosyncratic).

A COMPUTATIONAL EXPERIMENT

It is useful, in this rather conceptual paper, to hint whether it is possible to estimate the model, and what aspects of travel patterns reasonably can be modelled. To this purpose, we will focus on departure time choice. In fact, the proposed approach will be used in a project evaluating the Stockholm congestion pricing scheme that probably will be implemented in August 2005. Since the congestion pricing scheme is time differentiated, with a higher charge during peak hours, it will be interesting to test the hypothesis that the congestion pricing will shift time departure choice, both for travel to and from the workplace. Our model ensures that a delayed departure time to work in the morning may result in a delayed departure time from work in the afternoon.

To test whether the model is flexible enough to capture a shift in departure time choice as a response to a congestion charge in the morning peak hour, we will here briefly report the results of an experiment (for more computer experiments, see Jonsson and Karlstrom, 2005). We consider here an individual that lives 10 km north of Stockholm and have workplace in the city center. The travel time to work is on average 19 minutes during the morning peak hour (and approximately the same by public transport). The congestion charge will be enforced from 8 a.m., but the peak hour is from 6 - 9 a.m. The individual does not have hard constraints in terms of flex hours or shopping activities. That is, the individual does not necessarily have to stay at work for full 8 hours, and does not necessarily have to shop. However, the individual has to leave children at the day care in the morning, and pick them up before 6 p.m. The individual has to be at work at least from 9 a.m. to 3 p.m.

Figure 2.1 shows the probability of departure from home for this individual. First, note that the individual is certainly leaving home no later than 8:06 a.m. This is in fact the latest possible time of departure, if he will be able to make it to work before 9 a.m. Second, note that the departure probability is very similar for the first half hour. Since there are few hard constraints in the schedule for this individual, the individual may stay at home or go to work (and come home early). So, there is little difference between going to work, or staying at home early in the morning, since peak hour has not yet arrived. The small difference in choice probabilities early is due to the fact that leaving home early gives a little more flexibility in adjusting the schedule later, if some improbable events occur. However, if the individual has stayed longer at home, constraints will become active. These constraints are clearly visible in the diagram, indicated by peaks in the departure probabilities. For instance, the optimal schedule for the individual is to depart at 7:42, but if the individual does not choose to do so, he can just as well stay home a little longer. The reason is that if the individual starts from home earlier than 7:42, he can follow a schedule that allows him to shop that day (and other peaks in the departure probabilities are also directly derived from time constraints).

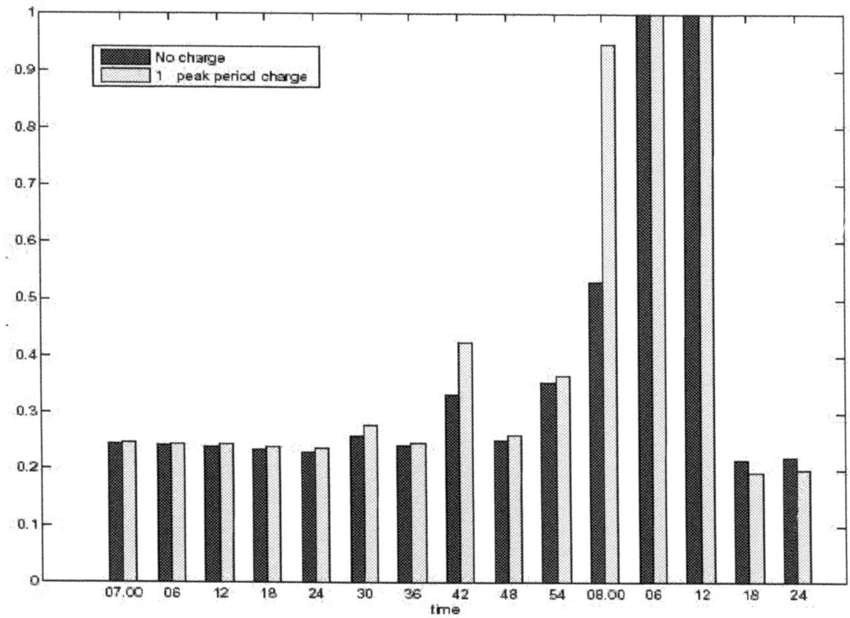

Figure 2.1
The Probability of Departing from Home Before and After a Change from 8 a.m.

After 7:42, it is a bad choice to go shopping that day, and the shopping activity is posponed until the next day. This illustrates the fact that the shopping activity is modelled endogeneously, and is allowed to affect the activity pattern throughout the day for this particular individual.

Third, note that the departure time shifts from the peak hour. The probability is higher for all departure times earlier than 8 am, when the charge is implemented. There is a marked increased probability just before the charge is implemented (from 8 a.m.), but otherwise the pattern looks similar.

SUMMARY AND FURTHER WORK

In this chapter, we have argued that we need methodological development in order to achieve the following objectives. First, a microeconomic foundation is needed for any activity based model that is to be used for welfare evaluation of policies. Secondly, this microeconomic model should be dynamically consistent, and allow for uncertainties and sequential decision making. Thirdly, to the extent that bounded rational behavioural rules are consistent with microeconomic theory, unrealistic assumptions on computational capabilities (and behaviour) should be avoided.

The first two objectives were the main focus of this chapter. In particular, we showed that the activity scheduling problem for one individual can be solved in a consistent dynamic microeconomic framework. We will later report on our experience with estimating such a model. At the same time, we acknowledge that further development is needed in order to use these techniques in a *household* activity model. However, it may be the case that having efficient methods for the one-individual activity scheduling and generation problem will prove useful in the development of such a household model. Whether this conjecture is justified or not is left for future research. In particular, it would be useful to compare reinforcement learning (or function approximation) approaches with the more brutal force method described in this paper.

ACKNOWLEDGEMENT

Financial support from VINNOVA is gratefully acknowledged.

REFERENCES

Adda, J. and R. Cooper (2003). *Dynamic Economics.* MIT Press, Boston.

Aguirregabiria, V. and P. Mira (2002). Swapping the nested fixed point algorithm: a class of estimators for descrete markov decision models. *Econom.*, **70**, 1519–1543.

Arentze, T.A. and H.J.P. Timmermans (2000). Albatross: *A Learning - Based Transportation Oriented Simulation System.* European Institute of Retailing and Service Studies, Eindhoven.

Arentze, T.A. and H.J.P. Timmermans (2004). A theoretical framework for modeling activity-travel scheduling decisions in non-stationary environments under conditions of uncertainty and learning. *Proc. Int. Conf. on Activity-Based An.*, Maastricht, The Netherlands (CD-Rom).

Bertsekas, D.P. and J. Tsitsiklis (1996). *Neuro-Dynamic Programming.* Athena Scientific, Belmont, MA.

Bowman, J.L. and M.E. Ben-Akiva (2001). Activity-Based Travel Forecasting. US Department of Transportation. http://tmip.fhwa.dot.gov/clearinghouse/docs/abtf/bowman.stm

Butler, J.S., G. Gumus and R. Burkhauser (2003). Comparing Option Value and Dynamic Programming Models Estimates of Social Security Disability Insurance Application Timing, IZA-DP 941, Institute for the Study of Labor, Bonn.

Charypar, D., P. Graf and K. Nagel (2004). Q-learning for flexible learning of daily activity plans.

Proc. Swiss Transp. Res. Conf. (CD-Rom), www.strc.ch/2004.html.

Doherty, S.T., E.J. Miller, K.W. Axhausen and T. Gärling (2005). A conceptual model of the weekly household activity-travel scheduling process. In: *Travel Behaviour: Patterns, Implications and Modelling* (E. Stern, I. Salomon, and P. Bovy, eds.). Elgar Publishing, Cheltenham, to appear.

Ettema, D.F., A.W.J. Borgers and H.J.P. Timmermans (2000). A simulation model of activity scheduling heuristics: an empirical test. *Geo. & Env. Modelling*, **4**, 175-187.

French, E. (2001). The effects of health, wealth, and wages on labor supply and retirement behavior. Manuscript, Federal Reserve Bank of Chicago, Chicago.

Fudenberg, D. and D. Levine (1998). Learning in games. *Eur. Ec. Rev.*, **42**, 631–639.

Gärling, T. and W. Young (2000). Perspectives on travel behaviour: decision paradigms, In: *Transportation Research:The Leading Edge* (D.A. Hensher, ed.), pp. 189-196. Elsevier, Oxford.

Gliebe, J.P. and F.S. Koppelman (2005). Modeling household activity-travel interactions as parallel constrained choices. *Proc. 84th TRB Conf.,* Washington D.C. (CD-Rom).

Greenwald, A., E.J. Friedman and S. Shenker (2001). Learning in network contexts: experimental results from simulations. *Games and E. Beh.*, **35**, 80-123.

Joh, C.-H., T.A. Arentze and H.J.P. Timmermans (2002). Modeling individuals' activity-travel rescheduling heuristics: theory and numerical experiments. *Transpn. Res. Rec.*, **1807**, 16-25.

Joh, C.-H., T.A. Arentze and H.J.P. Timmermans (2003). Estimating non-linear utility functions of time use in the context of an activity schedule adaptation model. *Proc. 10th Int. Conf. Travel Beh. Res.,* Lucerne, Switserland CD-Rom).

Joh, C.-H., T.A. Arentze and H.J.P. Timmermans (2004). Activity-travel rescheduling decisions: empirical estimation of the Aurora model. *Transpn. Res. Rec.*, to appear.

Jonsson, D. and A. Karlstrom (2005). SCAPES – A dynamic microeconomic model of activity scheduling. Working paper. Royal Institute of Technology, Stockholm.

Karlstrom, A., M. Palme and I. Svensson (2004). A dynamic programming approach to model the retirement behavior of blue-collar workers in Sweden. *J. Appl. Econom.,* **19**, 795-807.

McNally, M.G. (2000). The Activity-Based Approach. Report UCI-ITS-AS-WP-00-4, University of Irvine, Irvine, CA.

Nagel, K. and F. Marchal, (2003). Computational methods for multi-agent simulations of travel behavior. *Proc. IATBR Conf.*, Lucerne, Switzerland (CD-Rom).

Nagel, K., M. Rickert, P. Simon and M. Pieck, (2000). The dynamics of iterated transportation simulations. Working Paper 00-02-012, Santa Fe Institute, Santa Fe.

Rust , J. (1987). Optimal replacement of GMC bus engines: an empirical model of Harold Zurcher. *Econom.*, **55**, 999–1023.

Rust, J. (1994). Structural estimation of Markov decision processes, In: *Handbook of Econometrics*

(R. Engle and D. McFadden, eds.), Vol. 4, pp. 3081-3143. North-Holland, Amsterdam.

Rust, J. (1996). Numerical dynamic programming in economics. In: *Handbook of Computational Economics* (H.M. Amman, D.A. Kendrick and J. Rust, eds.), Vol. I, pp. 620-729. Elsevier, Amsterdam.

Rust, J. (1997a). A comparison of policy iteration methods for solving continuous-state, infinite-horizon Markovian decision problems using random, quasi-random, and deterministic discretizations. Working paper. Yale University, New Haven.

Rust, J. (1997b). Using randomization to break the curse of dimensionality. *Econom.*, **65**, 487-516.

Rust, J. and C. Phelan (1997). How social security and medicare affect retirement behavior in a world of incomplete markets. *Econom.*, **65**, 781–831.

Srinivasan, S. and C.R. Bhat (2005). Modeling household interactions in daily in-home and out-of-home maintenance activity participation. *Proc. 84th TRB Conf.,* Washington D.C. (CD-Rom).

Stock, J.H. and D.A. Wise (1990). Pension, the option value to work and retirement. *Econom.*, **58**, 1151–1180.

Sun, Z., T.A. Arentze and H.J.P. Timmermans (2005). Modeling the impact of travel information on activity-travel rescheduling decisions under conditions of travel time uncertainty. *Proc. 84th TRB Conf.,* Washington D.C. (CD-Rom).

Timmermans, H.J.P. (2000). Theory and models of activity patterns. In: *Albatross: A Learning-Based Transportation Oriented Simulation System* (T.A. Arentze and H.J.P. Timmermans, eds.), pp. 6–20. EIRASS, Eindhoven.

Williams, N. (2001). Stability and long run equilibrium in stochastic fictitious play. Working paper, Princeton University, Princeton, NJ.

3

SIMULATION OF DAILY ACTIVITY PATTERNS

Ondrej Pribyl, Czech Technical University, Prague, Czech Republic

Konstadinos G. Goulias, University of California, Santa Barbara, USA

THE PROBLEM

The traditional goal of modelling travel demand analysis is to estimate the volume of traffic on particular roads in a transportation network (Ortuzar and Willumsen, 1994). Recent policy propositions and actions (introduction of new technologies, taxation and congestion pricing) and market trends (market penetration of mobile telecommunication technologies) also motivate the study of impacts not only on traffic volumes but also on car ownership, trip consolidation into chains, departure times, and more general shifts in spatial and temporal aspects of travel demand. For this reason, many researchers and practitioners create models that address travel behaviour in a more comprehensive way, looking at schedules of activities to assess the impact of policy actions. The required precision and level of detail of such models have also changed dramatically over the years. An estimate of only daily volumes was sufficient in the early stages of modelling and regional simulation. However, the current objective of research is to find a model that can estimate traffic volumes at much finer time scales and also estimate other changes in travel behaviour such as departure time shifts affecting peak spreading and changes in time allocation from weekdays to weekends that may create unseen congestion types. It is clear that this is not an easy task and the original four-step model developed in the early 1950s is not sufficient (e.g., McNally, 2000).

A way to overcome these limitations is using activity-based approaches to travel demand analysis (e.g., McNally, 2000; Goulias, 2003). The idea behind these approaches is that travel demand is derived from the demand for activities. The models attempt to estimate the sequence of activities an

Progress in Activity-Based Analysis edited by H. Timmermans

individual follows in a day, called a *synthetic schedule* or *activity pattern*. Once we know the schedule, deriving volumes on particular roads in the network is a rather straightforward algebraic operation. Much effort has been spent on this problem, and knowledge in the field has developed significantly in just a few years (for an overview, see Arentze and Timmermans, 2000). Many different models have been proposed. However, there is no single model that considers all of the issues that a human being considers during her/his activity planning process. Moreover, the interaction among persons in activity scheduling is only recently receiving attention.

OBJECTIVES

A micro-simulation model that generates activity patterns of individuals in a given study area is described and evaluated in this chapter. The proposed algorithm belongs to the field of activity-based approaches to travel demand analysis. It aims to replicate the observed patterns that implicitly include the constraints and outcomes of the decision making processes underlying a person's time allocation in a day with other persons and alone. By replicating the entirety of a person's activity-travel pattern in a day, a feasible and robust solution that consists of timing and sequencing of activities is provided. These synthetically generated schedules are also linked to individual's and household's characteristics (such as income, number of cars, number of children, age, gender, and others).

The main contribution in travel behaviour research of the proposed approach is in capturing the interactions among household members. For example, in a family with a child of school age, one of the parents has to adjust her/his schedule in order to be able to drop off the child in the morning to school. Another example can be a joint dinner of both parents. Both must adjust their schedules in order to meet at the same time and at the same restaurant. These examples show the importance of such a model. The objectives defined at the early stages of research were met. As expected and to make the work tractable, however, many issues are left as future tasks and are reviewed at the end of this chapter.

THE DATA

The data set used to test the model in this study was obtained from the CentreSIM survey, conducted between November 23, 2002 and May 30, 2003 (Patten and Goulias, 2004). The activity and travel data collected from each person in the household using the two-day complete record of the activities not only includes the types of activities but also each person engaged and the different transportation options taken. The respondents were asked to record start and end time of each episode, its purpose, but also questions such as "With whom did you do the activity" and "For whom did you do the activity". For trips, the respondents reported the travel mode used for the trip, if they drove on this trip, start and end point of the trip, and they also estimated the distance of each trip and reported it. All these questions were included in order to capture a variety of contexts in the decision-making aspects of each household in activity and travel scheduling.

Data Representation

The described activity/travel data were organized in a structure in which each activity or trip corresponds to one record. However, a different structure is needed for the analysis. The entire daily activity/travel pattern, i.e., the sequence of all trips and activities will be recorded. Such a pattern is obtained by sampling the observed daily activity participation in regular time intervals, in this project every 10 minutes. The entire daily activity/travel pattern is thus represented as a vector of 144 values (24 hours * 60 minutes / 10 minutes). Each value of this vector is an integer that corresponds to the particular activity type combined with the information about joint/alone activity participation at a given time instant. In this project, we distinguish four activity purposes: **H**ome that denotes all-in home activities; **W**ork corresponds to all work related or school related activities; **M**aintenance that groups all shopping, dining out and other activities; and **D**iscretionary that focuses on leisure activities and recreation.

An important issue in the activity-based approaches to travel demand analysis is representing the within-household interactions. Different representations of within-household interactions were tested in Pribyl (2004). Here results of the most promising representation are summarized. The answers to the question "With whom did you do the activity?" are reduced to four different categories: **A**lone/Other (no other household member involved), with **S**pouse, with **C**hild(ren), and with **M**ultiple **F**amily (more than one household member). The activity type is combined with the joint/alone activity participation so that one vector is created. This implies that there are the following sixteen possible outcomes at every time instant:

Home-Alone, Home –Spouse, Home -Child, Home -multiple family, Work-Alone, Work -Spouse, Work -Child, Work -multiple family, Maintenance-Alone, Maintenance -Spouse, Maintenance -Child, Maintenance -multiple family, Discretionary-Alone, Discretionary -Spouse, Discretionary -Child, Discretionary -multiple family.

The cluster algorithm distinguishes all described values on the y-axis. When visualizing the patterns in figures, however, only two levels of joint/alone activity participation are distinguished, for simplicity: alone/other (includes also categories with child(ren) and with multiple family), and activity conducted with spouse.

An example of this representation of the activity patterns is provided in Figure 3.1. This individual is alone at home (H-A) until 8 a.m. when s/he leaves for work (W-A). At about 1:30 p.m. s/he goes home for lunch and after about 45 minutes/he returns to work again. S/he stays at work until 7 p.m. , and then returns home when s/he stays till the end of day (Note: The diagonal vertical lines in the figure only connect two activity types and do not have a real meaning).

The advantage of this representation is its ability to handle a daily activity pattern as a whole. The type of activities, their timing, duration, and sequencing, as well as the alone/joint activity participation are all embedded directly in this representation.

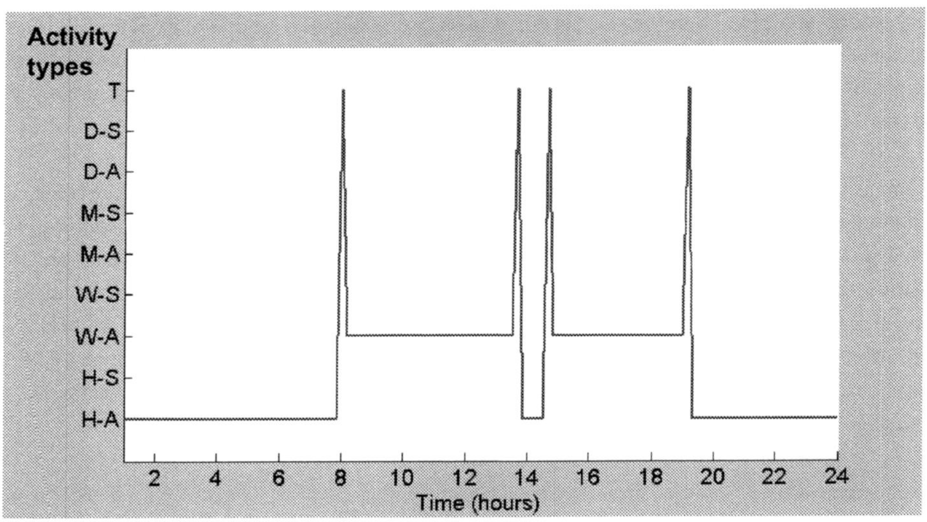

Figure 3.1
An Example of an Activity Pattern

The Data Sets

Since we are interested in representing the within-household interactions, the household is the proper level of analysis. The methodology was applied to adult members of the households only. For this reason, each object must represent the activity patterns of all adult household members of the household. It means that since the vector describing a one-adult household has a length of 144 (described above); the vector for a two-adult household has the length of 288 (schedule of the first adult plus of the second adult). In a two-adult household, we should be comparing only suitable individuals. For this reason we have to sort the adults in the household. The first person is called, for simplicity, the head of the household (even though this term is used with a high level of simplification). If one person is employed full-time and the second person has a different status, the first person is considered the head of the household. If both individuals are employed full-time, the male will be considered the head of the household. The definition of the head of the household is not essential as long as it is consistent for all households. It ensures that individuals having a similar role in the household are compared.

Past research dedicated to the cluster analysis of activity patterns still leaves one question open: what is the relationship between the activity patterns and household socio-demographic characteristics? This question is important since it enables the simulation of household activity patterns based on socio-demographic characteristics of the households.

One way to approach this task is to apply a cluster algorithm to the entire data set. The problem can be the high heterogeneity in the patterns. Wang (1996) therefore first specified six lifecycle groups and clustered the groups independently. The advantage of this method is that the found patterns are more homogenous. On the other hand, some of the found clusters can be redundant (similar or the same patterns can be found for different lifecycles). For the reasons stated above, a compromise of these two methods is used in this project. Prior to the cluster analysis, all households are split based on the number of adults in the household. The most common group, two-adult households, is further split based on the employment status of the head of the household – employed full time, and any other employment status. The analysis is performed for each split separately and this decreases the heterogeneity within the clusters. The variables used for splitting the data set were chosen based on empirical results and results of different papers (for example, Kulkarni and McNally, 2000). Because of low occurrences of households in which there are more than two-adults, in this analysis only households up to two adults will be used. The proposed methodology could be used without changes to any household size.

THE METHOD

This chapter describes an algorithm for generating daily activity-travel patterns of individuals in a household. The proposed approach is based on the belief that people with similar socio-demographic characteristics have similar travel behaviour (Bowman and Ben-Akiva, 1997; Stopher and Metcalf, 1999). In this model, first homogenous activity patterns are found, and then assigned to the households whose patterns are simulated. The probabilities of departing for particular activities by time of day are derived from the data belonging to each group. A micro-simulation model is used to assign each individual in a household to particular activities during the day. The output of the model is an estimate of individuals' activity-travel patterns on a detailed time span (10 minutes). The information whether each activity was conducted alone or together with other household members is also incorporated in the model.

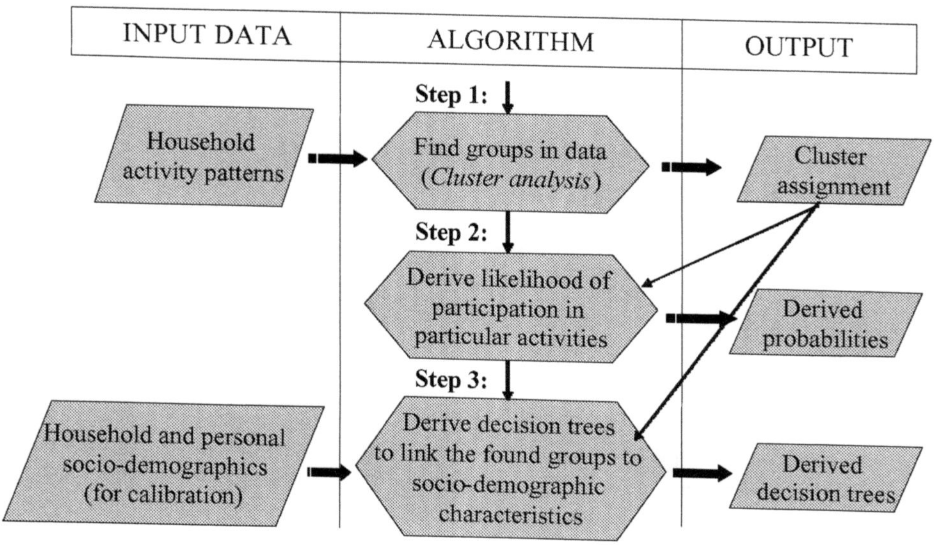

Figure 3.2
Overview of the Model Estimation Phase

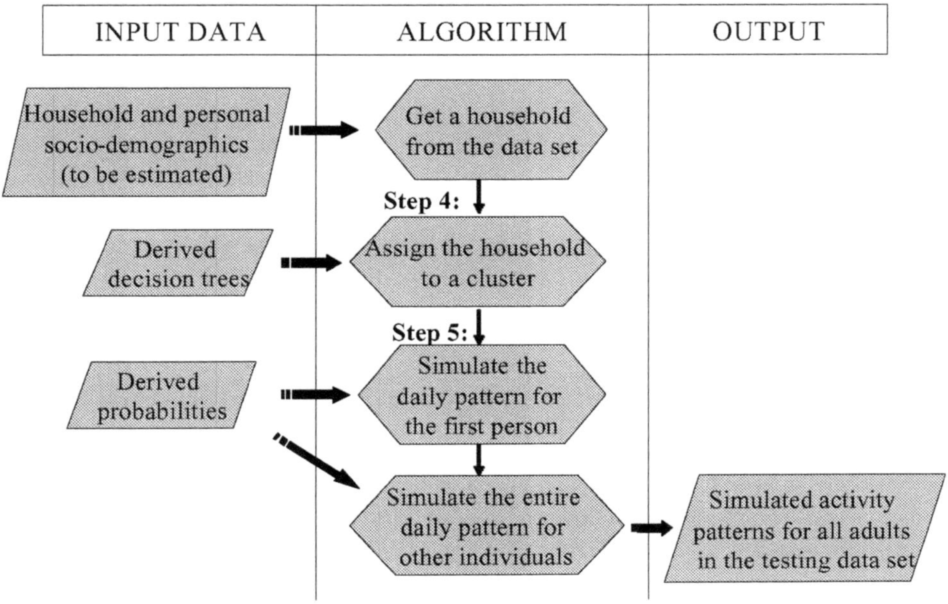

Figure 3.3
Overview of the Simulation Phase

The proposed algorithm is executed in several consecutive steps, which can be aggregated into two major phases: a *model estimation phase*, and a (micro)*simulation phase*. The task of the model estimation phase is to calibrate the proposed model, given a training data set that is known prior to analysis. The simulation phase aims to replicate the entire daily activity pattern for a given household using the results from previous phase. The schematic structure of this model is provided in Figure 3.2 and 3.3. These figures also show the input for each step as well as its output.

Model Estimation Phase

Step 1. The objective of the first step is to find groups of households with similar activity patterns, taking into consideration interactions within the family. This task will be accomplished using cluster analysis applied to the entire pattern. It should be noted that sequencing of activities is not explicitly modelled here, but this could be established applying sequence alignment measures (Joh *et al.*, 2001a,b,c,d). Many different clustering algorithms and heuristics have been developed over the years. The method used in this project is based on the so called *k-medoid clustering,* also called *partitioning around medoids* (PAM) (Kaufman and Rousseeuw, 1990). Consider a set $S=\{S_1, \ldots, S_N\}$ of N objects (in our case activity patterns). Each object S_i is a vector, containing L integer values describing an activity type at a particular time instant. Our objective is to find K objects, $m_1, \ldots, m_K$, which represent all objects in the data set. These representative objects are called *medoids*, because we expect them to be located in the centre of each cluster. The remaining objects are then assigned to the nearest representative object, using a given dissimilarity measure. Each object in the data set belongs exactly to one medoid. All objects belonging to the same medoid form a *cluster*. The objects in each cluster are clearly more similar (based on given dissimilarity measure) to each other than to objects in any other group.

Mathematically, we are looking for a set of K objects, $K<<N$, which minimize the *objective function* $F_K = \sum_{i=1}^{N} \min_{t=1,\ldots,K} d(S_i, m_t)$, where F_K is the sum of dissimilarities of all objects S_i to their nearest medoid. We are seeking the medoids so that this sum is minimal. An advantage of this method is that it uses only a dissimilarity matrix and not the original data. This implies that this method can be used for any type of data as long as we know how to measure dissimilarities among objects in the dataset.

A very important issue in any clustering algorithm is the definition of dissimilarity between two objects. We cannot use the common Euclidean distance, because we deal with objects of a categorical type (activity types). The distance measure used in this chapter is defined in Huang (1997) as follows: The dissimilarity measure between two objects S_1, and S_2 reflects the total number of mismatches of activity types at a corresponding time index. This can be written as:

$$d(S_1, S_2) = \sum_{j=1}^{|S_i|} \delta(S_1(j), S_2(j)), \tag{1}$$

where,

$$\delta(S_1(j), S_2(j)) = \begin{cases} 0 & (S_1(j) = S_2(j)) \\ 1 & (S_1(j) \neq S_2(j)) \end{cases} \tag{2}$$

and $|S_i|$ denotes the length of a schedule (in our case it equals to 144). Therefore, the value of the dissimilarity measure between two patterns can range in case of one-adult households from 0 (the two patterns are the same) to 144 (totally different); in case of two-adult households from 0 to 288. The developed algorithm requires the number of clusters, K, to be known. In order to find the optimal number of clusters, the algorithm is usually repeated for different numbers of clusters. The optimal number is chosen off-line, after the clustering problem for different K will be performed.

There are many different measures that can be used (see overview in Fridlyand, 2001). Based on existing literature (Maulik and Bandyopadhyay, 2000; Fridlyand, 2001) and author's preliminarily tests (Pribyl, 2004), the use of the silhouette coefficient (Kaufman and Rousseeuw, 1990) is the most promising. For every cluster, X_j, $j = 1,\ldots,K$, the silhouette technique assigns the ith member (x_{ij}, $i = 1,\ldots,n_j$) of cluster X_j a quality measure (silhouette width):

$$s_{ij} = \frac{b_i - a_i}{\max\{a_i, b_i\}}, \tag{3}$$

where, a_i is the average distance where between x_{ij} and all other members in X_j and b_i denotes the minimum of a_i, $i = 1, 2, \ldots, n_j$, where n_j is the number of patterns in cluster X_j and naturally, $n_1 + \ldots + n_k$ equals n if each pattern belongs to one and only one cluster, n is the number of patterns to be clustered. The silhouette is defined for each object i in the interval $-1 \leq s_i \leq 1$. Out of these silhouette values we can also compute another entity called average silhouette width for the entire data set, $\overline{s}(K)$. This coefficient is very important since it helps to determine the optimal number of clusters K^*. The clustering algorithm will be used on the same data for different numbers of clusters and for each setting the $\overline{s}(K)$ will be computed. The number of clusters will be determined based on the *silhouette coefficient* (SC) that is computed according to the following equation

$$SC = \max_K \overline{s}(K) \tag{4}$$

where, the maximum is taken over all K for which the silhouettes can be computed. The optimal number of clusters is taken as the argument of this maximum.

In order to avoid some limitations of the algorithm such as finding only local extremes, genetic algorithms are used for implementing the problem. The most common clustering techniques, such as partitioning methods are so-called greedy algorithms, i.e. they look for the largest improvement at each step. This does not ensure finding the global optima. Also their performance is very sensitive to the initial partitioning, that is in most cases done randomly. Using a genetic algorithm also does not ensure reaching the global optimum. The result depends mostly on the setting of its parameters, for example the selection mechanism, probability and type of recombination. However, the existing literature demonstrates that genetic algorithms are well suited for clustering and with the right parameter setting outperform standard clustering algorithms (Maulik and Bandyopadhyay, 2000; Lucasius *et al.*, 1993). In this study, an algorithm developed in Pribyl (2003) is used. It implements partitioning around medoids using genetic algorithms. The algorithm is rather robust and not too sensitive to its parameter settings, which is a significant advantage.

Step 2. The objective of this step is to find probabilities that an individual starts a particular activity at every time instant and its duration. For every cluster and every time step, the relative frequencies of leaving for a particular activity are derived. Also for each time instant, average duration and standard deviation of duration are computed for each activity type and travel. These probabilities will be essential for the simulation of daily activity patterns.

Step 3. In this step, the identified clusters are linked to the persons in the data set (this is also called action-assignment), based on their personal socio-demographic characteristics as well as characteristics of their entire households. There are several methods that can be used to link the household and personal socio-demographic characteristics to the resulting clusters found in the previous step, for example *multidimensional cross-tabulation*, *multinomial logit model* (MNL), or *decision trees* (DT). Each of these methods has advantages and drawbacks (discussed for example in Ortuzar and Willumsen, 1994; Bowman and Ben-Akiva, 1997; Arentze and Timmermans, 2000). Previous research does not show any evidence that any of these methods clearly outperforms the others (Lim *et al.*, 1998; Wets *et al.*, 2000). However, the decision tree algorithms have some potential advantage over the MNL model in terms of interpretability. They find context-sensitive rules that are well suited for discontinuous behavioural changes. Compared to MNL, the decision trees are also theory free. We do not make any assumptions about rational behaviour of individuals and about their maximizing of utility. All results are directly derived from the data. Also these algorithms are less sensitive to outliers and multicollinearity. These issues are also very important. In contrast, logit models (and in general all parametric models) have an advantage in predicting the size of the impact of particular explanatory variables on the probabilities of the outcome. This is not a major concern in the case of the simulation model proposed here. Based on findings from Arentze and Timmermans (2000), a decision-tree algorithm called CHAID (*Chi-square Automatic Interaction Detection*) is used in this model (Kass, 1980).

Simulation Phase

Step 4. In the next step the decision trees are used to link these groups to socio-demographic characteristics. The results from decision trees must allow generalization for some other data set of similar features to a testing data set. The question is how to assign one group of synthetic schedules (response variable) to a person (case) described by socio-demographic characteristics. In most applications, the used action-assignment is purely deterministic. A typical example is the *plurality action-assignment* rule. This rule assigns the modal response at a leaf node after training to every new case assigned to this leaf node. Clearly, it describes well the average values, but fails to describe the variance present in the sample.

In this model, we therefore use a probabilistic action-assignment rule (Arentze and Timmermans, 2003). The choice probabilities are mathematically expressed in the following equation:

$$p_{ij} = \frac{f_{ji}}{n_j}, \tag{5}$$

where, f_{ji} is the frequency of choice i at leaf node j and n_j is the total number of cases at leaf node j in the training data. In their work, Arentze and Timmermans (2003b) derived a more general form of the probabilistic assignment rule, which takes into account the possibility that for a given case one or more choice alternatives are not feasible or available. Since there are no constraints of this type in the problem as formulated here, the simple probabilistic rule as described in the previous equation is used.

Step 5. Once we know to which cluster each household belongs, we proceed with simulating its daily activity pattern. In this step, the variance in activity behaviour must be reintroduced to the data. The activity patterns consist of the sequence of activities, each with their start time, their duration, and also the within household interactions. The simulation model used in this chapter is similar to Kulkarni and McNally (2000), but some modifications are introduced to reflect the different representation of the data. The patterns are generated conditional on the probability distributions derived from each cluster and each time step. The simulation is performed sequentially. Starting at the first time step, the procedure assigns the type of the first activity, and its duration according to the derived distributions. The activity type is assigned probabilistically, based on the proportions in the training data set engaging in particular activities. Once the activity type is known, its duration is computed as a normally distributed random number with a mean value and standard deviation corresponding to the activity type and time instant. The assumption of the normal distribution of the duration of particular activities is based on the work of Kulkarni and McNally (2000). At the end of the first simulated activity, a new activity and its duration are selected accordingly.

In order to decrease the sensitivity of the model to the number of households contributing to each cluster, a time window of $t \pm 30$ min is used in the estimation phase (in the case where we have more households contributing to the estimation of the model, the time window can be decreased.) The model is constructed for each time step from the proportion of cluster members that start each particular activity within half hour on either side of the time step in question. It means that if an activity ends at time t, we look at the probabilities with which the next activity starts in the interval *(t-30 min, t+30 min)*. The travel in this model is not treated as a separate activity (as within the cluster algorithm), but rather as an indivisible part of each activity. A normally distributed random number with the mean value and standard deviation obtained from the sample for a particular activity type at every time instant will be used.

Another important issue that needs to be addressed is the simulation of alone or joint activity participation in multi-adult households. We expect that all joint-activities will be the same in both patterns. The patterns of all adult household members will be simulated sequentially. First, the pattern of the first person in the household (head of the household), similarly to the previous example, will be simulated. In case some activity was estimated as a joint activity with spouse, the schedule of the spouse will define an exact part of her/his schedule. The remainder of the schedule will be simulated conditioned on the derived probabilities and the joint parts of the schedule. The probability of an activity to start at the end of the joint activity will be used; similarly the probability of an activity to end at the beginning of the joint activity will be used. The schedule will be simulated starting at the edges of the joint activities. In this way, we implicitly account for Hägestrand's coupling constraints and introduce a schedule hierarchy by imposing conditionality of one person's schedule on the first simulated schedule.

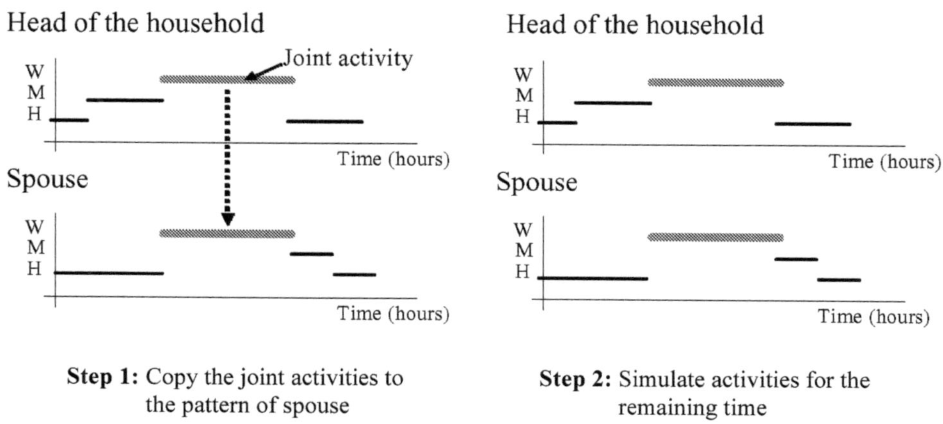

Figure 3.4
The Principle of Simulating Joint Activities for Multiple-Adult Households

OVERALL PERFORMANCE OF THE MODEL

The overview of the micro-simulation model of activity patterns was described above. Here, the methodology is applied to the data collected in Centre County, Pennsylvania and the results of the model are presented. The simulation does not attempt to mimic one particular activity pattern, rather it focuses on probabilities of the entire cluster. For this reason, the evaluation should be executed at the aggregate level, which means the averages of all patterns in a particular cluster will be compared.

Three ways of comparing the results are followed. First, the *activity profiles* (percentage involvement in particular activity types at every time instant) for each group in the observed and simulated data will be compared. Second, *the time spent in activities* by particular type will be evaluated. Third, the *numbers of episodes* (number of activities) simulated and observed per day will also be compared.

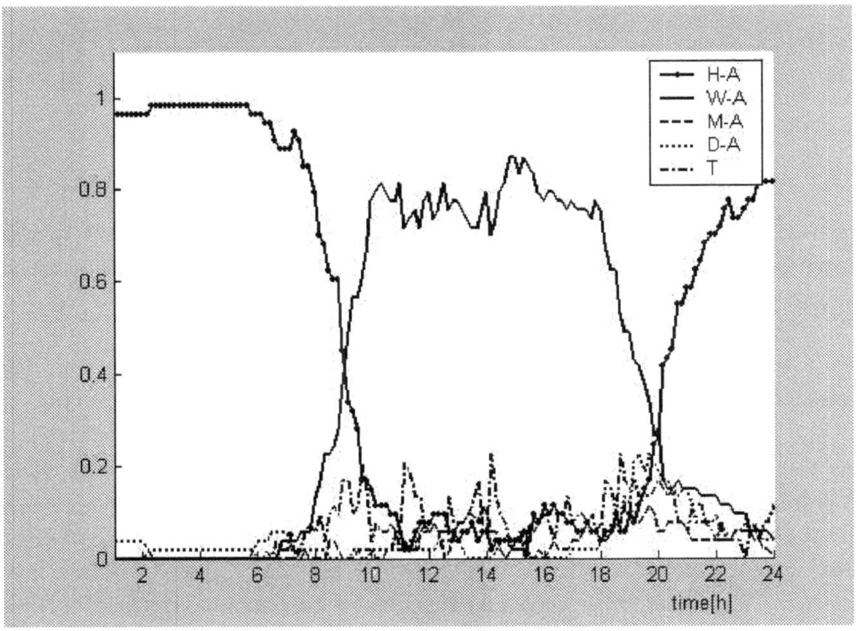

Figure 3.5
An Observed Activity Profile for One-Adult Household

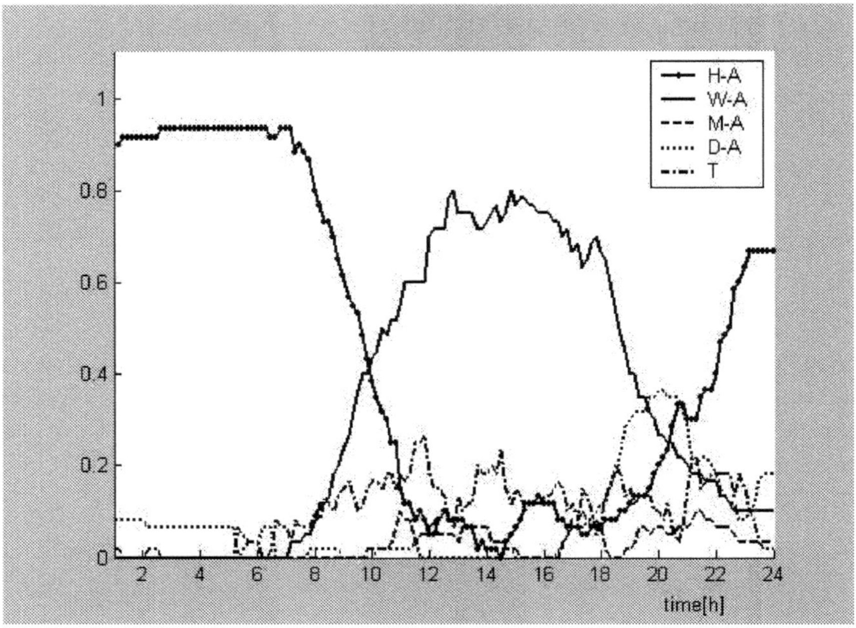

Figure 3.6
A Simulated Activity Profile for One-Adult Household

Legend: *H-A*...home alone, *H-S*...home with spouse, *W-A*...work alone, *W-S*...work with spouse, *M-A*...maintenance alone, *M-S*...maintenance with spouse, *D-A*...discretionary alone, *D-S*...discretionary with spouse, *T*...travel.

Evaluation of Activity Profiles

The first evaluation is based on comparisons of the so-called *activity profile*s of each cluster. An activity profile is an aggregate pattern of the entire cluster. In every time step, the percentage of individuals participating in different activity types is computed. Examples of observed and simulated activity patterns are shown in Figures 3.5 and 3.6. Figure 3.5 shows that until approximately 6 a.m. about 98 percent of all individuals in this group stayed "alone" at home. The remaining part of the individuals participated alone in some discretionary activity. Starting at about 8 a.m. , the percentage of individuals who are at work increases, and between 10 a.m. and 6 p.m. it fluctuates around 80 percent, only at about noon, there is a peak of maintenance activity (lunch).

Table 3.1
The Average MSE and P-Values for the Evaluation of Activities Profiles

		Evaluation of Activity Profiles	Evaluation of Time Spent on Activities
Activity Type		**Average MSE**	**Average P-Value**
H-A	home alone	0.0267	0.3080
H-S	home with spouse	0.0130	0.3426
W-A	work alone	0.0073	0.3786
W-S	work with spouse	0	0.4115
M-A	Maintenance alone	0.0023	0.4592
M-S	Maintenance with spouse	0.0010	0.4473
D-A	Discretionary alone	0.0053	0.5046
D-S	Discretionary with spouse	0.0010	0.4358
T	Travel	0.0117	0.2790
	MEAN	0.0100	0.3934

At about 8 p.m., there is a peak in discretionary activity participation. Late at night most individuals return home (up to 90 percent).

The activity profiles of the simulated patterns have similar trends; however, the numbers differ. In order to provide a numerical evaluation, the mean square error (MSE) between the observed and simulated activity profiles will be computed (Ma, 1997). This measure is applied to each particular activity type. The results are summarized in the third column of Table 3.1.

Evaluation of Time Spent on Activities

The second comparison focuses on the time spent in particular activities. In order to have more insight into the time-aspects of the results, the time spent in particular activities for particular groups will be compared for the entire day as well as for parts of the day. For the observed as well as simulated patterns we compute the average time spent in activities of all different types during the following time periods: (i) Morning hours (6 a.m. – 10 a.m.); (ii) Noon (10 a.m. – 2 p.m.); (iii) Afternoon hours (2 p.m. – 6 p.m.), and (iv) Evening hours (6 p.m. – 12 a.m.). There is not much activity before 6 a.m., so we will not focus on this time period. There are two parts to this evaluation; a) evaluation of the time spent in activities using the p-value and b) evaluation of the time spent in activities using a regression model.

Table 3.2

Strength of Relationship Between the Observed and Simulated Patterns

Model	R	R Square	Adjusted R Square	Std. Error of the Estimate
1	0.956	0.914	0.914	16.398

Evaluation of the Time Spent in Activities Using the P-Value. Once having the measured values of average time spent in activities, a statistical test should be performed to evaluate if the simulated patterns are actually similar to the observed patterns. There are different tests with different assumptions that could be used. Since we do not have any additional information on the values of the standard deviations, we assume that the standard deviations are not the same. Consequently, a *t-test for two unknown means and standard deviation* will be performed (Bhattacharyya and Johnson, 1977). The *p*-values of the null hypothesis that the values of time spent in activities in the simulated patterns are drawn from a distribution with the same mean as the sample of observed patterns are provided in the fourth column of Table 3.1. The lowest *p*-value (which implies the weakest estimate) was obtained for both in-home activities and travel activity.

Evaluation of the Time Spent in Activities Using a Regression Model. There is another comparison that will be performed to compare the time spent in activities. The time spent in particular activities for different clusters and different data sets can be written in two vectors; first corresponding to the observed data and the second representing the simulated data (i.e., we lose the table format, but we have corresponding values of observed and simulated patterns next to each other). These two vectors will be used in a simple regression model - the observed values as the dependent variable, and the simulated values as an explanatory variable. We use the R-square of the regression equation as a measure of the strength of the relationship between the vectors. The results of the regression model presented in this section are presented in Table 3.2. The value of 0.914 shows a very strong relationship between observed and simulated values.

Evaluation with Respect to the Number of Episodes

The last comparison focuses on the average number of activities in the simulated patterns as compared to those observed. It is an important indicator of the quality of the simulation especially for its practical use. The proposed model aims to replace the trip generation step in Urban Transportation Planning System. For this reason, the most important output of the model is the number of episodes in the simulated patterns. Figure 3.7 compares the average number of activities in the observed and simulated patterns within the entire activity patterns (during the simulated 24 hours).

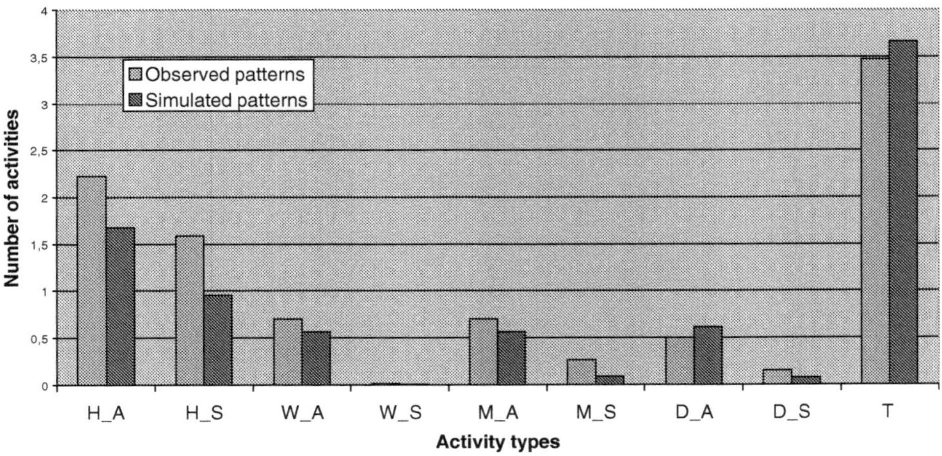

Figure 3.7
Comparison of the Average Number of Episodes Between Observed and Simulated Patterns (24 hours)

Legend: *H-A*...home alone, *H-S*...home with spouse, *W-A*...work alone, *W-S*...work with spouse, *M-A*...maintenance alone, *M-S*...maintenance with spouse, *D-A*...discretionary alone, *D-S*...discretionary with spouse, *T*...travel.

The values depicted in the figure show similar trends for all different activity types. It tends to underestimate the average number of most activities, except for discretionary activities conducted alone, and the travel activity; however, overall the results seem promising.

One of the drawbacks of traditional methods to travel demand analysis is that they are not accurate in estimating the number of trips for peak hours (see for example discussions provided at http://tmip.fhwa.dot.gov/clearinghouse/docs/amos/ch2.stm). For this reason, comparisons of the number of episodes between observed and simulated patterns during peak hours are provided. The four following figures compare the number of trips in the following time intervals: from 7 a.m. to 8 a.m., from 8 a.m. to 9 a.m., from 5 p.m. to 6 p.m., and finally from 6 p.m. to 7 p.m. The results are similar for all four time intervals. The largest discrepancies are obtained for home activities, both conducted alone or jointly. The average number of trips is also underestimated in all cases. The difference ranges from 0.1 to 0.2 trips per hour, which is still satisfactory for the majority of applications. The results of Pearson chi-squared statistic (Fienberg, 1981) are also provided in Table 3.3.

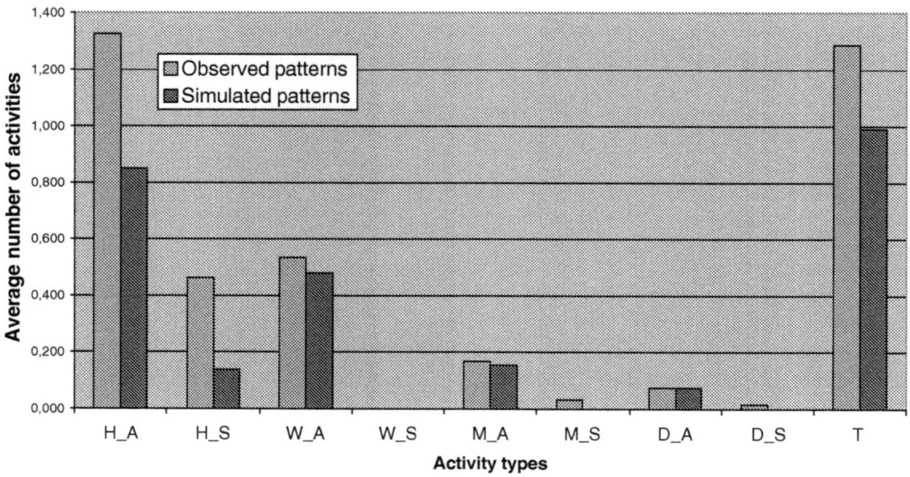

Figure 3.8
Comparison of Number of Episodes Between Observed and Simulated Patterns (Conducted between 7 a.m. and 8 a.m.)

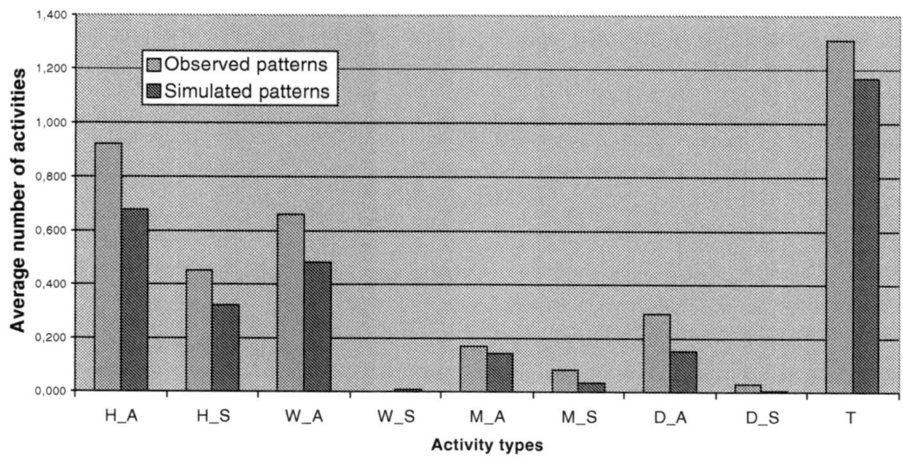

Figure 3.9
Comparison of Number of Episodes between Observed and Simulated Patterns (Conducted between 8 a.m. and 9 a.m.)

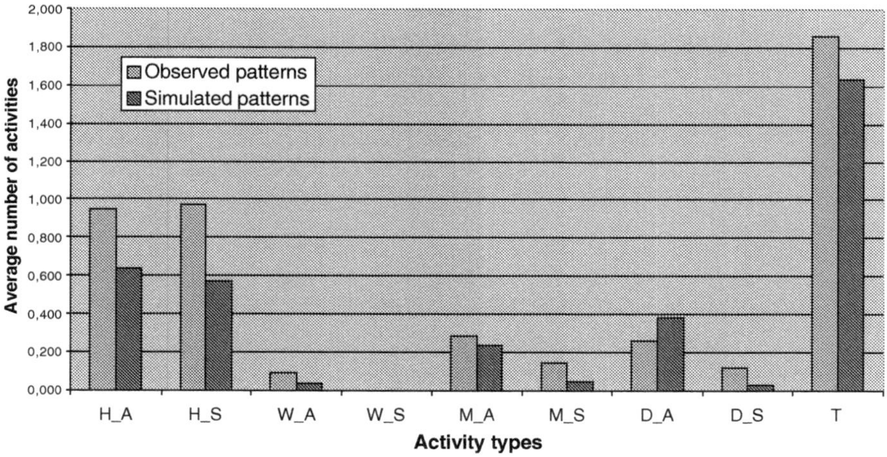

Figure 3.10
Comparison of Number of Episodes Between Observed and Simulated Patterns (Conducted between 5 p.m. and 6 p.m.)

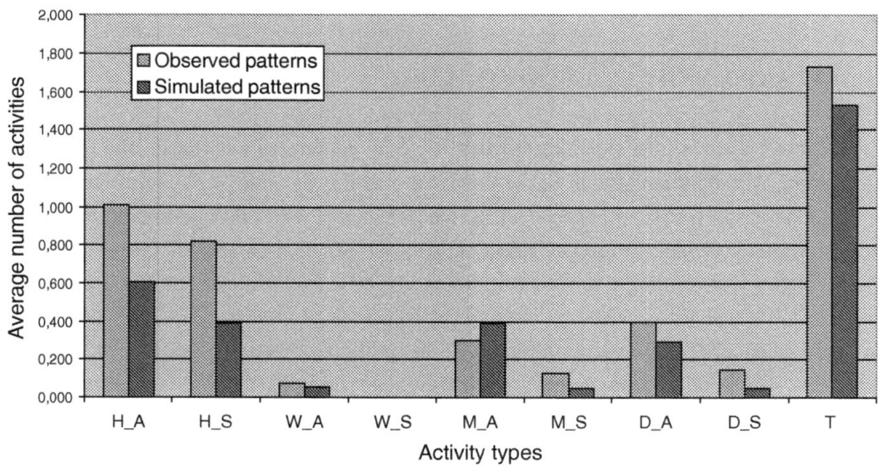

Figure 3.11
Comparison of Number of Episodes Between Observed and Simulated Patterns (Conducted between 6 p.m. and 7 p.m.)

Table 3.3
Similarity of the Frequency of the Number of Activities Between Observed and Simulated Patterns

Statistics	Entire Day	Morning Peak Hours		Afternoon Peak Hours	
		6 a.m. – 7 a.m.	7 a.m. – 8 a.m.	5 p.m. – 6 p.m.	6 p.m. – 7 p.m.
Test statistics - total $\div^2$	0.636	0.521	0.283	0.517	0.586
Degrees of freedom	8	7	7	7	7
Critical value *	15.51	14.07	14.07	14.07	14.07
Asymptotic significance	0.9996	0.9994	0.99996	0.9993	0.9991

* Critical value is computed for level of significance alpha = 0.05.

The values *Test statistics - total* $\div^2$ show the computed test statistics. In case of all peak hours (columns three to six in the table), there are no activities of the type "*work with spouse*" in the observed activity patterns. We have to exclude this activity type from the analysis and the number of degrees of freedom equals seven. The critical value is computed for level of significance alpha = 0.05. In all cases the critical values are much higher than values of the test statistics. Asymptotic significance is the estimated probability of obtaining a chi-square value greater than or equal to test statistics. The high significance values suggest that there is no statistical evidence that the number of trips in the observed and simulated patterns differ. This test really strongly supports the validity of the proposed model with respect to the number of activities.

CONCLUSIONS AND FUTURE STEPS

In this chapter a model that simulates the daily activity patterns of individuals and their households is presented. An important feature of the proposed model is its ability to simulate activities that are conducted alone as well as those that are conducted jointly with other household members. These interactions are essential for models based on activity participation and affect other features of the model, for example, vehicle availability or others (Arentze and Timmermans, 2000).

The performance of the entire model (not its particular steps) is evaluated from different viewpoints. The output of this model is each individual's activity pattern. In general, the simulated patterns are feasible and reasonable. The evaluation of performance of the model is performed at an aggregate level, however. Overall the performance of the model is encouraging. The simulated patterns seem to be reasonable and the simulated activity profiles are similar to those observed with respect to all compared characteristics. The model replicates the overall distribution of participation in particular activities, taking implicitly into consideration temporal constraints and constraints imposed by other household members. It also provides the level of detail required for advanced applications.

Although the results of the model are promising, there is indisputably room for improvement in terms of prediction accuracy, but also to address some additional issues and constraints. The most important issues that could lead to better performance of the model are summarized here. This model focused on the temporal aspects of activity patterns. The main objective was to find a model that could substitute and enhance the trip generation step of the UTPS model. However, an important issue in an activity-based model is the treatment of spatial characteristics of activity patterns. The proposed methodology could be easily adjusted to address this problem. Probabilistic tables of distances from home and distances from the previous activity could be derived in the second step of this model, similar to the probabilistic tables of activity participation and activity duration. This enhancement would not require any additional modifications of the algorithm. In the simulation step, these two distances would be assigned to a particular activity, based on the derived probabilities. In connection with GIS software, these distances would determine the traffic analysis zone to which the particular trip leads. Similarly, we could determine the travel mode for each trip, for example. It could be assigned based on probabilistic tables that would be derived in the second step of this model.

REFERENCES

Arentze, T.A. and H.J.P. Timmermans (2000). Albatross: *A Learning-Based Transportation Oriented Simulation System.* European Institute of Retailing and Services Studies. Eindhoven.

Arentze, T.A. and H.J.P. Timmermans (2003). Measuring impacts of condition variables in rule-based models of space-time choice behavior: method and empirical illustration. *Geo. An.*, **35**, 24-45.

Bhattacharyya, G.K. and R.A. Johnson (1977). *Statistical Concepts and Methods.* John Wiley & Sons, New York.

Bowman, J.L. and M.E. Ben-Akiva (1997). Activity-based travel forecasting. In: *Activity-Based Travel Forecasting Conference* (L.J. Engelke, ed.), pp. 33-36. Texas Transportation Institute, Austin.

Fienberg, S.E. (1981). *The Analysis of Cross-Classified Categorical Data.* The MIT Press, Cambridge.

Fridlyand, J. (2001). Resampling Methods for Variable Selection and Classification: Applications to Genomics. Ph.D. thesis, UC Berkeley.

Goulias, K.G. (2004). Transportation planning and modeling, In: *Handbook of Transportation Engineering* (M. Kutz, ed.), pp. 29.3-29.25. McGraw-Hill, New York.

Huang, Z. (1997). Clustering large data sets with mixed numeric and categorical values. *Proc. First Pacific-Asia Conf. Knowl. Disc. & Data Mining*. World Scientific, Singapore, 21-34.

Joh, C.-H, T.A. Arentze and H.J.P. Timmermans (2001a). A position-sensitive sequence alignment method illustrated for space-time activity diary data. *Env. and Plann. A,* **31**, 313-338.

Joh, C.-H, Arentze, T.A. and Timmermans, H.J.P. (2001b). Activity-travel pattern similarity: a multidimensional alignment method, *Transpn. Res. B,* **36**, 385-403.

Joh, C.-H, T.A. Arentze and H.J.P. Timmermans (2001c). Multidimensional sequence alignment methods for activity-travel pattern analysis: a comparison of dynamic programming and genetic algorithms, *Geo. An.,* **33**, 247-270.

Joh, C.-H., T.A. Arentze and H.J.P. Timmermans (2001d). Pattern recognition in complex activity-travel patterns: a comparison of Euclidean distance, signal processing theoretical, and multidimensional sequence alignment methods. *Transpn. Res. Rec.*, **1752**, 16-23.

Jones, P., F.S. Koppelman and J.-P. Orfeuil (1990). Activity analysis: state-of-the-art and future directions. In: *Developments in Dynamic and Activity-Based Approaches to Travel Analysis* (P. Jones, ed.), pp. 34-55. Gower, Aldershot.

Kass, G.V. (1980). An exploratory technique for investigating large quantities of categorical data. *Appl. Stat.*, **29**, 119-127.

Kaufman, L. and P.J. Rousseeuw (1990). *Finding Groups in Data, An Introduction to Cluster Analysis*. John Willey & Sons, New York.

Kulkarni, A.A. and M.G. McNally (2000). A Microsimulation of Daily Activity Patterns. Institute of Transportation Studies, University of California, Irvine.

Lim, T.-S., W.-Y. Loh and Y.-S. Shih (1998). An Empirical Comparison of Decision Trees and Other Classification Methods. Technical Report 979, Department of Statistics, University of Wisconsin, Madison.

Lucasius, C.B., A.D. Dane and G. Kateman (1993). On k-medoid clustering of large data sets with the aid of a genetic algorithm: background, feasibility and comparison. *Anal. Chim. Acta*, **282**, 647-669.

Ma, J. (1997). An Activity-Based Approach and Micro-simulated Travel Forecasting System: A Pragmatic Synthetic Scheduling Approach. PhD Thesis, The Pennsylvania State University, Department of Civil and Environmental Engineering, University Park, Pennsylvania.

Maulik, U. and S. Bandyopadhyay (2000). Genetic algorithm-based clustering technique. *Patt. Recogn.*, **33**, 1455-1465.

McNally, M.G. (2000). The activity-based approach. In: *Handbook of Transport Modelling* (D.A. Hensher and K.J. Button, eds.), pp. 53-68. Elsevier Science, Oxford.

Ortuzar, J. de D. and L.G. Willumsen (1994). *Modelling Transport.* John Wiley & Sons, New York.

Patten, M.L. and K.G. Goulias (2004). Integrated survey design for a household activity-travel survey in centre county, Pennsylvania. *Proc. 83rd Ann. Meet. Transpn. Res. B.,* Washington, D.C. (CD-Rom).

Pribyl, O. (2003). Clustering of activity patterns using genetic algorithms. Paper presented at the 8th On-line World Conference on Soft Computing in Industrial Applications.

Pribyl, O. (2004). A Microsimulation Model of Activity Patterns and Within Household Interactions. PhD Thesis, The Pennsylvania State University, Department of Civil and Environmental Engineering, University Park, Pennsylvania.

Stopher, P.R. and H.M.A. Metcalf (1999). Household activities, life cycle, and role allocation tests on data sets from Boston and Salt Lake City. *Transpn. Res. Rec.*, **1676**, 95-102.

Wang, R. (1996). An Activity-Based Microsimulation Model. Ph.D. Dissertation. UC Irvine, Irvine, CA.

Wets, G., K. Vanhoof, T.A. Arentze and H.J.P. Timmermans (2000). Identifying decision structures underlying activity patterns: an exploration of data mining algorithms. *Transpn. Res. Rec.*, **1718**, 1-9.

4

SIMULATING DAILY ACTIVITY PATTERNS THROUGH THE IDENTIFICATION OF SEQUENTIAL DEPENDENCIES

Davy Janssens, Limburgs Universitair Centrum, Diepenbeek, Belgium

Geert Wets, Limburgs Universitair Centrum, Diepenbeek, Belgium

Tom Brijs, Limburgs Universitair Centrum, Diepenbeek, Belgium

Koen Vanhoof, Limburgs Universitair Centrum, Diepenbeek, Belgium

INTRODUCTION

For the last decade, activity-based transportation models have set the standard for modelling travel demand. The research community has witnessed a multitude of modelling approaches, which can mainly be divided into utility-based transportation models and computational process models. Utility-based models typically use econometric techniques and assume that individuals evaluate a number of complete, one- or several-day activity-travel patterns and choose the particular pattern that maximises their utility. Examples of these models are the Daily Activity Schedule Model (Ben-Akiva *et al.*, 1996), Wen and Koppelman (1999), CEMDAP (Bhat *et al.*, 2004) and many others.

Several scholars however have argued that people do not necessarily arrive at "optimal" choices, but rather use heuristics that may be context-dependent. In their most simple form, the resulting computational process models are based on a set of IF-THEN rules, which take on the following form: IF (condition=X) THEN (perform action Y). Dependent on the context or the situation an individual faces, another outcome or another decision is taken. The first conceptual framework,

Progress in Activity-Based Analysis edited by H. Timmermans

which is based on these decision heuristics is the SCHEDULER model (Gärling *et al.*, 1989). Another model system that resembles computational process modelling is AMOS, a dynamic micro-simulator of household activities and travel over time and space (Pendyala *et al.*, 1998). The AMOS model uses micro-simulation techniques to predict a traveller's adaptation from a baseline activity-travel pattern. Another important advanced operational process model is Albatross: a learning-based transportation oriented simulation system (Arentze and Timmermans, 2000, 2004; Arentze *et al.*, 2003).

While activity-based models can be differentiated into one of these two categories, both approaches try to find the most accurate representation of observed distributions of travel characteristics, consistent with their underlying theory about human decision-making. Based on this most accurate representation, activity patterns are then generated. Given however a set of variables that are assumed to influence (aspects of) activity-travel behaviour, and assuming that these variables are known for the target population, micro-simulation can equally be used to predict activity-travel patterns. For instance, McNally (1995), Kulkarni and McNally (2000, 2001), and Recker *et al.* (1986a, 1986b), proposed the use of representative activity patterns (RAP), and suggested to simulate activity facets such as purpose and duration by drawing from the distributions that are associated with the target pattern. In the identification of RAP's, segmentation (clustering) approaches can be adopted to derive more homogeneous and hence more representative activity patterns, based on variables that are assumed to influence the activity-travel pattern (for instance socio-demographic variables). The identified RAP's are then used for simulating and predicting new activity patterns.

While the representativeness of such representative patterns has been questioned due to fluctuations over longer periods of time (Clarke, 1985) and regional differences (Veldhuisen *et al.*, 2000), simulation approaches are also susceptible to the criticism that they do not account for sequential information and sequential dependencies in the identification of representative activity patterns. However, it has been shown that the inclusion of sequential information in deriving a segmentation of activity-travel patterns produces better results (Joh *et al.*, 2002). Hence, the omission of sequential information in the identification of activity-travel patterns may be a deficiency, especially because some skeleton activity structure is often assumed first. For instance, Vaughn *et al.* (1997), emphasized the need for a skeletal structure which imposes time-space constraints and simplifies the simulation of the remaining facets of the activity-travel pattern. Similarly, Kitamura and Kermanshan (1983) introduced a technique in which the different characteristics of the set of activities can be generated sequentially using a Markov approach.

The aim of this chapter is to explicitly incorporate sequential information and dependencies in the identification of representative activity patterns. In addition, a test which determines the optimal historical (sequential) information that needs to be taken into account is incorporated in the framework. Once the skeleton activity-pattern has been identified, additional facets such as location and time (duration) information are generated. The simulation framework that is proposed in this chapter is able to generate these facets in an iterative heuristic manner. The presented approach is an extension of previous studies (Janssens *et al.*, 2004, 2005) in which it is assumed that each activity pattern only consists of a set of correlated successive observations of activities and travel characteristics (transport modes). The heuristic nature of the proposed simulation processes and the fact that all sequential information is immediately derived from the data itself, facilitate the immediate applicability of the simulation procedures in other study areas or regional settings, at least when regional data is available. Indeed, existing simulation and computational process models often use procedures or detailed variable-specific information that is tuned towards a particular regional setting, which makes transferability sometimes difficult, due to the lack of this specific information or due to the infeasibility of these procedures in other regions. The next section of this chapter portrays the general outline of the simulation framework.

OVERVIEW OF THE SIMULATION FRAMEWORK

It can be seen from Figure 4.1 that the transportation model presented in this chapter is a sequential process that first involves the simulation of a sequence of in-home and out-of-home activities along with a particular transport mode during a single day. Next, the general skeleton of the activity pattern is used as the basis for simulating time and location facets.

With respect to the simulation of activities and transport modes, it is assumed that each sequence in the activity pattern consists of a set of correlated successive observations of a random variable. To this end, a discrete random variable X_t is considered, taking values from the finite set $\{1,\ldots, m\}$, where each value represents an activity that occurs in a persons' activity pattern. Travelling is considered as an activity as well, however the transport mode is added as an additional attribute in this case. For this reason, the parameter m contains the number of non-travel activities and transport modes that occur in an activity pattern. The index t represents the position of the activity in the activity pattern.

In this first step, the goal is to generate (predict) the value taken of X_t as a function of the values taken by previous observations of this variable. Obviously, the most important question here is to investigate which number of previous observations can best explain the current observation in the activity pattern. In the limit, the current value taken by X_t can be entirely explained by the previous observation (Activity t-1), which is also referred to as the first lag. Analogously, in the limit, it might only be possible to accurately explain the current value of X_t by the last k-1 observations

(Activity *t-1*, Activity *t-2*, …, Activity *k -1*) (i.e. $k\text{-}1^{th}$ lag) in which *k* represents the length of the activity pattern. However, when the current value can only be explained by a relative large number of previous observations, it is unlikely that the information that is identified is suitable for generalisation (prediction) purposes. On the other hand, a low number of previous observations might not be sufficient to explain the current observation either. This trade-off between accuracy versus generalisation will be validated later in the chapter. It is assumed that an optimal skeleton (i.e. the skeleton with that lag that achieves the best match to the observed patterns) is able to generate better start and end times and location information than a suboptimal skeleton (Figure 4.1). The general description about how this dependency and correlation information can be taken into account is described in the first part of the methodological section. Once the optimal skeleton is selected, duration probability distributions from the sample dataset are used to generate the start and end times of each activity in the skeleton. The procedure for doing this is shown in the second part of the methodological section. Finally, origin-destination information from the sample data is used to generate the location where each activity in the skeleton is performed. The methodology is presented in the final part of that particular section. The chapter concludes with a description of the experimental results and with conclusions and topics for future research.

It should be emphasized that in the simulation framework, the dataset is split up into a training and a validation set. The training set is used for generating the information, whereas the test set is used to validate the generated patterns. Splitting the dataset is widely used to account for overfitting, which occurs when high "accuracy" of the simulation model is achieved on the training set, while rather low "accuracy" is obtained on the unseen test set.

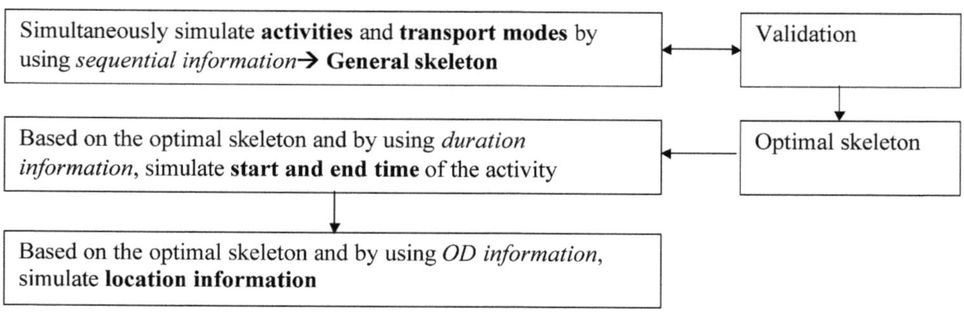

Figure 4.1
Overview of the Simulation Framework

METHODOLOGY

Generate the General Skeleton

Introducing Transition Probability Matrices. Markov chains are probabilistic models, which are commonly used to model dependencies in data (Stewart, 1995). While Markov chains and the corresponding transition probabilities have been used in a number of discrete choice models (Lerman, 1979; Kitamura and Kermanshah, 1983; van der Hoorn, 1983), their full and unmodified application turned out to be infeasible in our simulation framework. The most important drawback is that the number of independent parameters increases exponentially and becomes too large to be estimated. Moreover, it is important to notice that in calculating transition probabilities in Markov Chains, the independent character of each activity pattern in the data is in fact ignored, which may result in estimates that are seriously biased by specific combinations that may appear in one particular sequence. A more detailed discussion is given in Janssens *et al.* (2005). In order to cope with both these drawbacks, an adapted methodology to calculate transition probabilities and to generate activity patterns was developed and empirically tested.

We will discusses different approaches for storing sequential information (sequences of activities) in 'activity bundles', a term which is introduced to reflect that the information which is kept here represents low- and high-order combinations of activities that typically sequentially occur in one particular activity pattern. Activity bundles are constructed *per activity pattern* (i.e. for each respondent). Aggregating activity bundles will result in transition probability matrices, which give an idea about the sequential information for the whole *sample population.* In this sense, the activity bundles need to be interpreted as an intermediate but crucial step before building transition probability matrices.

As indicated before, it is possible that the current value taken by X_t can be entirely explained by the previous observation (Activity *t-1*), with the index t representing the current position of the activity in the activity pattern. In this case, it is in fact assumed that:

$$P(X_t = i_0 \, | X_0 = i_t, \ldots, X_{t-1} = i_1) = P(X_t = i_0 \, | X_{t-1} = i_1) = q_{i_1 i_0}(t), \text{ where } i_t, \ldots, i_0 \in \{1, \ldots, m\} \quad (1)$$

Each value in the set {1,..., m} represents a non-travel activity or transport mode that occurs in a persons activity pattern. Considering all combinations of i_1 and i_0, we can now construct a probability matrix Q, each of whose rows sums to 1.

$$Q = \begin{array}{c|cccc|} & & X_t & & \\ X_{t-1} & 1 & \cdots & \cdots & m \\ \hline 1 & q_{11} & \cdots & \cdots & q_{1m} \\ \vdots & \vdots & \ddots & & \vdots \\ \vdots & \vdots & & \ddots & \vdots \\ m & q_{m1} & \cdots & \cdots & q_{mm} \end{array} \qquad (2)$$

Suppose now that we wish to take into account that the current value of X_t is not only explained by the first lag, but also by an additional second lag. To illustrate, assume that the variable X_t originally takes the values in the state space {1,2,3}, where "1" stands for instance for "Sleeping", "2" stands for "Eating" and "3" stands for "Working". When one wishes to take into account that X_t is not only explained by the first lag (t-1) but also by a second lag (t-2), the state space can be redefined as {(1,1)}, indicating that this person was Sleeping at moment t-2 and at moment t-1, and with {(1,2), (1,3), (2,1), (2,2), (2,3), (3,1), (3,2), (3,3)} defined similarly. In this case, the number of previous observations that are taken into account equals 2, which is also referred to as the order (ℓ) of the transition probability matrix. The number of activities that occurs in this person's activity pattern (m) equals 3. The corresponding transition matrix Q for $\ell = 2$ and $m = 3$ is then:

$$Q = \begin{array}{cc|ccc|} & & & X_t & \\ X_{t-1} & X_{t-2} & 1 & 2 & 3 \\ \hline 1 & 1 & q_{111} & q_{112} & q_{113} \\ 2 & 1 & q_{211} & q_{212} & q_{213} \\ 3 & 1 & q_{311} & q_{312} & q_{313} \\ 1 & 2 & q_{121} & q_{122} & q_{123} \\ 2 & 2 & q_{221} & q_{222} & q_{223} \\ 3 & 2 & q_{321} & q_{322} & q_{323} \\ 1 & 3 & q_{131} & q_{132} & q_{133} \\ 2 & 3 & q_{231} & q_{232} & q_{233} \\ 3 & 3 & q_{331} & q_{332} & q_{333} \end{array} \qquad (3)$$

More general, in an ℓ-th order transition probability matrix it is implicitly assumed that

$$P(X_t = i_0 \mid X_0 = i_t, \ldots, X_{t-1} = i_1) = P(X_t = i_0 \mid X_{t-\ell} = i_\ell, \ldots, X_{t-1} = i_1) = q_{i_\ell,\ldots,i_0}(t) \qquad (4)$$

Obviously, the way in which the transition probabilities $q_{i_\ell \ldots i_0}$ are calculated determines the quality of the transition probability matrix. In order to avoid that transition probabilities are calculated for all sequences at once, as it is often done in Markov Chains, the idea of calculating probabilities for each respondent by means of activity bundles was developed.

From Activity Bundles to Transition Probability Matrices. The approach presented here estimates the probabilities based on the frequencies, observed in the data. These frequencies are maximum likelihood estimates, calculated as $N_{ij}/N_{i.}$, where N_{ij} equals the number of transitions from state i to state j in each activity pattern and $N_{i.}$ represents the number of transitions starting from state i in the pattern. The algorithm, which is used to construct an activity bundle of the ℓ-th-order, is displayed in Figure 4.2. The algorithm starts with first-order activity bundles. To this end, the unique elements (activities) in the activity pattern are identified. Next, the number of times that an activity bundle starts with that unique element in a particular activity sequence, is calculated (n_i).

```
Set k:=length of the sequence (diary)
Set ℓ:=1                                  // ℓ is the order of the activity bundles
Do while an ℓ-th activity bundle can still be constructed (ℓ <k)
Begin
If ℓ =1 then
  begin
        identify all unique elements u, with u ∈ {1,..., m}
        for each unique element u do
          begin
            set i:=current unique element
            calculate the number of transitions that start from state i in the activity pattern (n_i.)
            identify all the elements ij which follow immediately after i, with j ∈ {1,..., m}
            for each ij do count the number of times that ij occurs (n_ij)
           store each ij and each weight (n_ij/n_i.) in the first-order activity bundle
          end
  end
else
  begin
        read the (ℓ-1)-th order activity bundle
        for each combination A in the (ℓ-1)-th order activity bundle do
          begin
            set A:=current combination
            calculate the number of transitions that start from A in the activity pattern (n_A)
            identify all the elements Aj which follow immediately after A, with j ∈ {1,..., m}
            for each Aj count the number of times that Aj occurs (n_Aj)
           store each Aj and each weight (n_Aj/ n_A) in the ℓ-th-order activity bundle
          end
  end
ℓ:=ℓ+1
end
```

Figure 4.2

The Construction of Activity Bundles with Maximum Likelihood Estimates

Also, the frequency that an activity immediately follows after the current unique activity is stored (n_{ij}). Then, the weight of the unique activity is calculated, dividing n_{ij} by $n_{i.}$. This procedure is repeated for every unique element. When dealing with activity bundles of a higher order (> 1), the algorithm continues with the activity bundle that immediately precedes the current activity bundle, while the procedure for calculating the weights remains similar. It can thus be seen from Figure 4.2 that the construction of higher-order activity bundles is based upon the activity bundle which immediately precedes the current higher-order activity bundle. By doing this, the bundles can be built in a more efficient manner. Consider the following example to illustrate the algorithm, where T_c= **T**ransportation, with **c**ar as transport mode, F=visit **F**amily, E=**E**at, and R=**R**ead.

Activity pattern 1: T_cFFFFFFFFFFFFFFFFE
Activity pattern 2: T_cEEFREREERFT$_c$FT$_c$FFT$_c$FET$_c$F
Activity pattern 3: RREFEFEET$_c$T$_c$R
Activity pattern 4: EEFFT$_c$FT$_c$FRRT$_c$T$_c$RT$_c$RR
Activity pattern 5: FFT$_c$FFRE
Activity pattern 6: EET$_c$FRRE

The first-order activity bundles of these activity patterns are shown in Table 1, with weights shown in brackets. Based on these activity bundles, the final first-order transition probability matrix can be constructed for this example by aggregating the same bundles of activities across the different activity patterns (Table 2). Normalizing Table 2 such that each row sums to one, gives a first-order transition probability matrix as defined previously. After the transition probability matrices are generated (for different orders), one can adopt this information to generate in-home, out-home and travel activities in the pattern. This procedure is described in the next section.

Table 4.1
A First-order Example of Activity Bundles

Seq. number	First-order Combinations
Activity pattern 1	T$_c$-F (1); F-F (0.94); F-E (0.06)
Activity pattern 2	T$_c$-E (0.2); T$_c$-F (0.8); E-E (0.33); E-F (0.17); E-R (0.33); E-T$_c$ (0.17); F-R (0.17); F-T$_c$ (0.5); F-F (0.17) ; F-E (0.17) ; R-E (0.67); R-F (0.33)
Activity pattern 3	R-R (0.5); R-E (0.5); E-F (0.5); E-E (0.25);E-T$_c$ (0.25); F-E (1); T$_c$-T$_c$ (0.5); T$_c$-R (0.5)
Activity pattern 4	E-E (0.5); E-F (0.5); F-F (0.25); F-T$_c$ (0.5); F-R (0.25); T$_c$-F (0.4); T$_c$-T$_c$ (0.2); T$_c$-R (0.4); R-R (0.5); R-T$_c$ (0.5)
Activity pattern 5	F-F (0.5); F-T$_c$ (0.25); F-R (0.25); T$_c$-F(1); R-E (1)
Activity pattern 6	E-E (0.5) ; E-T$_c$ (0.5) ; T$_c$-F (1) ; F-R (1) ; R-R (0.5); R-E (0.5)

Table 4.2
The Final First-order Transition Probability Matrix

		X_t			
	X_{t-1}	T_c	E	R	F
	T_c	0.70	0.20	0.90	4.20
$Q=$	E	0.92	1.58	0.33	1.17
	R	0 .50	2.67	1.50	0.33
	F	1 .25	1.23	1.67	1.85

Generate the General Skeleton of an Activity Pattern. The aim of the simulation procedure is to predict the value taken by X_t as a function of the values taken by previous observations of this variable. The procedure for simulating the activity patterns is described in Figure 4.3. The left part of this figure shows the different steps of the procedure; the right part shows the real outcome of these steps by means of an example. The procedure starts by initialising the values of the indexes t and "diarypointer". The index t is preferably interpreted as the position of the activity in the activity pattern, whereas the "diarypointer" is a kind of technical index that keeps track of the lag, which is used in the simulations. The diarypointer is always initialised at position zero; the index t is variable and is set equal to the order of the transition probability matrix. The second order transition probability matrix is considered as an example (see right part).

Reading the transition probability matrix is the first logical step. The length of the activity pattern is generated using Monte Carlo simulation. This implies for this step that a random number is generated, based on a given sample distribution of the length of the activity pattern. The decision was made to incorporate this dimension into the generated activity patterns since some people fill out their diaries carefully (or simply perform more activities), while others are more imprecise. In our example, it is assumed that 15 activities will be generated.

Next, the first ℓ elements in the activity patterns are generated. The initial sequential probability distributions in the sample data are used for the Monte Carlo simulation. This means that the first ℓ elements of the sequence are generated from the prior probability distributions, which are in the data and not from the empirically constructed transition matrices. Assume that a sleep and an eat-activity are the first two elements, which are simulated. The diarypointer can now be augmented from zero to one in order to keep track of the two lags that are used in the example. Note that these and the subsequent steps of the simulation procedure will only occur when the order of the activity bundles (ℓ) does not exceed the simulated length.

The next step is to search for the combinations of elements in the transition probability matrix in the interval [diarypointer..t]. This means for our example that the Sleep-Eat-combination is looked up in the transition matrix and that the distribution which is in this row of the table is used as a constraint for simulating the next activity. If no combination of elements is found, the procedure stops simulating elements for this particular activity pattern (not shown in Figure 4.3).

After the "diarypointer" and the index t are augmented, the element ("transportation by car" in our example) is stored at position t (i.e. 3) in the activity pattern. The simulation procedure is repetitive, i.e. when the prediction of the value X_t is based on two lags, then the next value to be predicted becomes X_{t+1}, which is based on X_t (predicted in previous step) and on X_{t-1}. This repetition continues until the generated activity pattern equals the simulated length of the pattern. This procedure is repeated for every activity pattern in the data set. Next, the simulated skeleton of every activity pattern is used to generate time and location information (see also Kulkarni and McNally, 2001). A validation procedure is first executed to select the most appropriate lag. However, for the sake of clarity, this step is only described in the empirical section. In addition to the selection of the most appropriate lag, the accuracy of the generated general skeleton may be improved by using transition probability matrices, dependent on time of day. This is a topic for future research.

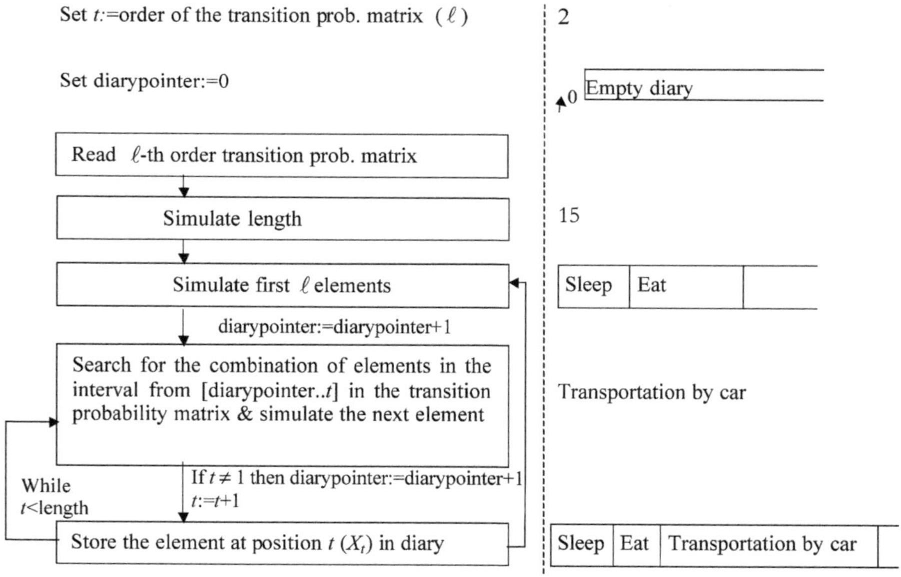

Figure 4.3
Description of the Simulation Procedure for Generating a General Skeleton

Simulating Time Information

Consistent with the generation of the general skeleton of activities and transport modes, the generation of time information is also based on empirically constructed sample probability distributions. To this end, in a first step, when simulating start and end times, probability distributions about the duration of activities need to be derived. It is assumed that the length of the general skeleton has a major impact on the duration of the activities. To give an example, if the generated skeleton is short in terms of number of activities, this may indicate that people have not very carefully filled out their diary and supposedly have aggregated some activities into one category, which lasts much longer. After all, diaries are collected during a 24-hour period, and thus logically when the number of activities in the diary is smaller the duration per activity on average is larger.The derived duration probability distribution is split up according to the length of the generated general skeleton. Skeletons are discretized into a number of intervals using a simple equal frequency discretization method, which categorizes the length of the activity pattern into a fixed number of intervals, each interval containing an "equal" number of observations. Duration probability distributions are constructed for every discretized interval.

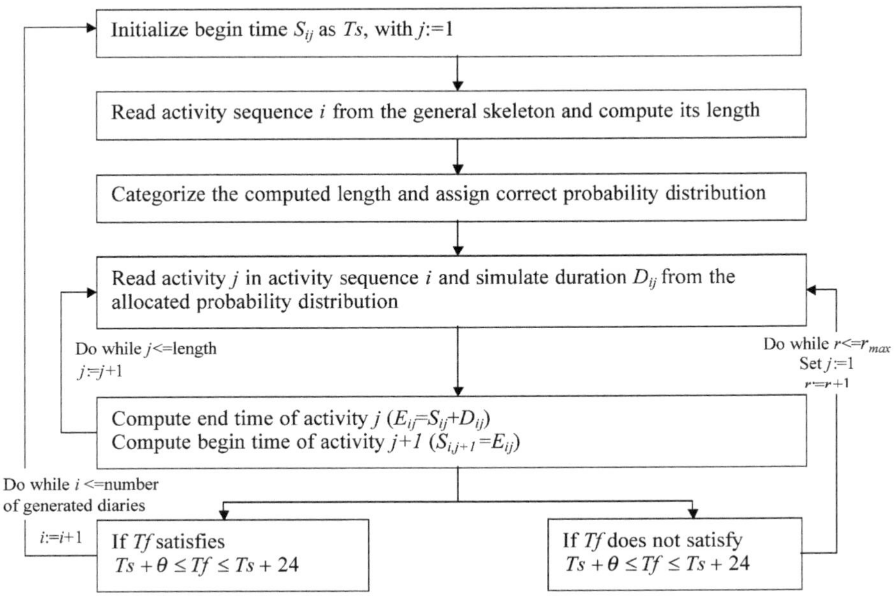

Figure 4.4
Description of the Simulation Framework for Generating Time Information

The framework for simulating start and end times is shown in Figure 4.4. The procedure first starts with an initialisation of the start time of the first activity (*j*:=1) in the activity sequence (*i*), which is defined as *Ts*. *Ts* is a parameter which can be tuned in each simulation, however it is kept constant in the experiments reported later. After this step, the length of the general skeleton of activities is computed, categorized into one of the intervals that were defined before and the probability distribution, which is associated with this interval is assigned accordingly. The duration of each activity in the generated general skeleton is then generated using Monte Carlo simulation from the previously allocated probability distribution for this sequence. Next, the end time of the present activity is simply computed by adding the simulated duration to the start time of this activity. Since it is assumed that there are no gaps in the activity patterns, computing the end time of the current activity simply gives us the begin time of the next activity. This loop procedure is repeated for the full length of the activity pattern.

The final step in the simulation framework in Figure 4.4 performs a check to control whether the generated start and end times are simulated within an acceptable error range. Therefore, the end time of the final activity in a persons' diary is defined as *Tf*. If *Tf* is within a time window that is defined as $Ts + \theta \leq Tf \leq Ts + 24$, the simulation is considered to be reliable under the assumed error parameter θ. Indeed, this procedure only checks the end time of the final activity, but since every begin and end time is connected with each other, it gives us a quite reliable estimate of the overall quality of the time simulation in the patterns. If considered to be satisfactory, the index i is augmented and the time simulation of a new activity sequence can start. Alternatively, if *Tf* does not pass the test, the full time information simulation is repeated for the first activity in the sequence. In this case, a parameter $r_{\max}$ is defined to account for exceptions (i.e., time simulations which are beyond the error range). It should be clear that there is no distinction in this simulation between in-home, out-home or travelling. Every transport mode (travel) has its own probability distribution which is derived from the training data, just as the in-home and out-home activities.

Simulating Location Information

The last step in the simulation framework deals with the generation of location information. To this end, and consistent with the general concept presented in the chapter, origin-destination matrices are derived from the training data set. To establish a link with the time information that is generated by means of the procedure that was described previously, it is assumed that the duration of the trip (with a particular transport mode) has an important impact on the (origin and) destination of activities. Indeed, common sense lets us believe that the longer the duration of the trip, the more likely it is that locations are visited further away from the start location (i.e. number of out-of-area locations is likely to increase). To this end, origins and destinations are coupled with the duration of the travelling which is needed to get there. That is, the derived origin and destination matrices are split up according to the duration of the trip. As before, the equal frequency interval method is used to categorize durations of the trip. The framework for simulating locations is shown in Figure 4.5. The procedure first starts with a random assignment of a home location to the activity sequence under consideration, based on a list of locations in the study area and a category: "out-of-area". Then, the length of the activity sequence is computed. Next, an element in the sequence is read by means of a loop procedure until a transport mode is encountered, and it is evaluated whether this is the first transport mode in the sequence or not. In the first case, the home location is assigned to the activity that immediately comes before the trip ($X_{i,j-1}$). This implies that the simulation framework assumes that every first travel episode in an activity pattern starts at the home location.

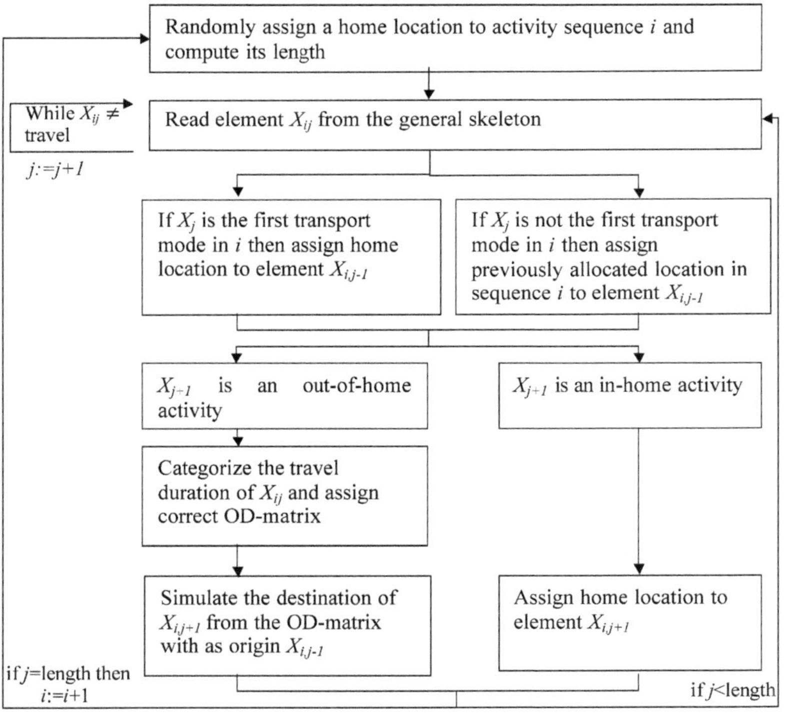

Figure 4.5
Description of the Simulation Framework for Generating Location Information

A small experiment supported this assumption. In the latter case, the location that was previously allocated in the sequence, is assigned to the activity that immediately comes before the trip ($X_{i,j-1}$). By doing this, the simulation does not allow errors, which means that somebody cannot perform an activity at a particular location, and immediately after this perform another activity at another location, without first reporting transportation to get there. In the next step, it is evaluated whether the activity that follows immediately after the trip (X_{j+1}) is an out-of-home activity or not. In case it is an in-home activity, the simulation of the location is simple; i.e. the home location of the sequence is allocated to element $X_{i,j+1}$. On the other hand, in case of an out-of-home activity, the duration of the transport mode is categorized into one of the intervals that were defined before and the OD-matrix that is associated with this interval is assigned accordingly. After this, the destination location $X_{i,j+1}$ is generated by means of Monte Carlo simulation in which the different number of trips that are present in the particular row of element $X_{i,j-1}$ of the origin-destination matrix are used as a probability distribution that serves as a constraint for simulating the next location.

EMPIRICAL APPLICATION

The Data

The activity diary data used in this study were collected in the municipalities of Hendrik-Ido-Ambacht and Zwijndrecht in the Netherlands (South Rotterdam region) to develop the Albatross model system (Arentze and Timmermans, 2000). The data involve a full activity diary, implying that both in-home and out-of-home activities were reported. Respondents were asked, for each successive activity, to provide information about the nature of the activity, the day, start and end time, the location where the activity took place, the transport mode and the travel time. A pre-coded scheme was used for activity reporting. Eighteen different activity classes and five different transport modes were distinguished. The activity categories are work or study in-home, bring or get persons or goods, daily shopping, non-daily shopping, service activity, medical visit, eating or drinking, sleeping, out-of-home leisure, in-home leisure, in-home non-leisure (household tasks), out-of-home non-leisure, receive social visit, bring social visit, work or study out-of-home, return home (e.g. drop bags), "other" and "missing" activities. The transport modes, which respondents could report were car (as driver or as car-passenger), walk, bike and public transport. Instead of using intervals, users were asked to report exact start, end and travel times. The sample, which was used in this study contains 1847 person-day diaries. In order to be able to test the transition matrices on a holdout sample, only 75 percent of these diaries (1385) were used as the training set for building transition matrices, OD matrices and duration probability distributions. Given the large number of observations, the 25 percent subset (462) was judged to be sufficiently large for a reliable validation set.

Generating the General Skeleton

Deriving Transition Probability Matrices. The first step in the generation of a general skeleton of activities and transport modes is the development of transition probabilities by means of the algorithm that is shown in Figure 4.2. There are 23 first-order transition probability matrices as 23 different activity and travel categories were distinguished in the activity patterns. The sequential information that is revealed seems logical and useful for low-order combinations. Table 3 provides an example of a 5 x 5 second-order probability matrix, which is randomly selected from the full 325 x 23 matrix. Indeed, the identified combinations seem credible. For example, the combination "Eat/drink – In-home Leisure – Work (out-of-home)" never occurs. Also, people first need transportation to arrive at work. The reader may also note that this matrix is the result of the aggregation of equal activity bundles and accordingly still has to be normalized.

Once all these transition probability matrices for each lag in the training data are built, the next step is to apply the simulation procedure described in Figure 4.4. In order to make a fair comparison between observed and generated activity patterns, 1385 activity patterns were predicted for the training data set. Predictions for the training data set measure how well the simulation framework is capable of generating data. In addition, 462 activity patters were predicted for the test data set. Predictions for the test data set allow us to measure how well transition matrices are capable of simulating new unseen data and give an indication about the reliability of the results. As shown in Figure 4.1, once the general skeleton is predicted, the aim of the next step is to derive an "optimal" skeleton (i.e. that particular skeleton with a lag that achieves the best match to the observed pattern) from the general skeleton. This is likely to result in more accurate results for the time and location facets of the simulation.

Table 4.3
An Example of a 5 x 5 Second-Order Probability Matrix

			X_t					
	X_{t-2}	X_{t-1}	**Sleep**	**Eat/ Drink**	**Transport (Car)**	**Work (Out-of-home)**	**In-Home Leisure**	...
	Sleep	Eat/drink	45	3	211.5	0.5	214.5	...
	Eat/drink	Transport (car)	1.5	3	0	236.8	6.1	...
$Q =$	Transport(car)	Work	0	47.5	539.4	3	0	...
	Work	Transport (car)	16.3	163.4	0	31.7	202.6	...
	Eat/drink	In-home Leisure	33.4	37.1	142.9	0	371.4	...
	...	...	...	...	...	...	...	...

Validating the General Skeleton at Pattern Level. At an aggregate level, pattern level attributes are used to evaluate goodness-of-fit. The goodness-of-fit for the generated patterns can be measured by comparing the generated activity patterns with the observed patterns in the training and the test data set. The mean number of tours in the observed and the generated patterns is used as the evaluation measure at the pattern level. A tour is defined as a subsequence of activities that starts and ends at the same base location. Since location information is not yet incorporated at this stage in the simulation, a tour refers to any appearance/subsequence of (out-of-home) activity(ies) between two trips. The z-test for equal population means with known variances was used to test for significant differences. The results are shown in Table 4. It can be seen from this table that high-order combinations are not very successful in generating reliable patterns of activities. This seems counter-intuitive at first sight. Indeed, one might expect that reliability increases with more sequential information being incorporated in the transition probability matrices. This turned out to be only true to some extent (only until $\ell=7$). Indeed, recall Figure 4.3, where it was explained that one of the first steps is to draw the first ℓ elements from the prior sample distributions. This means for high numbers of ℓ that the generated activity patterns are much larger than the activity patterns in the sample data, which damages the accuracy of the results. Activity bundles with orders ranging from 5 to 7, turned out to generate no significant differences with respect to the training data set.

Table 4.4
Comparing the Observed and Predicted Mean Number of Tours by Differentiating Between the Order of Activity Bundles

Training dataset			Test dataset		
Observed Tours (mean)	**Order of Q**	**Predicted Tours (mean)**	**Observed Tours (mean)**	**Order of Q**	**Predicted Tours (mean)**
	$\ell=1$	1.722*		$\ell=1$	1.621*
	$\ell=2$	1.975*		$\ell=2$	1.954*
	$\ell=3$	1.949*		$\ell=3$	2.128*
	$\ell=4$	2.563*		$\ell=4$	2.316
2.801	$\ell=5$	2.732	2.435	$\ell=5$	2.424
	$\ell=6$	2.779		$\ell=6$	2.621*
	$\ell=7$	2.821		$\ell=7$	2.730*
	$\ell=8$	2.262*		$\ell=8$	2.003*
	$\ell=9$	1.951*		$\ell=9$	1.621*
	$\ell=10$	1.312*		$\ell=10$	1.222*

* Statistically significant difference in means (observed vs. predicted) at the 95 percent level of confidence

When we compare these results with the data which is generated for the test set, it appears that 6th and 7th order activity bundles slightly overfit the training data, i.e. the good performance on the training data could not be kept on the unseen test data. The 4th and 5th order activity bundles seem to generate the best fit. Based on this validation, five lags were considered most appropriate for generating (optimal) skeletons for this data set.

Generating Time Information

As explained previously, in a first step, probability distributions for the duration of the different activities are derived from the training data. To this end, the length of the generated activity patterns was discretized into three intervals, i.e. activity patterns less than 12 activities, patterns between 12 and 15 activities, and patterns equal or more than 16 activities, respectively containing 422, 485 and 478 activity sequences. After this discretization, duration probability distributions were constructed for every discretized interval. This was done for 5 randomly chosen activities (for the sake of clarity) in Figure 4.6. Probability distributions were aggregated into 15-minutes intervals to improve the readability of the figure.

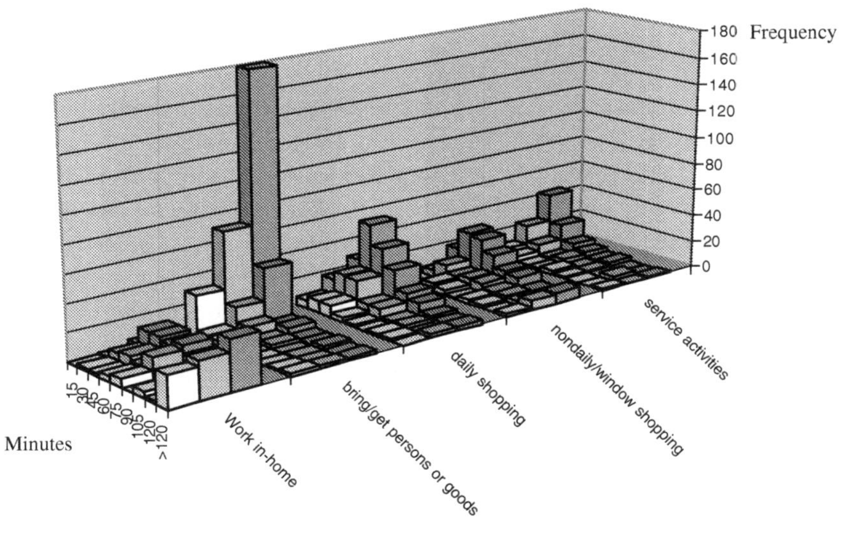

Figure 4.6
Difference in Duration Probability Distributions for Three Discretized Intervals (5 Randomly Chosen Activities)

Table 4.5
Comparing Observed and Simulation Probability Distributions of Departure Times for the Training and Test Set

	Training Set		Test Set	
	Observed	**Simulation**	**Observed**	**Simulation**
<10 a.m.	0.278	0.269	0.254	0.281*
10 a.m. – 12 a.m.	0.141	0.125	0.148	0.138
12 a.m. – 2 p.m.	0.168	0.151	0.166	0.159
2 p.m. – 4 p.m.	0.146	0.117*	0.148	0.121*
4 p.m. – 6 p.m.	0.109	0.121	0.110	0.132
> 6 p.m.	0.158	0.217*	0.174	0.169

* Statistically significant difference at the 95 percent level of confidence

It is obvious from this figure that the activity pattern length has an important impact on the duration probability distributions. Especially the difference in distributions with activity patterns of length smaller than 11 and those larger than 16 is significant. It can be seen that people who report more activities also provide significantly more detailed information, as especially the reported frequency of short time (15 and 30 minute intervals) activities is significantly larger. A similar conclusion was reached for frequency distributions of other in-home, out-of-home activities and trips (not shown here).

Once the probability distributions were built, the begin time of the first activity in all activity patterns was initialized at 3 am (*Ts*). This start time decision is consistent with previous data collection efforts. The idea behind this decision is that researchers look for data that consist of activities that form an entity and "belong to each other". Obviously, when the start time is set at 3 am this is more likely to be the case than when the start time was set at 12 pm (people may often read a book until 12:05 pm for example). Another parameter, which needs to be set, is the error parameter θ. In the experiments, θ was set to 21. With *Ts* set equal to 3, this implies that *Tf* needs to be between 24 and 27 to pass the reliability test defined in Figure 4.4. Note that in this time notation 24 is considered to be midnight, 25 is defined as 1 am, and 26 and 27 are defined accordingly. Finally, the parameter r_{max} is arbitrarily set at 5, which means that the simulation procedure tries at most 5 times to get the end time of the final activity in the activity pattern situated within the time window. With these parameters, start and end times of activities and trips can be generated. The results of these simulations are reported in Table 5. Recall that 5th order transition matrices were used to generate the general skeleton as the basis for the columns labelled as "simulation" in this table. The values reported in this table represent distributions instead of means (Table 4). The Kolmogorov-Smirnov-test (Siegel, 1956) was used to test the level of significance.

The simulation results of the training set give an indication about how well the framework is capable of capturing and simulating the time information, which is incorporated in the training data. Except for the last time category (> 6 pm), which is overestimated in the detriment of the fourth time category, the results are satisfactory. Simulated results for the test set are also encouraging.

Generating Location Information

For the generation of location information, origin-destination matrices first need to be derived from the training set, as explained before. The distinct locations in the dataset are defined by means of different zip codes and are labelled as 'Rot-Noord 1-3'; 'Rot-Zuid 4-6'; 'Hido A7-Hido A9'; 'Zwijn10-Zwijn15', 'Drecht16-Drecht19'; 'Outarea' and 'Missing'. A derived origin-destination matrix is shown in Figure 4.7 in a bubble chart. The locations on the X-axis represent the origin of the trip; destinations are shown on the Y-axis. The diameter of the circle gives an indication of the frequency that a particular origin-destination pair is observed. Except for the category 'Outarea', that could use more differentiation, the results do not suffer much from the spatial aggregation, which is used in the generation of location information. In order to improve the readability of the figure, the "missing" category is not shown. The dark grey circles represent the general OD-matrix across all different transport modes. The light grey and white circles only represent the locations in which respectively the car and slow modes were used and in which a significant difference could be

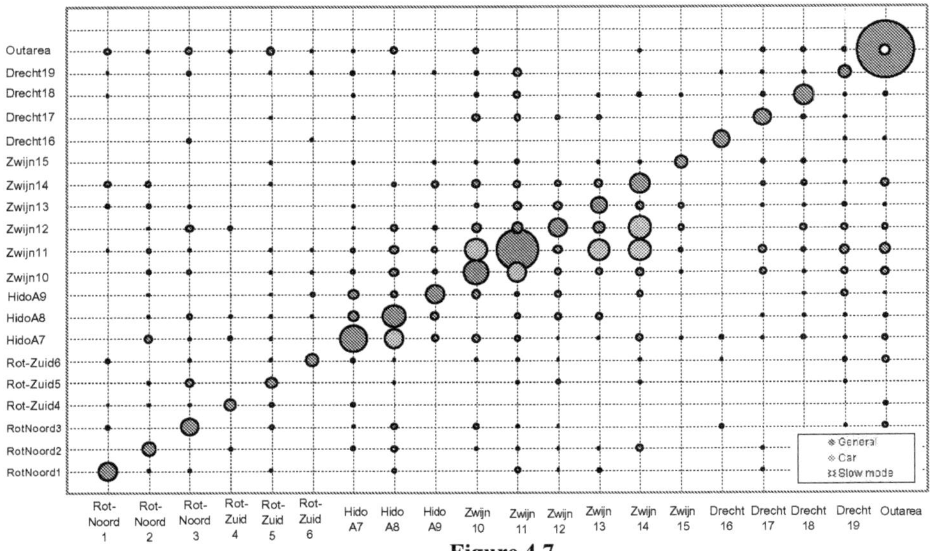

Figure 4.7

Origin-destination Bubble Chart for the Training Data

perceived compared to the general OD-matrix. The only slow mode, which is shown on the figure is located in the upper right part. Given the small differences, transport mode-specific OD matrices were not calculated in the simulations. However, experiments discovered that there was a stronger relationship between origins and destinations and the duration of the trip. Especially the relationship between out-of-area locations and long travel durations appeared to be significant.

The validation of the simulation results between predicted and observed origin-destination matrices was assessed in terms of correlation coefficients. The correlation coefficient between the observed and the predicted origins and destinations of the training set was 0.885, while it was 0.843 for the test set. Both coefficients were judged to be satisfactory.

Conclusion and Future Research

In this chapter, it was assumed that each activity pattern consists of a set of correlated successive observations. A methodology for simulating activity patterns and for storing sequential information was developed and empirically tested. In a first step, the simulation framework generates a sequence of in-home and out-of-home activities along with transport modes. The notion of "activity bundles" is used to store the information of low- and high-order combinations that typically sequentially occur in one particular activity sequence. Based on these activity bundles, transition matrices can be calculated, which in turn are used for simulating the general skeleton. The general skeleton was evaluated at the pattern level by calculating the mean number of tours. An optimal number of lags was selected based on this criterion. Obviously, the simulation of activity patterns is multidimensional. This means that also time and location information needs to be generated. Procedures for doing this were described in this chapter.

The methodology described here is novel, especially with respect to explicitly capturing sequential information in data and with respect to using this information for generating a general skeleton. The procedures for generating time and location information are straightforward when compared to other methodologies that exist in the literature.

Despite the promising results presented in this chapter, it is important to understand that the presented framework needs to be augmented with additional procedures and functionalities. First and foremost, the conditional and causal relationships between the different facets of the simulation approach need to be strengthened by adopting additional constraints and rules that make the generated patterns more realistic. Secondly, previous research efforts have emphasized the need for modelling procedures that are capable of capturing the correlation information (for instance Greaves and Stopher, 2000) between socio-demographics and travel or activity characteristics. Ideally, transition probability matrices need to be clustered according to socio-demographics in order to generate more homogeneous groups with the same skeleton. The integration with this socio-demographic information is currently being implemented and has already generated some initial promising results, but needs to be validated further. This is an important topic for further research.

Using transition probabilities that can be made time-dependent is another interesting area to explore. It is also important to use the presented transportation model for policy evaluation. Policy measures such as for instance an increase of part-time workers or changes in start times for work can be evaluated by re-calibrating the model for the particular subset of respondents under evaluation (in the case of an increase of part time workers) or by imposing constraints on the sequence and timing of activities which are generated (in the case of a change in the start time for work). A more thorough empirical comparison with other existing transportation models will then become feasible in the future.

ACKNOWLEDGEMENT

The authors would like to thank the Urban Planning Group in The Netherlands for providing the data that were used in the empirical section of the chapter.

REFERENCES

Arentze, T.A., F. Hofman and H.J.P. Timmermans (2003). Re-induction of Albatross' decision rules using pooled activity-travel diary data and an extended set of land use and costs-related condition states. *Proc. 82nd TRB Conf.*, Washington D.C. (CD-Rom).

Arentze, T.A., R. Katoshevski and H.J.P. Timmermans (2001). A micro-simulation model of activity-travel behavior of individuals in urban environments. *Proc. 7th CUPUM Conf.* Honolulu, USA (CD-Rom).

Arentze, T.A. and H.J.P. Timmermans (2000). *Albatross: A Learning-Based Transportation Oriented Simulation System.* European Institute of Retailing and Services Studies, Eindhoven, The Netherlands.

Arentze, T.A. and H.J.P. Timmermans (2004). A learning-based transportation oriented simulation system. *Transpn. Res.B*, **38**, 613-633.

Ben-Akiva, M.E., J. Bowman and D. Gopinath (1996). Travel demand model system for the information area. *Transpn.*, **25**, 241-266.

Bhat, C.R., J. Guo, S. Srinivasan and A. Sivakumar (2004). Comprehensive econometric microsimulator for daily activity-travel patterns. *Proc. 83rd TRB Conf.,* Washington, D.C., USA. (CD-Rom).

Clarke, M.I. (1985). Activity modeling – a research tool or a practical planning technique. Paper presented at *Beh. Res. for Transp. Pol., Int. Conf. on Trav. Beh.,* Noordwijk, The Netherlands.

Gärling, T., K. Brännäs, J. Garvill, R.G. Golledge, S. Gopal, E Holm and E. Lindberg (1989). Household activity scheduling. In: *Selected Proc. WCTR Conf.*, Vol. 4, pp. 82-91. Western Periodical, Ventura, CA.

Greaves, S.P. and P.R. Stopher (2000). Creating a synthetic household travel and activity survey: rationale and feasibility analysis. *Transpn. Res. Rec.*, **1706**, 82-91.

Janssens, D., G. Wets, T. Brijs and K. Vanhoof (2004). Simulating activity diary data by means of sequential probability information: development and evaluation of an initial framework. *Proc. 83rd TRB Conf.,* Washington, D.C., USA (CD-Rom).

Janssens, D., G. Wets, T. Brijs and K. Vanhoof (2005). The development of an adapted Markov Chain modelling heuristic and simulation framework in the context of transportation research. *Expert Syst. Appl.*, **28**, 105-117.

Kitamura R. and M. Kermanshah (1983). Identifying time and history dependencies of activity choice. *Transpn. Res. Rec.,* **944**, 22-30.

Koppelman, F.S. (1981). Non-linear utility functions in models of travel choice behavior. *Transpn.,* **10**, 127-146.

Kulkarni, A.A. and M.G. McNally (2000). An activity-based travel pattern generation model. *Proc. Western Reg. Sci. Assoc. Conf.*, Kauai, Hawaii (CD-Rom).

Kulkarni, A.A. and M.G. McNally (2001). A microsimulation of daily activity patterns. *Proc. 80th TRB Conf.*, Washington, D.C., USA (CD-Rom).

Lerman, S.R. (1979). The use of disaggregate choice models in semi-Markov process models of trip chaining behavior. *Transpn. Sc.,* **13**, 273-291.

McNally, M.G. (1995). An activity-based microsimulation model for travel demand forecasting. In: *Activity-Based Approaches to Travel Analysis* (D.F. Ettema and H.J.P. Timmermans, eds.), pp. 37-54. Pergamon, London.

Pendyala, R.M., R. Kitamura and D.V.G.P. Reddy (1998). Application of an activity-based travel demand model incorporating a rule-based algorithm. *Env. & Plan. B*, **25**, 753-772.

Recker, W.W., M.G. McNally and G.S. Root (1986a). A model of complex travel behavior: Part I theoretical development. *Transpn. Res. A*, **20**, 307-318.

Recker, W.W., M.G. McNally and G.S. Root (1986b). A model of complex travel behavior: Part 2: an operational model. *Transpn. Res. A*, **20**, 319-330.

Siegel, S. (1956). *Nonparametric Statistics for the Behavioral Sciences*. McGraw-Hill , New York.

Stewart, W.J. (1995). *Introduction to the Numerical Solution of Markov Chains*. Princeton University Press, Princeton.

van der Hoorn, T. (1983). Experiments with an activity-based travel model. *Transpn.,* **12**, 61-77.

Vaughn, K.M., P. Speckman and E.I. Pas (1997). Generating household activity-travel patterns (HATPs) for synthetic populations. *Proc. 76th TRB Conf.*, Washington, D.C., USA (Cd-Rom).

Veldhuisen, J., L. Kapoen and H.J.P. Timmermans (2000). Ramblas: A regional planning model based on the micro-simulation of daily activity travel patterns. *Env. & Plan. A*, **32**, 427-443.

Wen, C-H. and F.S. Koppelman (1999). An integrated system of stop generation and tour formation for the analysis of activity and travel patterns. *Transpn. Res. Rec.*, **1676**, 136-144.

5

Adjustments of Activity Timing and Duration in an Agent-Based Traffic Flow Simulation

Michael Balmer, ETH Zurich, Zurich, Switzerland

Bryan Raney, ETH Zurich, Zurich, Switzerland

Kai Nagel, TU Berlin, Berlin, Germany

Introduction

One possibility to bring activity-based demand generation into the transportation planning processes is to use it to replace the first two (or three) steps of the conventional four step process. Activity-based demand generation would produce a standard origin-destination (OD) matrix, which would then be fed into the existing assignment model. Both the OD matrix and the assignment model could be time-dependent. The advantage of this approach is that it ties in with the arguably most sophisticated and best understood part of the four step process: route assignment. Yet, some of these advantages disappear when the OD matrices are time-dependent. In that situation, very few of the mathematical results of static assignment carry over. In addition, coupling activity-based demand generation to network assignment through an OD matrix disrupts the connection between individuals and their performance in the simulated traffic system. Any iterative feedback from the traffic system performance to the activity generation can only be based on aggregate measures, such as link travel times, not on individual performance of the traveller. An obvious case where the coupling through the OD matrix goes wrong is when it is possible for a person to complete an activity even before he/she has arrived at the destination where the activity will be conducted.

Progress in Activity-Based Analysis edited by H. Timmermans

To avoid such situations, we propose to use a truly agent-based representation of the traffic system and the assignment process. In a truly agent-based representation, each person remains individually identifiable throughout the whole simulation process. In particular, the traffic micro-simulation assumes the role of a realistic representation of the physical system, including explicit modelling of persons walking to the bus stop, or of a bus being stuck in traffic. Also, in terms of analysis such a system offers enormous advantages. It is, for example, possible to obtain the demographic characteristics of all drivers being stuck in a particular traffic jam. It is also possible to make each traveller react individually to exactly the conditions that this traveller has experienced, rather than to aggregated conditions.

This multi-agent concept consists of basically two parts: (i) the simulation of the "physical" properties of the system, and (ii) the generation of the agents' strategies. The simulation of the physical system is the place where the agents interact with each other—car drivers produce congestion, traffic lights change their intervals dependent on the amount of traffic, pedestrians wait for the next train to catch, and so on. The agents make their strategies based on what they experienced in the physical simulation—car drivers try other routes to avoid congestion, pedestrians need to leave earlier to catch the train, traffic lights favour the main streets to maximize the throughput of an intersection, etc.

We are in the process of implementing such a multi-agent simulation for the whole of Switzerland. This paper concentrates on the Zurich area, with about 260,000 agents that cross this region. The challenges with such an implementation are many: availability and quality of input data, computational implementation and computational performance, conceptual understanding of agent learning, and validation. In previous research (Raney *et al.*, 2003), we have reported the first results based on typical transportation planning data: standard origin-destination matrices; the transportation planning network from the corresponding Swiss federal planning authority; and performed route assignment (dynamic traffic assignment or DTA) based on these input data. The two main differences with other DTA systems, such as DYNASMART (http://mit.edu/its) or DYNAMIT (http://www.dynasmart.com), were that our system uses individual route plans for each agent while standard DTA systems store the routing decisions in the network, and that our system was run on really large scale scenarios with several millions of travellers. A newer version of DYNASMART, however, now also uses individual routes, and other systems also move towards increasingly large scales. In contrast to TRANSIMS (www.transims.net), which has used individual routes and large scales for many years now, we used a so-called agent database, which keeps track of several plans for each agent.

This chapter goes further by now also internalizing the time structure of the input data. In other words, it is possible for the simulation system to predict when agents start and end their main activities. The main result is that it is possible to completely ignore the time structure of the time-dependent OD matrices without compromising predictive power. This is similar to the approach used and results obtained with METROPOLIS (De Palma and Marchal, 2002). The main difference is that our implementation uses complete daily activity chains, whereas METROPOLIS only uses trips. We believe that our method, while computationally more demanding, opens the door to more flexible transportation planning models.

This chapter will continue with an outline of the general simulation structure, where we also describe in more detail the modules we will use. Next, we introduce the network and the scenario. The results of the different setups of the scenario are compared with traffic count data. Some computational issues are discussed next. The paper is concluded by a section on future work.

SIMULATION STRUCTURE

Overview

As pointed out before, our simulation is constructed around the notion of agents that make independent decisions about their actions. Each traveller of the real system is modelled as an individual agent in our simulation. The overall approach consists of three important pieces:

1. Each agent independently generates a so-called *plan*, which encodes its intentions during a certain time period, typically a day. As this is an application to traffic forecasting, a plan contains the itinerary of activities the agent wishes to perform during the day, plus the trips the agent must take to travel between activities. An agent's plan details the order, type, location, duration and other time constraints of each activity, and the mode, route and expected departure and travel times of each leg.

2. All agents' plans are simultaneously executed in the simulation of the physical system. In this chapter, this is a *traffic flow simulation*. In other publications, we use the term *mobility simulation* in order to emphasize that the simulation of the physical system can go beyond traffic.

3. There is a mechanism that allows agents to *learn*. In our implementation, the system iterates between plan generation and traffic flow simulation. The system remembers several plans for each agent, and scores the performance of each plan. Agents normally choose the plan with the highest score, sometimes re-evaluate plans with bad scores, and sometimes obtain new plans. Further details will be given below.

This chapter concentrates on "home" and "work" as the only activities, and "car" as the only mode. We do not distinguish between a trip (between two activities) and a leg (a part of a trip which uses exactly one mode), since the "mode change" can also be defined as an activity (which has a specified location, i.e. a train station). Each of the details described in the plan, such as activity duration, is a decision that must be made by the agent. These decisions are mutually dependent, but the decisions made by one agent are independent of those made by another. We divide the task of generating a plan into sets of closely related decisions, and each set is assigned to a separate *module*. An agent strings together calls to various modules in order to build up a complete plan. To support this "stringing", the input to a given module is a (possibly incomplete) plan, and the output is a plan with some of the decisions updated. Some possible modules are:

Activity Pattern Generator: Decides which activities an agent actually wishes to perform during the day, and in what order. At present, this module is not used, but we have a fixed "home-work-home" pattern for all agents.

Activity Location Generator: Determines where the agent will perform a particular activity. At present, this module is not used, but we have a fixed location for each agent's "home" and "work" activity.

Activity Time Allocator: Determines the timing attributes the agent will utilize for each activity in a plan. Activities have two possible timing attributes: "activity duration" and "activity end time". After starting an activity, an agent performs the activity either for the length of "duration", or until the "activity end time", whichever comes first. Activities cannot overlap in time.

Router: Determines which route and which mode the agent chooses for each trip leg that connects activities at different locations.

A special feature of our approach is that users can choose any number and type of these modules as long as they generate some information that contributes to a plan. For that reason, it is easy to combine for example activity and mode choice into a single module or to add residential or workplace choice. This application will employ two modules only: "activity time allocator" and "router". Other modules will be the topic of future work.

Once the agent's plan has been constructed, it can be fed into the traffic flow simulation module. This module executes all agents' plans simultaneously on the network, allowing agents to interact with one another, and provides output describing what happened to the agents during the execution of their plans. The modules produce dependencies. The outcome of the traffic flow simulation module (e.g., congestion) depends on the planning decisions made by the decision-making modules.

However, those modules can base their decisions on the output of the traffic flow simulation (e.g., knowledge of congestion). This creates an interdependency ("chicken and egg") problem between the decision-making modules and the traffic flow simulation module. We need these modules to be consistent with one another, and therefore we introduced feedback into the traffic flow simulation structure (Kaufman *et al.*, 1991; Nagel, 1995; Bottom, 2000). This involves an iteration cycle which runs the traffic flow simulation with specific plans for the agents, then uses the time allocator and the router to update the plans, and these changed plans are again fed into the traffic flow simulation, etc., until consistency between modules is reached.

The feedback cycle is controlled by the agent database, which also keeps track of multiple plans generated by each agent, allowing agents to reuse those plans at will. The repetition of the iteration cycle coupled with the agent database enables the agents to learn how to improve their plans over many iterations. In the following sections we describe the modules in more detail.

Activity Time Allocator

This module is called to change the timing of an agent's plan. At this point, a very simple approach is used which just applies a random mutation to the duration and end time of an agent's activities. More precisely, for the first activity, the activity end time is the only attribute that is specified and thus mutated, while for all other activities, the duration is what is specified and mutated. For each such attribute of each activity in an agent's plan, this module picks a random time from the uniform distribution [-30 min, +30 min] and adds it to the attribute. Any negative duration is reset to zero; any activity end time before 00:00 a.m. is reset to 00:00 a.m.. The entire plan is returned to the agent, with only the time attributes modified.

Although this approach is not very sophisticated, it is sufficient to obtain useful results. This is consistent with our overall assumption that, to a certain extent, simple modules can be used in conjunction with a large number of learning iterations (e.g., Nagel *et al.*, 2004). Since each module is implemented as a "plug-in", this module can be replaced by an enhanced implementation if desired.

Router

The router is implemented as a time-dependent Dijkstra algorithm. It first calculates link travel times from the events output of the previous traffic flow simulation. The link travel times are aggregated into 15 minute time bins, and then used as the weights of the links in the network graph.

Apart from relatively small but essential technical details, the implementation of such an algorithm is straightforward (Jacob *et al.*, 1999). With the knowledge about activity chains, it computes the fastest path from each activity to the next one in the sequence as a function in time. It returns the entire plan, completed with updated paths, to be used by the agents for the next run of the traffic flow simulation.

TRAFFIC FLOW SIMULATION

The traffic flow simulation simulates the physical world. It is implemented as a queue simulation (Gawron, 1998; Cetin and Nagel, 2003), which means that each street (link) is represented as a FIFO (first-in first-out) queue with three restrictions. First, each agent has to remain for a certain time on the link, corresponding to the free speed travel time. Second, a link storage capacity is defined which limits the number of agents on the link. If this capacity has been reached, no more agents can enter this link. Third, there is a flow capacity, which limits the number of vehicles that can leave the link in any given time step.

Even though this structure is indeed very simple, it produces traffic as expected and it can run directly using the data typically available for transportation planning purposes. On the other hand, there are some limitations compared to reality, i.e., the number of lanes, weaving lanes, turn connectivities across intersections or signal schedules cannot be included into this model. The output that the traffic flow simulation produces is a list of events for each agent, such as entering/leaving link, left/arrived at activity, and so on. Data for an event includes which agent experienced it, what happened, at what time it happened, and where (link/node) the event occurred. With this data it is easy to produce different kinds of information and indicators such as link travel time, trip travel time, trip length, percentage of congestion, and so on.

AGENT DATABASE — FEEDBACK

As mentioned above, the feedback mechanism is important for making the modules consistent with one another, and for enabling agents to learn how to improve their plans. In order to achieve this improvement, agents need to be able to try out different plans and to tell when one plan is "better" than another. The iteration cycle of the feedback mechanism allows agents to try out multiple plans. To compare plans, the agents assign each plan a "score" based on how it performed in the traffic flow simulation. Essentially, each agent is running its own classifier system (e.g. Holland, 1992; Palmer *et al.*, 1994). It is very important to note that our framework always uses actual plan performance for the score. This is in contrast to all other similar approaches that we are aware of which typically feedback some aggregated quantity such as link travel times and reconstruct performance based on those (e.g., URBANSIM—www.urbansim.org; Ettema *et al.*, 2004). Because of unavoidable aggregation errors, such an approach can fail rather badly in the sense that the performance information derived from the aggregated information may be rather different from the performance that the agent in fact experienced (Raney and Nagel, 2003). The procedure of the feedback and learning mechanism is as follows:

Initial Conditions: Start with a plan file that specifies one complete plan for each agent. The agent database loads these plan files into the memory of the agents. Each agent marks its initial plan as the "selected" plan.

Simulate: The agent database sends the set of "selected" plans (one for each agent) to the traffic flow simulation. The simulation executes the plans simultaneously and outputs events.

Process Events: The agent database reads the events that are output by the traffic flow simulation and sends each one to the agent identified within it. Each agent uses its events to calculate the score of its "selected" plan—the one it most recently sent to the traffic flow simulation.

Plan Pruning: The number of plans kept in an agent's memory for reuse can be limited to N plans to conserve memory. If N is defined, each agent that has $P > N$ plans deletes its lowest-scoring $P - N$ plans in this step. Note that when an agent that has N plans generates a new one, it temporarily keeps $N + 1$ plans until the new plan has been scored. Then, in this step, it deletes the worst plan (even if it is the newest one).

Select Plans: Each agent decides which plan to select for execution by the next traffic flow simulation. It chooses from the following selection options, according to the indicated probabilities:

- *(10 %) New Plan, Routes Only*: The agent sends an existing plan (chosen with equal probability among all plans in memory) to the router. The router calculates new routes in that plan based on the link travel times calculated from the events data from the most recent traffic flow simulation, and returns the updated plan. The new plan is added to the agent's memory and marked as "selected".

- *(10 %) New Plan, Times and Routes*: The agent sends an existing plan (chosen with equal probability among all plans in memory) to the activity time allocation module. This module "mutates" the durations and/or end times of all activities in the plan and returns the updated plan. The returned plan is also sent to the router for route re-planning. When it comes back from the route re-planner, it is added to the agent's memory and marked as "selected". (Note that now 20 % of agents will have new routes, while only 10 % will have new times).

- *(10 %) Random Selection*: The agent picks an existing plan, chosen with equal probability among all plans in memory, without regard to their scores. This plan is marked as "selected".

- *(Rest) Probabilistic Selection*: The agent picks an existing plan from memory, choosing according to probabilities based on the scores of the plans. The probabilities are of the form

$$p \propto e^{bS_j} \tag{1}$$

where, S_j is the score of plan j, and β is an empirical constant. This is equal to a logit model from discrete choice theory. The chosen plan is marked as "selected".

The cycle returns to step 2 (simulate), and continues until the system has reached a relaxed state. At this point, there is no quantitative measure of when the system is "relaxed"; we just allow the cycle to continue until the outcome seems stable. Note that when an agent reuses an existing plan, its previous score is not forgotten, but averaged with its new score:

$$S = (1-\alpha)S_{old} + \alpha S_{new} \tag{2}$$

with the blending factor α. This allows the agent to base plan selection on the plans' history and not only on the last iteration. With $\alpha = 0$ no score will be updated and the agents will not learn. With $\alpha = 1$ the history of a plan is neglected. Score averaging requires all plans to have an S_{old}, so when a new plan is generated, it is optimistically given a preliminary score equal to the score of the agent's best plan. More sophisticated approaches to agent learning are discussed in Timmermans *et al.* (2003).

SCORES FOR PLANS

In order to compare plans, it is necessary to assign a quantitative score to the performance of each plan. In principle, arbitrary scoring schemes can be used (e.g., prospect theory by Avineri and Prashker, 2003). We used a simple utility-based approach, which is related to the Vickrey bottleneck model (Arnott *et al.*, 1993), but needs to be modified to be consistent with our approach based on complete daily plans (Charypar and Nagel, 2003; Raney and Nagel, in press). The total score of a plan is computed as the sum of individual contributions:

$$U_{total} = \sum_{i=1}^{n} U_{perf,i} + \sum_{i=1}^{n} U_{late,i} + \sum_{i=1}^{n} U_{travel,i} \tag{3}$$

where U_{total} is the total utility for a given plan; n is the number of activities/trips; $U_{perf,i}$ is the (positive) utility earned for performing activity i; $U_{late,i}$ is the (negative) utility earned for arriving late at activity i; and $U_{travel,i}$ is the (negative) utility earned for travelling during trip i. In order to work in plausible real-world units, utilities are measured in Euro.

A logarithmic form is used for the positive utility earned by performing an activity (e.g., Axhausen, 1990b):

$$U_{perf,i}(t_{perf,i}) = \beta_{perf} \cdot t_i^* \cdot \ln\left(\frac{t_{perf,i}}{t_{0,i}}\right) \tag{4}$$

where, $t_{perf,i}$ is the actual performed duration of the activity, t_i^* is the "typical" duration of an activity, and β_{perf} is the marginal utility of an activity at its typical duration. β_{perf} is the same for all activities, since in equilibrium all activities at their typical duration need to have the same marginal utility. $t_{0,i}$ is a scaling parameter that is related both to the minimum duration and to the importance of an activity.

If the actual duration falls below $t_{0,i}$, then the utility contribution of the activity becomes negative, implying that the agent should completely drop that activity. A $t_{0,i}$ only slightly less than t_i^* means that the utility of activity i rapidly decreases with decreasing $t_{perf,i}$, implying that the agent should rather cut short other activities where the utility does not decrease as quickly when reducing their duration. In this application, we use

$$t_{0,i} = t_i^* \cdot e^{-\varsigma/(p \cdot t_i^*)} \tag{5}$$

where ς is a scaling constant set to 10 hours, and p is a priority indicator, here set uniformly to one. Note that with this specific form, $U_{perf,i}(t_i^*) = \beta_{perf} \cdot \varsigma$, independent of the activity type. This "consequence" is actually the motivation for the specific mathematical form of the activity performance utility contribution, which was used because no better argument was available (Charypar and Nagel, in press); future research should lead to better versions.

The (dis)utility of being late is defined as:

$$U_{late,i} = \beta_{late} \cdot t_{late,i} \tag{6}$$

where, $\beta_{late} \leq 0$ is the marginal utility (in Euro/h) for being late, and $t_{late,i}$ is the number of hours late for activity i. To be able to calculate the utility of being late, a starting time window for the activities has to be given. The (dis)utility of travelling is defined as:

$$U_{travel,i} = \beta_{travel} \cdot t_{travel,i} \tag{7}$$

where $\beta_{travel} \leq 0$ is the marginal utility (in Euro/h) for travel, and $t_{travel,i}$ is the number of hours spent travelling during trip *i*.

At this point, our traffic flow simulation does not differentiate between "being at an activity location" (which potentially includes waiting) and "performing an activity". Consequently, the simulation makes the agent stay at the activity location for the length of "duration", no matter whether the agent can perform the activity or not. For example, when work starts at 8 a.m. but the agent arrives at 7 a.m. with a duration of 8 hours, then the agent will depart from the activity location at 7 a.m. plus 8 hours = 3 p.m.. The utility function, however, differentiates between "arrival time" and "activity start time". The "work" activity has a particular starting time, and arriving before this time causes the agent to wait until then before actually starting the activity. This means that arriving early to an activity does not gain an agent any activity performance utility.

VERIFICATION OF IMPLEMENTATION

We have verified that the simulation structure as described above works as we intended by running it on a simple test scenario consisting of a circular network with 2,000 agents going back and forth between home and work. All agents have the same "home" location on one side of the circle and the same "work" location on the other side. Nine routes are available between home and work, and one route is available between work and home. We ran three setups with various combinations of decision-making modules enabled:

New plan, routes only: The agents are only allowed to use the router module. They may do so with a 10 % probability.

New plan, times only: The agents are only allowed to use the activity time allocation module. They may do so with a 10 % probability.

New plan, times and routes: Agents may use the router module with a 10 % probability, or both modules with a 10 % probability (see p. 97 Select plans).

The results from these three scenarios were as expected (Raney and Nagel, in press).

INPUT DATA AND SCENARIO

Network

The street network that is used was originally developed for the Swiss regional planning authority (Bundesamt fuer Raumentwicklung), and covered Switzerland. It was extended with the major European transit corridors for a railway-related study (Vrtic *et al.*, 1999). Some further modifications, in particular a capacity increase inside the Zurich city area, are described in Raney *et al.* (2003). The resulting network has the fairly typical size of 10,564 nodes and 28,624 links (Figure 5.1). Also fairly typical, the major attributes on these links are type, length, speed, and capacity.

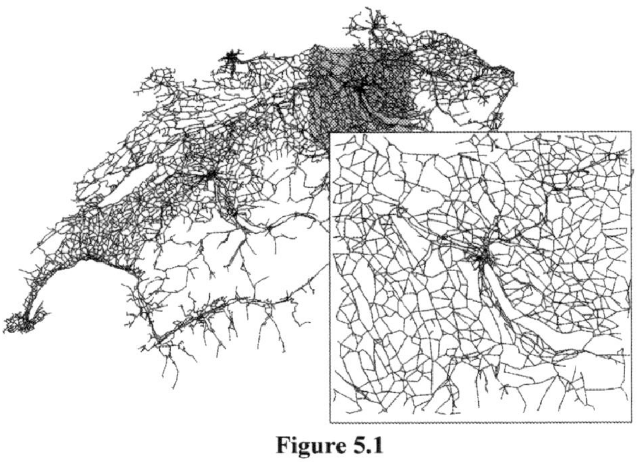

Figure 5.1
Switzerland Network

ZURICH AREA SCENARIO

The full Switzerland scenario demand generation is based on 24-hour origin-destination matrices from the Swiss regional planning authority (Bundesamt fuer Raumentwicklung). The original 24-hour matrix was converted into 24 one-hour matrices using a three step heuristic (Vrtic and Axhausen, 2002). The first step employed departure time probabilities by population size of origin zone, population size of destination zone and network distance. These were calculated using the 1994 Swiss National Travel Survey (BfS, 1996). The resulting 24 initial matrices were then corrected (calibrated) against available hourly counts using the OD-matrix estimation module of VISUM (www.ptv.de). Hourly traffic count data are available from the counting stations on the national motorway system. Finally, the hourly matrices were rescaled so that the totals over 24 hours match the original 24h matrix. VISUM assignment of the matrices showed that the patterns of congestion over time are realistic and consistent with the known patterns.

For the multi-agent simulation, these hourly matrices were then disaggregated into individual trips. That is, we generated individual trips such that summing up the trips would again result in the given OD matrix. The starting time for each trip was randomly selected between the starting and the ending time of the validity of the OD matrix. The OD matrices assume traffic analysis zones (TAZs) while in our simulations trips start on links. We converted traffic analysis zones to links by the following heuristic. First, the geographic location of the zone is found via the geographical coordinate of its centroid given by the database. Next, a circle with radius 3 km is drawn around the centroid. Finally, each link starting within this circle is now a possible starting link for the trips. One of these links is randomly selected and the trip start or end is assigned. This led to a list of approximately 5 million trips, or about 1 million trips between 6 a.m. and 9 a.m.. Since the origin-destination matrices are given on an hourly basis, these trips reflect the daily dynamics. Intra-zonal trips are not included in those matrices, as by tradition.

Since an agent should keep more than one plan during the iteration process, the memory requirements of one million agents exceeded the available memory. So we restricted our interests to the Zurich Area only. This was done with the following steps: (i) all trips are routed using free flow travel times; (ii) we define the area of interest as a circle of 26 km radius around the center ("Bellevue") of Zurich City, and (iii) each trip that does not cross this area is removed. This results in 260,275 trips between 6 a.m. and 9 a.m.. All trips are now identified with an agent. The "origin" location for the morning trip is assigned to the home activity, and the "destination" location is assigned to the work activity. The end time of the home activity is set to the departure time of the original trip. The daily patterns "home-work" are then extended to the "home-work-home" pattern, where the two homes are at the same location. The duration of the "work" activity is set to 8 hours, with no fixed activity end time. At the end we get 260,275 agents that have an initial day plan.

TRAFFIC COUNT DATA

There are about 230 automatic counting stations registered with the Swiss Federal Roads Authority (Bundesamt fuer Strassen). Of those, we had hourly traffic count data for 75 stations. A total of 33 of these could be located unequivocally on our network. Unfortunately there are only 6 useful bi-directional counting stations left in the Zurich area, implying we can compare 12 links with reality.

SIMULATION PARAMETERS

The maximum number of plans per agent, N, was set to 5 plans. The value of the empirical constant β used to convert plan scores to selection probabilities is $2.0/Euro$. We use the following values for the marginal utilities of the utility function used for calculating scores:

$$\beta_{perf} = +6Euro/h, \ \beta_{travel} = -6Euro/h \text{ and } \beta_{late} = -18Euro/h$$

Although it is not obvious at first glance, these values mirror the standard values of the Vickrey scenario (Arnott *et al.*, 1993): An agent that arrives early to an activity must wait for the activity to start. During this time, the agent cannot perform *any* activity and therefore forgoes the $\beta_{perf} = +6Euro/h$ that it could accumulate instead (opportunity cost). An agent that travels fore-goes the same amount, *plus* a loss of $6Euro/h$ for travelling. And finally, an agent that arrives late receives a penalty of $18Euro$ per hour late, but is not losing (or gaining) any time elsewhere by being late. We only look at daily activity chains that consist of one home and one work activity. The "typical" times were set to $t_h^* = 16\ hours$ and $t_w^* = 8\ hours$. With these assumptions, the maximum score is 120 *Euro* (60 *Euro* per activity). For the work activity a starting time window is defined between 7:08 a.m. and 8:52 a.m.. The blending factor α is set to 0.1. This is a useful compromise between zero learning and overreaction. We expect that changes in α will mostly affect the speed of relaxation; this may be a topic of future research.

RESULTS

Overview

We present the results of four different setups, which result from two different initial conditions and from using time re-planning or not. The two initial conditions are:

Initial departure times given externally: Here, the activity end times from the home activity are generated as described earlier. When the home activity ends, agents immediately depart and drive to work, where they stay for 8 hours, and then return. We will call the two setups where agents initially use externally defined times *times-routes-initial-times-extern* and *routes-only-initial-times-extern* when times re-planning is enabled and disabled, respectively.

All agents depart home at 6 a.m.: Once departed, agents drive to work, where they work for 8 hours, and then return. These initial conditions are used to have a scenario where the simulation starts with a clearly implausible situation. The question that is tested is whether it will recover to a realistic solution by itself. We will call the two setups where all agents depart at 6 a.m. *times-routes-initial-times-all6a.m.* and *routes-only-initial-times-all6a.m.* when times re-planning is enabled and disabled, respectively.

Note that when times re-planning is disabled, only 10 % of agents perform route re-planning, but when it is enabled, a total of 20 % of agents perform route re-planning, with half of those also performing times re-planning. We compare the results with the following indicators: (i) *Average travel time*: The average travel time across all agents plans for each iteration; (ii) *Average score*: The average score across all agents for each iteration; (iii) *Departure and arrival time histograms*: The number of agents that arrive/depart from an activity over time during a certain iteration; (iv) *Traffic count data comparison*: Mean bias and error of the simulations compared to the counting data described above.

Initial Plans with Externally Defined Departure Times

This setup tests whether or not the learning, once time re-planning is switched on, drifts away from the time structure given by the external data. Since these initial plans are based on realistic time distributions, one would assume that the time re-planning will not affect the result that much. Re-routing alone should decrease the average travel time and congestion. Figure 5.2 compares the average travel times over the iterations. The routes-only iteration (Figure 5.2a) quickly gets to a stable result because re-routing is the only part, which has to be optimized. The small fluctuations are due to the fact that some percentage of the agents always re-plans, and that the traffic flow simulation is stochastic.

The iterations where time re-planning is switched on (Figure 5.2b) behave in a similar way, but the average travel time is slightly higher than routes-only and also it fluctuates more. However, the scores of the times-routes setup are not worse than the scores of the routes-only setup. This indicates that the agents are "trading off" travel time for other parts of their utility. In other words, by adjusting their activity times (i.e., the times they make their trips) they make up for the fact that trips are longer by arriving at a more suitable time to work. The higher fluctuations can be attributed to the fact that there are now two re-planning parts, which have to be optimized.

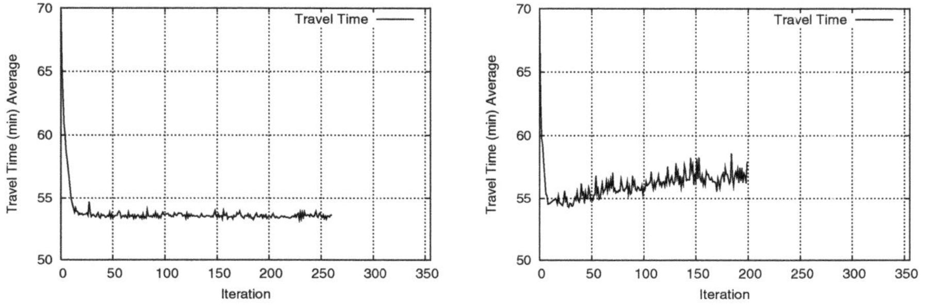

(a) Average Travel Times per Iteration for Routes-Only-Initial-Times-Extern Setup

(b) Average Travel Times per Iteration for Times-Routes-Initial-Times-Extern Setup

Figure 5.2
Average Travel Times of Routes-Only-Initial-Times-Extern and Times-Routes-Initial-Times-Extern

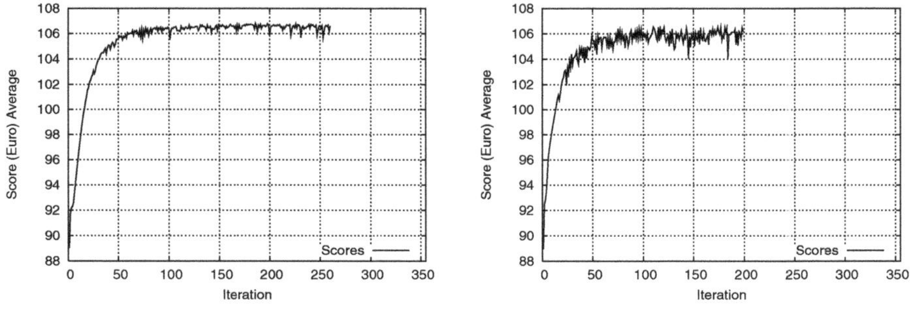

(a) Average Scores per Iteration for Routes-Only-Initial-Times-Extern Setup

(b) Average Scores per Iteration for Times-Routes-Initial-Times-Extern Setup

Figure 5.3
Average Scores of Routes-Only-Initial-Times-Extern and Times-Routes-Initial-Times-Extern

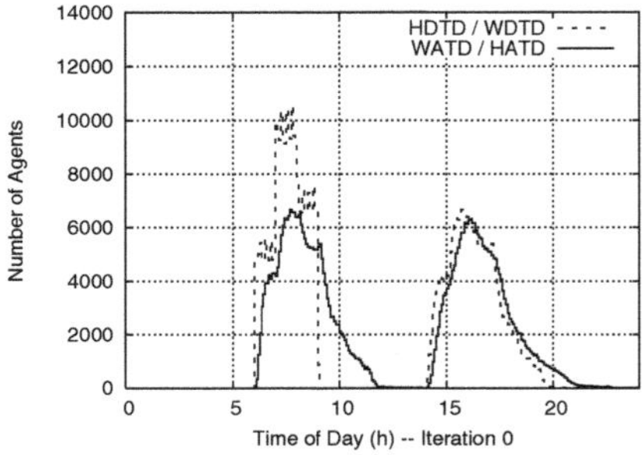

(a) Arrival and Departure Histograms (5 min Time Bins) of Iteration 0 with "Plausible" Initial Activity Times

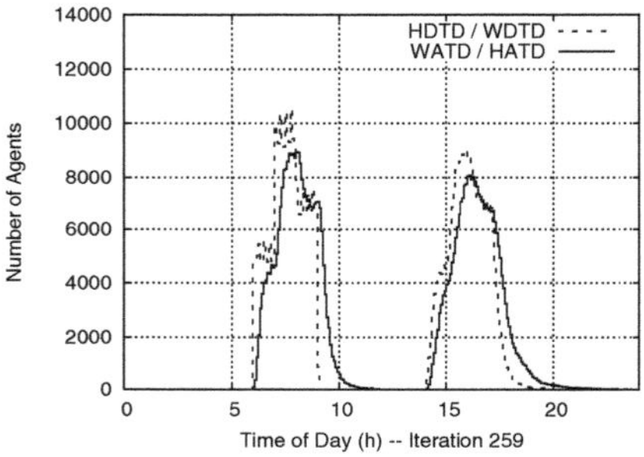

(b) Arrival and Departure Histograms (5 min Time Bins) of Iteration 260 with Time Re-Planning Switched off

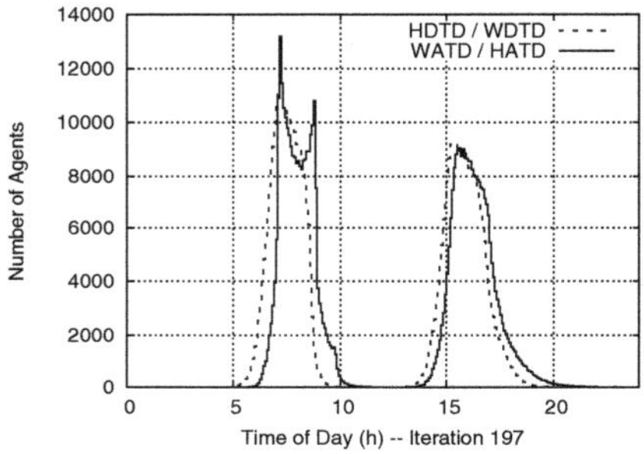

(c) Arrival and Departure Histograms (5 min Time Bins) of Iteration 200 with Time Re-Planning Switched on

Figure 5.4
Arrival and Departure Histograms when the Initial Plans have "Plausible" Departure Times

Figure 5.3 shows the scores for each iteration of both setups. They are once more similar to each other, and once more the routes-only setup (Figure 5.3a) shows less fluctuation than the setup with time re-planning (Figure 5.3b). The reason is the same as described above. Comparing to Figure 5.2, one can see that in both setups, the average scores relax considerably more slowly than the average travel times. This is due to the score averaging in the agent database.

The histograms (Figure 5.4) show how the re-planning affects the agents. Starting with the same configuration (Figure 5.4a), the routes-only iteration only tries to minimize travel times, so that the periods of arrivals decreases (see bold graph of Figure 5.4b), while departure from home stays the same (see dotted graph of Figure 5.4b). Switching on time re-planning changes also the dotted graph (see Figure 5.4c). The two peaks of the arrival (bold) graph are at 7:08 a.m. and 8:52 a.m., which is the border of the time window we defined for these scenarios.

Table 5.1
Bias and Error of Routes-Only-Initial-Times-Extern and Times-Routes-Initial-Times-Extern Compared to Field Data at 7-8 a.m.

	Initial-Times-Extern	
Mean / Bias	**Routes-Only 7-8 a.m.**	**Times-Routes 7-8 a.m.**
Mean Absolute Bias:	+331.403	+306.320
Mean Relative. Bias:	+19,6 %	+25.3 %
Mean Absolute Error:	533.553	503.768
Mean Relative Error:	37.5 %	35.4 %

The reason for that is the fact that agents, which are too late or too early at work try to "squeeze" into this time window. Once they are inside the time window they will more or less stay at this plan if they succeeded. Since an "optimal" plan for an agent is still to have short travel times, more and more agents try to arrive earlier in the defined time window. That is why the left peak is higher than the right one.

Finally, we look at the traffic count data. Figure 5.5 shows the relations of the two setups and the real data given by the already mentioned 12 links. As expected, the two results do not differ very much, and they are comparable to reality. Also the quantitative measures of bias and errors are similar (Table 5.1).

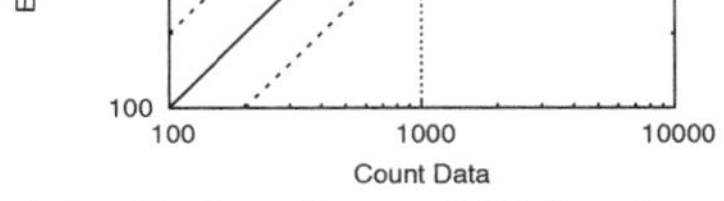

(a) Traffic Count Data vs. 260th Iteration of Routes-Only-Initial-Times-Extern Setup at 7-8 a.m.

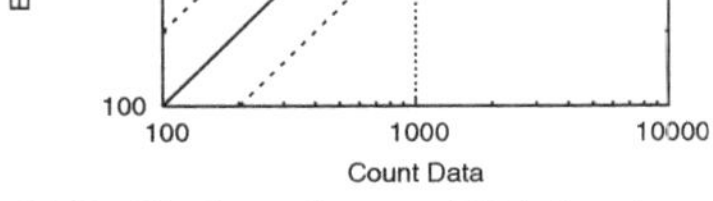

(b) Traffic Count Data vs. 200th Iteration of Times-Routes-Initial-Times-Extern Setup at 7-8 a.m.

Figure 5.5
Initial Plans with Externally Defined Departure Times: Comparison to Traffic Count Data

Initial Plans with Departure Time at 6 a.m. for all Agents

The previous section demonstrated that the results both with respect to the time structure and with respect to validation do not (at least) become worse when time re-planning is switched on. However, the initial condition was still based on the externally given time structure. The experiments in this section will test in how far a realistic time structure can be generated even when starting from a clearly implausible initial condition. For this purpose, all initial plans will be modified so that all agents initially depart at 6 a.m.. Apart from that, the initial plans are the same as before.

Figure 5.6 shows again the average of travel times for both setups. We see that this time, the routes-only setup decreases travel time more slowly than before because it is harder to avoid congestion when all agents start travelling at the same time. Of course, at the end the average travel time will be higher. With time re-planning switched on, average travel times decrease rather quickly, because agents are now allowed to change their departure time, too. Also average scores without time re-planning (Figure 5.7a) show only little improvement. Only optimizing routes does not help very much because a major part of the agents will then arrive at work too early which does not increase scores (Figure 5.8b). When the time re-planning module is also switched on, agents are now able to have short travel times and still arrive at work within the given time window. Figure 5.7b shows that the average score slowly increases to the same level as in Figure 5.3b.

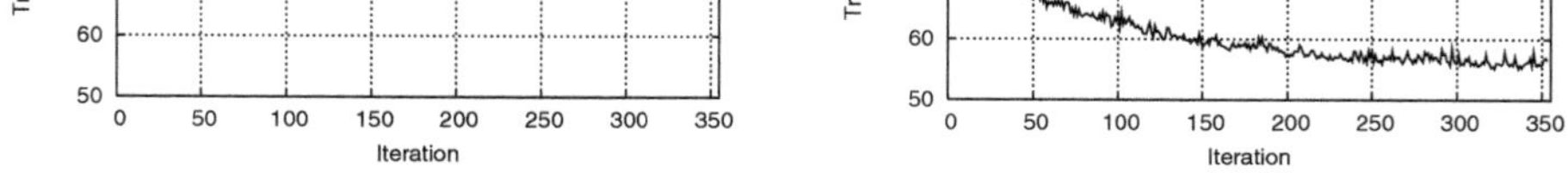

(a) Average Travel Times per Iteration for Routes-Only-Initial-Times-All6a.m. Setup

(b) Average Travel Times per Iteration for Times-Routes-Initial-Times-All6a.m. Setup

Figure 5.6
Average Travel Times of Routes-Only-Initial-Times-All6a.m. and Times-Routes-Initial-Times-All6a.m.

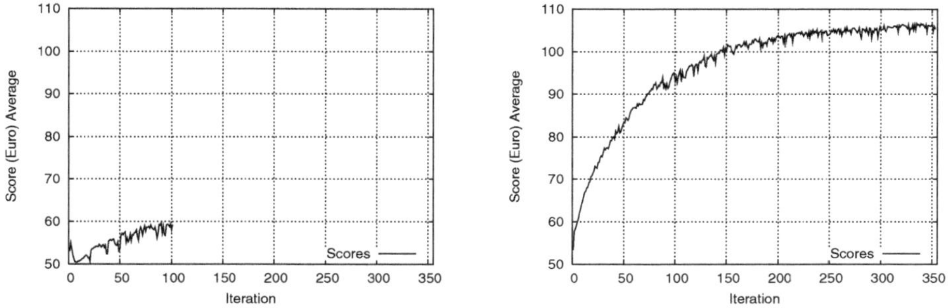

(a) Average Scores per Iteration for Routes-Only-Initial-Times-All6a.m. Setup

(b) Average Scores per Iteration for Times-Routes-Initial-Times-All6a.m. Setup

Figure 5.7
Average Scores of Routes-Only-Initial-Times-All6a.m. and Times-Routes-Initial-Times-all6a.m.

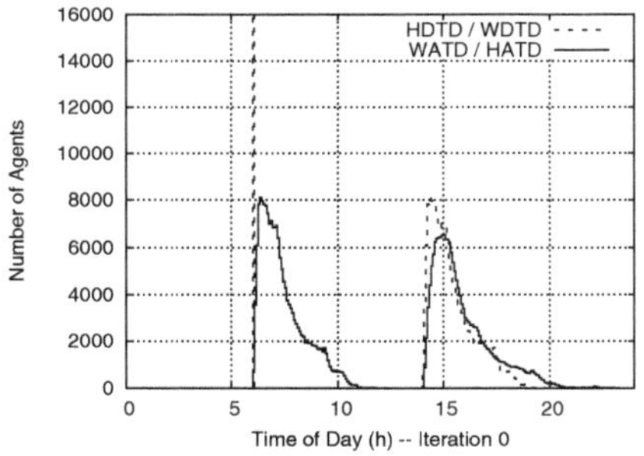

(a) Arrival and Departure Histograms (5 min Time Bins) of Iteration 0 with Initial Departure Ttime 6 a.m.

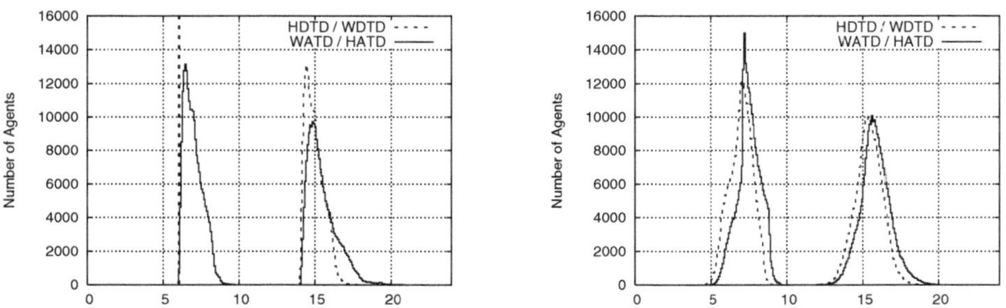

(b) Arrival and Departure Histograms (5 min Time Bins) of Iteration 100 with Time Re-Planning Switched off

(c) Arrival and Departure Histograms (5 min Time Bins) of Iteration 350 with Time Re-Planning

Figure 5.8
Arrival and Departure Histograms when in the Initial Plans Everybody Departs at 6 a.m.

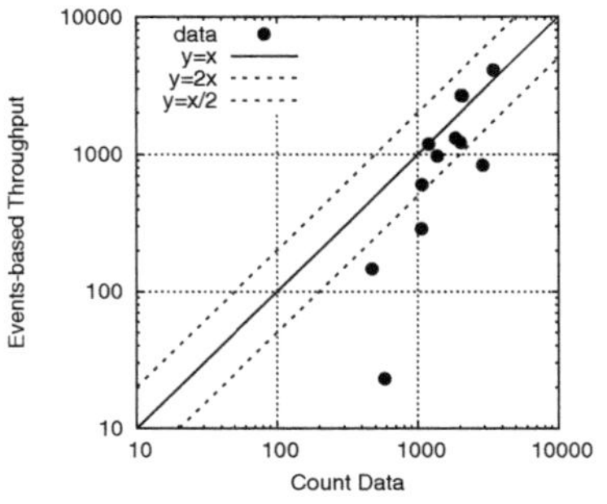

(a) Traffic Count Data vs. 100th Iteration of Routes-Only-Initial-Times-All6a.m. Setup at 7-8 a.m.

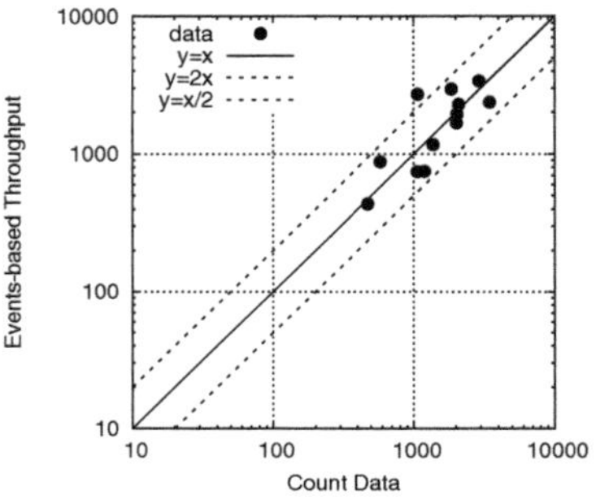

(b) Traffic Count Data vs. 350th Iteration of Times-Routes-Initial-Times-All6a.m. Setup at 7-8 a.m.

Figure 5.9
Departure Time 6 a.m. Plans: Comparison to Traffic Count Data

The histograms (Figure 5.8) also show those facts. There are many more people who arrive between 6 and 7 a.m. in the routes-only setup (Figure 5.8b) than in the times-routes setup (Figure 5.8c). The peak of the departure time (dotted) graph of Figure 5.8c moved toward the same time as shown in Figure 5.4c of the previous section.

Comparing the results with real word data shows a high discrepancy between the two setups. In the routes-only setup almost everybody starts too early. So it underestimates the throughput between 7 and 8 a.m. (Figure 5.9a). In the times-routes setup (Figure 5.9b), agents slowly move to more appropriate departure times which—at the end—will converge to similar results as obtained before. Of course, the calculation of the bias and the error (Table 5.2) now produces completely different results for the routes-only setup.

Table 5.2
Bias and Error of Routes-Only-Initial-Times-All6a.m .
and Times-Routes-Initial-Times-All6a.m. Compared to Field Data at 7-8 a.m.

	Initial-Times-All6a.m.	
Mean / Bias	**Routes-only 7-8 a.m.**	**Times-Routes 7-8 a.m.**
Mean Abs. Bias:	-344.764	+99.236
Mean Rel. Bias:	-31.3 %	+12.4 %
Mean Abs. Error:	644.107	520.256
Mean Rel. Error:	43.8 %	36.1 %

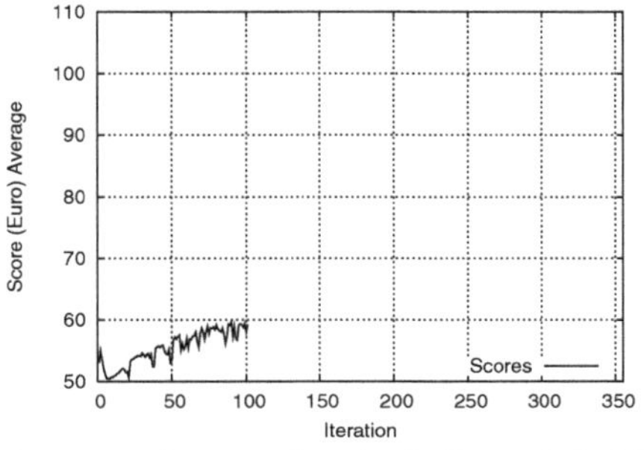

(a) Average Scores per Iteration for Routes-Only-Initial-Times-All6a.m. Setup

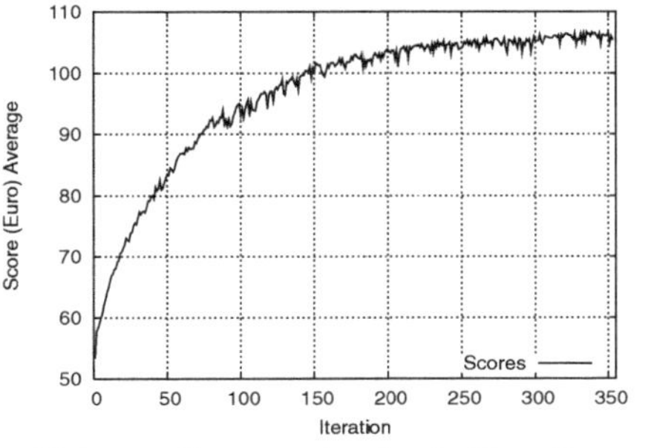

(b) Average Scores per Iteration for Times-Routes-Initial-Times-All6a.m. Setup

Figure 5.7
Average Scores of Routes-Only-Initial-Times-All6a.m. and Times-Routes-Initial-Times-all6a.m.

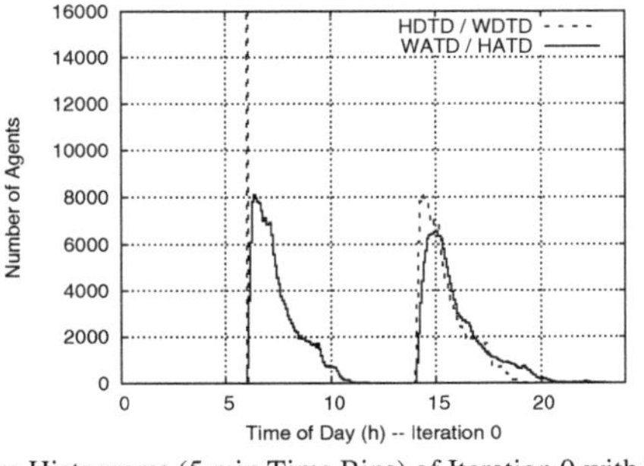

(a) Arrival and Departure Histograms (5 min Time Bins) of Iteration 0 with Initial Departure Ttime 6 a.m.

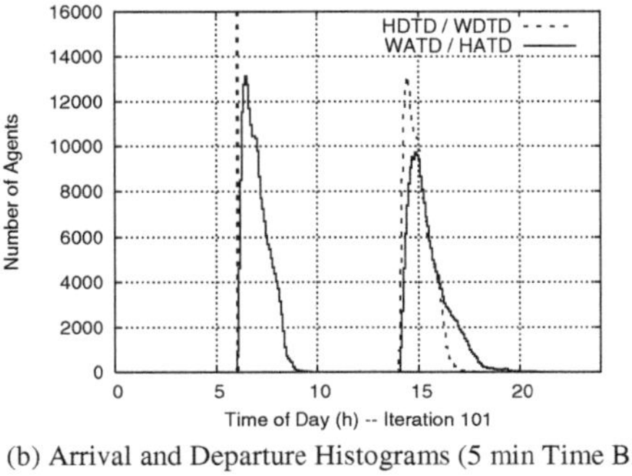

(b) Arrival and Departure Histograms (5 min Time Bins) of Iteration 100 with Time Re-Planning Switched off

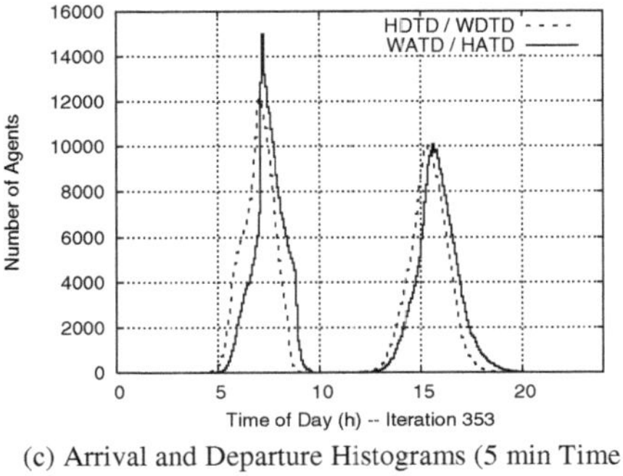

(c) Arrival and Departure Histograms (5 min Time Bins) of Iteration 350 with Time Re-Planning

Figure 5.8
Arrival and Departure Histograms when in the Initial Plans Everybody Departs at 6 a.m.

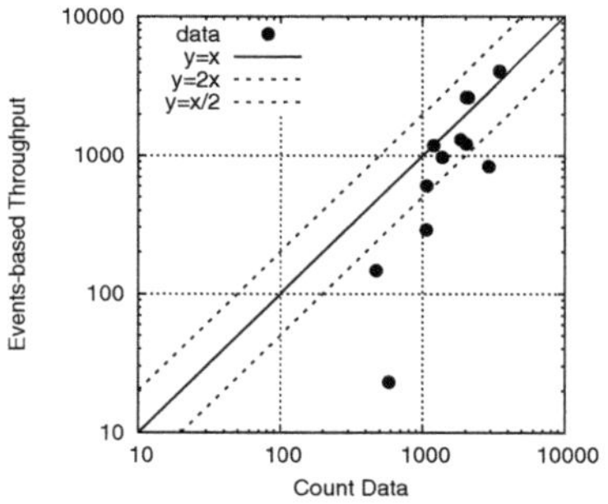

(a) Traffic Count Data vs. 100th Iteration of Routes-Only-Initial-Times-All6a.m. Setup at 7-8 a.m.

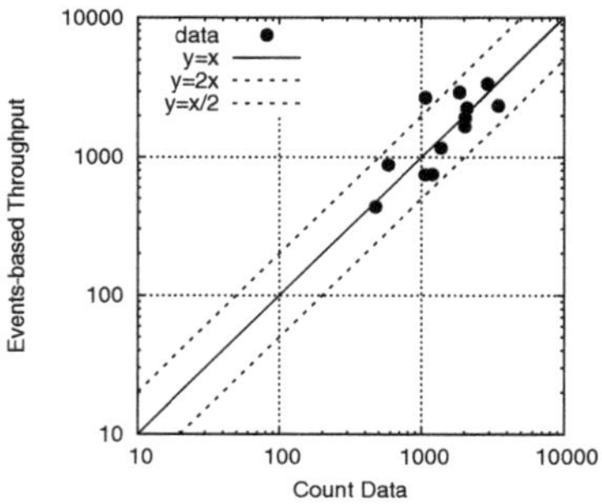

(b) Traffic Count Data vs. 350th Iteration of Times-Routes-Initial-Times-All6a.m. Setup at 7-8 a.m.

Figure 5.9
Departure Time 6 a.m. Plans: Comparison to Traffic Count Data

The histograms (Figure 5.8) also show those facts. There are many more people who arrive between 6 and 7 a.m. in the routes-only setup (Figure 5.8b) than in the times-routes setup (Figure 5.8c). The peak of the departure time (dotted) graph of Figure 5.8c moved toward the same time as shown in Figure 5.4c of the previous section.

Comparing the results with real word data shows a high discrepancy between the two setups. In the routes-only setup almost everybody starts too early. So it underestimates the throughput between 7 and 8 a.m. (Figure 5.9a). In the times-routes setup (Figure 5.9b), agents slowly move to more appropriate departure times which—at the end—will converge to similar results as obtained before. Of course, the calculation of the bias and the error (Table 5.2) now produces completely different results for the routes-only setup.

Table 5.2
Bias and Error of Routes-Only-Initial-Times-All6a.m .
and Times-Routes-Initial-Times-All6a.m. Compared to Field Data at 7-8 a.m.

	Initial-Times-All6a.m.	
Mean / Bias	**Routes-only 7-8 a.m.**	**Times-Routes 7-8 a.m.**
Mean Abs. Bias:	-344.764	+99.236
Mean Rel. Bias:	-31.3 %	+12.4 %
Mean Abs. Error:	644.107	520.256
Mean Rel. Error:	43.8 %	36.1 %

COMPUTATIONAL ISSUES

Performance: One iteration takes on average about 102 minutes. The average duration of sub-steps of an iteration are: about 18 sec for the departure time allocation module; about 23 min for the router module, of which about 19 min is spent reading the events; about 39 min for the traffic flow simulation module, including file input and output (Cetin and Nagel, 2003); about 11 respectively 16 min for sorting events and reading processing them into scores; about 105 sec for writing the new plans; the remaining time is used for other I/O processes/bottlenecks, communication and data preparations. These times allow the calculation of 15 to 20 iterations per day.

Disk Usage: A complete data set generated by *one* iteration produces about 280 MB of data (when compressed). These will be kept for the first and the last 5 iterations and also for every 10th iteration. For each of the other iterations only about 40 MB are kept.

Memory Usage: Since we are simulating about 260,000 Agents with most 5 different plans and each of them needs about 700 Bytes of memory plus some overhead, we end up with a requirement of about 1 GB of memory. The router module also needs about 200 MB of memory. Higher resolution networks will need more memory which might become a problem in the future.

FUTURE WORK

At present we only model the "primary" activities "home" and "work". We are working on adding "secondary" activities, such as shopping and leisure to the system. This requires the addition of two more modules: the activity pattern generator and the activity location generator. Another module we are interested in adding is a *Population generation* module, which would disaggregate demographic data to obtain individual households and individual household members, with certain characteristics, such as a street address, car ownership or household income (Beckman *et al.*, 1996; Frick, 2004). The population would not match reality, but would result in the same statistics. These modules should also be implemented as "plug-ins". We are also investigating other travel modes such as public transport or pedestrian mode.

Another issue of interest is the possibility that agents could also learn during the day. They could re-route while they are stuck in congestion, drop an activity because they are already too late, and so on. This "within day re-planning" (e.g., Axhausen, 1990; Cascetta and Cantarella, 1991) should help to improve their strategies faster than only "day-by-day re-planning", and the interaction with other entities (like traffic lights, changing traffic signs and other ITS entities) can be added to the traffic flow simulation. Within-day learning is more realistic since some types of decisions are made on time scales much shorter than a day (Doherty and Axhausen, 1998). However, within-day re-planning is at odds with the parallel computing approach to the traffic flow simulation (Nagel and Marchal, 2003), which is the reason why it is currently not used in our project.

Simulation speedup can also be improved by elimination of performance bottlenecks. At the moment the agent database keeps track of *all* agents of the simulation. Since recalculating an agents' strategies is completely independent to other agents, it would be useful to introduce parallelism into this. These "multiple agent databases" should then be controlled by a separate module, which keeps track of the feedback. This leads us to a clear separation of "agent databases" and "feedback".

The activity time allocation module itself could be improved, too. It should recognize when agents are too early or too late, so the adaptation to a more realistic departure time should be done with fewer iterations. High resolution networks are another issue, especially if there is more precise information available about locations. The main goal will be that each agent has its home location at a street with a house number, possibly a ramp to its garage, a private pedestrian path to the next tram station, and so on. Last but not least, high resolution scenarios are indeed a computational challenge. Quite in general, more precise traffic count data is required. There is some effort to extract information of the raw data of the Kanton Zurich, which gives more precise information about local traffic situations.

ACKNOWLEDGMENTS

The ETH-sponsored 192-CPU Beowulf cluster "Xibalba" performed most of the computations. Marc Schmitt is doing a great job maintaining the computational system. Nurhan Cetin provided the traffic flow simulation. The work also benefited from discussions with Fabrice Marchal. Michael Balmer and Bryan Raney are funded in part by the ETH project "Large scale multi-agent simulation of travel behavior and traffic flow"; Bryan Raney is also funded in part by the ETH project "A unified approach for agent-based learning with application in architecture and in transportation planning".

REFERENCES

Arnott, R., A. De Palma and R. Lindsey (1993). A structural model of peak-period congestion: a traffic bottleneck with elastic demand. *Am. Ec. Rev.*, **83**, 161.

Avineri, E. and J.N. Prashker (2003). Sensitivity to uncertainty: need for paradigm shift. *Proc. 82nd Ann. Meet. Transpn. Res. B.,* Washington, D.C. (CD-Rom).

Axhausen, K.W. (1990a). A simultaneous simulation of activity chains. In: *New Approaches in Dynamic and Activity-based Approaches to Travel Analysis* (P. M. Jones, ed.), pp. 206–225. Avebury, Aldershot.

Axhausen, K.W. (1990b). Judging the day: a synthesis of the literature on measuring the utility of activity patterns. Working Paper 561, Transport Studies Unit, University of Oxford, Oxford.

Beckman, R.J., K.A. Baggerly and M.D. McKay (1996). Creating synthetic base-line populations. *Transpn Res. A*, **30**, 415–429.

BfS—Bundesamt fuer Statistik und Dienst fuer Gesamtverkehrsfragen (1996). Verkehrsverhalten in der Schweiz 1994. *Mikrozensus Verkehr 1994*, Bern. See also www.statistik.admin.ch/-news/archiv96/dp96036.htm.

Bottom, J.A. (2000). Consistent Anticipatory Route Guidance. PhD thesis. Massachusetts Institute of Technology, Cambridge, MA.

Cascetta, E. and C. Cantarella (1991). A day-to-day and within day dynamic stochastic assignment model. *Transpn. Res. A*, **25**, 277–291.

Cetin, N. and K. Nagel (2003). A large-scale agent-based traffic microsimulation based on queue model. *Proc. Swiss Transp. Res. Conf.,* Monte Verita, Switzerland. See also www.strc.ch.

Charypar, D. and K. Nagel (in press). Generating complete all-day activity plans with genetic algorithms. *Transpn.*

De Palma, A. and F. Marchal (2002). Real case applications of the fully dynamic METROPOLIS tool-box: an advocacy for large-scale mesoscopic transportation systems. *Netw. & Spat. Ec.*, **2**, 347–369.

Doherty, S.T. and K.W. Axhausen (1998). The development of a unified modeling framework for the household activity-travel scheduling process. In: *Verkehr und Mobilitaet* (K.-J. Beckmann, ed.), pp. 45-59. Stadt Region Land No. 66, Institut fuer Stadtbauwesen, Technical University, Aachen, Germany.

Ettema, D.F., H.J.P. Timmermans and T.A. Arentze (2004). Modelling perception updating of travel times in the context of departure time choice under ITS. *Intell. Transpn. Syst.*, **8**, 33-43.

Frick, M.A. (2004). Generating synthetic populations using IPF and Monte Carlo techniques: some new results. *Proc. Swiss Transp. Res. Conf.*, Monte Verita, Switzerland. See also www.strc.ch.

Gawron, C. (1998). An iterative algorithm to determine the dynamic user equilibrium in a traffic simulation model. *Int. J. Modern Phys. C*, **9**, 393-407.

Holland, J.D. (1992). *Adaptation in Natural and Artificial Systems*. Bradford Books, New York.

Jacob, R.R., M.V. Marathe and K. Nagel (1999). A computational study of routing algorithms for realistic transportation networks. *ACM J. Exp. Algorithms*, **4**/1999es/6 (electronic journal).

Kaufman, D.E., K.E. Wunderlich and R.L. Smith (1991). An Iterative Routing/Assignment Method for Anticipatory Real-Time Route Guidance. IVHS, Technical Report 91-02, University of Michigan, Department of Industrial and Operations Engineering, Ann Arbor, MI.

Nagel, K. (1995). High-Speed Microsimulations of Traffic Flow, PhD thesis, University of Cologne. See also www.inf.ethz.ch/~nagel/papers or www.zaik.uni-koeln.de/~paper.

Nagel, K. and F. Marchal (2003). Computational methods for multi-agent simulations of travel behavior. *Proc. Int. Assoc. Trav. Beh. Res.*, Lucerne, Switzerland. See www.ivt.baum.ethz.ch/allgemein/iatbr2003.html.

Nagel, K., M. Strauss and M. Shubik (2004). The importance of timescales: simple models for economic markets. *Phys. A*, **340**, 668-677.

Palmer, R.G., W.B. Arthur, J.H. Holland, B. LeBaron and P. Tayler (1994). Artificial economic life: a simple model of a stockmarket. *Phys. D,* **75**, 264-274.

Raney, B., N. Cetin, A. Voellmy, M. Vrtic, K. Axhausen and K. Nagel (2003). An agent-based microsimulation model of Swiss travel: first results. *Netw. & Spat. Ec.*, **3**, 23-41.

Raney, B. and K. Nagel (2003). Truly agent-based strategy selection for transportation simulations. *Proc. 82nd Ann. Meet. Transpn. Res. B.,* Washington, D.C. (CD-Rom).

Raney, B. and K. Nagel (in press). An improved framework for large-scale multi-agent simulations of travel behavior. In: *Towards Better Performing European Transportation Systems* (P. Riedveld, B. Jourquin and K. Westin (eds.).

Timmermans, H.J.P., T.A. Arentze and D.F. Ettema (2003). Learning and adaptation behaviour: empirical evidence and modelling issues. *Proc. ITS Conf.*, Eindhoven, The Netherlands (CD-Rom).

Vrtic, M. and K.W. Axhausen (2002). Experiment mit einem dynamischen Umlegungsverfahren. *Strassenverkehrstechnik: Arbeitsberichte Verkehrs- und Raumplanung,* No. 138. See also www.ivt.baug.ethz.ch.

Vrtic, M., R. Koblo, and M. Voedisch (1999). Entwicklung bimodales Personenverkehrsmodell als Grundlage fuer Bahn2000, 2. Etappe, Auftrag 1. Report to the Swiss National Railway and Dienst fuer Gesamtverkehrsfragen, Prognos AG, Basel, Switzerland. See also www.ivt.baug.ethz.ch/vrp/ab115.pdf for a related report.

Weiss, G. (1999). *Multiagent Systems. A Modern Approach to Distributed Artificial Intelligence.* MIT Press, Cambridge, MA.

6

The Scheduling Agent - Using Sesam to Implement a Generator of Activity Programs

Guido Rindsfüser, Emch+Berger AG, Bern, Switzerland

Franziska Klügl, Universität Würzburg, Germany

Introduction

Several reviews of the activity-based approach have underscored the importance of travellers' activities as objects of examination in transport research and the shift from vehicles to people (households) as the behavioural units (e.g., Jones *et al*,. 1983; Kurani and Lee-Gosselin, 1996). Activity scheduling (e.g., Damm and Lerman, 1981; Doherty, 2000; Timmermans, 2001), a current research field within the framework of the activity-based approach (e.g. Axhausen and Herz, 1989; Jones, 1990; Ettema and Timmermans, 1997; Arentze and Timermans, 2000) aims at understanding the underlying behavioural mechanisms that give rise to activity sequencing over the course of a day or a week. Activity scheduling surveys brought new insights into individual scheduling behaviour (Ettema *et al.*, 1994; Rindsfüser *et al.*, 2003; Lee and McNally, 2003; Doherty *et al.* 2004). The first models of activity scheduling have been developed and some have been tested recently (e.g., Ettema *et al.*, 1993, 2000; Arentze and Timmermans, 2000; Gärling *et al.*, 2001; Joh *et al.*, 2002, 2003, 2004; Miller and Roorda 2003). Some of these models can be viewed as simple multi-agent models. It is believed that these models of activity scheduling behaviour can improve transport demand modelling because of their behavioural underpinnings.

Progress in Activity-Based Analysis edited by H. Timmermans

The focus of these models of activity-scheduling behaviour has primarily been on the formulation of these models. Less attention has been paid to the implementation of the estimated models in real-time simulations. Multi-Agent technology has much promise in this regard (Klügl, 2001). The concept "agent" is especially useful for developing solutions in highly dynamic problem domains. Thus, in simulation applications the multi-agent view enables the specification and simulation of variable structure models where the components of the models may interact with each other. This allows for models of human decision making that are less abstract than any traditional paradigm. The multi-agent paradigm provides interesting modelling options for traffic and transportation systems.

The remainder of the paper is organized as follows. In the next section, the objectives of the presented approach are summarized. This is followed by a brief overview of current research related to the use of agent technology in transport applications. Next, we will give a short introduction to the modelling and experimenting environment "SeSAm", followed by a description of the application of a scheduling agent. Conclusions will close this chapter.

OBJECTIVES

The objective of this chapter is to present an emerging activity scheduling process simulation designed as a multi-agent simulation – briefly describing the overall model concept, the simulation modules, and first experiments. The focus is on demonstrating the feasibility and benefits of this approach, and not that much on a detailed description of the behavioural models, which are and which have to be included in the simulation. During the development of the simulation system a number of problems, related to both information technology and transportation occurred. Therefore, a second objective of this chapter is to discuss these difficulties and provide hints for further research.

AGENTS IN TRANSPORT

The Agent Paradigm

Before discussing characteristics and examples of agents in transportation modelling, the meaning of the term software agent has to be clarified. An examination of the literature suggested that no single definition was available that is widely acceptable. Franklin and Graesser (1997) collected a variety of definitions of the concept "agent". Their definition only focused on the situational context of an agent in an environment, and did not explicitly explain other important properties like autonomy, flexibility and responsiveness, pro-activeness or ability for interactions with other agents or users. Wooldridge and Jennings (1995) on the other hand as probably the most cited paper in the context of agent properties did identify those properties.

AgentLink, the European Network of Excellence for agent-based systems (www.agentlink.org) called "autonomous, problem-solving computational entities capable of effective operation in dynamic and open environments" agents (Luck *et al.*, 2003). The key property is autonomy – although the definition of this property turns out to be difficult as well. Autonomy is what distinguishes an agent from an object, as an agent should be capable of choosing its actions and interactions itself. For implementation, object-oriented techniques might be used.

Multi-Agent Systems

A multi-agent system can be seen as a collection of agents. The following broad characteristics are associated with multi-agent systems:

- Each agent has incomplete information or capabilities for solving the problem, as it is only capable of perceiving part of the problem space or part of its environment (locality).
- There is no global system control.
- Data is decentralized.
- Computation is asynchronous.

Multi-agent system technology may form an ideal basis for traffic and transportation modelling because these properties can also be ascribed to traffic systems with all their individual participants that are processing information and modify their environment based on this information.

Existing Applications

Applications of multi-agent systems in traffic and transportation modelling can be distinguished into three domains and levels of application:

- Use of agent technology as a basis for sophisticated traffic control and/or traffic management strategies. A prominent and early example is the OASIS system (Ljungberg and Lucas, 1992), where air traffic was modelled as agents that are negotiating for scheduling take-off and landing time slots. This system was actually tested at Sydney Airport. Other examples in transportation management can be found in Fischer *et al.* (2000), Dijkstra and Timmermans (2002), Hernández *et al.* (2002), and Adler and Blue (2002).

- Use of agent technology for abstract modelling of phenomena. This consists mostly of basic research into information technologies with a game-theoretic background (e.g., Klügl and Bazzan, 2004)
- Use of agent technology for more realistic traffic simulation. For example, Dia (2002) proposed an agent simulation for route choice behaviour. Microscopic traffic flow models - many of them originally based on cellular automata - are now developed as agent-based models resulting in more interesting travel behaviour due to the decision making abilities (e.g., Nagel and Marchal, 2003). Another example is an agent system to model lane changing within a micro simulation (Hidas, 2002). These simulations are not only based on predefined behaviour schemes but also use "intelligent" and adaptive activity generation (e.g. Charypa and Nagel, 2003; Balmer *et al.*, 2004).

A fourth category concerns the simulation of individual activity scheduling processes, which is a relatively new research field within the activity-based approach. The problem of modelling individual demand and scheduling of activities and as a result the generation of activity patterns seems to call for an agent-based simulation where a single agent represents a single person. Due to the fact that the society is becoming more and more heterogenous, activities and observed activity patterns increasingly diversify. Thus, new models for the estimation of single and individual decision processes for activity participation and derived transport demand are required. Traditional models are able to depict only a small range of variation in behaviour (e.g., Kitamura, 1996; Recker *et al.*, 1986). More current models such as for example the sophisticated *Albatross* model system (Arentze and Timmermans, 2000) produce more variations and more complex patterns while taking into account several constraints influencing individual behaviour. The use of several rules and heuristics underscore the new approaches as agent systems. Various agents perform different jobs to assign activities to decision units.

With a multi-agent approach in the sense of having one agent for one person of the represented real system (plus agents for facilities and other constructs like households, resources or the world), one should be able to simulate activity patterns with more behavioural soundness, covering a broader range of behavioural change due to new transport measures. The proposed approach is a first step towards a multi-agent simulation system to generate individual activity programs.

SeSAm

The model and experiments described in this chapter are implemented and conducted using the multi-agent simulation environment SeSAm. This "Shell for Simulated Agent Systems" provides a generic environment for modelling and experimenting with agent-based simulations. Its development was specially focused on providing a comfortable tool for the construction of complex agent-based models (Klügl, 2001). In the following, we want to present relevant properties of SeSAm as a short introduction to the simulation system. More information about SeSAm can obtained from http://www.simsesam.de.

Basic Agent Architecture

One of the most important characteristics is the underlying agent architecture, as it determines the potential abilities of the agents based on it. SeSAm-Agents consist of two parts: a body that describes the current state and beliefs of the agent and a "brain" that is responsible for selecting the agents' actions. The body is represented by a set of state variables that may not only contain numeric information but also nested symbolic data. Every variable may have some associated dynamics that may be used for unconscious changes in this variable, like aging, etc. The brain interprets a behaviour network – described based on enhanced UML Activity Graphs. The nodes of the behaviour network can be seen as scripts that are initiated and terminated by firing rules. An agent is always found in one behaviour node in each network (there can be more than one graph responsible for behaviour representation, in this case an agent concurrently is in more than one behavioural state).

A set of predefined primitive actions and perceptions can be used for filling the nodes and rules for "implementing" the concrete agent behaviour. This simple and transparent agent architecture resembles a computationally complete agent programming language (Oechslein *et al.*, 2001). Thus, it can also be used for BDI (Belief Desire Intention (Klügl, 2001)) agent models or other intelligent behaviour generation, like the activity scheduling approach described in this chapter.

Representation of the Agents' Environment

In SeSAm different forms of spatially explicit simulations are possible. The basic space representation is a continuous map (can be mapped to a discrete grid). Possible movement directions are arbitrarily values for movement angles (0-360°). On the map, agents and resources can be positioned. Resources are passive objects that also possess a body, but no behaviour. They can store arbitrary complex state descriptions but may not manipulate their environment actively. The environment for an agent consists not only of other agents and resources that are localized on the map, but also on some properties on the global level, e.g. overall temperature or time of day. However, all agents may not perceive this information in the same way.

Simulation Routine

For technical reasons the simulation routine and update order is relevant, especially as SeSAm does not support real agent parallelism. The simulation is time-stepped, that means at every time point, all agents and all other objects containing some update formula in their state variable descriptions are updated. After one update round the global time step is incremented. For emulating agent parallelism their update happens in random order, which leads to the system property that the modeller cannot rely on one agent being updated before a particular other one. In detail the update routine is as follows:

1. Generate start situation and initialize simulation clock.
2. While not terminating condition true
    a. Update all simulation components in a random sequence
       For every object in the situation do
        i. Update body variables
        ii. If object is an agent
            For every reasoning engine
            1. Execute Action Sequence
            2. If current behaviour node is terminating → Select new node
               (Execute termination and starting sequence resp.)
            3. Repeat 2+3 until a time consuming node is executed
    b. Update World behaviour

Consequently, there is only one possible synchronization point, namely the defined update time of the world. All other synchronization and coordination may use more or less traditional approaches known from distributed and multi-agent systems.

The SeSAm System

Based on the representation concept described above, both a specification language and a software environment for modelling and simulation were developed. The specification framework focuses on the representation of agent behaviour especially in relation to other agents, resources or the general environment. In SeSAm, the actual modelling and simulation environment user is able to design visually the behaviour of agent classes. Analogous mechanisms are provided for specifying the static structures of an agent, an agent system or the environmental elements and their configuration. Thus, a complete agent-based simulation model can be implemented visually without programming in a traditional programming language. Due to the provision of abstract data structures and diverse modularization and abstraction instruments, the SeSAm also allows to handle complex models.

Plugins are provided that also support real-world simulations, like database connectivity or sophisticated spatial enhancements. Support for simulation cannot end when the model is implemented completely. Testing and executing experiments are also very demanding and effortful tasks. As it provides freely configurable instruments for gathering data and scripting options for constructing simulation experiments on several computers concurrently, SeSAm is a highly valuable tool even for large-scale agent based simulations.

THE SCHEDULIING AGENT

"The Scheduling Agent" is a complex system of models of individual behaviour related to the scheduling processes realized as a multi-agent system. It aims at the generation of activity patterns of individuals. The multi-agent model was (it is still in progress) designed using SeSAm. This chapter describes the first implementation and tests. Therefore, it was the major objective to bring the two approaches, the transport related scheduling process approach and the IT related multi-agent approach, together and to demonstrate the feasibility and advantages. Because of the complexity, the amount of data required, the models of individual behaviour and the poor available empirical data, the behavioural models are kept relatively simple.

Simulation Concept

The basic concept goes back to the ideas of a unified modelling framework for the scheduling process proposed by Doherty and Axhausen (1999). Following this framework, Rindsfüser and Doherty (2000) elaborated the original idea and conducted some data analyses for single facets of the concept. In particular, the concept of The Scheduling Agent system is based on the simulation components displayed in Figure 6.1.

The input is a data set describing a synthetic population, a synthetic city and a set of behavioural parameters. In a first modelling step, an "activity repertoire" and a "habitual program" are generated. The repertoire is a list of activities with various sets of distributions for every attribute value, as for example the start time. For different combinations of individual socio-demographic attributes an assigned set of activity attribute value distributions can be found. The habitual program is a set of routine/habitual activities with already assigned attributes. The second modelling step concerns the simulation of the scheduling processes. It starts with an initial "current" program, generated from the habitual program while adding and fixing values of some variables. This current program can be seen as a skeleton program and can be modified during scheduling.

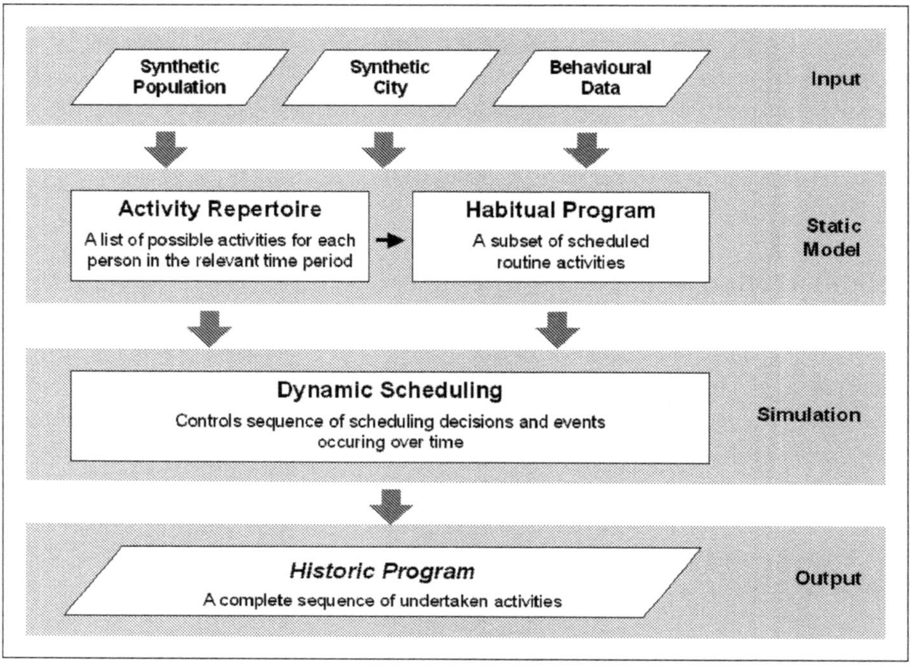

Figure 6.1
Schematic Representation of the Main Structure and Components

The simulated scheduling may happen at every time step during the execution of the current program (a day, a week, and so on). The output of the overall simulation is a "historic program", a list (sequence) of activities and trips the individual has undertaken during the simulation run including activity and trip attributes as there are start time, duration, visited location and used mean of transport at present. There is no route choice realized until now. In this first implementation agents move on direct line to the destination with an average speed (depending on the used mode).

The general simulation framework of a scheduling process is displayed in Figure 6.2. One of the major benefits of using agents is the possibility to integrate the situational context of each individual into the model. As displayed in Figure 6.2, an agent can act in reaction to a modification in his environment or act depending on his own status or the status of other agents at every time step during the simulation. All actions of the simulated agents result in changes in attributes of the environment, agents' own attributes or attributes of the other agents.

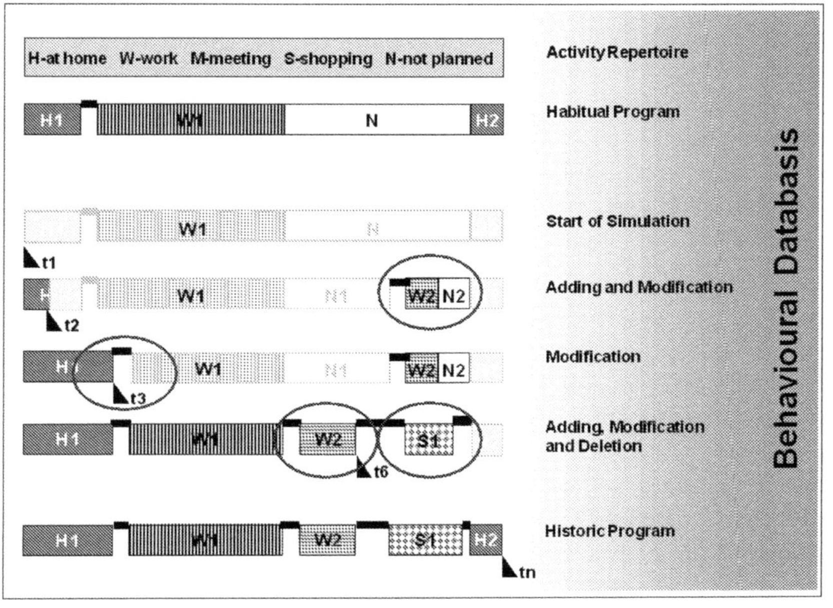

Figure 6.2
Schematic Representation of the Simulation Concept

Data

Modelling the scheduling processes at the individual level requires not only a lot of data on socio-demographics, but also new, additional data about the scheduling process (Doherty, 2001). To develop an executable simulation at the microscopic level, empirical data (where possible), synthetic data (where needed) and assumptions/rules to generate additional data (where necessary) had to be used:

- The population (modelled as agents):
  A synthetic population, created for the project ILUMASS (Strauch *et al.*, 2004), was used to specify different agents. A description of the generation of the synthetic population is given in Moeckel *et al.* (2003). These data sets do not include any choice variables.
- The agents' behaviour:
  To generate input data for the various components of the agents' behaviour model mainly the Mobidrive dataset (Axhausen *et al.*, 2002) was used. For example, a measure of similarity of activity patterns was calculated using the sequence alignment method (e.g., Wilson, 1998; Joh *et al.*, 2000, 2001a,b,c; Rindsfüser, 2001) besides several other parameters including

flexibility measures, routine measures, etc. These and socio-demographic attributes of the individuals are used to (rule based and stochastically) choose activities to perform, choose start times and durations as well as to serve as a basis for the decisions during fitting the chosen activities into the schedule. Some additional findings from data produced with CHASE (Litwin *et al.*, 2004) and data from the time budget study in Germany (Statistisches Bundesamt, 1997) were used as a supplement for statistical distributions of activity start times, durations, etc.

- The "world" and the facilities (modelled as agents):
  A synthetic city, created for the project ILUMASS was used to specify different facilities (residences and places to undertake an activity). Some properties of the facilities, as for example opening hours, were created using general and simplified assumptions, for example opening hours are not differentiated between facilities.

An example of observed scheduling behaviour using EX-ACT (Rindsfüser *et al.*, 2003) in Dortmund is shown in Figure 6.3. For a single person and a single (here out-of-home) activity the whole planning/scheduling process is displayed. In the middle of day six, the execution of this single activity starts. All gathered information after this time concerns decisions made during the execution of the activity. With EX-ACT all activities (30 categories) were observed.

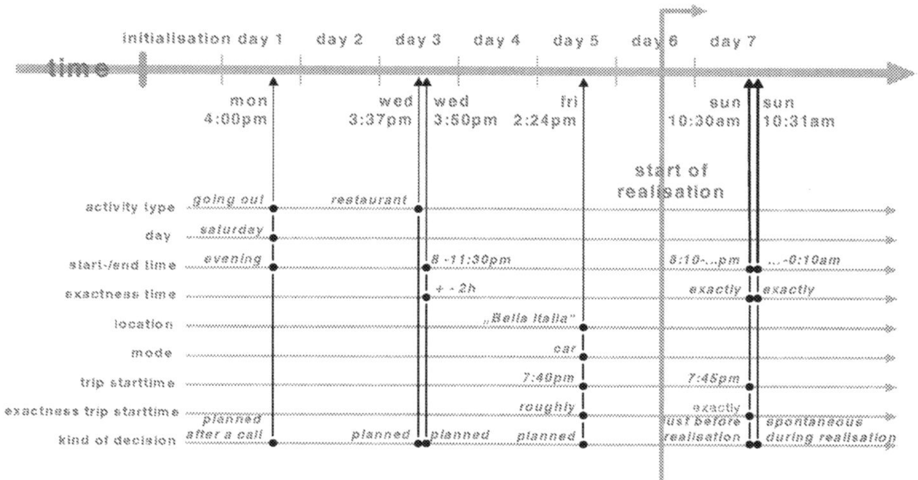

Figure 6.3
Schematic Representation of Observed Scheduling Behaviour of a Single Person for a Single Activity

These data will support this simulation system in the future. For the moment these data are used to determine decision (scheduling) horizons or decision (scheduling) points and scheduling sequences, which choice is made first or last before realisation.

Implementation

Situational Context "The World". The "world" - the environment in which the agents act - is an agent too in SeSAM, to ease the specification of the environment. Therefore, the world can possess "behaviour" (can have a behaviour graph). The world, as every agent, can have attributes, which can be modified by other agents. The behaviour of the world can also influence the other agents. In the current application, the world is used to provide the current time, day and week. With every simulation step, one minute is added to the current time. Additionally, the world is used to generate aggregated simulation data used for the generation of output files.

Agents and "Reasoning Engines". The main objective during the design process of a scheduling agent in SeSAm was to create a universal agent and not to design different classes of agent groups. The individuality of the agents should be represented by the individual set of attributes of the agents - resulting in maximum individuality. Meta-level behaviour - scheduling and executing activities - was defined for only one agent class. Therefore, the simulation is based on generated, different agents as instances of the designed agent. Consequently, all agents have the same generic behaviour. Their individual behaviour results from individual attribute values, from the assignment rules based on these variations in personality and therefore from individual tasks generated from their individual context. A "Task" is one single future activity or trip, which is chosen, specified and carried out during simulation.

The generic behaviour of the agents is represented in the "reasoning engine", one part of the agent specification. The behaviour network is the graphical representation of the rules and functions, which determine the agents' behaviour. At first glance, the graph seems to be quite simple (Figure 6.4). The main components are the states of an agent. An agent can be found either in an activity or in a trip. Is a person in one of these states, and only then, time elapses. All other actions are only containers of more or less inner processes occurring during an activity or a trip excluding the behaviour node "out of simulation", which is only implemented to end the simulation.

An instrument to generate some kind of real situational context is the use of triggers. Some of the most frequently observed reasons for scheduling are used in the simulation to cause a transition from "during activity" or "during trip" to "schedule". In the following, the single components of the behaviour graph and these triggers and steps in the processes will be explained in more detail.

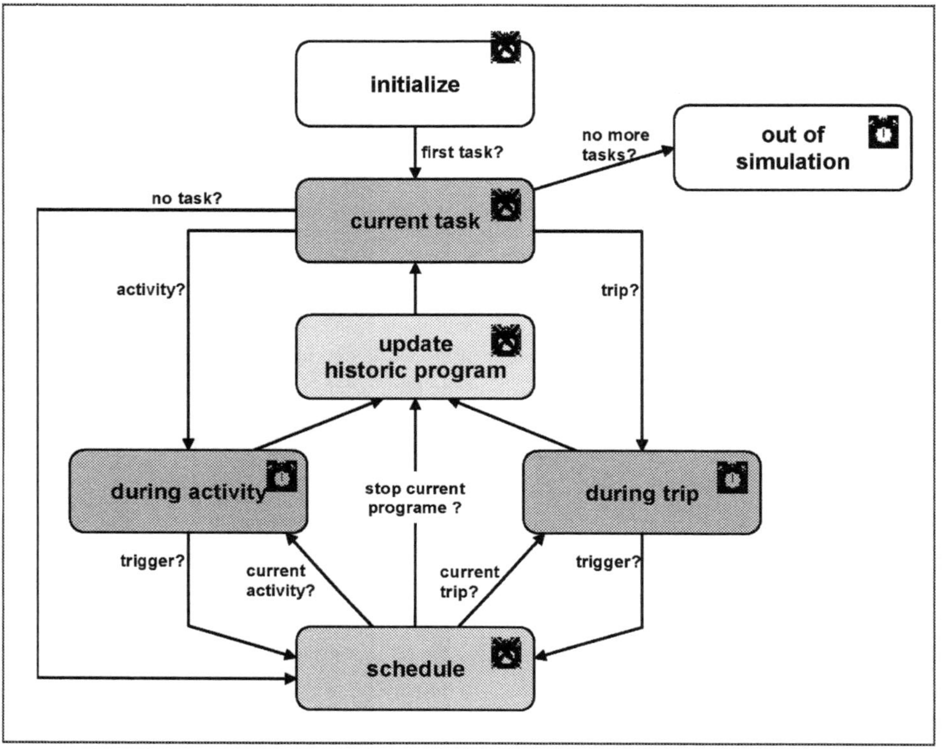

Figure 6.4
The Agents' Activity Graph

- Initialize:
  Before starting the actual simulation an initialisation step is needed. All assignments and variable value settings required to generate a start situation and all modelling which is not part of the inner scheduling are made during initialisation:

  - For each facility present in the situation a list of possible related activities is generated.
  - The individual dwelling (originally coming with the household file) is assigned to the agents.
  - The initial current program of the individual for the simulation period is generated from the habitual programs and assigned to the agents.
  - The starting position of the agents in the world is assigned depending on the location of their first activity in the current program.

- For each activity in the current program (until this time only activities are generated) a trip (a new task) is generated when the location of the inspected activity differs from the location of the previous activity. The distance (direct line) is calculated. Depending on the distance and individual characteristics a transport mode is chosen. Based on the mean speed for the chosen mode and the distance, the expected trip duration is calculated. The end time of the trip is set to the start time of the following activity.

• Current Task:
The determination of the current task of an agent plays a key role. Depending on the current task the next actions of the agent are determined. But also other functions and rules are located in this node of the behaviour graph:

- Using the time of the world, the current task is determined out of the agents' current program as the task with start time <= the current time and with end time >= the current time.
- If the current task is an activity, a transition to "during activity" is made.
- Is the current task a trip, the agent's state in the next simulation step is "during trip".
- If the current task is empty the agent continues with the node "schedule".

• During Activity:
The agent is in this state during an activity. The simulation time is advancing until the activity end time is reached and the agent leaves this behaviour node. The agent can act or react in every simulation step due to modifications of the situational context. The agent is continuously updating his variables and his knowledge of the environment with every simulation step. The current time for example is a public attribute from the world, which can be read by the other agents, as well as all public attributes. Another advantage is that agents can communicate, in a single form through changing of attributes, which the other (specified) agents can read.

- In the state "during activity" the agent stays at the current position (facility) to carry out the current task (an activity), the agent is not moving.
- It may happen due to some triggers (personal, from communication with other agents or related to modifications of the environment) that the agent needs to do some scheduling. If so, he leaves the node "during activity" and changes to the node "schedule". After finishing his scheduling behaviour, the agent updates his current program and continues

with his current activity or reaches the state "update historic program". This happens if the current task is to stop spontaneously and a new task is to start directly.

- During Trip:
  The agent is in this state during a trip. The simulation time is advancing until the destination is arrived and the agent leaves this behaviour node. As in the node "during activity" the agent can act or react in every simulation step due to modifications of the situational context. The agent is continuously updating his variables and his knowledge every simulation step.

  - In the state "during trip" the agent moves in the environment to reach the given (chosen) destination where the next task has to be carried out.
  - It may happen due to some triggers (personal, from communication with other agents or related to modifications of the environment) that the agent needs to do some scheduling. If so, he leaves the node "during trip" and changes to the node "schedule". After finishing his scheduling behaviour, the agent updates his current program and continues with his current trip or reaches the state "update historic program". This happens if the current task is to stop spontaneously and a new task is to start directly.

  The current realization of the movement is a direct moving from one position to another with a calculated mean speed (depending on the used mode). In the future, it will be possible to let the agents move along the road network, imported from a GIS.

- Schedule:
  This state can be seen as a collection of behavioural models. Thus, it is the most complex module of the simulation and still under development. All scheduling related processes are simulated within this behaviour node. These processes are additions of new activities, modifications or deletions of already scheduled activities. Because it is possible that an already scheduled activity has to be modified again for some reason, this concept and realisation in this simulation includes rescheduling of activities and trips. For all these decisions, a reasonable number of constraints and intentions needs to be taken into consideration (as for example location and mode choice or temporal overlapping with other, already scheduled, activities). For this first simple implementation some rules derived from CHASE and EX-ACT data are used, as for example, which activity types are more likely to be moved in time or deleted, or which activities are to what amount shortened. Again depending on the own attributes these values are chosen stochastically. Constraints like opening hours and time budgets are (simplified) considered.

This node can be reached from every other node, comparable to real life, where planning and scheduling is not fixed to specific times of the day. This is a major benefit of the using multi-agent simulation compared to "traditional" modelling techniques. Depending on the received triggers an agent needs to add a new task, modify or delete an already scheduled or the current task or their variables.

- New tasks:
  The need to choose a new task (during simulation run) is given if the agent has no current task (just finished a previous task) or if a trigger (e.g. external demand from another agent) is recognized during a just executed task (activity or trip). Several steps, new task generation, activity type choice, start time choice, duration choice, location choice and fitting into existing schedule are performed based on simple algorithms and distributions depending on the very own characteristics of the agents attribute values and the situation. Several more sophisticated choice procedures are known and well documented (e.g., Arentze and Timmermans 2000). The difference is that in the presented simulation each agent uses these procedures to act (and react) depending on the state of his own attributes and all other public attributes from other agents depending on the specific situation during simulation is running.

  - A new task (an activity or trip) is generated based on choosing an activity type out of the agents' repertoire. The possibility to choose a specific activity depends on the individual characteristics and the time of the day. In the future, some kind of learning (e.g., Arentze and Timmermans, 2002, 2004; Charypar *et al.*, 2004) should be implemented to enable the agent to compare with his activity history or choose from the history to simplify modelling and to narrow real behaviour. In addition, an individual start time and duration is chosen from specific distributions.
  - Depending on the activity type a location (a facility) is chosen. In the current model, an appropriate facility nearest to the current position is chosen, with some randomness (so that in 10% of the choices a facility further away is taken into consideration). The agent is also considering his home position and the next task to check whether it is possible to find a location on "the way home".
  - A new task, a trip, is generated (see above).
  - The new task (tasks) is fitted into the current program. Several context sensitive rules are used. Due to the poor availability of empirical data, many assumptions had to made for prioritizing activities (which task has to be moved in time, to be shortened or to be deleted, which task cannot be moved, etc.), change start and end times, etc. For example, if a new task leisure is to be scheduled and the time window between possible start time of the planned leisure task and the next task, lets say work, is not long enough, it is only to some

extend possible to move the start time from the work task to fit the leisure task into the program. The amount of movement is chosen again from a distribution derived from time budget studies.

- Modifications and/or deletions:
  Based on similar/same rules, the current or future task is modified or deleted.

- Update Historic Program:
  This status is not time consuming, and is entered when the end time of an activity or a trip is reached (including the case, that the current task is to stop spontaneously).
  - The activity or trip carried out is written at the end of a historic program (a list), which is consequently a sequence of activities and trips already carried out.
  - The agents' current task is set to zero.

- Out of Simulation:
  This behaviour node is needed, because the simulation is set up to stop when the very last activity or trip is carried out. Due to the fact that during generation of the current program the individual programs vary and a simulation end per se not exists one of the future tasks can be marked as last one. An agent with no next task after the marked one moves to the "out of simulation" state where he is doing nothing until the simulation stops (the last agent reached this node). In addition the simulation can be stopped immediately.
  - After all agents have reached the out of simulation status, the realized programs of all agents are written to a file. This is done in a XML format, corresponding to an XSD schema and an XSL stylesheet, so that the results can be displayed and inspected within a standard browser (e.g., Internet Explorer).

As indicated before, the objective of this project was to demonstrate the feasibility of using SeSAM as a shell for implementing a scheduling based activity generation using a multi-agent system. Therefore, the implemented behavioural rules and models are kept simple so that the general framework could be build and the advantages be demonstrated. In terms of sophisticated behavioural models, other work is available and documented. Nevertheless, the approach differs from most from existing work due to the fact that it is a step towards (re-)scheduling and simulation of individual behaviour within the running execution of activities and trips. Only Gärling *et al.* (1998) and Aurora (Joh *et al.*, 2003, 2004) have a similar focus.

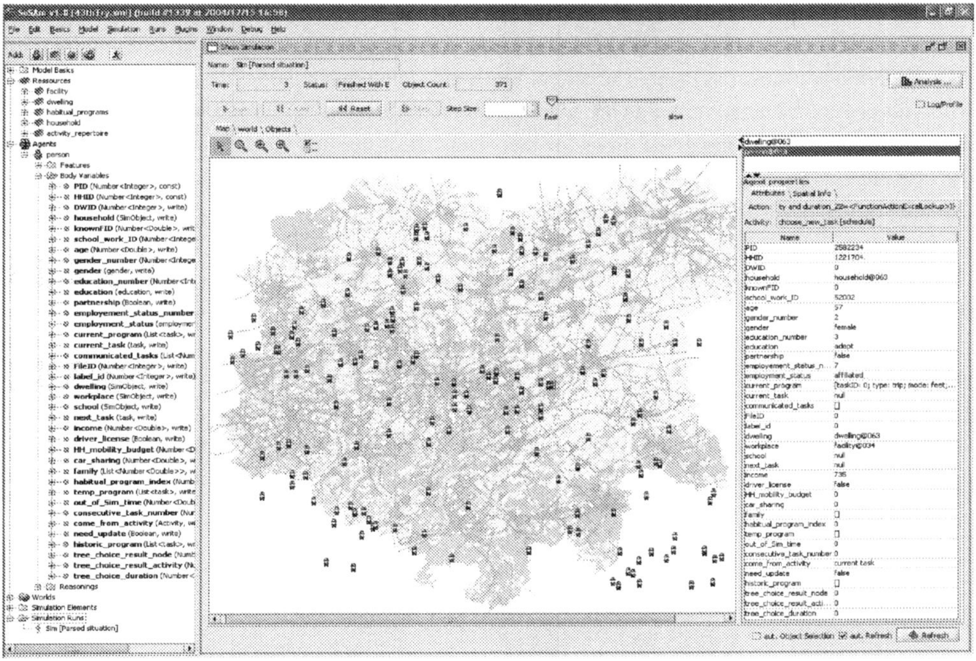

Figure 6.5
A Situation within a Simulation in SeSAm (Screenshot)

Generating a Situation. Before starting a simulation run, a situation has to be created. Generating a situation means generating so many instances of the designed agent that the desired number of individual agents exists. Also, the different facilities must be generated, meaning that there is a simulation object for each facility with a position in the environment. The data concerning the synthetic population is stored in two files: a household file with all the household-related characteristics and a person file, including ~3 million persons with their attributes. Two other files contain data related to the residences and the facilities. Pre-processing of these data is necessary to convert the ASCII coded files into SeSAm readable data. A converter is used to produce one SeSAm description file and another file with general preferences for the situation. In the latter, for example a background map can be determined to be loaded into the situation. New advances in SeSAm make it possible to load geo-referenced data directly. A plugin for SeSAm generates a specific situation directly out of the read data via a parser and an interface. For the first test simulation runs different numbers of agents, households and facilities were used, but until now not more than 200 agents due to the processing speed. This is still a major problem simulating such models with such numbers of calculations. The screenshot displayed in Figure 6.5 displays an example of a generated situation within a simulation and some of the SeSAm control windows. On the left hand the structure of the model is displayed. For example the variables of an agent used in the simulation are shown in the tree structure. The main simulation window shows the agents in the world (at present a bitmap out of a GIS is used). On the right hand the current values of the chosen agents attributes can be inspected. Some of the agents (facilities are agents too) are found outside the map. This is because the map only shows the area of the city Dortmund, whereas the simulation the more or less (transport) modelling relevant areas around the city are considered.

Simulation. Using SeSAm, it is possible to run and test simulations almost after every model programming step (Figure 6.5). Therefore, the design of the agent and the environment was an iterative process of implementing the rules and functions and testing variations (specifications) of the agents' behaviour and the implementation into SeSAm. Agents have their individual current program, fill the remaining time windows by adding new activities out of a list of possible activities with a start time and duration, try to carry out the scheduled tasks considering the environment and do some problem solving (overlapping of new scheduled tasks with existing tasks). This results in generated activity sequences.

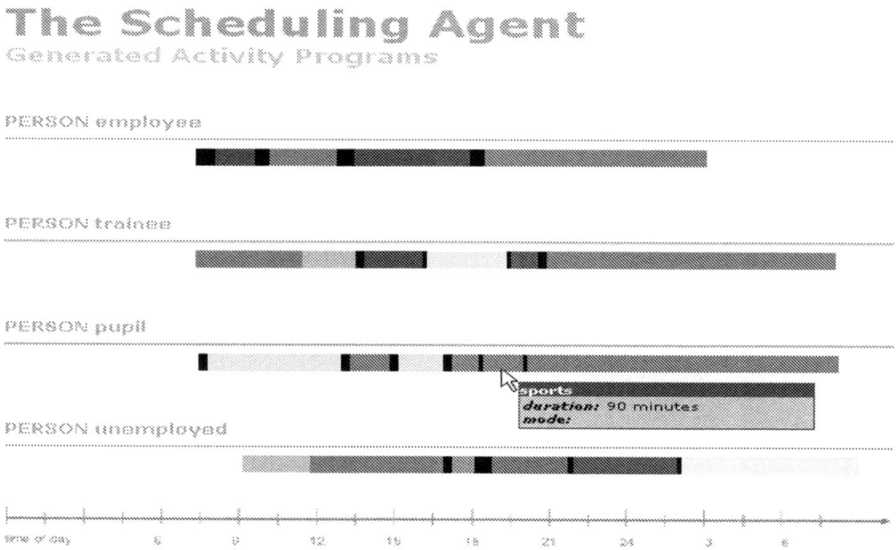

Figure 6.6
Examples of Generated Activity Patterns (Screenshot from Browser)

Figure 6.6 shows the result for a somewhat detailed activity categorization and a test for including simple communications between the agents to test external triggers. Communications are actually used only to generate triggers. In Figures 6.6 activity/trip sequences of individuals (two for each for two days) are displayed. Starting at the left side at midnight, time is displayed on the x-axis. The first light grey bar indicates an activity at home, continued from previous day. Black bars indicate trips. Variations of grey (in original: colours) indicate different activities. To display the night sleep (or other activities lasting over midnight) as a complete activity, the time axis is longer than 24 hours. When moving the computer mouse over a segment, additional information is displayed (for example type of activity, duration, mode). As can be seen, the simulation results are individual sequences of activities and trips with day-to-day-variations. The result depends strongly on variations in activity categories, input data, and behavioural rules.

CONCLUSIONS

A multi-agent simulation including the generation of individual activity programs, the simulation of these programs under ongoing planning and modification of future tasks based on the idea simulating the scheduling processes was developed and tested. Such a simulation on a microscopic level requires data about the scheduling processes and related behaviour in addition to the usual data about infrastructure and the temporal organisation of the physical environment. It is necessary to find the right balance between the possible level of modelling detail and the availability of or demand for empirical data. The results suggest that with some rules based on expert knowledge and empirical data - where possible - a satisfying result can be obtained.

The major benefits of the proposed work concern the researcher's ability to enrich the agents' behaviour iteratively, and in doing so learn and understand scheduling behaviour and derived activity-travel patterns. The relatively rules used in the present illustration can be replaced by more sophisticated models. The presented approach aims at simulating the individual persons (represented as an agent) decision during execution of his and all others "life" in the simulation. The agent cannot only react but act based on sensing the environment and for that incorporate much more information for his decision. The need to derive behavioural rules and appropriate data is unquestionable. With further research on analysing decisions in time and space, this approach can be promising in terms of improved behavioural underpinnings in integrated transportation modelling. Nevertheless, whether such improved behavioural realism will results in better forecasts and outweighs additional data acquisition remains a question of further research. Overall this combination of a highly sophisticated simulation shell and a complex and behaviourally rich modelling of individual scheduling behaviour could improve both our understanding of behavioural mechanisms underlying mobility and the predictive ability of transport demand models.

REFERENCES

Adler, J.L. and V.J. Blue (2002). A cooperative multi-agent transportation management and route guidance system. *Transpn. Res. C*, **10**, 433-454.

Arentze, T.A. and H.J.P. Timmermans (2000). *Albatross: A Learning - Based Transportation Oriented Simulation System*. European Institute of Retailing and Service Studies, Eindhoven.

Arentze, T.A. and Timmermans, H.J.P. (2002). Modeling learning and adaptation processes in activity-travel choice: a framework and numerical experiments. *Transpn.*, **30**, 37-62.

Arentze, T.A. and H.J.P. Timmermans (2004). A learning-based transportation oriented simulation system. *Transpn. Res. B*, **38**, 613-633.

Axhausen, K.W. and R. Herz (1989). Simulating activity chains: a German approach. *J. Transpn. Eng.*, **115**, 316-325.

Axhausen, K.W., A. Zimmermann, S. Schönfelder, G. Rindsfüser and T. Haupt (2002). Observing the rhythms of daily life: a six-week travel diary. *Transpn.*, **29**, 95-124.

Balmer, M., K. Nagel and B. Raney (2004). Large scale multi-agent simulations for transportation applications. *Proc. Beh. Responses to ITC*, Eindhoven (CD-Rom).

Charypa, D. and K. Nagel (2003). Generating complete all-day activity plans with genetic algorithms. *Proc.10th Int. Conf. Travel Beh. Res.*, Lucerne, Switzerland (CD-Rom).

Charypar, D., P. Graf and K. Nagel (2004). Q-learning for flexible learning of daily activity plans. *Proc. 4th Swiss Transp. Res. Conf.*, Ascona, Switzerland (CD-Rom).

Damm, D. and S.R. Lerman (1981). A theory of activity scheduling behaviour. *Env. & Plan. A*, **13**, 703-718.

Dia, H. (2002). An agent-based approach to modelling driver route choice behaviour under the influence of real-time information. *Transpn. Res. C*, **10**, 331-349.

Dijkstra, J. and H.J.P. Timmermans (2002). Towards a multi-agent model for visualizing simulated user behavior to support the assessment of design performance. *Autom. in Constr.*, **11**, 135-145.

Doherty, S.T. (2000). An activity scheduling process approach to understanding travel behaviour. *Proc. 79th Ann. Meet. Transpn. Res. B.*, Washington, D.C. (CD-Rom).

Doherty, S.T. (2001). Meeting the data needs of activity scheduling process modelling and analysis. *Proc. 801thth Ann. Meet. Transpn. Res. B.*, Washington, D.C. (CD-Rom).

Doherty, S.T. and K.W. Axhausen (1999). The development of a unified modeling framework for the household activity-travel scheduling process. In: *Traffic and Mobility: Simulation - Economics - Environment,* W. Brilon, F. Huber, M. Schreckenberg and H. Wallentowitz (eds.), pp. 35-56. Springer Verlag, Berlin.

Doherty, S.T., E. Nemeth, M. Roorda and E.J. Miller (2004). Design and assessment of the Toronto area computerized household activity scheduling survey. *Proc. 83rd Ann. Meet. Transpn. Res. B.,* Washington, D.C. (CD-Rom).

Ettema, D.F., A.W.J. Borgers and H.J.P. Timmermans (1993). A simulation model of activity scheduling behaviour. *Transpn. Res. Rec.*, **1413**, 1-11.

Ettema, D., A.W.J. Borgers and H.J.P. Timmermans (1994). Using interactive computer experiments for identifying activity scheduling heuristics. *Proc. 7th Int. Conf. Travel Beh. Res.*, Valle Nevado, Chile.

Ettema, D.F., A.W.J. Borgers and H.J.P. Timmermans (2000). A simulation model of activity scheduling heuristics: an empirical test. *Geo. & Env. Modelling*, **4**, 175-187.

Ettema, D.F. and H.J.P. Timmermans eds. (1997). *Activity-Based Approaches to Travel Analysis*. Pergamon, Oxford.

Fischer, K., H.-J. Bürckert and G. Vierke (2000). Holonic transport scheduling with TELETRUCK. *J. Appl. Artif. Intell.*, **14**, 697-725.

Franklin, S. and A.C. Graesser (1997). Is it an agent, or just a program?: a taxonomy for autonomous agents. In: *Intelligent Agents III: Agent Theories, Architectures and Languages* (J.P. Muller, M.J. Wooldridge and N.R. Jennings (eds.), pp. 21-35. Springer-Verlag, Berlin.

Gärling, T., S. Fujii and O. Boe (2001). Empirical tests of a model of determinants of script-based driving choice. *Transpn. Res. F*, **4**, 89-102.

Gärling, T, R. Gillholm and W. Montgomery (1999). The role of anticipated time pressure in activity scheduling. *Transpn.*, **26**, 173-192.

Gärling, T., T. Kalén, J. Romanus and M. Selart (1998). Computer simulation of household activity scheduling. *Env. & Plan. A*, **30**, 665-679.

Gärling, T., M.-P. Kwan and R.G. Golledge (1994). Computational-process modeling of household travel activity scheduling. *Transpn. Res. B*, **25**, 355-364.

Hernández, J.Z., S. Ossowski and A. Garcia-Serrano (2002). Multiagent architectures for intelligent management systems. *Transpn. Res. C*, **10**, 473-506.

Hidas, P. (2002). Modelling lane changing and merging in microscopic traffic simulation. *Transpn. Res. C*, **10**, 351-371.

Joh, C.-H., T.A. Arentze and H.J.P. Timmermans (2000). Sequence alignment method. In: *Albatross: A Learning-Based Transportation Oriented Simulation System* (T.A. Arentze and H.J.P. Timmermans (eds.), pp. 227-250. European Institute of Retailing and Services Studies, Eindhoven.

Joh, C.-H., T.A. Arentze and H.J.P. Timmermans (2001a). Pattern recognition in complex activity-travel patterns: a comparison of Euclidean distance, signal processing theoretical, and multidimensional sequence alignment methods. *Transpn. Res. Rec.*, **1752**, 16-22.

Joh, C.-H., T.A. Arentze and H.J.P. Timmermans (2001b). Multidimensional sequence alignment methods for activity-travel pattern analysis: a comparison of dynamic programming and genetic algorithms. *Geo. An.*, **33**, 247-270.

Joh, C.H., T.A. Arentze and H.J.P. Timmermans (2001c). A position-sensitive sequence alignment method illustrated for space-time activity-diary data. *Env. & Plan. A*, **33**, 313-338.

Joh, C.-H., T.A. Arentze and H.J.P. Timmermans (2002). Modeling individuals' activity-travel rescheduling heuristics: theory and numerical experiments. *Transpn. Res. Rec.*, **1807**, 16-25.

Joh, C.-H., T.A. Arentze and H.J.P. Timmermans (2003). Estimating non-linear utility functions of time use in the context of an activity schedule adaptation model. *Proc. 10th Int. Conf. Travel Beh. Res.,* Lucerne, Switserland CD-Rom).

Joh, C.-H., T.A. Arentze and H.J.P. Timmermans (2004). Activity-travel rescheduling decisions: empirical estimation of the *Aurora* model. *Transpn. Res. Rec.*, to appear.

Jones, P.M. (1990). *Developments in Dynamic and Activity-based Approaches to Travel Analysis.* Avebury, Aldershot, England.

Jones, P.M., M.C. Dix, M.I. Clarke and I.G. Heggie (1983). *Understanding Travel Behaviour.* Gower, Aldershot.

Kitamura, R. (1996). Applications of models of activity behaviour for activity based demand forecasting. *Proc. Activity-Based Trav. Forec. Conf.*, New Orleans.

Klügl, F. (2001). *Multiagentensimulation.* Addison-Wesley, München.

Klügl, F. and A.L. Bazzan (2004). Route decision behaviour in a commuting scenario: simple heuristics adaptation and effect of traffic forecast. *J. Artif. Societ. and Soc. Sc. Sim.*, **7**, 1.

Kurani, K.S. and M.E.H. Lee-Gosselin (1996). Synthesis of past activity analysis applications. *Proc. TMIP - Conference on Activity-Based Travel Forecasting*, US Department of Transportation, New Orleans, Louisiana.

Lee, M.S., J-H. Chung and M.G. McNally (2003). An empirical investigation of the underlying behavioral process of trip chaining. *Proc. 82nd Ann. Meet. Transpn. Res. B.,* Washington, D.C. (CD-Rom).

Litwin, M.S., S.T. Doherty and E.J. Miller (2004). Investigating competition patterns in household activity scheduling processes. *Proc. 83rd Ann. Meet. Transpn. Res. B.,* Washington, D.C. (CD-Rom).

Ljungberg, M. and A. Lucas (1992). The oasis air-traffic management system. *Proc. 2nd Pac. Rim Int. Conf. on Artif. Intell.,* Seoul, Korea.

Luck, M., P. McBurney and C. Preist (2003). *Agent Technology: Enabling Next Generation Computing (A Roadmap for Agent Based Computing).* AgentLink, New York.

Miller, E.J. and M.J. Roorda (2003). A prototype model of household activity/travel scheduling. *Proc. 82nd Ann. Meet. Transpn. Res. B.,* Washington, D.C. (CD-Rom).

Moeckel, R., K. Spiekermann and M. Wegener (2003). Creating a synthetic population. *Proc. 8th Int. Conf. Comp. Urb. Plan. & Urb. Mngmnt*, Sendai (Cd-Rom).

Nagel, K. and F. Marchal (2003). Computational methods for multi-agent simulations of travel behavior, *Proc. 10th Int. Conf. Trav. Beh. Res*., Lucerne, Switzerland (Cd-Rom).

Oechslein, C., F. Klügl, R. Herrler and F. Puppe (2001). UML for behavior-oriented multi-agent simulations. *Proc. 2nd Int. Workshop of Central and Eastern Europe on Multi-Agent Systems*, Cracow, Poland.

Oi, W.Y. and P.W. Shuldiner (1962). *An Analysis of Urban Travel Demands*. Northwestern University Press, Evanston, IL.

Raney, B. and K. Nagel (2005). An improved framework for large-scale multi-agent simulations of travel behavior. In: *Towards Better Performing European Transportation Systems* (P. Rietveld, B. Jourquin and K. Westin, eds.), to appear.

Recker, W.W., M.G. McNally and G. S. Root (1986). A model of complex travel behavior: Part I - theoretical development. *Transpn. Res. A*, **20**, 307-318.

Rindsfüser, G. (2001). Analysis of multi-week activity patterns using sequence alignment methods. *Proc. 80th Ann. Meet. Transpn. Res. B.*, Washington, D.C. (CD-Rom).

Rindsfüser, G. and S.T. Doherty (2000). Konzept, Module und Datenerfordernisse für SMART - Simulationsmodell des Aktivitäten-(Re)Planungsprozesses. In: *Stadt, Regio, Land* No. 69, pp. 109-129. Institut für Stadtbauwesen und Stadtverkehr, RWTH Aachen, Aachen.

Rindsfüser, G., H. Mühlhans, S.T. Doherty and K.-J. Beckmann (2003). Tracing the planning and executing attributes of activities and travel - design and application of a hand-held scheduling process survey. *Proc. 10th Int. Conf. Trav. Beh. Res.*, Lucerne, Switzerland (CD-Rom).

Statistisches Bundesamt (1997). Wo bleibt die Zeit? Zeitverwendung der Bevölkerung in Deutschland. *Scientific Use File der Zeitbudgeterhebung 91/92.* Version 1.1, Wiesbaden.

Strauch, D., R. Moeckel, M. Wegener, J. Gräfe, H. Mühlhans, G. Rindsfüser and K.-J. Beckmann (2004). Linking transport and land use planning: the microscopic dynamic simulation model ILUMASS. In: *GeoDynamics* (P. Atkinson, G. Foody, S. Darby and F. Wu (eds.), CRC Press, Boca Raton, to appear.

Timmermans, H.J.P. (2001). Models of activity scheduling behaviour. *Stadt, Region, Land*, No. 71, pp. 33-47. Institut für Stadtbauwesen und Stadtverkehr, RWTH Aachen, Aachen.

Wilson, W.C. (1998). Activity pattern analysis by means of sequence-alignment methods, *Env. & Plan. A*, **30**, 1017-1038

Wooldridge, M. and N.R. Jennings (1995). Intelligent agents: theory and practice. *Knowl. Eng. Rev.*, **10**, 115-152.

7

ACTIVITY-BASED ANALYSIS OF TRAVEL DEMAND USING COGNITIVE AGENTS

Rosaldo J. F. Rossetti, Universidade Atlântica, Barcarena, Portugal

Ronghui Liu, University of Leeds, Leeds, UK

INTRODUCTION

Transportation systems are subject of much concern as they play an important role in contemporary society and have long been attracting great interest among technical and scientific communities. Analysing and understanding traffic phenomena has historically been approached from two major points of view (McNally, 2000a). In a longer-term perspective, transportation analysis seeks to improve network and service infrastructures either to accommodate a forecasted future demand or to increase capacity as resources become obsolete. A shorter-term approach, on the other hand, is rather oriented at the evaluation of immediate application of management policies that are somehow expected to directly influence travel behaviour. Not surprisingly, the physical modification of infrastructures and the improvement of control systems are some attempts to tackle the problem of traffic congestion with moderate success. When similar measures were used to try and deal with the demand side, the results have been less positive. Nonetheless, recent advances in communication and computer technologies have encouraged the advent of Intelligent Transportation Systems (ITS). One prime goal of ITS-based solutions is to ensure productivity and efficiency of existing transportation systems through the application of distributed solutions that handle users' needs on an individual basis (Chatterjee and McDonald, 1999). Heterogeneity, uncertainty, and dynamics are key elements in this scenario. So, how can one assess the complex impact of these new technologies

Progress in Activity-Based Analysis edited by H. Timmermans

through modelling? At the very least, this implies the need for more powerful tools of analysis that are capable of coping with the representation of the complexity inherent in human behaviour and interactions. Existing microscopic approaches suffer from various shortcomings and the need for more expressive modelling approaches is frequently recognised (Watling, 1994; Ettema *et al.*, 2003). In this way, the activity-based approach emerges as an attempt at addressing the inability of more conventional trip-based models to reflect the underlying behaviour of travellers (McNally, 2000b). According to Ben-Akiva *et al.* (1998) the fundamental problem facing demand modelling is the trade-off between behavioural realism and complexity. Representing human behaviour has received special attention from technical and scientific communities. Much effort has been devoted to adapt traditional models to meet ITS requirements and significantly contributed to building up the roadmap towards the development of new generation traffic network models, which explicitly incorporate behavioural realism of demand (Arnott *et al.*, 1991; Mahmassani and Jayakrishnan, 1991; Cantarella and Cascetta, 1995; Liu *et al.*, 1995; Bazzan *et al.*, 2000). Recently, agent-based techniques have been increasingly applied in that way. Schleiffer (2000) claims that modelling heterogeneity at a microscopic level is a key step towards the understanding of macroscopic behaviour. The author suggests that the use of artificial agents to represent simple fundamental individual mechanisms is the tool to better comprehend highly complex and dynamic collective behaviour of traffic.

In this research project, we seek to incorporate behavioural realism in travel demand forecasting by means of featuring travellers with cognitive abilities. The use of predicate logics through a BDI (**b**eliefs, **d**esires, and **i**ntentions) architecture allows for an expressive way to specify and implement activity-based travel behaviours. Some examples are identified in the literature, which apply the concept of autonomous agent and multi-agent system to address different issues in transport analysis. Travel demand is presented from a multi-agent system point of view, where travellers are agents that autonomously plan their journeys according to their activities' needs and parameters. Different behaviours for departure time selection are proposed accounting for activity arrival time constraints, and a microscopic simulation framework is set up. Some experimental scenarios are simulated in order to support the methodological approach presented in this chapter. Finally, some conclusions are drawn and next steps for further research on this topic are presented.

INTELLIGENT AGENTS AND THEIR APPLICATION IN TRANSPORT SYSTEMS

Multi-agent systems (MAS) are under the umbrella of the Distributed Artificial Intelligence (DAI) and have triggered increasing interest among scientists from different knowledge fields. The rapid evolution in computational resources, both in hardware and in software, has contributed a great deal to its development. Basically, there are two major ways through which agent-based solutions have been proposed and effectively applied. First, real agents are playing an important role in contemporary society. Not only robotics has profited from such a technology but also the Internet environment experiences the presence of software agents that are frequently interacting with human users. Second, agent-based models become a natural metaphor to represent domains where a number of intelligent and autonomous entities interact with each other and with the environment. These models are being increasingly based on elaborate frameworks of analysis as an effective tool to aid the understanding of complex and stochastic phenomena. Traffic and transportation systems have profited from these approaches and have also stimulated much research on and development of agent-based technologies.

The main premise of multi-agent systems is to interpret the real world in terms of agents that exhibit intelligence, autonomy, and some degree of interaction with other agents and with their environment. Other characteristics of agents include, for example, reactivity, adaptability, pro-activity, and the ability to communicate and to behave socially. The basic structure of an agent features sensors through which it can gather information from the environment, and effectors through which it can act and behave according to its objectives (Russell and Norvig, 1995). This structure can feature both reactive and cognitive abilities, and a mixture of both, to mimic human behaviour in a wide range of applications. Steels (1990) suggests that each single agent possibly having a very simple structure can contribute to a more complex and efficient behaviour of the system as a whole. If the behaviour of such a single agent can be backtracked, then this can be used to aid the understanding of the more complex behaviours at the aggregate level, such as social phenomena for instance.

To the best of our knowledge, former attempts to apply agent-based techniques to address transportation issues date back to the 90's. For instance, Haugeneder and Steiner (1994) proposed a co-operative agent-based architecture as a means of improving traffic management and control, where agents were implemented in the MAI^{2}L language (Steiner *et al.*, 1993). Not surprisingly, that was when much research and controversies were going on to define the actual scope of agents (e.g., Wooldridge and Jennings, 1995). For instance, many people from different fields in Computer Science (CS) and even in Artificial Intelligence (AI) were trying to decide whether agents were different from objects, as in the object-oriented perspective, or from autonomous processes, considering operating systems and network points of views. Whereas in the beginning people from AI community benefited from the complex and dynamic nature intrinsic in transportation systems to devise and support agent theory, transportation engineers and practitioners have now started to recognise the natural ability of the multi-agent metaphor to model traffic phenomena. Owing to their characteristics and concepts, multi-agent systems have a natural aptitude to cope with a wide range of issues in contemporary traffic and transportation scenarios (Schleiffer, 2002). Not amazingly, most publications report on the application of agent-based techniques to control systems and traffic management to make those systems more autonomous and responsive to recurrent traffic demand (e.g., Hernández *et al.*, 2002). The analysis of ITS systems through this approximation has also been investigated (e.g., Rickert and Nagel, 1997; Wahle *et al.*, 2002), and some other publications report on applications to freight transport and the optimisation of resource use (e.g., Adler and Blue, 2002). A research project has been recently presented, which provides a survey on the application of agent-based approaches to transport logistics (Davidsson *et al.*, 2004). Nevertheless, the challenging issue of modelling the decision-making process underlying travellers' behaviour in a more realistic way has encouraged the increasing use of agents for such a purpose. For example, drivers are endowed with cognitive abilities to plan a trip accounting for a mental model of the world and an expectation of the utility their choices would bring about (e.g., Dia, 2002; Nagel and Marchal, 2002; Rossetti *et al.*, 2002a).

In this same direction, agent concepts have also proved to be very useful in fostering the improvement of an activity-based analysis of travel demand. *Albatross* (Arentze and Timmermans, 2000) is one of the most comprehensive and operational models in this respect and uses a decision tree induction approach to derive choice heuristics. Rindt *et al.* (2003) conceptualised a reactive architecture on the basis of a series of interrelated sub-modules that implement different aspects of agent behaviour, such as assessment, interpretation, decision processes, and learning abilities. With such a structure the authors seek to explore the dynamic nature of the activities' interactions in the formulation of demand. Raney and Nagel (2004) used activities parameters to generate a plan that drives the decision-making of each agent of the population. Each agent may have several plans, which are associated with a score that keeps a record of performance evaluation that is used by the agent when choosing a plan to follow. Marchal and Nagel (2004) devised a model that is rather focused on addressing performance issues in the selection of secondary activities' location in large data sets with gains in computation workload. Rindsfüser *et al.* (2004) formulated a model in which daily activities schedules result from the combination of the interaction of different behavioural processes defined within an agent architecture implemented in SeSAm (Klügl, 2001).

In this research project, we focus on the cognitive process that underlies the planning of activity journeys rather than building daily activity schedules. All activities characteristics are assumed to be exogenous with regard to the planning process and are assimilated as beliefs through perception from an outer level of interaction, as suggested in Rindt *et al.* (2003). Therefore, beliefs on activity parameters will trigger possible courses of actions that are expected to produce the best journey options that satisfy time constraints for an activity.

THE AGENT-BASED DEMAND APPROACH

The Driver Agent

The transportation domain is approached from a multi-agent system point of view, according to the framework presented in Rossetti *et al.* (2000), and Rossetti and Liu (2004). If the natural phenomenon of traffic in urban areas is already complex enough in its own right, the different levels of interactions implied by contemporary intelligent transportation solutions bring about additional modelling challenges. Thus, in order to ease the task of identifying which components of the system we might model as agents, we use a simple rule: entities make decisions in an autonomous fashion are considered to be autonomous decision entities (ADE) and therefore are potential agents.

To avoid going too much into detailed specifications, the task of identifying agents within a system is basically reduced to the identification of ADEs. We have devised an agent shell to structure the way agents can be implemented and inserted into the environment. Such a structure is very flexible in the sense that it is only defined at the meta-level, comprising sensors through which the agent can perceive the world and effectors through which it can effectively act on the environment. It is also featured with a reasoning kernel that drives the decision-making processes. It is important to notice that this meta-level agent shell only specifies the basic structure for the ADEs, allowing the definition of different kinds of agents with different reasoning capabilities, skills, and goals. Communication among agents is simply considered to be acting/sensing behaviour. Then, all messages are issued as actions, trough effectors, and received as perceptions, through sensors as further explained in Rossetti *et al.* (2002a). The environment is basically formed by a network topology and parameterises all the information shared by the inhabitant agents. Traffic signs and basic rules are considered part of the environment structure and dynamics.

With this conceptualisation, it is possible to virtually represent all aspects involved in contemporary traffic scenarios: drivers are agents in the sense they make their decisions en-route and at the time of departure; travellers are agents as they have to choose among transport modes and activities; each level of decision in an advanced traffic management system could be an agent that interacts directly or indirectly with the others in order to optimise overall traffic performance; in the same way, traveller information systems could be agents interacting with drivers or travellers so as to optimise individual performance levels; and so forth. Albeit all these ADEs are encapsulated into agent shells, they may be internally different, implementing distinct reasoning approaches and having different knowledge representations. Basically, the shell defines the common interface shared by all these agents within the transportation system.

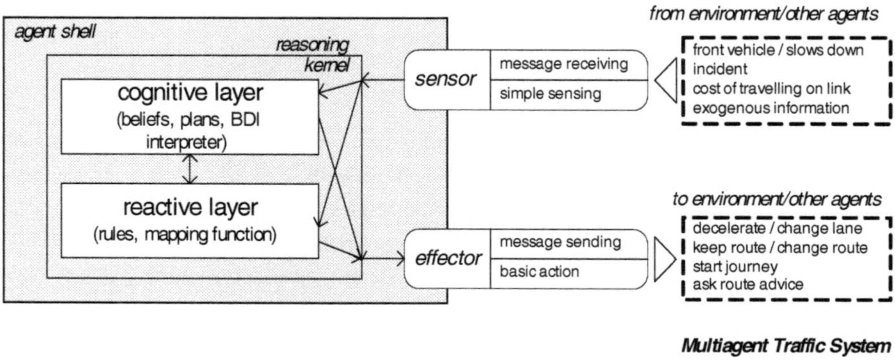

Figure 7.1
A Two-Layered Architecture for the Driver Agent

To demonstrate our approach, we have started by modelling the driver agent whose structure is depicted in Figure 7.1. This is a simplification adopted in the first stage of our research in activity-based travel analysis, as we believe the traveller is a more adequate trip-maker unit to be considered at this level. We have designed a two-layered reasoning kernel to base the driver model so that it is able to exhibit both reactive and cognitive behaviours to some extent. The reactive layer relies on a simple set of rules that map perceptions to actions and is basically used to implement rather instinctive behaviours. Individual's driving abilities, in terms of car-following and lane-changing behaviours, are performed in this layer. The more complex decisions, such as selecting the next activity and which itinerary to follow, are addressed in the cognitive layer.

The Cognitive Layer

The cognitive approach for the BDI architecture, devised by Rao and Georgeff (1991), basically relies on the mental states of **b**eliefs, **d**esires, and **i**ntentions, and on their relationship. This approach was inspired by Bratman's (1987) philosophical work, which deals with intentions as an important element for rational reasoning. The BDI architecture makes a clear distinction between the choices an agent has over the actions it can perform and the different outcomes an action can possibly bring about. Thus, an agent might consider that a certain action would produce an expected effect, however environmental conditions actually dictate the results of executing that action. This allows for the overall system dynamics and non-determinism, which are important characteristics to feature models of complex domains such as transport systems.

The conceptual architecture of a BDI agent is depicted in Figure 7.2, and is briefly explained in (Wooldridge, 1999) as follows. With every perception from the environment, the agent's set of base beliefs is updated. The new configuration of beliefs is established by a belief revision function (BRF), which is responsible for preserving the consistence of the agent's beliefs. An *options* function determines the options available to the agent, which are its desires. This function receives as inputs the current configuration of the beliefs set (**B**), as well as the agent's current intentions (**I**). As further discussed in Georgeff and Lansky (1987), and Georgeff and Rao (1996), an agent is equipped with a library of plans that are used to perform means-ends reasoning. Deliberation is achieved on the basis of instantiating the meta-descriptions of plans, which generates the agent's options and are able to modify its intention structure dynamically during run time. The desires represent possible course of actions available to the agent, and a simplification is generally made in the sense that conflicting desires are discarded and only non-conflicting ones (the goals) are considered. A filter function representing the deliberation process determines new intentions on the basis of the agent's current beliefs, non-conflicting desires (goals), and the intentions currently being performed. The intentions represent those states of affairs that an agent has committed trying to bring about. An action selection function then executes the next action the agent must perform on the basis of its current intention.

The cognitive layer of our agent is represented by the tuple $\langle \mathbf{E}, \mathbf{B}, \mathbf{P}, \mathbf{I}, \mathbf{A}, S_E, S_O, S_I \rangle$, as initially proposed in Rossetti *et al.* (2002a), according to the AgentSpeak(L) language specification (Rao, 1996), where **E** is the set of events, **P** is the set of non-instantiated plans, and **A** is the set of basic actions. S_E, S_O, and S_I denote the selection functions for events, options, and intentions, respectively. As intentions are generated dynamically during the lifecycle of an agent in the system, and events result either from perception or during the iteration of current intentions, the modelling task is reduced to identifying base beliefs, non-instantiated plans, and the actions an agent can perform, as mentioned before. Rather than using the basic approach of aggregating all the factors that can in any way influence drivers decisions on travel solely within the concept of trip, we have opted to use the activity metaphor intrinsic in human behaviour. Driver agents make their decisions on activities, rather than on actions or individual parameters of trips. Thus, the basic structure of a trip is defined as the tuple $\langle i, j, \mathbf{R}, d \rangle$, where i is the origin, j is the destination, **R** is the set of links from i to j giving the itinerary chosen for the trip, and d is the departure time at which the journey is to be started. An activity is given by tuple $\langle p, s, t \rangle$, where p is the activity purpose, s is the site where the activity is to be performed, and t is the time the activity is expected to start. In order to evaluate the next journey on a certain day, the agent needs to seek among every activity belief in its base beliefs set for the next activity on the base of the starting time term, t, of the activity belief entry. Once another activity is selected for execution the agent proceeds with the journey planning, which encompasses departure time and route selection.

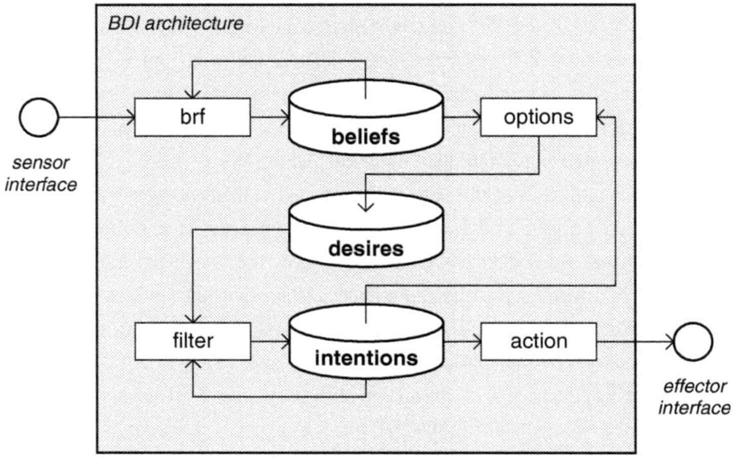

Figure 7.2
The Basic BDI Architecture (Wooldridge, 1999)

ANALYSING SCHEDULED ARRIVAL OF ACTIVITIES: A FIRST EXERCISE

The driver plans each trip on a next-activity basis. Recalling that an activity has been basically characterised by its purpose, site at which it is performed, and time at which it is expected to begin, the driver then proceeds with the selection of the time to start the journey and its itinerary based on the desired arrival at the activity site. The departure time and route selection behaviours are proposed on the basis of the habitual choices as initially implemented in Liu *et al.* (1995).

Habitual Behaviour

For the original habitual behaviour, departure time is chosen in response to the traveller's previous experiences and preferred arrival time, so that the activity can be started as expected. The absolute delay for a driver m travelling from certain origin i to a destination j on day k is given in Equation 1, where $d_{ijm}^{(k)}$ is the departure time, $t_{ijm}^{(k)}$ is the travel time, and $a_{ijm}^{(k)}$ is the desired arrival time. Drivers are also assumed to be indifferent to a delay of $\varepsilon_m \times t_{ijm}^{(k)}$ (relative to the travel time experienced). Equation 2 represents the lateness actually perceived by individuals.

$$\delta_{ijm}^{(k)} = d_{ijm}^{(k)} + t_{ijm}^{(k)} - a_{ijm}^{(k)} \tag{1}$$

$$\Delta_{ijm}^{(k)} = \delta_{ijm}^{(k)} - \varepsilon_m \times t_{ijm}^{(k)} \tag{2}$$

As suggested in Mahmassani *et al.* (1997), drivers are likely to be indifferent to early arrivals. Accounting for that fact, we consider that users only adjust their departure time for a future journey in the case of $\Delta_{ijm}^{(k)} > 0$, otherwise they will keep the same departure time. That is:

$$d_{ijm}^{(k+1)} = \begin{cases} d_{ijm}^{(k)} & , if \ \Delta_{ijm}^{(k)} \leq 0 \\ d_{ijm}^{(k)} - \Delta_{ijm}^{(k)} & , if \ \Delta_{ijm}^{(k)} > 0 \end{cases} \tag{3}$$

The route choice model is based on bounded rational behaviour, as suggested in Mahmassani and Jayakrishnan (1991). Drivers are assumed to use their habit routes, unless the cost expected for the minimum cost route is significantly better. Thus, a driver will use the same route unless $C_{p_1} - C_{p_2} > \max(\eta \times C_{p_1}, \tau)$, where C_{p_1} and C_{p_2} are the costs along the habit and the minimum cost routes, respectively. The parameters η and τ, representing the relative and the absolute cost improvement required for a route switch, are associated with the activity belief.

This model seems to be quite flexible as lateness tolerance is evaluated with respect to the travel time experienced (as will be seen later in the simulation results). This means that the longer the trip lasts, the more tolerant the driver will be with regard to being late. Moreover, the model completely disregards early arrivals. This may constitute a problem when the activity is being performed within a one-day activity chain. In this case, previous activities, work for instance, cannot just be interrupted for the traveller to adjust the departure time much earlier for the next activity, shopping for example. Therefore, two extensions to the initial structure of the habitual behaviour are suggested in order to support the definition of earliness-lateness tolerance windows that will depend on the nature of the activity. They both differ from one another basically in terms of how lateness and earliness thresholds are identified. In the first extension, limits are drawn from the total travel time experienced, whereas in the second one boundaries are given in absolute terms.

The Relative Tolerance Window

The first extension considers an earliness threshold in addition to the lateness constraint, both relative to travel time. Leisure and shopping activities, for instance, might have associated such thresholds. Indeed, a family travelling to the beach on a summer holiday may have an expected arrival time but may also be very keen to tolerate early or late arrivals depending on the access conditions to the place they are going to. So, let λ_{ijm} be the earliness tolerance factor as ε_{ijm} still represents the lateness tolerance factor, both related to a driver *m*. As in the original habitual behaviour, perceived lateness and earliness will be drawn from trip cost as $\varepsilon_m \times t_{ijm}^{(k)}$ and $\lambda_m \times t_{ijm}^{(k)}$, respectively. The term $t_{ijm}^{(k)}$ refers to the total travel time from *i* to *j* on day *k*. The sign of the absolute delay $\delta_{ijm}^{(k)}$ (as defined in Equation 1) is also important as it allows one to identify whether the driver has arrived earlier or later. Bearing in mind the definition for perceived lateness $\Delta_{ijm}^{(k)}$ (see Equation 2), let $\Theta_{ijm}^{(k)}$ be the perceived earliness, as given in Expression 4.

$$\Theta_{ijm}^{(k)} = \left|\delta_{ijm}^{(k)}\right| - \lambda_{ijm} \times t_{ijm}^{(k)} \tag{4}$$

One should notice that the absolute value of $\delta_{ijm}^{(k)}$ is used instead as its sign is negative meaning the agent was earlier. Thus, the departure time on the next day $k+1$ is adjusted according to the following criterion.

$$d_{ijm}^{(k+1)} = \begin{cases} d_{ijm}^{(k)} - \Delta_{ijm}^{(k)}, \text{if } \delta_{ijm}^{(k)} > 0 \text{ and } \Delta_{ijm}^{(k)} > 0 \\ d_{ijm}^{(k)} + \Theta_{ijm}^{(k)}, \text{if } \delta_{ijm}^{(k)} < 0 \text{ and } \Theta_{ijm}^{(k)} > 0 \\ d_{ijm}^{(k)} \qquad\qquad, \text{otherwise} \end{cases} \tag{5}$$

The Absolute Tolerance Window

A similar approach is used for the absolute lateness-earliness window. For other activities, such as work, travellers are unlikely to exhibit such a flexible behaviour with respect to either early or late arrival. In this case we consider an absolute top lateness and an absolute bottom earliness thresholds, within which no adjustment to departure is required. In turn, any arrival experience perceived outside these bounds should be considered in future journeys. So, let ι_{ijm} be the absolute lateness whereas ν_{ijm} represents the absolute earliness tolerances. Then, the perceived lateness $\Delta_{ijm}^{(k)}$ and the perceived earliness $\Theta_{ijm}^{(k)}$ should be redefined as follows.

$$\Delta_{ijm}^{(k)} = \delta_{ijm}^{(k)} - \iota_{ijm} \quad (6)$$

$$\Theta_{ijm}^{(k)} = \left|\delta_{ijm}^{(k)}\right| - \nu_{ijm} \quad (7)$$

The adjustment for departure time on day $k+1$ happens on the same conditions as in the case of the relative lateness-earliness tolerance window, in Equation 5. It is important to notice that both relative and absolute lateness and earliness factors are very likely to depend on the trip purpose rather than being global parameters. In this sense, all these factors are associated with the activity belief in the agent's base beliefs set. Other types of behaviour were also suggested and specified in AgentSpeak(L), where all the plans and base beliefs were presented (Rossetti *et al.*, 2002a).

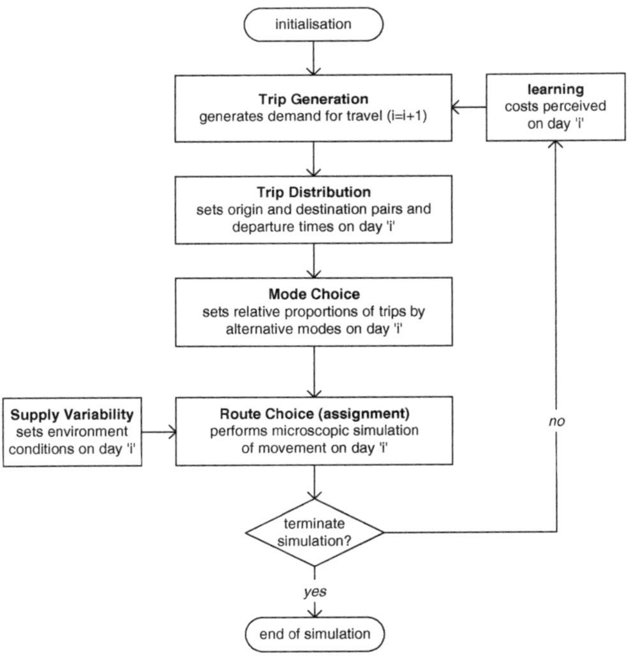

Figure 7.3
A Basic Transport Analysis Framework

THE SIMULATION FRAMEWORK AND SOME EXPERIMENT RESULTS

Madam+Dracula

Some transport analysis frameworks (e.g., Chatterjee and McDonald, 1999) implement a structure that resembles the original four-step model, as depicted in Figure 7.3. In a *trip generation* module, measures of trip frequency are developed, which provide the propensity of a certain trip unity to make a trip. Then, origins and destinations are associated in the *trip distribution* module. These tables are generally related to a certain period of the day. Trips are distributed among transport modes (*mode choice*), whereas routes are assigned to mode-specific networks (*route choice*). Generally, the assignment part is connected to a *supply variability* module capable of emulating the supply dynamic nature (giving link capacity as a function of weather, for instance).

In this research project, we extend the demand formulation of the DRACULA original structure in order to allow for autonomous and cognitive behaviour of trip makers, as suggested in Rossetti *et al.* (2000). DRACULA is a microscopic network simulator that has been developed in the Institute for Transport Studies, at the University of Leeds (Liu *et al.*, 1995). It comprises basically a demand and a supply model whose integration gives rise to the main premises in DRACULA, namely the within-day decision-making process and the day-to-day dynamics, as depicted in Figure 7.4. These are two important concepts that deserve special attention in modelling traffic systems with regard to users' behaviour. The within-day formulation focuses on the travel choices made by individuals. These choices are made with regard to each specific journey to take place at a given time on a given day. All trip preferences, such as travel goals and purpose, travel needs and mode, and other traveller parameters, such as perceptions, behavioural tendencies, and cognitive abilities that influence the decision-making process are reflective of the traveller's mental state at the instant the choice is being undertaken. The dynamic formulation, on the other hand, is concerned with modelling how the state of the network changes from one day to the other and evolves over time.

In addition, the spatial knowledge of travellers is constantly evolving in response to trips made throughout the network, in terms of the travel cost experienced for each journey (which may be associated to travel time, for instance, or any other performance measure). Such a structure has been used as an attempt at improving the representation and simulation of the complexity and the uncertainty inherent in the transportation domain. Rather than executing separate modules to set variable states from a probability distribution, demand emerges from the aggregate combination of individual choices carried out by autonomous decision entities. In such a demand model, coined MADAM, the cognitive kernel of each agent is implemented in JAM (Huber, 1999) and encapsulated into a Java-based ADE (Rossetti *et al.*, 2002b; Rossetti and Liu, 2004).

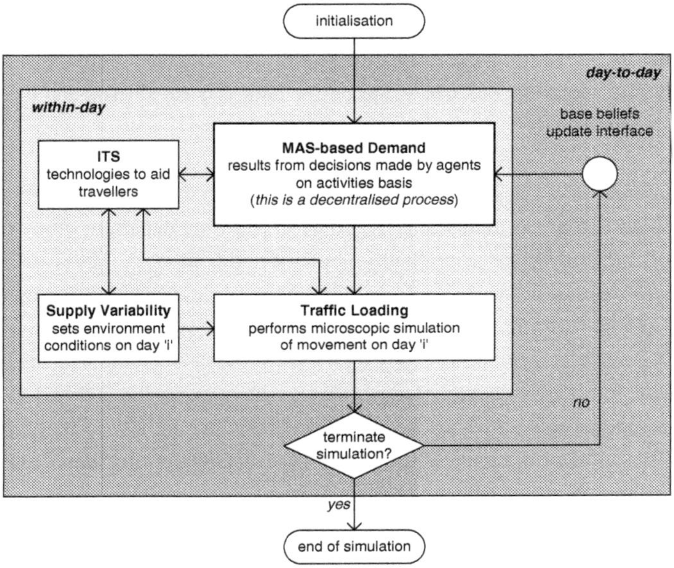

Figure 7.4
MADAM+DRACULA Simulation Framework

Thus, contrary to the approaches based on fixed matrix structures, the demand stage predicts the level of individual demand on a certain day from a full population of potential traveller agents. In the supply side, trips are individually simulated from their origins throughout the network to their destinations, and travellers are able to gather information about trip conditions to improve future choices as their base beliefs sets are constantly being updated as new perceptions are gathered. Such a structure will very likely reflect back in the demand side as travellers may present different abilities of learning with their previous experiences, thus influencing their decision-making process in future journeys.

Simulation Scenario Set-Up

Some experiments were carried out in order to demonstrate the methodological approach presented in this chapter. A small network with 54 links (uni-directional road segments) connected through 14 junctions was selected for this purpose. Most road junctions follow a priority regime, whereas two of which are governed by traffic signals. In this simple example, demand is generated from a population of 2,323 driver agents who wish to travel in a one-hour morning peak period, and their day-to-day choices on next-activity basis are simulated. The agents can perform their trips to/from 11 zones, *i.e.* there are 11 zones generating traffic onto and 11 zones draining traffic from the network. A hypothetical morning peak period starting at hour 8 is considered and the simulation is carried out from day 0 to day 100.

In order to test the ability of our approach to cope with the activity-based analysis of travel demand, we started by considering the constraints of arrival time in choosing departure time of activities. As discussed above, the habitual driver as originally implemented (Liu *et al.*, 1995) seems to be quite flexible with regard to earliness, which may not be realistic for every kind of activity. Thus, two extensions to the habitual behaviour were proposed in order to allow for both relative and absolute earliness-lateness tolerance at arrival. For studying the different behaviours presented in this work, different values were considered for the tolerance factors, as presented in Table 7.1. As discussed earlier, these values are considered to depend on the nature of the activity to be performed by the traveller. Some belief entries in a typical traveller's base beliefs set are illustrated in Table 7.2 alongside their terms and an instance example.

The predicate *today* denotes the current *day* on which the agent should make *activity* selections to carry out. We are currently considering a week horizon for the agent memory so that *day* can be any value in {*sun, mon, tue, wed, thu, fri, sat*}. Daily activities, such as *work* are inserted in the base beliefs set at the beginning of each day, whereas other activities are assigned to the agent as the result of activity interactions, at the household level, as mentioned before. Thus, whenever the *day* term in the *today* predicate binds with the day predicate in the *activity* predicate, it is possible for the agent to identify what activities are to be performed. The belief *timeNow*(*time*) is updated periodically, for instance every minute, according to the environment clock.

Table 7.1
Tolerances at Arrival Time

Original Behaviour (Relative to Travel Time)	Relative Tolerance Windows (Relative to Travel Time)		Absolute Tolerance Windows (in Minutes)	
ε	ε	λ	ι	ν
0.20	0.20	0.20	5	5
	0.20	0.30	5	10
	0.20	0.50	5	20
	0.20	1.00	5	30

Table 7.2
Base Belief Entries

Belief Predicate	Terms	Example
today	current day	today(wed)
timeNow	current time	timeNow(0815)
activity	day, purpose, time, location	activity(wed, work, 0900, 109)
tripRoute	origin, destination, link list	tripRoute(109, 105, [1 5 ...35])
tripDeparture	origin, destination, time	TripDeparture(109, 105, 0837)
location	current location	location(105)
route	origin, destination, mode, duration, link list	route(101, 106, car, 32, [14 15...3])
routeSwitch	origin, destination, η, τ	RouteSwitch(103, 109, 0.20, 5)
rwScheduledDelay	purpose, ε, λ	rwScheduledDelay(shop, 0.10, 0.20)
awScheduledDelay	purpose, ι, ν	awScheduledDelay(work, 5, 10)

The predicates *tripRoute*(*origin*, *destination*, *links*) and *tripDeparture*(*origin*, *destination*, *time*) result from the planning of an activity journey through the *links* within the itinerary to be followed. When the current time in *timeNow*(*time*) is found to be the departure time for the next activity in *tripDeparture*(*origin*, *destination*, *time*), then the agent starts the journey. A number of beliefs *route*(*origin*, *destination*, *mode*, *duration*, *links*) gives the alternatives for an agent to move from its current *location*(*agent_location*), resulting from the binding between the terms *agent_location* and *origin*, to the selected activity location, as the term *activity_location* in *activity*(*day*, *purpose*, *time*, *activity_location*) binds with *destination*. As many *route* alternatives may be applicable, the term *duration* in the belief predicate *route* is used to denote an expected travel time, which is updated whenever the option is selected. The route selection will consequently be constrained by the values of the terms η and τ in the belief predicates *routeSwitch*(*origin*, *destination*, η, τ), which are related to trips made from a certain *origin* to a certain *destination*. The activity journey departure *time* in *tripDeparture*, on the other hand, is constrained by the *purpose* of the activity and respective values for the terms ε and λ, in *rwScheduledDelay*, and the values of ι and ν, in *awScheduledDelay*. Contrary to considering global values for these parameters as originally assumed in Liu *et al.* (1995), we now associate these values to different kinds of activities.

Results and Comments

As expected for the original habitual behaviour, drivers can experience a smoother arrival time after few days from the start of the simulation as observed in graph of Figure 7.5. The departure choice remains the same unless new delay beyond what is tolerable by the driver is perceived.

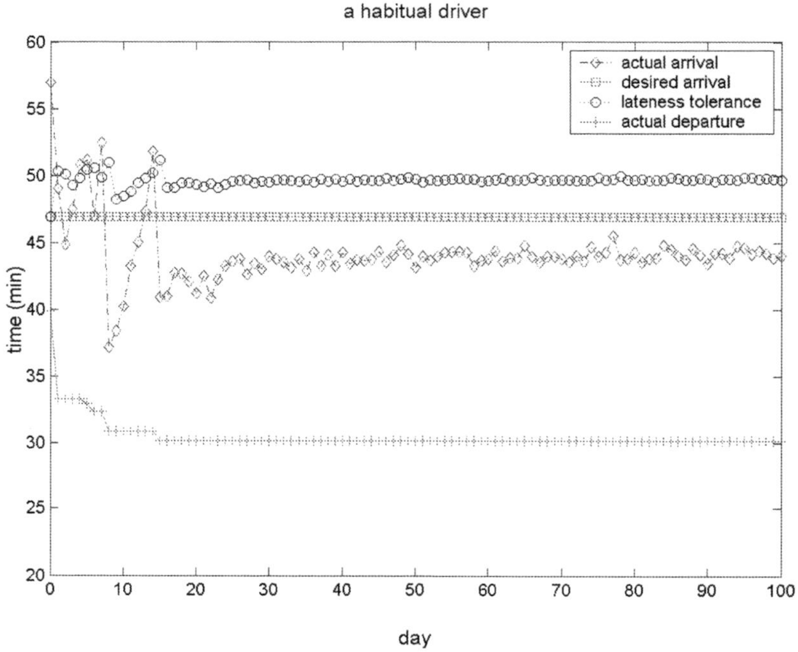

Figure 7.5
Activity Journeys with Lateness Thresholds

This means that stabilising arrival much earlier than deadline results in keeping the same departure choice. Although this behaviour can be representative of some kind of activities, such as for leisure, featuring agents with such an unlimited earliness tolerance may not be exactly the case for all activities. Indeed, real commuters may not be so tolerant to early arrivals at work, for instance, especially during morning journeys. Some kind of activities may also pose some sort of penalty for early arrival.

Then a first extension to the habitual behaviour is suggested, which considers an earliness threshold with respect to the travel time in addition to the lateness tolerance. The behaviour of single instances of travellers presenting a relative lateness-earliness tolerance to desired arrival time is depicted in Figure 7.6. It is interesting to note that the fluctuation of the agent's arrival time can be considerably high, and this has also been observed even for some agents in the population when testing activities with the [20%, 100%] relative window.

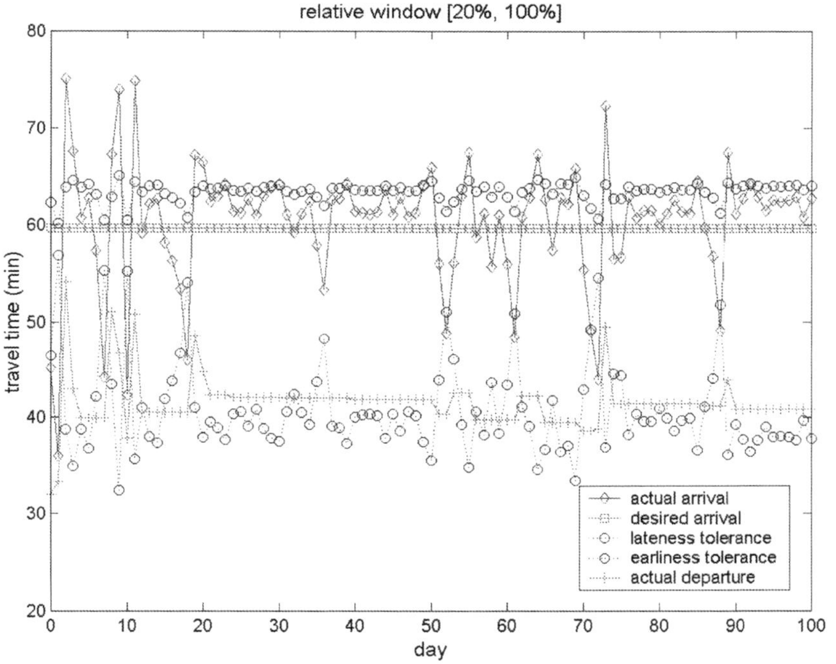

Figure 7.6
Activity Journeys with both Relative Earliness and Lateness Thresholds

This specific behaviour may result in a considerably shorter journey yielding a very restrictive tolerance. As thresholds may vary on a daily basis, even if the driver sometimes experiences longer journeys, it seems to be very difficult to reach a steady state and wide fluctuations have been observed in all window sizes simulated. A similar experiment was carried out for the suggested extension, taking into consideration that certain activities pose an absolute lateness-earliness tolerance window. In this case, earliness and lateness thresholds of arrival time window are kept constant, according to the tolerance parameters ι and ν as listed in Table 7.1. Simulation results, as exemplified in graph of Figure 7.7, have shown that it is easier for the traveller to meet its lateness-earliness thresholds after a number of iterations and to keep this state for a longer period of time. The wider the distance between the upper and lower boundaries, the more tolerable the agent will be in absolute terms. What remains to be analysed is the departure time behaviour of travellers when activities present mixed thresholds, that means either relative earliness and absolute lateness or absolute earliness and relative lateness tolerances.

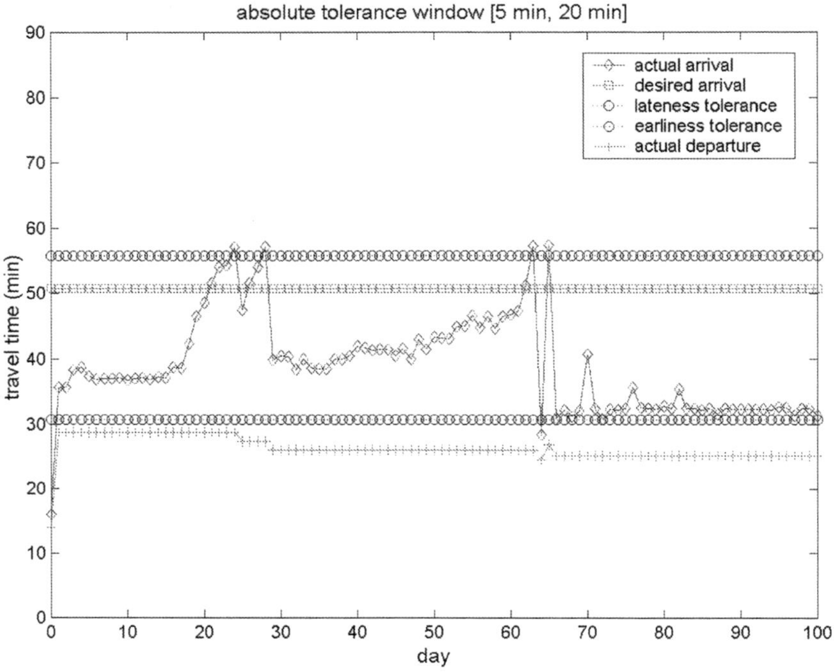

Figure 7.7
Activity Journeys with both Absolute Earliness and Lateness Thresholds

CONCLUSIONS

In this research project, we used a cognitive approach to achieve behavioural realism. The expressiveness provided by the BDI approach adopted allows for a straight match between modelling language constructors and the mental attitudes that play a role in deliberation processes of travellers. This fosters the activity-based analysis of travel demand as activity parameters can be accounted for on an individual and scalable basis. For example, the first experiments focused on the effect of activity arrival time constraints on departure time selection where all tolerance factors were associated with the kind of activity rather than being treated as global parameters. After extending the original habitual behaviour to consider both earliness and lateness tolerance at arrival time, it was observed that drivers are very likely to meet deadline window for activities with absolute thresholds, even for the lowest values of ι and ν. For those activities with thresholds depending on travel time, drivers experienced high fluctuation in arrival time, meeting thresholds occasionally, even for the highest values for ε and λ, as unpredictability of travel time dictates the tolerance to be considered.

Next steps include the development of a methodological approach for validation and calibration of activity-based demand models based on cognitive agents. Dia and Purchase (1999) and Dia (2002) have proposed a survey of driver behaviours to provide useful insight into the characteristics of commuters, their preferences and thresholds. Results from such a survey could be easily specified in terms of AgentSpeak(L) constructors to enrich agents' base beliefs and plan options. The use of virtual environments, such as the one implemented in Vladimir (Bonsall *et al.*, 1997), and the use of stated preference experiments (McNally, 2000a) could also serve this purpose.

Unfortunately, a high computation has been observed while simulating the simple scenarios herein presented, and this is an important issue to overcome for larger populations. In our simulation set-up, a population of 2,323 BDI agents took approximately 10 to 12 hours' CPU time, running sequentially in a PC featured with an AMD Athlon™ (a trademark of Advanced Micro Devices, Inc) processor at 1.1 GHz. In fact, cognitive approaches, such as the one adopted in this work, have been found to be very suitable from a psychological point of view, as all mental attitudes are accounted for when modelling the complex nature of human reasoning. However, only systems with a reduced number of decision-making entities have been represented in terms of cognitive models. On the other hand, reactive solutions relying on the overall behaviour emerging from the interaction of simpler agent structures have proven to be very effective when applied to larger data sets (Balmer *et al.*, 2003; Gloor *et al.*, 2004). Thus, a coupling of both behaviours within the layered architecture we have used could be the basis for profiting from the qualities of the reactive and the cognitive approaches. Indeed, the frequent exercise of heavy deliberation performed for some decision-makings is very likely to be transformed into a natural aptitude in a longer-term, allowing the individual to behave more instinctively as the result of a learning process, through which new rules might be assimilated. This being so, complex driver behaviours could be modelled and analysed on a individual basis in a cognitive level, while the simpler reactive behaviour could be used to assess the overall system performance in complex and highly dynamic scenarios on aggregate basis.

REFERENCES

Adler, J.L. and V.J. Blue (2002). A cooperative multi-agent management and route-guidance system. *Transpn. Res. C*, **10**, 433-454.

Arentze, T.A. and H.J.P. Timmermans (2000). *Albatross: A Learning Based Transportation Oriented Simulation System*. EIRASS, Eindhoven, The Netherlands.

Arnott, R., A. de Palma and R. Lindsey (1991). Does providing information to drivers reduce traffic congestion? *Transpn. Res. A*, **25**, 309-318.

Balmer, M., K. Nagel and B. Raneya (2003). Large scale multi-agent simulations for transportation applications. *Proc. Workshop on Behavioral Responses to ITS*, Eindhoven, The Netherlands, (CD-Rom).

Bazzan, A.L.C., R.H. Bordini, G.K. Andriotti, R.M. Vicari and J. Wahle (2000). Wayward agents in a commuting scenario (personalities in the minority game). *Proc. 4th Int. Conf. on Multi-Agent Systems*, pp. 55-62. IEEE Comp. Soc., Washington.

Ben-Akiva, M., J. Bowman, S. Ramming and J. Walker (1998). Behavioral realism in urban transportation planning models. In: *Transportation Models in the Policy-Making Process: Uses, Misuses and Lessons for the Future* (T. Goldman, ed.), pp. 41-45. ITS/University of California, Berkeley.

Bonsall, P.W., P.E. Firmin, M. Anderson, I. Palmer and P. Balmforth (1997). Validating the results of a route choice simulator. *Transpn. Res. C*, **5**, 371-387.

Bratman, M.E. (1987). *Intentions, Plans, and Practical Reason*. Harvard University Press, Massachusetts.

Cantarella, G.E. and E. Cascetta (1995). Dynamic process and equilibrium in transportation networks: towards a unifying theory. *Transpn. Sc.*, **29**, 305-329.

Chatterjee, K. and M. McDonald (1999). Modelling the impacts of transport telematics: current limitations and future developments. *Transp. Rev.*, **19**, 57-80.

Davidsson, P., L. Henesey, L. Ramstedt, J. Törnquist and F. Wernstedt (2004). Agent-based approaches to transport logistics. *Proc. 3rd Workshop on Agents in Traffic and Transpn.*, pp. 14-24, AAMAS, New York.

Dia, H. (2002). An agent-based approach to modelling driver route choice behaviour under the influence of real-time information. *Transpn. Res. C*, **10**, 331-349.

Dia, H. and H. Purchase (1999). Modelling the impacts of advanced traveler information systems using intelligent agents. *Road and Transpn. Res.*, **8**, 68-73.

Ettema, D.F., G. Tamminga and H.J.P. Timmermans (2003). A micro-simulation model system of departure time and route choice under travel time uncertainty. *Proc. 10th IATBR Conf.*, Lucerne, Switzerland (CD-Rom).

Georgeff, M.P. and A.L. Lansky (1987). Reactive reasoning and planning. *Proc. 6th Nat. Conf. on Artif. Intell.*, pp. 677-682. AAAI Press, Menlo Park.

Georgeff, M.P. and A.S. Rao (1996). A profile of the Australian AI Institute. *IEEE Exp.*, **11**, 89-92.

Gloor, C., D. Cavens and K. Nagel (2004). A message-based framework for real-world mobility simulations. *Proc. 3rd Works. Agents in Traff. & Transpn.*, pp. 25-34, AAMAS, New York.

Haugeneder, H. and D. Steiner (1994). A multi-agent approach to cooperation in urban traffic. *Proc. Workshop of the Special Interest Group on Coop. Knowl. Based Systems*, pp. 83-99. Dake Centre, Keele.

Hernández, J.Z., S. Ossowski and A.G. Serrano (2002). Multiagent architectures for intelligent traffic management systems. *Transpn. Res. C*, **10**, 473-503.

Huber, M.J. (1999). JAM: A BDI-theoretic mobile agent architecture. *Proc. 3rd Int. Conf. on Autonomous Agents*, pp. 236-243. ACM Press, New York.

Klügl, F. (2001). *Multiagentensimulation.* Addison-Wesley, München.

Liu, R., D. Van Vliet and D.P. Watling (1995). DRACULA: dynamic route assignment combining user learning and microsimulation. *Proc. 23rd Eur. Transp. Forum*, Vol. E, pp. 143-152. PTRC, London.

Mahmassani, H.S., S.G. Hatcher and C.G. Caplice (1997). Daily variation of trip chaining, scheduling, and path selection behaviour of work commuters. In: *Understanding Travel Behaviour in an Era of Change* (P.R. Stopher and M.E.H. Lee-Gosselin, eds.), pp. 351-379. Elsevier, Oxford.

Mahmassani, H.S. and R. Jayakrishnan (1991). System performance and user response under real-time information in a congested traffic corridor. *Transpn. Res. A*, **25**, 293-308.

Marchal, F. and K. Nagel (2004). Computation of location choice of secondary activities in transportation models with cooperative agents. *Proc. 3rd Workshop on Agents in Traff. & Transpn.,* pp. 56-61. AAMAS, New York.

McNally, M.G. (2000a). The four step model. In: *Handbook of Transport Modelling* (D.A. Hensher and K.J. Button, eds.), pp.35-52. Pergamon Press, Oxford.

McNally, M.G. (2000b). The activity-based approach. In: *Handbook of Transport Modelling* (D.A. Hensher and K.J. Button, eds.), pp.53-70. Pergamon Press, Oxford.

Nagel, K. and F. Marchal (2002). Computational methods for multi-agent simulations of travel behaviour. *Proc. 10th IATBR Conf.*, Lucerne, Switzerland (CD-Rom).

Raney, B. and K. Nagel (2004). An improved framework for large-scale multi-agent simulations of travel behaviour. *Proc. 4th Swiss Transpn. Res. Conf.*, Monte Verità, Ascona, (available from ⟨http://www.strc.ch⟩, accessed 25 Jan. 2005).

Rao, A.S. (1996). BDI agents speak out in a logical computable language. *Proc. 7th Eur. Workshop on Modelling Autonomous Agents in a Multi-Agent World*, pp. 42-55. Springer Verlag, Heidelberg.

Rao, A.S. and M.P. Georgeff (1991). Modeling rational agents within a BDI architecture. *Proc. Int. Conf. on Principles of Knowedge Representation and Reasoning,* pp. 473-484. Morgan Kaufmann, San Mateo.

Rickert, M. and K. Nagel (1997). Experiences with a simplified microsimulation for the Dallas/Fort Worth area. *Int. J. Modern Phys. C*, **8**, 483-504.

Rindsfüser, G., F. Klügl and J. Freudenstein (2004). AI and transport demand modelling: a multi-agent system simulation for the generation of activity programs. *Proc. 3rd Workshop on Agents in Traff. & Transpn.*, pp. 79-87. AAMAS, New York.

Rindt, C.R., J.E. Marca and M.G. McNally (2003). Toward dynamic, longitudinal, agent-based microsimulation models of human activity in urban settings. *Proc. 82nd Ann. Meet. Transpn. Res. B.,* Washington, D.C. (CD-Rom).

Rossetti, R.J.F., S. Bampi, R. Liu, D. Van Vliet and H.B.B. Cybis (2000). An agent-based framework for the assessment of drivers' decision-making. *Proc. 3rd Ann. IEEE Conf. on Intell. Transpn. Syst.*, pp. 387-392. IEEE, Piscataway.

Rossetti R.J.F., R.H. Bordini, A.L.C. Bazzan, S. Bampi, R. Liu and D. Van Vliet (2002a). Using BDI agents to improve driver modelling in a commuter scenario. *Transpn.Res. C*, **10**, 373-398.

Rossetti, R.J.F. and R. Liu (2004). A dynamic network simulation model based on multi-agent systems. *Proc. 3rd Works. Agents in Traff. and Transpn.,* pp. 88-93. AAMAS, New York.

Rossetti R.J.F., R. Liu, H.B.B. Cybis and S. Bampi (2002b). A multi-agent demand model. *Proc. 13th Mini-Euro Conf. and 9th Meeting of the Euro Working Group Transpn.*, pp. 193-198. Polytechnic University of Bari, Bari.

Russell, S.J. and P. Norvig (1995). *Artificial Intelligence: A Modern Approach.* Prentice-Hall, Englewood Cliffs.

Schleiffer, R. (2000). Traffic itself is simple – just analyzing it is not. *Proc. 33rd Ann. Hawaii Int. Conf. on Syst. Sc.* (R.H. Sprague, ed.), Vol. 2. IEEE Comp. Soc., Washington, (CD-Rom).

Schleiffer, R. (2002). Intelligent agents in traffic and transportation. *Transpn. Res. C*, **10**, 325-329.

Steels, L. (1990). Cooperating between distributed agents through self-organisation. In: *Decentralized A.I.* (Y. Demazeau and J. P. Muller, eds.), pp. 175-196. North-Holland, Amsterdam.

Steiner, D., A.D. Burt, M. Kolb and C. Leri (1993). The conceptual framework of MAI^2L. *Proc. 5th Eur. Workshop on Modelling Autonomous Agents in a Multi-Agent World*, pp. 217-230. Springer Verlag, Berlin.

Wahle, J., A.L.C. Bazzan, F. Klügl and M. Schreckenberg (2002). The impact of real-time information in a two-route scenario using agent-based simulation. *Transpn. Res. C*, **10**, 399-417.

Watling, D.P. (1994). Urban traffic network models and dynamic driver information systems. *Transpn. Rev.*, **14**, 219-246.

Wooldridge, M. (1999). Intelligent agents. In: *Multiagent Systems: A Modern Approach to Distributed Artificial Intelligence* (G. Weiss, ed.), pp. 27-78. MIT Press, Cambridge.

Wooldridge, M. and N.R. Jennings (1995). Intelligent agents: theory and practice. *Knowl. Eng. Rev.*, **10**, 115-152.

8

Adaptation of Time Use Patterns to Simulated Travel Times in a Travel Demand Model

Georg Hertkorn, German Aerospace Centre, Berlin, Germany

Peter Wagner, German Aerospace Centre, Berlin, Germany

Introduction

The theoretical number of possible alternative activity patterns for a day is huge. A varying number of various activities can be sequenced in different order. The activity patterns then can be combined with a great number of locations for activity participation. Another dimension is the transport mode, which is used to get to these locations. The imitation of the processes that leads to the activity travel patterns observed in reality is still challenging in terms of theoretical frameworks and computing time (Arentze and Timmermans, 2000; Bowman and Ben-Akiva, 2001). The model discussed in this paper employs observed time use diaries to determine activity patterns for travel demand estimation. Thus, an efficient microscopic model can be implemented that takes spatio-temporal and logical constraints into account. Travel demand models are often used to evaluate the consequences of changes in the transportation system. Such changes in the transportation system usually cause changes in travel times and accessibility perceived by the travellers. Travel demand modelling has to provide insights into the likely effects of such changes for activity-travel patterns, destination choice and mode choice. Therefore, the model presented in this paper has been applied in combination with a traffic flow simulation model in order to generate a travel demand that is consistent with travel times in the network. The effect of the feedback between travel demand and traffic network performance is studied using data from the City of Cologne, Germany. A scenario in

Progress in Activity-Based Analysis edited by H. Timmermans

which an important link in the traffic network is interrupted is studied. The discussion of the scenario simulation is preceded by a description of the model structure and by a section about the pre-processing of input data.

STRUCTURE OF THE MODEL

The model uses four different types of input data: a synthetic population, a synthetic town, time use data, and mode choice probabilities. The synthetic population consists of individuals with home locations and socio-demographic characteristics, and it is obtained from marginal distributions using iterative proportional fitting (Beckman *et al.*, 1996). For the test case of this paper, the process is described in detail in Hertkorn (2002). To explain the algorithms of the model, we call a member of the synthetic population representing an individual in reality an agent. The synthetic town is made up of the locations for activities, e.g. working places, shops, gyms. For activities like shopping with a great number of possible locations, the data are based on zones. For other activities, the coordinates of every single possible location are known, e.g. for theatres and cinemas (Rindsfüser *et al.*, 2002).

Time use patterns are the third type of input. They are the result of a two-step classification process of diary data. In the first step diaries are compared and 24 classes are formed. Within these classes the variation of starting times of similar episodes is determined. The data were collected by the Federal Statistical Office in Germany. The analysis of the time use data is described in the next section. The travel demand model deals with one agent at a time. The first step is to select a time use pattern on the basis of the socio-demographic characteristics of the agent and the associated household (see Figure 8.1). The trips in the pattern need to be completed with information about destination, transport mode and departure time. The choice of destination and transport mode are interrelated because it is assumed that destination choice is sensitive to the travel times, which in turn depend strongly on the means of transport.

We call the activity-travel pattern, complemented with destinations and modes, a schedule. For every schedule it has to be decided whether it can be realised in the environment in which the agent lives. Usually the travel times that the agent experiences in the model will differ from the travel times in the activity pattern. This induces a time stress in the schedule, depending on the magnitude of the differences in travel times and on the rigidity of the episodes in the pattern. The next step aims at minimizing the time stress. This is achieved by shifting the starting time of the episodes. The magnitude of the shifts depends on the flexibility parameters of each episode. If after this equilibration the value of total time stress still exceeds a threshold value, it is assumed that the pattern is not likely to occur in this form and new locations and modes are selected.

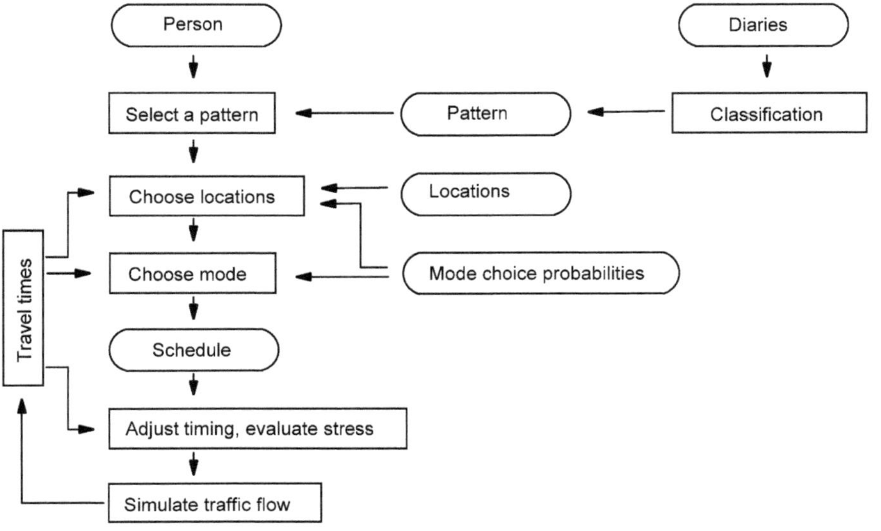

Figure 8.1
Structure of the Model

Finally, origin, destination and starting time of the car trips are written to a trip table that can be used as input for a traffic flow simulation. The traffic flow simulation results in travel times on the network. In an iterative process they are used for the next run of the travel demand model. This cycle is repeated until a self-consistent situation is obtained. This is the case when the actual travel times that the agents encounter during their trips are very similar to those that they expected when they planned their trips.

Mode Choice

Mode choice depends on many other aspects of the trip and the attributes of the traveller. Influencing variables to be considered are trip purpose, number of cars in the household, gender and age of the person, and trip distance. The full combination of all of these variables would lead to a huge number of cases to be differentiated. It would be expensive to provide an empirical data set big enough to estimate each case in a reliable way. The CHAID-Algorithm (Chi-square automatic interaction detection) (Kass, 1980) allows one to reduce the number of combinations to those that are significantly different with respect to the dependent variable.

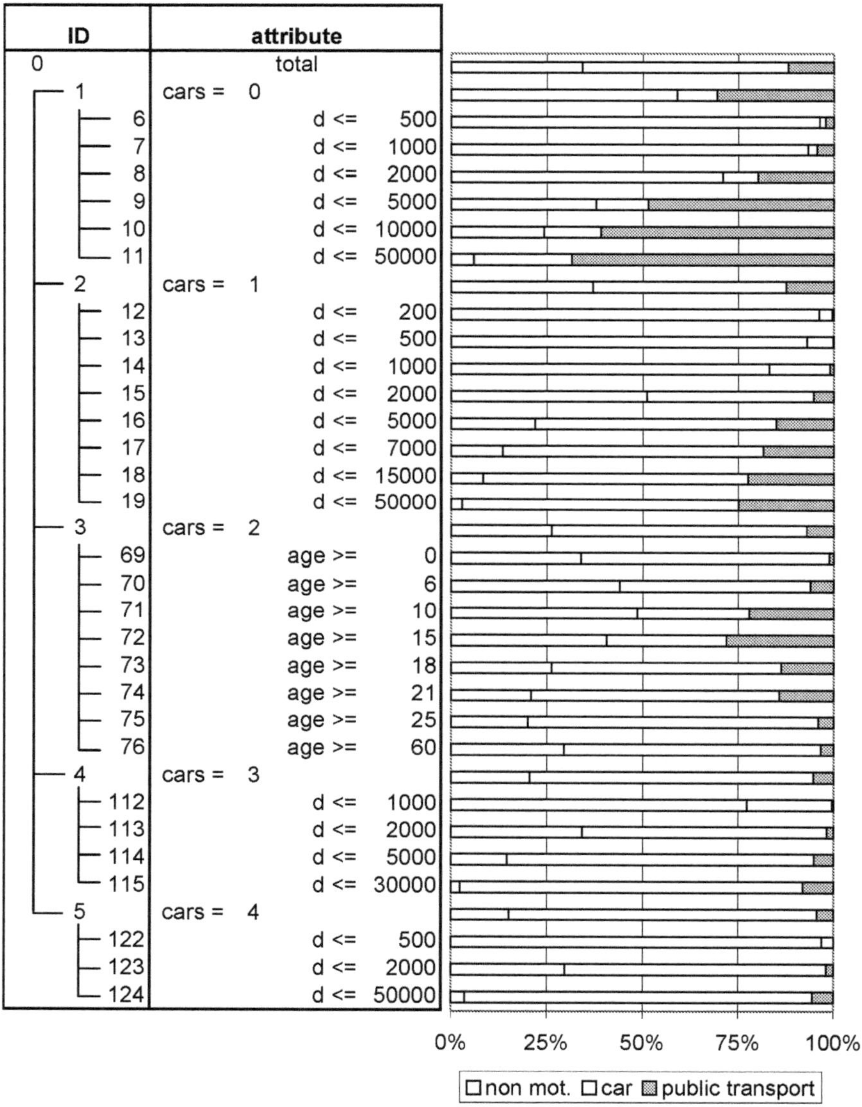

Figure 8.2
Nodes on the First Two Levels of the CHAID-Decision Tree with Mode Choice Probabilities

It builds up a tree structure: in each step the variable leading to the most significant subdivision of the data set, based on the χ^2-statistic, is determined. Often two or more variables provide a significant distinction at an arbitrarily small level of significance with respect to the χ^2-test. In these cases, the value of the Goodman-Kruskal-Tau (Ludwig-Mayerhofer, 1999) is used in this model to select the variable for a split. The data set for the estimation of the CHAID tree stems from a nationwide survey in Germany: Mobilität in Deutschland, MID (Bundesministerium für Verkehr, Bau-und Wohnungswesen, 2003; Clearingstelle für Verkehrsdaten und -modelle, 2003). 100 000 respondents were asked to report all their trips on a given day. The CHAID-tree is built on the basis of 44 000 trips. Only trips of those respondents, who live in a region of similar density compared to the area under investigation here, are taken into account.

Figure 8.2 lists the root node and the nodes at the first two levels of the CHAID-decision tree. Missing IDs belong to nodes of higher levels omitted in this table. Car trips include trips as car passenger. *Cars* is the number of cars in the household, *d* is distance in meters. On the first level, the set of trips is divided based on the variable "number of cars in the household". The value of this variable for nodes 1 to 5 can be found in the second column of the table included in Figure 8.2. Then, for example, the subset of trips for node number 1 is further split, based on variable "distance". Some of the nodes are further differentiated, e.g. node 14. However, due to the restricted space, nodes at higher levels are omitted in the table. In addition to the variables "number of cars in the household", "distance", and "age" the variables "purpose of the trip", and "gender" were found significant by the splitting procedure.

The CHAID-tree reveals that trip distance strongly influences transport mode choice. On the other hand, destination choice is based on travel time, which in turn depends on the transport mode. The next section explains how mode choice is integrated in the destination choice procedure.

Destination Choice

The basic idea of the destination choice algorithm in this model is the concept of intervening opportunities (Ortúzar and Willumsen, 1999). People use the closest destination for a given purpose unless there are reasons not to do so, such as for example the fact that a location is unknown to the traveller. Consequently, there is some probability q that the closest location will not be taken into account by the traveller or that it does not meet particular needs. The same holds for the second closest destination and all the others. The probability that location number i is chosen in a list of locations sorted by travel times is then equal to

$$f(i) = pq^{(i-1)}; \qquad p = 1 - q.$$

Two tasks have to be accomplished to apply this procedure. First, the possible destinations have to be sorted by travel times, and second, the appropriate value for q has to be determined. For the first task, the value of the travel time is computed as the weighted mean value for the modes that may be used by the traveller. The second task requires a survey where the home location and the job location of the respondent and the location of all other possible destinations for each trip are known. Usually such a data set is not available. As an alternative, the value of q for different groups of travellers can be calibrated indirectly using distance or travel time from usual travel surveys.

Temporal Adaptation of the Schedule

At this stage the schedule is complete. It is a sequence of either activity episodes or trips. For the trips, departure time, destination and mode are known. However, the process starts with a pattern in which for each episode and each trip a certain time span is reserved. The result of the location and mode choice usually yields travel times different from those in the genuine pattern. The deviations are used to calculate for episode l a local stress value

$$s_l = \alpha_l(\Delta t_l)^2 + \beta_l(\Delta d_l)^2,$$

where,

t_l : deviation of starting time,

d_l : deviation of duration,

α_l, β_l : parameters.

The starting time of the episodes are shifted such as to minimize the total stress $S = \sum_l s_l$. As illustrated in Figure 8.3 shifts in starting times can be both positive and negative. This means that the shifts are not regarded as spontaneous reactions of the travellers to disturbances in traffic flows. The shifts reflect the fact that travellers usually are aware of typical travel times. Therefore, they adapt their schedules to the situation they are used to live with. The second and the third trip require longer travel times than those in the original pattern. The rightmost column displays the new starting times, for example the work episode starts and ends earlier. The sleep episode in the evening is shortened.

The value of S is used to decide whether the schedule is still likely to occur. In the example shown in Figure 8.3, the shift of the working time may be acceptable only for people with flexible working hours. Thus, a threshold value is introduced. If the total stress is above this value, alternative locations and modes are considered.

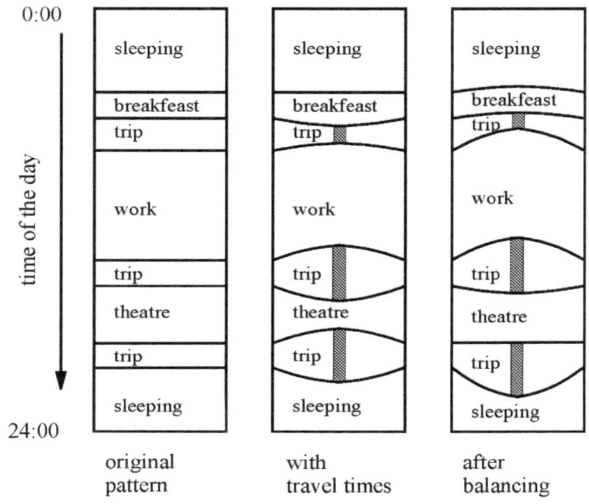

Figure 8.3
Adaptation of Starting Times and Durations to Actual Travel Times

Traffic Flow Simulation

The traffic flow simulation is based on a queuing algorithm described in Gawron (1998). It performs a stochastic dynamic traffic assignment to find a route for each driver that leads to a user equilibrium. The travel times computed with the traffic flow simulation are used to establish a feedback loop with the travel demand simulation. As described above, travel times are essential for the location choice algorithm. But they also have an impact on mode choice. If it is assumed that many locations can be reached within a short time by car, these locations get a higher ranking on the location selection list and it is likely that the car is used for the corresponding trip. Finally, travel times are the source for deviations of starting times in the schedules compared to the original activity pattern.

INPUT DATA

There are four groups of input data: synthetic population, activity patterns, locations and mode choice parameters. The population of the area under investigation is represented in the model as a synthetic population. The challenge is to assign all relevant attributes such as age, gender, principal occupation, and number of cars in the household to each agent, knowing only the marginal distributions of these variables for the set of zones. To this end, iterative proportional fitting was used. It starts with a sub-sample, in which the values of the variables are known simultaneously. Then, the share of each 'person type' is increased until the sums are consistent with the marginal distributions for the zones.

The activity patterns are derived from the diary data, using a nationwide time use survey in Germany. In this survey, respondents were asked to complete a diary for two consecutive days. The second part of the survey provides information about socio-demographics of the respondents and their households (Blanke *et al.*, 1996). The data set consists of 32 000 diaries from 7 200 households. Each diary is a sequence of 288 activity codes (one for a five minute interval). Sequence alignment was applied to determine the similarity of each pair of diaries (Sankoff and Kruskal, 1983; Wilson, 1998; Hertkorn *et al.*, 2001). The temporal dimension is included here, because not only the type of activity is known but also the duration of the episode by the number of intervals it occupies. The resulting distance table is used in agglomerative clustering leading to 24 time-use classes. These classes are essential for the next steps in the modelling process. First, they are used to discern different socio-demographic groups. according to the frequency of the time-use classes within these groups (Hertkorn and Kracht, 2002). The link between specific distributions of time-use classes within the socio-demographic groups is used to select an activity pattern in the first step of the travel demand simulation process. Second, within the time-use classes the episodes of the diaries are compared to estimate the shift parameters α and β, necessary for the adaptation of a schedule to the travel times experienced by the traveller in his/her environment. The comparison is restricted to the scope of the classes, because they represent the temporal context of their episodes.

The environment is given as the positions of possible locations where out-of-home activities can take place. For some activities, the density of locations was estimated on a zonal basis, e.g. for bakeries and food stores. For other locations, such as schools, universities, and theatres, the true position is given for every single location. For activities such as paid work and school, each location has a capacity, which is taken into account when the destinations of trips are assigned (Rindsfüser *et al.*, 2002). In addition, it is ensured that work places and schools are selected again if they are visited more than once during a single day.

Case Study: City of Cologne

The case study was designed to study the functioning of the model with real world data. Travel demand was estimated for the 1.1 million inhabitants of Cologne, Germany, and the feedback loop with the traffic flow simulation was realised. It turned out that after five iterations a steady state in terms of number of trips, modal split and distances travelled was reached. A short description of the geographic characteristics of Cologne is necessary to interpret the results. As can be seen in Figure 8.4, the population density is quite heterogeneous within the administrative borders of the city. The mediaeval city centre, Altstadt, on the left riverside of River Rhine is easily identified. The quarter boundaries follow the line of the ancient city wall. It was destroyed in 1881, and today a major road follows its course. The river flows from the south to the north. Four bridges are in the vicinity of the city centre. In addition, there are two motorways crossing the river, at a distance of 11 km and 5 km to the north and the south of the centre, respectively. Only during the last century some independent municipalities joined the City of Cologne. Some of them have a relatively high population density (e.g., Kalk on the left riverside, Porz to the south of Kalk, and Weiden at the western boundary of the map). The high density in the district of Chorweiler to the north of the city centre is due to the tower block housing that was built there in the 1960s.

The reference scenario A is a typical working day, i.e. the activity patterns refer to a Tuesday, Wednesday or a Thursday. As an example of major changes in the road network, scenario B assumes that one of the bridges in the city centre cannot be passed anymore. It is instructive to compare the effects of this intervention for the inhabitants of the two sides of River Rhine. Table 8.1 contains some figures about the travel demand for the two scenarios. Under scenario A, the number of trips per person is comparable to the reported number of 3.7 trips in the nationwide study MID (Bundesministerium für Verkehr, Bau-und Wohnungswesen, 2003). The predicted share of car trips is 5 percentage points lower on the western side compared to the eastern side, reflecting the fact that the western side has many locations for activity participation within or close to residential neighbourhoods.

Table 8.1

Travel Demand by Residence on the Western or Eastern side of Rhiver Rhine for Scenario A: Normal Traffic Network Conditions and Scenario B: Deutzer Brücke Closed

	A: Bridge open		B: Bridge closed	
	West	East	West	East
Trips per person	3.84	3.73	3.74	3.66
Total number of trips [$\times 10^6$]	2.10	1.10	2.10	1.10
Share of car trips	0.35	0.40	0.33	0.38
Distance per person and day [km]	15.5	20.5	15.4	20.5
Travel time per person and day [min]	65	70	65	71
Average car trip length [km]	5.3	6.9	5.7	7.1

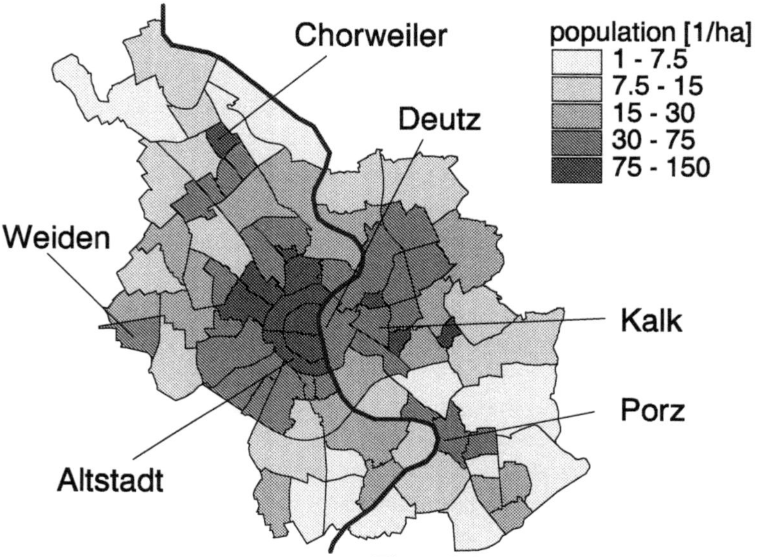

Figure 8.4
Population Density in the City of Cologne

As expected, this reduces the travel distance per person and day as well. Even for the eastern side, the value of 20.5 km per person and day is below the average value of 23.4 km reported in the MID study, which is likely due to the fact that very long trips cannot occur in the restricted area that is considered here. The city centre on the western side attracts many trips; therefore the average length of a car trip is longer for inhabitants of the eastern part of Cologne.

The effect of the bridge closure in the simulation is not very large with regard to the averages in Table 8.1. For inhabitants of both sides of the river, the number of trips and the share of car trips are both decreasing. Table 8.2 shows, however, that there are significant changes in the prediction of destination choice. The first row in each section contains the number of trips that start and end on the western side. The second row displays trips that start on the western side but end on the eastern side, and so on. As mentioned above, crossing the Rhine is much more important for people living on the eastern side than for those living on the western side. The share of trips crossing the river is only 2.5% for the latter, but 13% for the former. This share is reduced by 2 percentage points when the bridge is closed.

Table 8.2
Trips Crossing the Rhine by Residential Location

		A: Bridge Open		B: Bridge Closed	
Home	**Direction**	**Number [×10³]**	**Share**	**Number [×10³]**	**Share**
West	w → w	1987	94.6	1941	94.9
	w → e	54	2.6	50	2.4
	e → w	53	2.5	48	2.3
	e → e	6	0.3	6	0.3
East	w → w	37	3.2	29	2.5
	w → e	151	13.0	126	11.2
	e → w	153	13.2	127	11.3
	e → e	818	70.6	853	75.1

Considering only car trips crossing the river, the number per day drops from 179 000 to 136 000 (24%). As a consequence, the overall effect on the flows on the traffic network is reduced on most bridges, in the city centre, and on roads that lead from the bridges to the residential areas on the eastern side of the river (Fig. 5). Thus, according to the model it is easier for most travellers to abandon their trips to the city centre rather than to accept the longer travel times that would result from detours or the congestion caused by the concentration of traffic on the remaining bridges. The increase in the flow on the eastern Autobahnring may be caused by less interference with traffic in the direction of the city centre.

Figure 8.5
Left: Roads with Less Traffic; Right: Roads with More Traffic as a Consequence of the Bridge Closure

Cairns *et al.* (2002) studied more than seventy cases where road space was reallocated, "whether due to positively planned schemes, temporary road closure for maintenance or renewal of transport facilities, or natural disasters." In most of these cases, overall traffic was reduced by a noticeable amount. The median reduction was 11%, compared to the traffic on the road that was closed. In some cases, values above 100% were obtained. This occurred when not only the increase of traffic on alternative routes was smaller than the traffic on the missing road, but also when there was a decrease on the alternative routes as well, like in the simulation here. Traffic flow reduction related to the original flow on the bridge is 147%. However, comparing these values, it has to be kept in mind that they strongly depend on local conditions, e.g. on the definition of alternative routes.

CONCLUSION

The model discussed in this paper captures all aspects of a trip in an activity-based, microscopic framework. Time use patterns are derived from reported time use diaries and assigned to individuals according to their socio-demographic characteristics. The concept of intervening opportunities is applied to find the locations for out-of-home activities. A decision tree stores the mode choice probabilities depending on the characteristics of the trip. The starting times and durations of episodes can be adapted if travel times are different from the expected values. This is done in a way that takes into account the time use pattern of the day as a whole. The microscopic approach makes it possible to respect all spatio-temporal constraints and logical constraints related to the usage of individual modes of transport. The simulation can be computed fast enough on an ordinary PC to establish the feedback loop with a traffic flow simulation. Thus, consistency of travel demand with car travel times in the network is obtained.

The simulation results for a bridge closure scenario predict less traffic in the part of the network where the bridge is located. At first glance, one would expect an increasing number of vehicles crossing the other bridges, instead. However, the simulation results are in accordance with empirical findings about road space reallocations.

Except for some comparisons with aggregate models the results were not rigorously validated. Such validation requires the study of location and mode choice in different neighbourhoods and to check whether the model predicts the variability in these aspects of the trips in an adequate manner. The model will become more flexible if it is possible to relate the parameters of location choice to a behavioural concept rather than calibrating them using observed distances. Similarly, the integration of research results of how people allocate their time could lead to a better handling of the time stress within the schedules.

REFERENCES

Arentze, T.A. and H.J.P. Timmermans (2000). *Albatross — A Learning Based Transportation Oriented Simulation System*. EIRASS, Technische Universiteit Eindhoven, Eindhoven, The Netherlands.

Beckman, R.J., K.A. Baggerly and M.D. McKay (1996). Creating synthetic baseline populations. *Transpn. Res. A*, **30**, 415–429.

Blanke, K., M. Ehling and N. Schwarz (1996). Zeit im blickfeld: ergebnisse einer repräsentativen zeitbudgeterhebung. *Schriftenreihe des Bundesministeriums für Familie, Senioren, Frauen und Jugend*, Vol. 121. Kohlhammer, Stuttgart.

Bowman, J.L. and M.E. Ben-Akiva (2001). Activity-based disaggregate travel demand model system with activity schedules. *Transpn. Res. A*, **35**, 1–28.

Bundesministerium für Verkehr, Bau-und Wohnungswesen (2003). *Mobilität in Deutschland 2002 — Kontinuierliche Erhebung zum Verkehrsverhalten*. Endbericht 70.0681/2001, Bundesministerium für Verkehr, Bau-und Wohnungswesen der Bundesrepublik, Berlin.

Cairns, S., S. Atkins and P. Goodwin (2002). Disappearing traffic? the story so far. *Proc. Inst. Civ. Eng. – Mun. Eng.*, **151**, 13–22.

Clearingstelle für Verkehrsdaten und -modelle (2003). *KONTIV/Mobilität in Deutschland*. http://www.clearingstelle-verkehr.de, access: March 31st 2004.

Gawron, C. (1998). Simulation-Based Traffic Assignment, PhD thesis, Universität zu Köln.

Hertkorn, G. (2002). Synthetische bevölkerung. In: *SimVV, Mobilität Verstehen und Lenken — zu Einer Integrierten Quantitativen Gesamtsicht und Mikrosimulation von Verkehr* (K.-J. Beckmann, ed.), Part One. Ministerium für Schule, Wissenschaft und Forschung des Landes Nordrhein-Westfalen (MSWF), Düsseldorf.

Hertkorn, G. and M. Kracht (2002). Analysis of a large scale time use survey with respect to travel demand and regional aspects. In: *Work Time and Leisure Time — Dynamics and Convergence in Changing Contexts* (E. Fontainha and H. Perista, eds.), Int. Assoc. Time Use Res., Lisbon, Portugal, http://eprints.vf.ba.dlr.de/31/.

Hertkorn, G., P. Wagner and C. Hertfelder (2001). Klassifikation von zeitverwendungstagebüchern. In: *Zeitbudget in Deutschland — Erfahrungsberichte der Wissenschaft* (M. Ehling and J. Merz, eds.), Vol. 17, pp. 78–90. Metzeler Poeschel, Stuttgart.

Kass, G.V. (1980). An exploratory technique for investigating large quantities of categorical data. *App. Stats.*, **29**, 119–127.

Ludwig-Mayerhofer, W. (1999). *Internet-Lexikon der Methoden der Empirischen Sozialforschung, Tau (Goodman und Kruskal)*. http://www.lrz-muenchen.de/~wlm/ilmes.htm, access: December 5th 2003.

Ortúzar, J. and L.G. Willumsen (1999). *Modelling Transport*. John Wiley & Sons, Chichester.

Rindsfüser, G., J. Ansorge and H. Mühlhans (2002). Aktivitätenvorhaben. In: *SimVV, Mobilität Verstehen und Lenken — Zu Einer Integrierten Quantitativen Gesamtsicht und Mikrosimulation von Verkehr* (K.-J. Beckmann, ed.), Part Two, Ministerium für Schule, Wissenschaft und Forschung des Landes Nordrhein-Westfalen (MSWF), Düsseldorf.

Sankoff, D. and J.B. Kruskal, eds. (1983). *Time Warps, String Edits, and Macromolecules: The Theory and Practice of Sequence Comparison.* Addison-Wesley, Reading, MA.

Wilson, W.C. (1998). Activity pattern analysis by means of sequence-alignment methods. *Env. & Plan.* A, **30**, 1017–1038.

9

An Integrated Framework for Modelling Short- and Long-Run Household Decision-Making

Eric J. Miller, University of Toronto, Toronto, Ontario, Canada

Introduction

While the need for integrated models of transportation and land use is well understood, the conceptual framework for actually integrating relatively short-run activity/travel behaviour of households with their longer-run residential location and auto ownership choices (among other longer-run decisions) is not well developed. Current integrated models generally are rather *ad hoc* in their handling of this issue. In particular, while activity-based travel models continue to mature in both their theoretical foundations and their empirical implementation, a corresponding evolution of longer-term decision-making as a logical extension of the activity-based paradigm has not occurred to any significant extent (e.g., Miller *et al.*, 1998; Timmermans, 2003).

This chapter presents a conceptual framework that provides a unified, activity-based approach for dealing with household-based decision-making in both the short and the long run. It is unified in that the same framework "scales" from day-to-day to very long-term processes, and all decisions are viewed as occurring within a temporal continuum. The framework is explicitly activity-based, in that automobile ownership, residential location/tenure decisions, etc. are treated within the same "project" representation that is used to model activity/travel. The result is a comprehensive and consistent representation of household (and individual household member) planning/decision-making concerning the acquisition/allocation of personal and household resources (time, money,

Progress in Activity-Based Analysis edited by H. Timmermans

"durable goods" and information) over both the short and the long run.

The chapter is largely devoted to describing the proposed framework. It builds directly upon an earlier paper by the author, in which a comprehensive model of household decision-making was first sketched (Miller, 2004). Section 2 briefly summarizes some of the key concepts and definitions introduced in the earlier paper. Other important elements of the framework derive from recent contributions of Salvini (2003) and Litwin (2004), as well as from the seminal work of Maslow (1970). These elements are introduced in Sections 3 through 5, respectively. The sixth section then brings the various pieces together to construct an overall conceptual framework for household-based decision-making. The final section briefly summarises the chapter and discusses "next steps" in the elaboration and implementation of the framework.

BASIC CONCEPTS AND DEFINITIONS

Decision-Making Units (DMU)

A *decision-making unit* (DMU) is defined as an intelligent agent that is capable of perceiving the world around it, acquiring and using resources, scheduling its activities, and acting into the world (Miller and Salvini, 2001). Each of these four capabilities – perception, resource acquisition, scheduling, and action – are essential to characterizing human behaviour and be must explicitly incorporated into the conceptual framework. DMU's of interest within integrated urban models include persons, households, firms, and business establishments. Within this chapter, the focus is on the inter-related decision-making of persons and the households within which they live. The need to work simultaneously and consistently at both the person and household levels is well understood in the literature: it is persons who ultimately always act, but households are socio-economic units that constrain personal behaviour, coordinate the use of joint, household-level resources, generate joint activities among its members, and have "needs" of their own that can be different from and often supersede the needs of the individual household members.

Activity Episodes

A DMU acts into the world through a sequence of *activity episodes*, where each episode is a specific instance of an activity of some sort. Attributes of an episode include activity type, episode start time, episode duration, episode location, and episode mode. Start time, duration and location are self-evident in their definition. In Miller (2004), the concept of activity mode was introduced.

The idea is that, just as in the case of travel, undertaking **any** activity requires a certain combination of technology, culture and other contextual factors that determines the detailed nature, form, etc. of the activity episode. As a simple example, the generic activity "shopping" has "modes" for accomplishing the task that include in-store, mail-order, telephone and internet. With the addition of mode as a generic episode attribute, the mapping between "activity" and "travel" as simply two sub-classes of the episode class becomes complete, in that a trip (travel episode) also has start time, duration, location and mode.

Note that in speaking of an "activity episode" no distinction between the "short-run" and the "long-run" is made. The key difference between these two types of activities is that one occurs within a *fixed* set of resources and other constraints (a current number of household cars, a current job, a contracted child-care arrangement, etc.), while the other involves actions that *change* these resources or constraints. Thus these two types of activities are "short-run" and "long-run" in the classical economic definition of the terms; i.e., they depend on whether resources/technology (the "production function" of the household if you will) are fixed or not. The key point, however, is that decision-making processes concerning both types of activities (what to do today given available resources? can/should I change my available resources today?) "run" continuously and simultaneously within the DMU. They also interact, in that decisions about resources constrain day-to-day activities, while day-to-day experiences "feedback" to the DMU "resource manager" information concerning resource needs and opportunities.

Resources

The terms "resources", "technology" and "durable goods" are all used somewhat interchangeably within this chapter, although each term obviously carries somewhat different definition and connotations. What is meant collectively by these terms is the physical/technological/fiscal context within which day-to-day activity occurs. In particular, the engagement in any activity requires the expenditure of *resources* of various kinds. At a minimum, time must be expended by the person(s) involved. Knowledge of various kinds is also always involved (where the store is located, how to play the game, etc.). Very often money is required to engage in the activity (transit fare, admission price, cost of the goods purchased, etc.) and/or durable goods/technology or services of one sort or another (car, tools, etc.) are used. Without the required resources, the activity episode can not be executed as intended (no allowance left to spend on a movie; car isn't available this afternoon to go shopping; etc.), and it will either not be undertaken at all or it will have to be rescheduled for some point in time when the necessary resources will be available.

While activities consume resources, they also generate them. "Outputs" from activities can include money (paid labour; returns on investment, etc.), new or improved durable goods/technology and increased knowledge (e.g., now know how to play the game). The money, knowledge and/or durable goods acquired through this activity episode, of course, are added to the household's or person's stock of such resources and are available for use in subsequent activity episodes. In particular, note that it is through action that we learn (i.e., increase our *knowledge base*).

Utility is also generated by most activities, where "utility" is simply the label used to represent the benefit, pleasure, avoidance of adverse consequences, etc. derived from engaging in the given activity episode at a given point in time. Unlike the other activity outputs (which are all instrumental in nature; that is, they are means to achieve other ends -- their value lies in their ability to support new activities), utility is an "end" output of the activity. It is "consumed" as it is generated, and, while the memory of the event can be "stored" in the recipient's mind and thereby provide future utility (e.g., recollections of last summer's vacation), utility does not itself become an input into other activity episodes.

Resources in addition to time (knowledge, money, technology embedded in durable goods, contracted services) thus clearly play a central role in the life of a DMU, both as enablers of DMU action and as constraints on the scope of DMU activity. As such, the acquisition and maintenance of resources – *resource management* – is a major DMU activity in its own right that must be explicitly accounted for in any activity-based model of the DMU.

Projects

Axhausen (1998) defines a *project* as a coordinated set of activities, tied together by a common goal or outcome. Axhausen's simple example to illustrate the concept of the project is the dinner party, in which the logical set of inter-related activities might include: the decision to hold a dinner party, planning the meal, shopping for the food and drink, cleaning the house prior to the party, preparing the meal, the dinner party event itself, and cleaning up after the party. In Miller (2004) it is argued at length that **all** activities can be conceived of as being contained within (or generated by) projects. It is also hypothesized that a set of *primary projects* can be identified that form the basis for all specific projects undertaken by a given DMU. Without revisiting these arguments in detail within this chapter, the assumption that each DMU (person or household) possesses at any time a set of active projects within which all the DMU's activities are "generated" represents a fundamental component of this conceptual framework.

It must be stressed that the adoption of the project as a basic building block of DMU behaviour has both behavioural and computational motivations. Behaviourally, it is argued that the project is a reasonable organizing principle for dealing with complex human behaviour. This holds even in cases in which people may not explicitly "think in projects". For example, it is probable that a person does not think of his/her social activities as a "project". But these activities do have a "logic of their own" that is clearly distinct from other activity sets (work, home maintenance, etc.) in terms of its "rules", motivation, information base, etc. Social activities also do "spawn" other logically related, but "non-social" activities, as the simple dinner party example above illustrates that are difficult to "explain" outside of the project that is motivating them.

In operational, computational terms this ability to group activities into logical projects is at least as compelling. The project *qua* agent (i.e., an intelligent object) allows the modeller to encapsulate the logic of each project within its own object: in building the behaviour describing the work project, the modeller does not have to be concerned with how to model social networks (and *vice versa*). At the same time, since each project is a sub-class of the abstract project class, it benefits through the object-oriented (OO) properties of generalization and inheritance from common elements (attributes and behaviours) that are generic across all projects. As a modelling approach this is extremely powerful, since it allows the modeller to decompose an extraordinarily complex problem into more "bite size" pieces (the standard modelling approach for dealing with complexity) while still retaining a coherent, consistent representation of the "whole". That is, while the internal logic of each project typically will be very different, they all remain projects, and their generalized behaviour, as viewed "externally" by the rest of the modelling system (and, in particular, as viewed by the DMU agent that is "managing" these projects), is common (and hence computationally tractable).

Further, the decomposition of behaviour into projects helps reduce the combinatorics of dealing with the huge "decision space" potentially faced by DMU's. That is, each project need only concern itself with its own objectives and actions, and it can leave the consideration of "tradeoffs" among competing actions/objectives to "higher-level" agents that deal with these decisions. For example, I do not need to know the details of my wife's day to schedule my work day (and *vice versa*), except in the case where competition for fixed household resources might arise (e.g., we both need the only family car at the same time) or where collaborative action is required (e.g., who is going to pick the child up from day care). In such cases, it is argued that a higher-level DMU (i.e., the household) mediates/decides the resolution of the conflict and/or collaboration issues. Similarly, competing demands for activity engagement by two or more projects within a DMU (e.g., should I mow the lawn Saturday afternoon or play golf?), it is argued, are resolved at the "higher level" of the activity scheduler, whose task is specifically to resolve such issues. That is, it is far more efficient to have a scheduler agent that "sees" the "activity requests" from all competing projects and mediates among them, than to try to have individual projects try to mediate these issues among themselves. This does, however, imply the need for each project, regardless of its specific nature, to be able to supply to the scheduler a standard set of activity episode attributes so that the scheduler can intelligently choose from among the candidate episodes those that will actually be scheduled for execution.

The three key elements of a project are:

1. The project will have a *knowledge base* that defines the information available to support its decision-making, including information concerning its mandate, how to evaluate the accomplishment of its objectives, etc. This information base may well be updated over time as new information is obtained, or as the "rules of the game" change.

2. The "output" of the project is an *agenda*, which consists of one or more specific activity episodes that the project currently would like to have scheduled and executed. This agenda is provided to the activity scheduler, which will decide which, if any, of these episodes to insert within the provisional schedule for the current planning period. Thus, the project does **not** schedule its episodes. Rather, it generates candidate episodes that may or may not be scheduled and subsequently executed. Note that the agenda does **not** consist of all activities in which the DMU might conceivably engage within the context of this project. This totality of possible activities is embedded within the task manager. The agenda, rather, is the current, finite, specific set of activity episodes that the project is contemplating for execution within the current planning period. The agenda, thus, is a dynamic list of potential episodes that evolves over time as candidate episodes are added to the list by the project and then subsequently deleted, either because they have been executed or because the project decides to "cancel" them for one reason or another.

3. The heart of the project is the *task manager*. The task manager contains the "intelligence" and dynamics of the project agent. The responsibilities of the task manager include:

    - instantiate the project agenda when the project is first created;
    - delete the project agenda when the project is terminating;
    - instantiate episodes within the agenda, as required;
    - monitor the progress of the project (what activities need to be done next; how well is the project's mandate is being fulfilled; etc.);
    - delete agenda episodes if they are now longer required and/or feasible; and
    - communicate with collaborating agents.

Thus, once instantiated, a project is an autonomous agent capable of monitoring its own activities and receiving and sending information to other, collaborating agents (e.g., the scheduler). The project's "view" of the world around it, however, is limited to information and events pertaining to the project itself. For the reasons discussed above, one project is not "aware" of other projects, and does not ever communicate directly with these other projects.

The Activity Scheduler

As already discussed, the logic of the project design concept is that another agent must exist that undertakes the actual scheduling of candidate episodes within a *provisional schedule*. This schedule is labelled "provisional" since it is subject to change up to the point that the next episode is to be actually executed by the DMU – it is only at the instant of episode execution that commitment to undertaking the episode is realized and the episode attributes are confirmed. This agent is labelled the *activity scheduler*, or, more simply, the *scheduler*. Many detailed models of activity scheduling (both conceptual and at least semi-operational) exist (see Arentze and Timmermans (2000) or Roorda *et al.* (2004) for reviews of such models), and the activity scheduling task *per se* will not be discussed in detail here. The conceptual model being developed here differs to varying degrees from other models of activity scheduling in at least two key ways. First, it clearly demarcates between the project (that generates an agenda containing activity episodes) and the scheduler (that schedules these episodes, including resolving conflicts among competing episodes). This permits project-specific calculations to be encapsulated within each project, leaving the scheduler free to focus on the scheduling task *per se*. Second, "day-to-day" scheduling occurs within an explicit, continuous, "longer-term" process within which the resources (knowledge, technology, money, etc.) that constrain these day-to-day decisions evolve over time. Further, the projects "driving" the scheduling process may change over time (change jobs, have a baby, etc.), thereby changing the nature of the activity episodes to be scheduled.

Event-Driven, Continuous Time Scheduling

Litwin and Miller (2004a,b) argue that activity scheduling is a discrete event, continuous time process in that:

- The human mind is constantly aware of and responding to the world around it.
- We continuously make both "short-run" and "long-run" decisions at arbitrary points in time. The only difference between a "short-run" decision (what am I going to do in the next half

hour?) and a "long-run" decision (should I buy a new home?) is the timeframe over which the impacts of the decision are likely to be felt. Another way of putting this is that we have multiple *planning horizons* over which we are continuously and simultaneously planning – today, this week, this year, etc.

- Scheduling decisions are *events* that cause the provisional schedule to change in a discrete way. Similarly, all physical events, whether these are caused by our own actions (drive to work) or exogenously imposed upon us by the world around us (traffic jam) potentially "trigger" scheduling/rescheduling decisions.

Litwin identifies three temporal "spaces" of relevance to scheduling:

- *A-Space*: This is a multi-dimensional space in which both past memories and possible future actions lie. It is the "home" of projects (i.e., the things we might do); indeed, each project can be thought of as a dimension in A-Space.
- *C-Space:* This is the space of physical events (past and current). A key point in this space is the moving "NOW" – the point in physical time at which we are at "now".
- *B-Space:* B-Space consists of a set of lines in which each line connects NOW to a future planning horizon. Each such line contains a set of proposed activities in which the agent wishes to engage during this planning period, at least roughly ordered by "before/after" relationships. Thus, one line may exist for today's activities, another for the time between now and next summer's vacation, etc. Similarly, a line may exist for each project over the project's planning horizon, containing the project's agenda over this time period.

The extent to which the "full" concepts of A-Space, B-Space, etc. can or should be implemented within an operational model of household decision-making remains to be seen (e.g., in practical terms, how might we "observe" the evolution of B-Space?). Regardless, Litwin's conceptual model provides a very strong framework for a consistent and unified model of household decision-making. Key points that are applied within this chapter include:

- C-Space is explicitly implemented within the conceptual model as the set of events within the world at large that a DMU perceives and responds to.
- The NOW point is similarly implemented as the point in simulated clock time that triggers "current" decision processes.
- Multiple planning horizons are explicit within the model.
- The concept of mental simulation is explicitly incorporated as the means by which alternative courses of action are explored by decision-making agents.
- Although not explicit in the current framework, A-Space is viewed as the "home" of

projects and, indeed, resource management.

- Similarly, B-Space is implemented within the concept of project agendas.
- The event-driven nature of decision-making is taken as an axiomatic feature of the framework. Each event can potentially trigger a response (decision) since each event brings with it new information, a change in system and DMU states, etc. Note in particular that each decision is, itself, an event that might trigger a new decision.

Stress Management

Definitions and the Example of Residential Stress

The concept of *stress* as an explanatory variable in urban spatial processes dates at least to the seminal work of Rossi in the 1950's (Rossi, 1955). Loosely defined, stress arises when one's current state deviates from some alternative desired/expected/optimal state. The larger this deviation, it is hypothesized, the more likely one is to act in some way that attempts to reduce the stress; i.e., to attempt to move one's state closer to the alternative "target" state. While consistent with utility concepts, the notion of stress recognizes that people often are unable (or are at least unwilling) to "act at the margin"; i.e., in many situations people do not make continuous marginal adjustments to their state so as to maintain themselves at their "optimal" (utility-maximizing) state. This is particularly the case with respect to "large", discrete choices such as residential location, employment, etc. in which: marginal adjustments are generally not possible; significant transaction costs are involved in state changes (usually both monetary and psychic); information about alternatives is imperfect (the "target state" is a hypothesized one, based on limited information); and state transitions typically involve engaging in a market process involving an explicit search for actual alternatives, with the outcome of this search being uncertain at the time the search is initiated.

For all of these reasons, it is reasonable to hypothesize that people will tend to remain in their current state (same house, same job, etc.) when stress is low, and will only actively seek to change this state when stress exceeds some threshold value. Sarjeant, for example hypothesized a cusp-catastrophe model of residential search (Sarjeant, 1986; Miller and Sarjeant, 1987), illustrated in Figure 9.1, in which the "explanatory" variables of search behaviour are "present satisfaction" with one's current dwelling (PS) and an "expected alternative satisfaction" (EAS) that one anticipates could be achieved if one were to move now to another dwelling. Both PS and EAS change over time in response to changing conditions within the household (household structure, employment status, etc.) and external factors (housing prices, interest rates, neighbourhood characteristics, etc.). Thus, the household traces a trajectory within the PS-EAS plane over time.

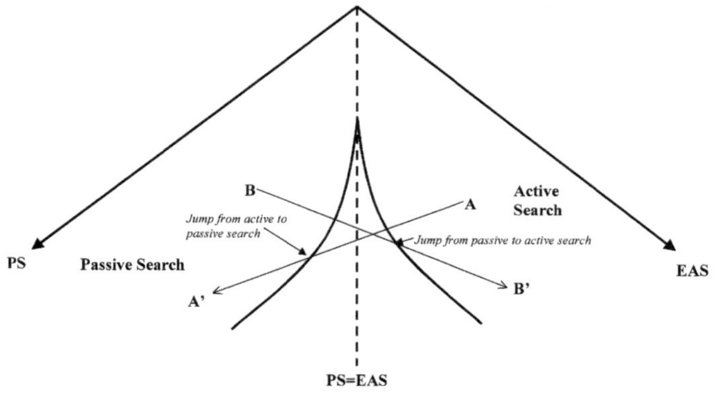

Figure 9.1
Stress-Based Search Process

Residential "search" is hypothesized to be a **continuous** process of information acquisition about the residential housing market. As sentient beings we are constantly exposed to some level of information about the housing market, which we filter in various ways and assimilate to varying degrees. Normally, this information acquisition process is an essentially *passive* one in that the information "comes to us" through our normal day-to-day activities (newspaper articles, noticing "for sale" signs as one drives through the neighbourhood, etc.); that is, we are not consciously, pro-actively seeking specific, detailed information about the housing market. As a result of this passive information acquisition process we each have some general (but certainly imperfect) information about the current state of the housing market, upon which we can form some judgement (again typically imperfect) about what our EAS would be if we were to move to a new dwelling at this point in time. It is also certainly the case that we have a good sense of our PS with our current dwelling, given that we are experiencing it on a day-to-day basis. Thus, in principle, both PS and EAS are "known" to the household at each point in time.

The difference between EAS and PS is assumed to define the household's *residential stress*. If (EAS- PS) < 0, then presumably the household is in an "optimal" location with respect to its current residential location. If, however, (EAS - PAS) > 0, then some amount of stress between its current and "desired" states exists. If (EAS - PS) > τ, where τ is some non-negative threshold value, then the household abruptly transitions into an *active* search state in which detailed, specific information about dwelling units that are currently available for purchase (or rent, as the case may be) is sought. A value of τ = 0 would correspond to a "classical" economic model of marginal, "utility maximizing" behaviour, in which as soon as a marginally better alternative exists, a transition to this alternative is attempted. As argued above, it is generally more reasonable to assume that $\tau > 0$ in value. τ and/or EAS and PS, of course, may well be random from the modeller's perspective.

This active state generally persists for only a relatively short period of time (a few weeks or months at the most), before the household reverts to a passive state again. This reversion to the passive state can arise in two ways. First, through the information gained within the active search, the household's EAS is revised (downward) to the point that the household no longer expects to (sufficiently) improve its satisfaction level by moving from its current location, and so it ceases to actively search in the market. As a result, the household "drops out" of active participation in the housing market. This corresponds to a very rapid trajectory in PS-EAS space, such as trajectory (A) in Figure 9.1. Second, the household is successful in finding an attractive new dwelling and in purchasing this dwelling. In this case, a sharp, discontinuous rise in PS occurs (associated with the new dwelling), which again returns the household to a passive state.

Stressors and Stress Management

The extended example above deals with only one source of stress (residential), and considers only one "stress resolution" mechanism (move or don't move). In general, DMU's face many sources of stress (congestion, dissatisfaction with one's current job, financial problems, etc.) and potentially have many options to relieve a specific stress (e.g., perhaps change jobs instead of move residence). Further, relief of one stress might trigger a "cascade" of responses (e.g., in order to accept an attractive job offer one may end up changing place of residence and/or buying a second car). In order to deal with this multi-dimensionality (in both sources of stress and stress resolution/management), Salvini introduced the concept of the *stress manager* (Salvini, 2003; Salvini and Miller, 2004). A stress manager is "simply" an object that "manages" stresses that arise from a variety of *stressors*, where a stressor is something that generates stress for the DMU.

It is argued further below that it is reasonable to assume that each project undertaken by a DMU has its own "stress". This stress is provisionally defined as the difference between the current project "state" and some perceived, feasible alternative (new job, expected/desired house, etc.). This difference can either represent a "deficit" ("my current job stinks") or an "opportunity" ("my current job is OK, but this job offer is too good to turn down"). Stress in one area generally will be resolved in the first instance within the area (i.e., within the project) itself, if this is possible. But stress, if large enough, will also be resolved through accommodations elsewhere: allocation of addition resources, modifications of other projects, etc.

The concept of stress is, inevitably, closely aligned with the concept of utility: a "natural" definition of stress is the difference between the current utility and some target utility level. This linkage between the two concepts is strengthened by the need to compare stresses in one area with another (e.g., a move to the suburb may create a stress due to loss of urban environment; this may be compensated for by more disposable income for vacations; greater stimulation and prestige at work; more green space for the kids; etc.). That is, it is not clear how to consistently evaluate stress management strategies and their outcomes within a general model outside of an explicit utility-based framework. This issue is returned to in the next section.

Similarly, the stress management concept supports a household base for these decisions. A move to the suburbs (which might be "triggered" by a job offer to one household member) impacts all household members. It is reasonable to argue that this decision must be made at the household level, with the household needing to mentally simulate the impacts of the proposed move on all household members, and to "add up" the utilities/stresses to see how they all balance out (e.g., "we can't move because we don't want the kids to have to change schools").

The very simple notion that underlies the concept of stress management is that for every stressor there is a logical set of responses that can be worked through in attempting to relieve the stress. For example, if travel time to work is too onerous then the person may try changing the trip route, start time and/or mode. Alternatively, the person might change jobs and/or residential location. In order to avoid the combinatorics of exploring all combinations/permutations of all such responses, it is suggested that sequential exploration of "increasingly major" changes (try change mode, if that doesn't work, try buying another car; if that doesn't work ...) may well represent a reasonable approach to stress management in most cases.

Certainly any stress management response will typically involve the "mental simulation" of alternative courses of action. In particular, the ramifications of any proposed change in "lower-level" and/or "collateral" projects should be considered. For example, every "move residence" assessment probably must trigger a mental simulation of auto ownership to see if the proposed change results in excessive new stress within this project. Every "change job" assessment probably should similarly trigger a "move residence" mental simulation – at least if the job location changes significantly (and "negatively" in terms of its stress implications) relative to the *status quo*. The job change, for example, may also trigger an independent change mode/change auto holdings assessment. "Smart" uses of stress calculations may help enormously here. If, for example, the job change does not involve a significant location change (i.e., no new residential stress) then there is no need to trigger a residential move change at this point in time. At the same time, however, commuting times may change, the need for a car may change, etc., thereby creating stress with respect to work mode and/or auto ownership that may need evaluation.

MASLOW'S HIERARCHY OF NEEDS

Basic Concepts and Definitions

Maslow's seminal 1970 book *Motivation and Personality* provides an in-depth exploration of the *motivated* nature of human behaviour. His best-known concept is that of the existence of a *hierarchy of needs* that influence our actions. Figure 9.2 provides a very simplistic summary of Maslow's hierarchy. A key point of this hierarchy is that lower-level needs are *prepotent* over higher-level needs; that is, lower-level needs take precedence over higher-level needs until they are satisfied to some level. A starving person is only interested in food; once a person has eaten he will be concerned with shelter and security; and so on. As a general principle, it is argued that the motivation for our behaviour "moves up" the hierarchy as lower-level needs are "satisfied". Having said this, however, it is important to note that Maslow stresses that the motivation for a given act generally is derived from many needs simultaneously, with different needs providing the motivation to varying degrees, depending on the context of the action. In addition to "needs" *per se*, Maslow identifies a number of *preconditions* for the satisfaction of needs. These preconditions can be prepotent in nature given their importance to the needs satisfaction process.

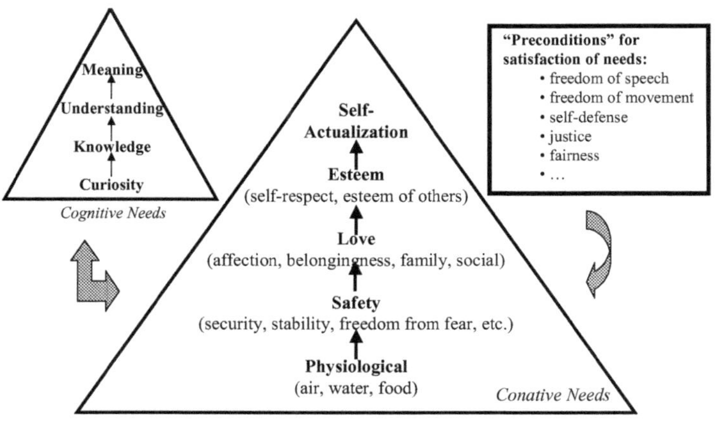

Figure 9.2
Maslow's Hierarchy of Needs

With respect to the conceptual framework being advanced within this chapter, Maslow's theory provides at least three important concepts. First, to the extent that one accepts Maslow's thesis that human behaviour is largely motivated in nature, it can be argued that the economist's concept of *utility* is a very simplified/abstracted approximation of motivation. We can never (certainly in ILUTE-type models) trace back to motivations in any meaningful or operational way, especially given their multiple/over-lapping nature within any given act. The "only" (but still useful) link here is that actions are viewed as being motivated, and hence utility would appear to be a reasonable "surrogate" to capture the benefits received from the action (i.e., the desired ends that motivated the act).

Second, Maslow's hierarchy of needs provides a sound foundation for the development of a set of "fundamental activities" or "primary projects" that are relatively universal across persons and households. The existence of such a universal set obviously would simplify the task of constructing a transferable, relatively "generic" model of household decision-making.

Third, the concepts of prepotency and preconditions may provide insights into the establishment of priority and precedence relationships among activities within a scheduling/decision-making process. It may also further strengthen the validity of the generic concept of a project as a set of logically connected actions; that is, some of the actions within the project are preconditions to others in the satisfaction of the needs motivating the given project.

Given these observations, it would seem very reasonable, as is often done in the literature, to speak of *place utility* as a measure of the level of satisfaction that a household possesses at any point of time with its place of residence, in terms of the extent to which these various needs and preconditions are being met. Note that in this definition place utility is an entity attached to the residence *per se*, independent of the daily "activities" that occur within it. Alternatively, one might argue that place utility is simply the sum of the expected utilities over all of the detailed activities that will go on within the residence (child-rearing, family entertainment, meals, personal maintenance, sleeping, socializing, gardening, etc.). This is clearly operationally a hopeless path to tread. But what is being argued here is that it is also behaviourally incorrect: one's home possesses utility because it directly addresses various conative needs. It also is a precondition for a wide variety of other needs. Indeed, it would be "double counting" to "integrate" over the expected utilities of these activities as part of the place utility *per se*, since these activities will possess utility in their own right that will accrue as they are executed. Further, when I buy my house I do not know all the activities that will occur within it or because of it. I do know, however, that it is of value to me for all of the reasons listed above.

As another example, consider the "mobility project". Project activities include: vehicle ownership and maintenance; obtain a driver's licence; purchase a transit pass; etc. Ownership of a car can address safety, love and esteem needs. The car is an important precondition for virtually all out-of-home activities, since it significantly increases the "action space" available to the car owner. The ability of the car to provide enhanced mobility (combined with the way we have typically built our cities) invests in the car a considerable amount of prepotency over many other investments, since the precondition of car availability for activity participation is extremely strong in many, many instances. Thus, as with the residence, the car possesses "mobility utility" that is separate from (and "prior to") the utility derived from travel *per se*.

To take a third (and final) case, the "work project" is different from the previous two discussed in that it generates income as well as utility. Work *qua* activity directly addresses both cognitive and conative (safety, love, esteem, self-actualization) needs directly (perhaps aesthetic needs as well). The income derived from work is an essential precondition to undertaking virtually all other activities and thereby achieving the full range of cognitive, conative and aesthetic needs (in particular all prepotent physiological needs – food, shelter, etc.). Thus, the work project generally has a very high level of utility/priority attached to it.

Other projects can also generate income. Projects also, in general, incur both capital costs (equity investments in dwelling units, purchase of autos, etc.) and on-going operating costs (out-of-pocket travel costs, activity participation costs, dwelling unit maintenance and utility costs, etc.). Ignoring capital costs for present purposes, one can on an "annualized basis" define for each project p:

U_p = the utility derived from project p
Y_p = the income derived from project p
C_p = the on-going costs for project p

One can bring these project-specific terms together to construct an overall DMU utility (U), which for current purposes of illustration will simply be written as:

$$U = \sum_p U_p + \alpha \log(\sum_p Y_p - \sum_p C_p) \tag{1}$$

It was previously argued that resource management is stress-driven. A natural definition of stress is the difference between current utility (U in total or U_p within a given project) and some expected utility from another possible "project target state" (different home, new car, etc.). Stress is also generated as $(\sum_p Y_p - \sum_p C_p)$ approaches zero, regardless of the values of U_p.

The stress manager is not necessarily trying to maximize U directly. Rather, it is assumed that it is trying to keep stresses within certain thresholds. That is, because most projects of interest are inherently lumpy things (home, car, job, marriage, etc.) with high transaction costs, etc., absolute utility maximization/stress minimization is not generally feasible or directly sought.

Without worrying how it might be computed for the moment (a big leap!), let us assume that for each project p there is an expected/desired "target" utility, T_p, that the decision-maker thinks might be achieved if the project was to change to the target state (change job, change eating habits, etc.). Define the project stress level, S_p, as:

$$S_p = T_p - U_p \qquad (2)$$

If $S_p \leq 0$ for all possible (or at least known) target states, then the current project state is optimal. As in the Sarjeant model, if $S_p \geq \tau$ for some threshold level τ, then the DMU will actively engage in trying to change the project state to improve its utility and reduce its stress. This stress resolution strategy may well impact other projects, and so utilities (and stresses) within other projects might be altered if the project state transition is implemented. Possible considerations for assessing whether to actually change project state may include:

$$\Delta U \geq 0 \qquad (3)$$

(any change in total utility must be positive; i.e., the DMU will not change project state if this change reduces overall utility)

$$S_p < \tau \qquad \forall p \qquad (4)$$

The Conceptual Framework

In this section, the previously developed concepts are combined into a unified conceptual framework of household-based decision-making. First, a general model of the decision-making unit (DMU) is presented. The DMU can be decomposed into several discrete *decision management agents* (DMA): the scheduler, the project and the stress manager. The general concept of the DMA is developed, followed by a brief revisit of the scheduler, project and stress manager within the DMA/DMU framework. Finally, the household as a DMU is discussed.

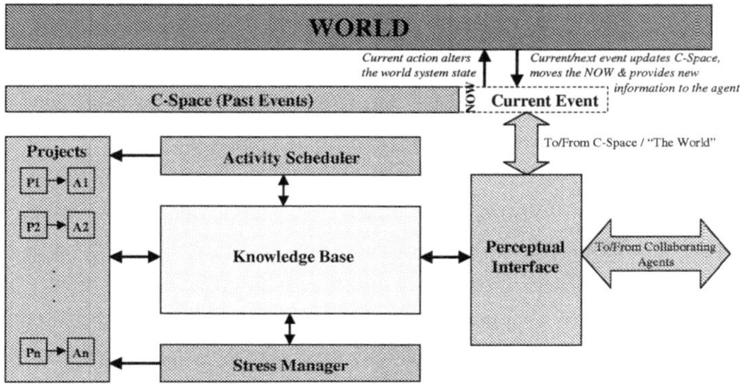

Figure 9.3
The Decision-Making Unit (DMU) Agent

The DMU Agent

Figure 9.3 "puts the pieces together" in terms of a provisional model of the DMU agent. It is argued that both persons and households represent sub-classes of this abstract agent class, and, hence, the DMU represents the "basic unit of analysis" for micro-simulation modelling of socio-economic processes.

At the heart of the DMU are three cognitive agents (the scheduler, the stress manager, and the set of DMU projects), interconnected by the DMU knowledge base. Each of these agents has been discussed in isolation in earlier sections of the chapter. Sub-sections below provide additional discussion of these agents and their roles/interactions within the DMU. It is noteworthy with respect to the DMU as a whole, however, that stress management is where "reflection" resides within the DMU. That is, it involves the agent observing its own condition and actions and "reflecting" upon them in terms of how things could be improved (would a new car help; is now the right time to move, etc.). Scheduling, on the other hand, is much less "conscious" an activity. That is, it is more mechanical in nature; it must go on due to the relentless need to "spend" each minute of the day. It is efficient in the programming sense, but also behaviourally plausible/sensible, to differentiate this activity from the reflective considerations of stress management. Also, the "B-Spaces" of the two activities are typically very different.

Allocation of task management within a given project also makes computational and behavioural sense. As has already been noted, it is where the "specialized knowledge" and the "specific management" of the project resides. Each project is different in its details from each other, and the encapsulation of these differences within the project (but with a common "interface" to the DMU) seems very sensible. And while task management of all projects is going on inside a person's head more or less simultaneously, it still makes sense to separate out projects in this way: we do "compartmentalize" within our own minds.

The DMU must be able to communicate with both the agents with which it is explicitly collaborating (other household members, members of a social network, etc.) and, more generally, the world within which the DMU exists. This interface with collaborating agents and the world at large is assumed to be handled by a *perceptual interface*. At a minimum, this interface handles the mechanics of information exchange. It may also be the case that some "perceptual filtering" occurs, in which "objective facts" about the world (e.g., current housing market prices) are modified to reflect the DMU's unique perception of these "facts". Care must be taken in the model design to ensure that this interface facilitates rather than hinders information exchange between agents. At the same time, the need to encapsulate each DMU's knowledge basis within each DMU object would appear to be a paramount design feature (and, hence, the need for an interface): I cannot "read the mind" of another person – it is only through their actions and their communications with me that I can obtain some understanding of their thoughts, tastes, etc.

The totality of all DMU's, all other *simulated objects* (buildings, roads, etc.), and all the events arising from DMU actions within the model is represented in Figure 9.3 as *the World*. In short, the World **is** the simulation model. Indeed, within ILUTE for example, the World exists as an explicit object which brings into existence all other objects within the model and which "releases" them to act (Salvini, 2003). Each DMU is, of course, contained within the World, is affected by the actions of other DMU's within the World, and, in turn, through its own actions affects others. In particular, it is through the individual actions of all the DMU's within the World that the "collective outcome" that is the "system state" of the World evolves over time.

It is beyond the scope of this chapter to deal in depth with the concepts of learning and adaptation. Clearly, however, a truly behavioural DMU representation is one that permits the DMU to learn and to adapt to changes in its surroundings over time. Adaptation presumably can occur in at least three ways within a DMU: changes in selected actions within a given set of resources and decision rules (e.g., changing work trip travel mode in response to new congestion levels, etc.); changes in the resources available to support activities within a given set of rules (purchasing an additional car for household use; acquiring a transit pass; moving residential location; etc.); and changes in the "rules" used to choose activities and/or resources (e.g., changes in tastes and preferences as a function of experience and socialisation; changes in the set of activities being considered). It is argued that the conceptual model of the DMU presented here explicitly deals with the first two of these processes. The extent to which the third process can be captured within this framework is discussed somewhat further below.

The Decision Management Agent (DMA)

In this preceding section three primary cognitive functions were identified for a DMU: project task management; activity scheduling; and resource/stress management. Each of these activities involves the intelligent use of DMU knowledge to arrive at decisions on an on-going, continuous basis. While all three of these activities obviously occur within the same mind, it is clear from the discussion above that these are three very different cognitive tasks. It is thus both practically and behaviourally reasonable to encapsulate each of these primary activities within its own object/agent for modelling purposes, which can be labelled the *project*, the *scheduler* and the *stress manager*, respectively. These, in turn, can be represented as sub-classes of an abstract class – the *decision management agent (DMA)* – as is illustrated in Figure 9.4.

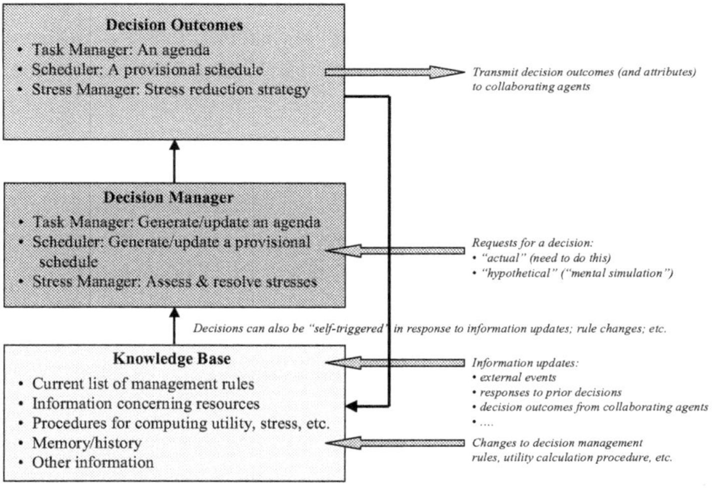

Figure 9.4
The Decision Management Agent

A DMA depends upon a knowledge base that provides it with information pertinent to its decision-making, as well as the rules/algorithms, tastes/preferences, etc. that are to be used by the DMA in its deliberations. Memory/history of past DMA decisions and their outcomes reside here as well. The knowledge base communicates with other agents and can thereby update itself over time. While one could assume that the knowledge base is intrinsic to the DMA *per se*, it is computationally far more efficient to assume that the DMA has access to appropriate portions of a common, DMU-level knowledge base, so as to avoid excessive and artificial "information requests" between DMA's. Or, in behavioural terms, while it is reasonable to divide decision processes among separate "agents" within the mind, it seems much less reasonable to divide the mind's knowledge (memory, tastes, etc.) among these agents. Thus, the primary means of communication among DMA's is through shared access to the common DMU knowledge base.

DMA's, however, can communicate directly with collaborating DMA's in terms of direct requests for a decision and the transmission of decision results. Direct DMA-DMA communications are discussed further below. In general, however, these consist of requests from a "higher-level" DMA (e.g., the stress manager) to a "lower-level" DMA (e.g., a project) for information (e.g., utility impacts of purchasing another car, as computed by the mobility project), and the passing of this information back to the requesting DMA. In general, direct DMA-DMA communications are not encouraged within the model for a variety of computational and behavioural reasons. In particular, note that projects do **not** communicate directly with one another: they are "self-contained" in their decision-making and neither "need" to know or "care" about information generated by other projects. That is, the interface between projects occurs at the "higher" levels of scheduling and stress management.

It is also the case that a DMA within one DMU does **not** communicate with any DMA belonging to another DMU (e.g., the household within which the person resides). All communications between DMU's must logically occur at the DMU level; that is, it is through the DMU agent that all information from "outside" the DMU passes (e.g., information from other household members). This information is then passed to projects and other within-DMU agents as required. In other words, a project does not interact with or act upon the world directly – its DMU agent does.

The DMA *per se* largely consists of the decision manager. The decision manager controls the execution of the DMA decision rules/algorithms and monitors the outcomes of its decisions. The decision manager runs continuously within (simulated) physical time. It is autonomous within the limits of its programming. That is, it can respond to new information as this becomes available, and it can change the outcome of its decisions (e.g., switch modes for the primary work chain) in the face of new circumstances.

Whether the DMA can "change its own programming" is a question for a more detailed model design than will be attempted within this chapter. It is also a function of the sophistication of our modelling and computational capabilities. It is conceivable, for example, that a project over time might alter its "utility function parameters" as the DMU's tastes and preferences change, or that a project "learns" over time how to execute its tasks more efficiently. Alternatively, it might be argued that a DMA's "programming" should only be changed by a higher-level agent (e.g., the stress manager might change a project's "rules" as a mechanism for reducing stress in a given area). Possible advantages of the latter approach may be improved computational efficiency (i.e., the lower-level agent is not constantly asking itself if it needs to change how it is doing things), as well as the ability for the higher-level agent to "consider" more factors in its decision-making (e.g., a change in one project's rules may be introduced so as to improve another project's performance -- something the individual project cannot, by definition "see").

The focus of this chapter is not on activity scheduling *per se*, and so relatively little will be said here concerning the scheduler DMA. Suffice to say, a number of activity scheduling models currently exist that fall within this category and are consistent with the conceptual framework presented in this chapter, whether or not they explicitly adopt the DMA class design. The key point is that scheduling is viewed as a dynamic, agent-based process in which activity episodes within some form of an agenda are inserted into a daily schedule for eventual execution.

The project concept has already been discussed in detail above. In terms of the overall conceptual structure, key points to note about the project DMA include:

- projects are instantiated by the resource manager;
- projects compute utilities and associated stresses, but the resource manager evaluates these stresses and acts upon them;
- projects generate activity episodes; these episodes are scheduled by the scheduler; and
- projects also must be able to respond to "what if" questions from both the scheduler and the resource manager when they are "mentally simulating" alternative courses of action (e.g., "What is the utility loss if this activity episode cannot be scheduled?").

Finally, key points to note about the stress manager include the following.

- Each project monitors its "stress" level, however stress is defined. These stresses are passed to the stress manager, which compares them with threshold values.
- The stress manager resides outside of the project, since stresses across the projects must be considered together, and action within one project may well alter the stress in another.

- It is the stress manager that instantiates (and terminates) all projects.
- The stress manager asks "what if" ("mental simulation") questions of projects ("What if we bought a new car?" "What if I took the new job?").
- The stress manager "authorizes" a search for new resources since it must consider the impacts of identified alternatives "across the board". That is, as with other projects, the stress manager must instantiate a "search project" in order for the DMU to participate in a given market activity. The search for and evaluation of options, however, must occur within the project, since it has the knowledge to undertake this.

Adaptation occurs at the stress manager level given that it is able to instantiate and terminate projects in response to changing conditions/stresses. One can argue, however, that, for "true" adaptation to occur, project rules, etc. need to be modifiable over time as well (in response to new information, adaptive learning, etc.). The stress manager is envisioned to have the "power" to change projects and to (perhaps) change scheduling rules. The question arises as to whether it is possible for the stress manager itself to adapt; that is, for its "programming" to change. If one accepts the argument that only a "higher-level" agent can change the programming of a (lower-level) agent, this would imply the need for yet a higher-level "consciousness" within the DMU that is capable of changing the stress manager over time. Alternatively, one might conceive of a stress manager that is sufficiently self-reflective that it might be able to "change itself". Given the rudimentary state of our current capabilities for modelling learning and adaptation, however, it is perhaps not necessary to pursue this train of thought too much further at this stage of model development! The important point for now is that the proposed framework is at least in a provisional way starting to be explicit concerning learning/adaptation and is conceivably extendable in this regard as need and modelling capabilities increase.

The Household DMU

To this point in the discussion relatively little has been said about the household *per se*, except to claim that the overall decision-making framework is "household-based", and to identify the household as a DMU sub-class. To a large extent the discussion concerning DMU's, and their DMA's (stress manager, scheduler, and projects), holds at the household level as well as the person level. Indeed, this is one of the strengths of the DMU concept, that it applies equally well at either level. The household is, however, also different from the person in numerous ways. Most notably, it is not a biological, sentient being, but rather a social unit – a collection of logically related individuals (often, but not always, a family). As such, households do not "do" anything; it is their members that undertake activities "on behalf" of the household.

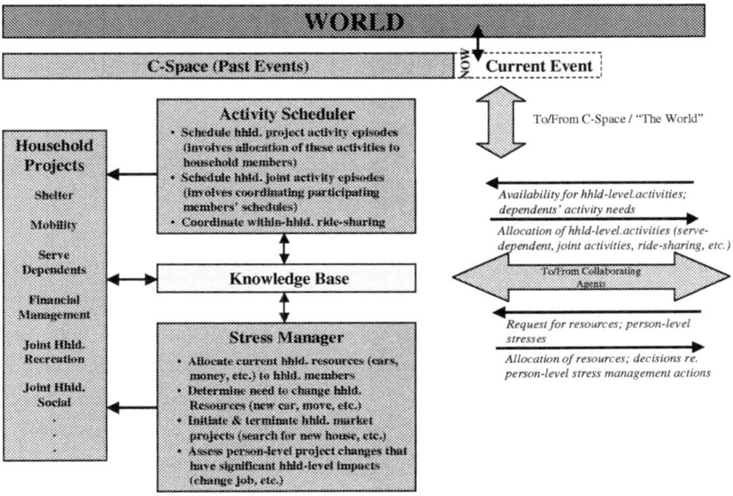

Figure 9.5
The Household DMU

Nevertheless, it is very convenient from a modelling point of view to treat the household as a sentient, wilful agent that is, in certain circumstances, able to "act" directly into the world, rather than have to model the actions of the individual household members who actually undertake these acts. A typical example of this is residential search. While it is individual household members who actually inspect vacancies, make bids on houses, sign purchase agreements, etc., it is far simpler, as well as sufficient for modelling purposes, to think of "the household" doing these activities directly.

Figure 9.5 sketches the household DMU and its key functions. As indicated in this figure, the household scheduler is responsible for controlling the scheduling of household-level activities (i.e., those activities episodes that are generated by household projects). This involves coordinating with the schedulers of household members to determine feasible allocations of these activities among the person-level provisional schedules. Thus, the household scheduler is quite different in its internal workings than the person scheduler. Types of household activities include: *joint* activities, in which two or more household members jointly participate; *serve-dependent* activities, in which an *independent* member (e.g., an adult) assists a *dependent* member (e.g., a child) undertake an activity; *ride-sharing*, in which one household member drives one or more other members to their activity location; and *individual* activities, in which a single household member undertakes a household-level activity (e.g., mow the lawn).

Most resource management of interest within an integrated urban model (houses, cars, etc.) occurs at the household level. Not only does the household allocate existing household resources (e.g., which driver gets the single family car in the case of a scheduling conflict), but it is the household that must decide when and how these commonly held resources should be changed in some way (e.g., move to a new place of residence). Further, certain person-level decisions that have significant impact on household-level resources (e.g., quit one's job) should be "vetted" at the household level in order to assess the overall ramifications of this decision.

Modelling household-level decision-making is not a well-developed field, especially in terms of modelling either inter-personal conflict resolution or inter-personal collaboration in joint activities. In the travel demand field the vast majority of models are person-based, with, at most, a few household-level variables included in the model in an attempt to capture (inevitably in a rather *ad hoc* fashion) at least some household-level effects. Examples of studies involving explicit household-level decision-making include: Miller and Rhamey (1987), Golob (1999), Scott (2001), Gliebe and Koppelman (2002), and Zhang *et al.* (2002). As indicated by the dates on most of these references, most of household-based work is very recent. In addition, residential location choice models within existing land use models are household-based, although in all such cases the household "agent" is not a very well defined object. A considerable literature in *household economics* also exists that might prove useful in developing operational models of household DMU's. Building upon the seminal work of Becker (1965), various researchers have attempted to build micro-economic models of household interactions and decision-making, recent examples of which include Manski (2000) and Ermish (2003). Overall, however, it is fair to say that much work remains to be done before satisfactory operational dynamic, household-level decision-making models of the sort envisioned within this chapter have been achieved.

Summary Comments and Future Work

This chapter has sketched a comprehensive model of an integrated model of person- and household-level decision-making. The model is continuous in time and deals with both "short-run" and "long-run" decisions and actions on the part of households and their individual members. It is comprehensive in that all person- and household-based actions are potentially included within its framework. The model is event-driven in that decisions always occur in response to events (including prior decisions which are, themselves, events). A fundamental organizing principle of the model is the concept of the project, which is defined as a logically connected set of activities. All activities are "generated", in the first instance, out of projects.

The fundamental unit of analysis is the decision-making unit (DMU). Both persons and households are DMU sub-classes. A DMU is an intelligent agent that is able to: perceive the world around it; acquire and use resources; schedule its activities; and act into the world. A DMU further sub-divides into three decision-making agents (DMA): an activity scheduler; a stress (resource) manager; and a set of project task managers. These DMA's interact to determine the DMU's behaviour. They share (and communicate through) a common DMU knowledge base. The DMU interacts with collaborating DMU's and the world at large via a perceptual interface that handles all inter-agent information exchange and filtering.

Research is ongoing at the University of Toronto to test this conceptual framework within operational models. The Travel and Activity Scheduler for Household Agents (TASHA) represents a prototype implementation of the activity scheduler DMA, as well as project-based activity agenda formation, that is undergoing continuous testing and elaboration (Miller and Roorda, 2003; Roorda *et al.*, 2004; Miller *et al.*, 2005). The Integrated Land Use, Transportation, Environment (ILUTE) model (Miller and Salvini, 2001; Salvini, 2003; Salvini and Miller, 2004) similarly represents an ongoing research effort to develop an integrated, comprehensive micro-simulation model of urban spatial processes (within which TASHA is but one component) using both person and household agents consistent with the conceptual framework presented in this chapter. In particular, stress-based resource management (residential mobility decisions; household auto transactions) is being explicitly explored and tested.

ACKNOWLEDGEMENTS

The research presented in this chapter was funded by a Social Sciences and Humanities Research Council (SSHRC) Major Collaborative Research Initiative (MCRI) grant. Contributions to the ideas expressed in the chapter from Juan Antonio Carrasco, Sean Doherty, Marek Litwin, Matthew Roorda, and Paul Salvini are gratefully acknowledged.

REFERENCES

Arentze, T.A. and H.J.P. Timmermans (2000). *Albatross: A Learning-Based Transportation Oriented Simulation System.* EIRASS, Eindhoven.

Axhausen, K. (1998). Can we ever obtain the data we would like to have?. In: *Theoretical Foundations of Travel Choice Modelling* (T. Gårling, T. Laitila and K. Westin, eds.), pp. 305-324. Pergamon, Oxford.

Becker, G.S. (1965). A theory of the allocation of time. *Ec. J.*, **75**, 493-517.

Ermisch, J.P. (2003). *An Economic Analysis of the Family*. Princeton University Press, Princeton.

Gliebe, J.P. and F.S. Koppelman (2002). A model of joint activity participation. *Transpn.*, **29**, 49-72.

Golob, T.F. (1999). A simultaneous model of household activity participation and trip chain generation. *Transpn. Res.*, **34B**, 355-376.

Litwin, M.S. and E.J. Miller (2004a). Event-driven time-series based dynamic model of decision making processes. *Proc. 82nd Ann. Meet. Transpn. Res. B.,* Washington, D.C. (CD-Rom).

Litwin, M.S. and E.J. Miller (2004b). Agenda formation: evolution of activity sequencing within an event-driven time-series based framework. *Proc. Conf. on Progress in Activity-Based Analysis*, Maastricht, The Netherlands (CD-Rom).

Manski, C.F. (2000). Economic analysis of social interactions. *J. Ec. Persp.*, **14**, 115-136.

Maslow, A.H. (1970). *Motivation and Personality*. Second Edition, Harper & Row, New York.

Miller, E.J. (2004). Propositions for modelling household decision-making. In: *Behavioural Foundations of Integrated Land-use and Transportation Models: Assumptions ad New Conceptual Frameworks* (M. Lee-Gosselin and S.T. Doherty, eds.), Elsevier, New York, forthcoming.

Miller, E.J., D.S. Kriger and J.D. Hunt (1998). Integrated Urban Models for Simulation of Transit and Land-Use Policies. Final Report, Transit Cooperative Research Project H-12. TRB, Washington, D.C., Web Document 9 at www4.nas.edu/trb/crp.nsf.

Miller, E.J. and W.R. Rhamey (1987). Modelling household mode choice. *Proc. 4th World Conf. Transp. Res.*, pp. 1551-1569.

Miller, E.J. and M.J. Roorda (2003). Prototype model of household activity-travel scheduling. *Transpn. Res. Rec.*, **1831**, 114-121.

Miller, E.J., M.J. Roorda and J.A. Carrasco (2005). A tour-based model of travel mode choice. *Transpn.*, forthcoming.

Miller, E.J. and P.A. Salvini (2001). The integrated land use, transportation, environment (Ilute) microsimulation modelling system: description and current status. In: *The Leading Edge in Travel Behaviour Research* (D. Hensher, ed.), pp 711-724. Pergamon, Amsterdam.

Miller, E.J. and P.M. Sarjeant (1987). A model of continuous intra-urban residential search. *Proc. 34th N. Am. Meet. Reg. Sc. Assoc.*, Baltimore.

Roorda, M.J., S.T. Doherty and E.J. Miller (2004). Operationalizing household activity scheduling models: addressing assumptions and the use of new sources of behavioural data. In: *Behavioural Foundations of Integrated Land-Use and Transportation Models: Assumptions and New Conceptual Frameworks* (M. Lee-Gosselin and S.T. Doherty, eds.), Elsevier, New York, forthcoming

Rossi, P.H. (1955). *Why Families Move*. Free Press, Glencoe, Ill.

Salvini, P.A. (2003). Design and Development of the ILUTE Operational Prototype: A

Comprehensive Micro-Simulation Model of Urban Systems. Ph.D. thesis, Department of Civil Engineering, University of Toronto, Toronto.

Salvini, P.A. and E.J. Miller (2004). ILUTE: an operational prototype of a comprehensive microsimulation model of urban systems. In: *Netw. Spat. Econ.*, forthcoming.

Sarjeant, P.M. (1986). A Continuous Search Model of Residential Mobility Decision Making. M.A.Sc. Thesis, Department of Civil Engineering, University of Toronto, Toronto.

Scott, D.M. (2001). Toward an Operational Model of Daily Household Activity-Travel Behavior. Ph.D. Thesis, School of Geography and Geology, McMaster University, Hamilton.

Timmermans, H.J.P. (2003). The saga of integrated land use-transport modelling. *Proc. 9th IATBR Conf.,* Lucerne (CD-Rom).

Zhang, J., H.J.P. Timmermans and A.W.J. Borgers (2002). Utility-maximizing model of household time use for independent, shard and allocated activities incorporating group decision mechanisms. *Transpn. Res. Rec.*, **1807**, 1-8.

10

STRATEGIES FOR RESOLVING ACTIVITY SCHEDULING CONFLICTS: AN EMPIRICAL ANALYSIS

Matthew Roorda, University of Toronto, Toronto, Canada

Eric J. Miller, University of Toronto, Toronto, Canada

INTRODUCTION

Modelling the process of activity scheduling has been difficult to do because there have been relatively few reliable data sources through which the day-to-day scheduling decision making process has been observed. Conflict resolution, the process of deciding what to do when multiple activity opportunities are available at the same time, requires information about the activity schedule before and after the conflicting opportunities arose. With new data from an activity scheduling panel survey in Toronto, we now have an opportunity to better understand conflict resolution outcomes, and use that understanding to improve models of this process.

There has been debate about the modelling of activity scheduling in the travel behaviour research community. Two methods of modelling activity schedules have emerged. The econometric modelling approach, which is almost always based on random utility maximization, has been applied to individual components of activity and travel behaviour, including activity frequency, task allocation, activity duration, departure time, travel party and trip chaining (see Arentze and Timmermans, 2000 for a comprehensive review). Models that are more comprehensive have also been developed based on random utility maximization, in which choices are made between

Progress in Activity-Based Analysis edited by H. Timmermans

alternative patterns of activity/travel (e.g. Jones *et al.*, 1983; Recker *et al.*, 1986a, 1986b; Kawakami and Isobe, 1990). These approaches have been criticised for their lack of an underlying behavioural theory (Gärling, 1994).

It has been argued that understanding the underlying process of activity scheduling is crucial for a more accurate prediction of activity travel patterns (Pas, 1985; Jones *et al.*, 1990; Axhausen and Gärling, 1992; Lee-Gosselin, 1996; Axhausen, 1998; Bhat and Lawton, 2000). In response to this view, computational process models have been proposed as an alternative to random utility maximization approaches. Computational process models conceptualize choices as an outcome of a set of rules, or heuristics. Examples of rule-based computational process models include SCHEDULER (Gärling *et al.*, 1989), SMASH (Ettema *et al.*, 1993), AMOS (Pendyala *et al.*, 1995, 1998) and *Albatross* (Arentze and Timmermans, 2000, 2004). The Travel-Activity Scheduling model for Household Agents (TASHA) is a computational model of activity scheduling that is the subject of research at University of Toronto (Miller and Roorda, 2003; Roorda *et al.*, 2005).

One of the key difficulties in developing a credible rule-based model of activity scheduling is that it is difficult to observe the scheduling process. Indeed, an observed pattern of activities and travel can be the outcome of countless long-term and short-term decisions, many of which are done subconsciously. Hence, rule bases for activity scheduling cannot adequately be assessed with traditional activity-based surveys that focus only on the characteristics of the final executed schedule. Observation of the *process* of activity scheduling is required to extract rules or strategies employed by individuals. A number of methods have been attempted to observe this process and to try to develop decision-making rules appropriate to the activity scheduling problem. Examples include MAGIC (Ettema *et al.*, 1994) and CHASE (Doherty and Miller, 2000).

This paper reports on an attempt to use data from the first wave of a multiple instrument panel survey conducted in Toronto (Roorda and Miller, 2004) to assess rules for activity rescheduling in response to scheduling conflicts. The CHASE survey instrument employed in the first wave (Doherty *et al.*, 2004) resulted in revealed activity rescheduling scenarios. Two types of rules are considered in the analysis. First, the concept of activity precedence is defined and analyzed; does activity precedence play an important role in resolving conflicts, is activity type a valid measure for activity precedence, and if so, what kinds of activities are more likely to be modified or deleted when a scheduling conflict occurs? Second, strategies for rescheduling of the activities are assessed. Given that a conflict has occurred, is the activity moved to another time in the same day, is it moved to another day, or is the activity skipped altogether? The intention of this analysis is to provide an empirical basis for enhancing the system of rules used in the prototype TASHA model of activity and travel scheduling.

CONCEPTS OF ACTIVITY PRIORITY, PRECEDENCE AND CONFLICT RESOLUTION FOR TASHA

TASHA models the process of schedule building by:

1. Generating activities with attributes based on empirical distributions,
2. Inserting those activities into project agendas, and
3. Constructing person schedules by moving activities from project agendas to be inserted into person schedules

The process of insertion in steps 2 and 3, above, can result in scheduling conflicts when two activities are generated at overlapping times. Conflicts arise when a "competing" activity, which is a new activity being inserted into the schedule, overlaps in time with an "original" activity, which already exists in the schedule. Three different classes of conflict arise, as shown in Figure 10.1:

Class 1 – A competing activity being added to the schedule is added within an original activity
Class 2 – A competing activity being added partially overlaps one or two original activities
Class 3 – A competing activity completely overlaps one or more shorter original activities

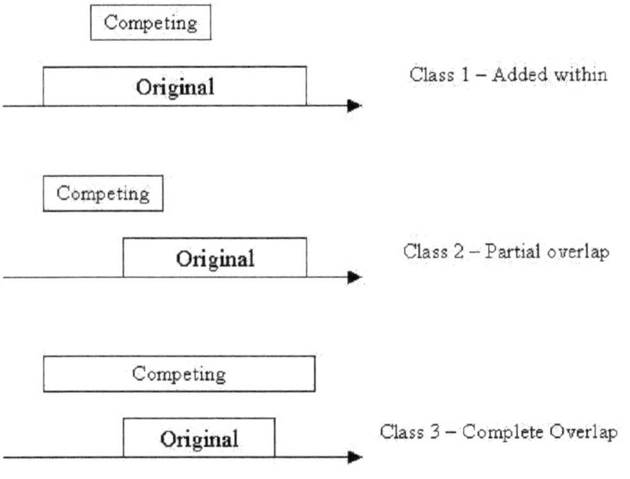

Figure 10.1
Classes of Activity Conflict

The concepts of "activity priority" and "activity precedence" play a large role in the current implementation of TASHA. Activities are moved from project agendas to form person schedules in order of priority/precedence (until the writing of this paper, the two terms had not been differentiated in TASHA), which were assumed to be linked to broad activity type. The order in which activities are added to the schedule clearly has an influence on the predicted activity schedule; the process is path-dependent.

Given our current goal of testing the assumptions made in TASHA, it is important to clarify our conceptualizations of priority and precedence. Priority is a term that holds connotations of importance, of "utility" (the satisfaction or benefit one obtains by participating in an activity) and of the degree of commitment to other parties. Precedence, on the other hand, is the degree to which an activity is planned at an earlier point in time than other activities. An activity's precedence may be related to its place in a "normal routine" (activities that are usually done at around the same time and place without thinking too much about their planning), and its "fixity" (the extent to which the attributes of the activity may not be changed, once they are planned), in addition to any influence of the activity's utility/importance.

Examples of activities with high and low levels of priority and precedence are shown in Table 10.1. It is noted that the fixity of an activity in time and space (i.e. the extent to which an activity is non-flexible) is **not** considered here to be an element of priority. In fact, high priority activities may be very flexible. Work might be have a high priority (important, involving commitment), yet one may indeed have the option to work at times and locations of one's choosing. A highly non-flexible activity might register very low in terms of priority if the activity is considered unimportant (for example, a one-time showing of a boring movie at 7:00pm at the theatre). Clearly flexibility can be related to commitments and contracts one has with another party regarding a particular activity, but it is not fixity *per se* that determines the priority of the activity.

Table 10.1
A Typology of Priority and Preference

	High Precedence	**Low Precedence**
High Priority	Preplanned, high utility activities, especially with commitment to others (e.g. doctors appointment)	Spontaneously planned, high utility activities, especially with commitment to others (e.g. pick up sick child from school)
Low Priority	Routine or preplanned activities with less utility, and less commitment to others (e.g. watch favourite TV show)	Spontaneously planned activities with less utility and less commitment to others (e.g. shopping at cornerstore for a magazine)

Similarly, the flexibility of an activity can be distinguished from its precedence, although the two may be related. One might have, as part of one's regular routine, a stop at the coffee shop on the way home from work. Yet, this may be an entirely flexible activity; one could just as easily go for coffee at a different time or location, with little consequence. When activities conflict, elements of priority, precedence and flexibility all play a part in determining the scheduling outcome. Intuitively, activities that have lower priority and higher flexibility are more likely to be adjusted than fixed, high priority activities. Furthermore, those activities that have higher precedence are more likely to be modified simply because they are already part of a schedule into which more spontaneous activities must fit. *Ceteris paribus,* activities with high precedence are more likely to be planned in advance, they are more likely to be the "original" activity in a scheduling conflict, and they are more likely to be modified or skipped, particularly if they are flexible or low priority activities. Priority of an activity is a latent variable. It is very difficult, if not impossible, to measure and no empirical analysis of activity priority can be provided in this analysis. Precedence is measurable using CHASE: we observe the time when each activity was entered into the schedule. However, measures of precedence are not available in traditional activity- or trip-based surveys, on which models such as TASHA are based. Therefore, we need to assess whether another activity attribute, such as broad activity type is a sufficient measure of precedence to explain the outcomes of scheduling conflicts (and thus is appropriate as clear rule for scheduling), or whether a more complex treatment is required.

TASHA, or any rule-based model which attempts to predict scheduling/rescheduling behaviour, must also represent the process of conflict resolution. Once it is decided which of two conflicting activities is displaced, a course of action must be chosen. Should the activity be moved to another time in the same day, should it be moved to another day, should the duration be shortened, should the activity be split, or should it be skipped altogether? The rules for activity rescheduling in the prototype implementation of TASHA are numerous and depend on the conflict class and the availability of nearby "gaps" in the schedule. In cases of conflict, the lower priority activity is shifted to an adjacent gap in the schedule, if one exists. If a gap does not exist, then one is created by shifting the adjacent activity provided there is sufficient "room" in the schedule to take place without significantly reducing activity durations. Once an activity is successfully added to a schedule in TASHA, it is never subsequently deleted or moved to another day.

DATA

The first wave of a three-wave panel survey undertaken in Toronto was designed to assess the mechanics of schedule building with particular focus on how scheduling conflicts are resolved as they arise. The panel survey was conducted with an initial sample of 270 households in the Toronto Area at one-year intervals over the period from 2002 to 2005. The panel survey is used to collect core data, which are common data gathered for every wave in the panel, and additional data elements, which are unique to each wave. Core data include information about the residence, the household structure, attributes of the people in the household, and the vehicles and other modes of transportation available to the household. In all waves, a minimum 2-day activity schedule is collected for multiple household members including:

- All activities completed for all participating adult household members,
- Activity description,
- Start time, duration, and location,
- Mode(s) of transportation, estimated travel time, and passengers in the vehicle,
- Other people involved in the activity, and
- Children under the respondent's care at the time of the activity

In the first wave, a seven-day computerized activity diary is self-reported by the respondent in customized scheduling software entitled CHASE© (Doherty and Miller, 2000). The application of the CHASE survey process to the first wave of the Toronto Area Panel Survey has been documented in detail by Doherty *et al.* (2004).

In short, the CHASE software records schedule adjustments made by respondents over the week, including activity additions, deletions and modifications. An activity addition in CHASE occurs when respondents enter an activity into their schedule for the first time. An activity modification occurs when one or more of the attributes of an activity in the schedule are changed, including changes to start time, duration, mode, location, or the other persons with home the activity is done. Deletions refer to activities that are removed from the 7-day schedule entirely. Deletions can include activities that are moved to another week or are done by another person.

ANALYSIS OF RESCHEDULING RULES

Summary of Activity Scheduling

The CHASE data can be summarized as follows:

Number of responding households:	264*
Number of responding persons:	423*

Total activity operations in 7-day schedule:	40756	(13.8 operations/person-day)
Total number of activity additions	35644	(87.5%)
Total number of activity deletions	753	(1.8%)
Total number of activity modifications	4359	(10.7%)
Total number of executed activities	34880	(11.8 activities/person-day)

* 6 households and 30 persons were eliminated from this analysis due to poor data quality

Of the deleted activities, 465 (62%) could be linked with an activity of the same type at the same location that was "added" to the schedule. If the addition was entered into the computer within one hour of "deletion" operation, the two activities are considered to be linked (and is treated as a modification).

One of the drawbacks of the CHASE data is that only a subset of possible activity conflicts and the resulting modifications and deletions are observed. Consider the following example. A friend calls to go for dinner with a CHASE respondent on Friday evening, but she had already planned to go to a movie with her brother at that time. Several solutions to this scheduling conflict exist, only some of which would be observed in the CHASE database. If the respondent decided not to go for dinner with the friend, then the activity would not have been entered into the CHASE database. If she decided to have dinner earlier than suggested, then the activity would have been entered into the CHASE database at a different time (likely into a gap in her schedule), but would not have been captured as a conflict. The conflict would only have been captured in CHASE if the movie was rescheduled or skipped completely because of the conflict and the dinner plan was entered into the schedule. Indeed, it would be very difficult to try to model all potential activity conflicts, since one is arguably screening a constant stream of possible opportunities subconsciously.

It is also important to note that not all activity modifications are made as the result of a conflict with another higher priority activity. Activities can be modified because of traffic delays, changes to plans made by other people, or simple adjustments in the attributes of the activity (e.g. "I had nothing else planned so I spent an extra hour studying econometrics at the library"). Yet, the concern for this research is on how scheduling conflicts are handled.

The identification of activity conflicts in the CHASE database is not straightforward, and requires some assumptions about what constitutes a conflict. For the purposes of this analysis, the following criteria are used to define a conflict:

- The competing activity overlaps in time with the original activity in a manner described by one of the conflict classes shown in Figure 10.1.

- The competing activity is entered by the respondent at a point in time after the initial entry of the original activity,
- A valid adjustment was made to the original activity such that the conflict was resolved. Valid adjustments included: moving the activity to another day, shifting the activity to another part of the day such that there was no overlap with the competing activity, or deleting the activity outright.
- The adjustment to the original activity is entered by the respondent not more than one hour before the entry of the competing activity.

The total number of conflicts discovered in the CHASE database is as follows:

Class 1 – (added within)	1023
Class 2 – (partial overlap)	445
Class 3 – (complete overlap)	449
Total	1917

Of these 1917 conflicts, there were cases where a competing activity conflicted with more than one original activity and cases where more than one competing activity conflicted with one original activity. The total number of activities represented in our database of conflicts is shown below:

Total number of competing activities	1678
Total number of original activities	1279
Total number of conflicts	1917

Table 10.2
Original Activities Ordered by Probability of Modification / Deletion

Activity Group	Total Activity Additions		Total Number of Conflicts		Total Number of Original Activities		Prob. of Displacement
Work/School	3390	9.5%	423	22.1%	225	17.6%	6.6%
Drop-off/Pick-up	1726	4.8%	91	4.7%	73	5.7%	4.2%
Recreation/Entertainment	7079	19.9%	402	21.0%	277	21.7%	3.9%
Household Obligations	4985	14.0%	227	11.8%	168	13.1%	3.4%
Social	1818	5.1%	115	6.0%	58	4.5%	3.2%
Services	805	2.3%	39	2.0%	25	2.0%	3.1%
Basic Needs	13906	39.0%	555	29.0%	405	31.7%	2.9%
Shopping	1240	3.5%	48	2.5%	35	2.7%	2.8%
Other	695	1.9%	17	0.9%	13	1.0%	1.9%
Total	35644	100.0%	1917	100.0%	1279	100.0%	3.6%

Table 10.2 shows the proportion of "original" activities that are displaced by another activity with which it is in conflict. It shows that work/school, drop-off/pickup and recreation/entertainment are the activities that are most likely to be displaced. Overall, 3.6% of activities were displaced by another activity (recognizing that this is an underestimate of all possible conflicts, as described on the previous page). These activity modifications or deletions represent about 1/3 of all activities that are modified at least once, which indicates that many modifications are made due to changing opportunities or constraints that do not relate to other conflicting activities.

Assessment of Activity Precedence

One way of assessing the precedence of an activity is to compare the number of conflicts in which activities of the same type are the competing activity (i.e. doing the displacing) to the number where they are the original activity (i.e. being displaced). Table 10.3 shows the number of scheduling conflicts that occur between competing and original activities, grouped by broad activity type. If activities are ordered from highest to lowest precedence in this matrix, then the entries above the main diagonal are conflicts where the lower precedence activity (which are entered later) displaces a higher precedence activity (which is entered earlier). Those elements below the main diagonal are conflicts where a higher precedence activity displaces one with lower precedence. If activity type was a perfect descriptor of precedence, then the activities types could be arranged such that the lower half of the matrix would be zeros. In this case, a simple rule could be developed for the order that activities are added into the schedule.

In Table 10.3, activities are ordered optimally, to maximize the sum of the elements above the diagonal. Yet, 531 (28%) of conflicting activities remain below the main diagonal, indicating that there is significant room for improvement in assessing precedence. We also recognize that for the 428 (22%) of conflicting activities on the diagonal, with two activities of the same type in conflict, we would need additional rules, using attributes other than activity type to specify which activity has higher precedence. In an attempt to improve this result, the matrix was cross-classified with sex, and subsequently with income. For each classification, optimal precedence rankings were developed and the number of "violations" was assessed, as shown in Table 10.4. Two observations can be made about this table. First, the precedence rankings change slightly for different groups of people, but major differences are not evident. Work/school does appear as the highest precedence activity for all groups of people regardless of income or gender. On the other hand, more "discretionary" activities such as shopping, services and other activities are consistently found to have low precedence. As the number of observations decreases within a particular group of persons the data within the matrix become more sparse, leading to less confidence in the resulting ranking.

Table 10.3
Optimal Precedence Ranking for Conflicting Activities

Rank	Activity Group – Original Activity	Activity Group - Competing Activity: Work/ School	Basic Needs	Recr/ Entert.	Drop-off/ Pick-up	Social	Household Obligations	Services	Other	Shop-ping	Total
1	Work/School	49	113	92	57	22	33	23	6	28	423
2	Basic Needs	43	215	115	13	48	68	24	20	9	555
3	Recr/Entert.	39	108	82	25	33	46	22	23	24	402
4	Drop-off/Pick-up	24	18	8	4	5	17	3	4	8	91
5	Social	9	30	24	2	18	22	2	4	4	115
6	Hhld Obligations	32	43	38	13	18	48	13	6	16	227
7	Services	4	5	3	2	2	8	6	1	8	39
8	Other	1	6	6		1	1	0	1	1	17
9	Shopping	6	7	14	1	3	8	3	1	5	48
	Total	207	545	382	117	150	251	96	66	103	1917

Shading indicates for each pair of activity groups, the "competing"/"original" ordering observed more frequently

Total number of entries where the lower precedence activity displaces the higher precedence activity	958	50.0%
Total number on the diagonal (same activity group)	428	22.3%
Total number of entries where the higher precedence activity displaces the lower precedence activity	531	27.7%

Some of the results that look somewhat anomalous for low-income groups (note the positions of household obligations and other activities) are partially explained by the low number of total conflict observations for low-income households (36 and 17 for these two activity types, respectively). Overall, evidence in Table 10.4 shows that, in terms of activity precedence rankings, the differences between people of different genders and incomes are minor and that the use of a single rule base for different people may be an appropriate simplification.

Second, with a single rule base for activity precedence, we observe violations at a rate of 27.7%. Yet, by disaggregating by sex or income, we can only obtain marginal improvements in this violation rate (to 27.4% and 25.7%, respectively). While the optimal rule base does explain the majority of choices, it is clear that precedence lists based only on broad activity type are not sufficient to fully predict the order in which activities are added to the schedule. To fully explain the outcome of scheduling conflicts, a more sophisticated measure for activity precedence is required. Two methods are suggested for further research based on these results.

(a) An improved measure of activity precedence could be developed that is a function of activity type and other key attributes that are elements of precedence. These elements could include the level of commitment to other people, the degree of pre-planning associated with this activity, the difficulty associated with rescheduling the activity, and so on. Such an improved measure could be used to develop better rules for predicting the outcome of scheduling conflicts.

Table 10.4
Optimal Precedence Rankings by Age and Household Income

		Gender		**Income**		
	All Conflicts	**Males**	**Females**	**Low Income**	**Medium Income**	**High Income**
				(<=$35K Can)	**($35-60K Can)**	**(>60K Can)**
Optimal Precedence Rankings (Rule: If activities higher in the list conflict with activities lower in the list, those lower in the list are modified or deleted)	Work/School	Work/School	Work/School	Work/School	Work/School	Work/School
	Basic Needs	Basic Needs	Recreation/ Entertainment	Recreation/ Entertainment	Recreation/ Entertainment	Basic Needs
	Recreation/ Entertainment	Recreation/ Entertainment	Drop-off/ Pick-up	Other	Drop-off/ Pick-up	Recreation/ Entertainment
	Drop-off/ Pick-up	Household Obligations	Basic Needs	Basic Needs	Basic Needs	Drop-off/ Pick-up
	Social	Drop-off/ Pick-up	Social	Drop-off/ Pick-up	Household Obligations	Social
	Household Obligations	Social	Household Obligations	Shopping	Social	Household Obligations
	Services	Shopping	Services	Social	Services	Other
	Other	Other	Shopping	Services	Shopping	Shopping
	Shopping	Services	Other	Household Obligations	Other	Services
Total Conflicts	1917	644	1228	327	573	950
Total Rule Violations	531	148	365	77	145	254
% Rule Violations	27.7%	23.0%	29.7%	23.5%	25.3%	26.7%
Overall % Rule Violations	27.7%	27.4%			25.7%	

b) There are some attributes of activities that have an influence on the activity's precedence that cannot be observed. Currently, with a simple specification of precedence based only on broad activity type, we are able to explain precedence with a rule violation rate of 27.7% (although we recognize that when two activities of the same type are in conflict, our rule base does not make any prediction about precedence). While the violation rate could be improved with a better specification of precedence, uncertainty will always exist. This uncertainty could be incorporated into the measure of precedence by means of an error term, such that the rule base for activity scheduling/rescheduling becomes more stochastic in nature.

Strategies Employed to Resolve Scheduling Conflicts

In the CHASE database, conflict resolution strategies could be determined by observing the modification or deletion of the original activity that was displaced by the competing activity. A summary of the strategies used to resolve each of the conflicts is shown in Table 10.5.

Table 10.5
Strategies Used to Resolve Conflicts

Description of Strategy	Total Number of Conflicts		Total Number of Original Activities	
Modify activity within the same day	1297	67.7%	888	69.4%
Shorten duration of activity	493	25.7%	330	25.8%
Shift activity to another part of the day	90	4.7%	75	5.9%
Shift and shorten duration of activity	204	10.6%	118	9.2%
Shift and lengthen duration of activity	110	5.7%	85	6.6%
Split the activity	400	20.9%	280	21.9%
Move activity to another day	233	12.2%	157	12.3%
Skip activity	387	20.2%	234	18.3%
Other	0	0.0%	0	0.0%
Total	1917	100%	1279	100%

It is noted that in some cases, some judgement was required to properly classify the conflict. The first source of ambiguity existed when a respondent deleted an activity, and some time later added another activity of the same type at the same location. If the respondent made the addition within one hour of the deletion, the two operations were considered a single modification; otherwise, the deletion and addition were assumed unrelated. Similarly, an activity was assumed to be "split" into two activities if the original activity was shortened, and a new activity of the same type was added at the same location such that the result resembled a split activity.

Table 10.6
Conflict Resolution Strategies by Conflict Class

Description of Strategy	Wave 1 (Revealed Response)							
	Class 1 - (added within)		Class 2 - (partial overlap)		Class 3 - (complete overlap)		Total	
Modify activity within the same day	831	81.2%	292	65.6%	174	38.8%	1297	67.7%
Shorten duration of activity	337	32.9%	156	35.1%	0	0.0%	493	25.7%
Shift activity to another part of the day	8	0.8%	24	5.4%	58	12.9%	90	4.7%
Shift and shorten duration of activity	95	9.3%	57	12.8%	52	11.6%	204	10.6%
Shift and lengthen duration of activity	23	2.2%	26	5.8%	61	13.6%	110	5.7%
Split the activity	368	36.0%	29	6.5%	3	0.7%	400	20.9%
Move activity to another day	63	6.2%	58	13.0%	112	24.9%	233	12.2%
Skip activity	129	12.6%	95	21.3%	163	36.3%	387	20.2%
Other	0	0.0%	0	0.0%	0	0.0%	0	0.0%
Total	1023	100%	445	100%	449	100%	1917	100%

Table 10.7
Conflict Resolution Strategies by Activity Type

Activity Group – Original Activity	Move activity within the same day		Move activity to another day		Skip Activity		Total
Basic Needs	423	76.1%	68	12.2%	65	11.7%	556
Drop-off/Pick-up	25	31.6%	19	24.1%	35	44.3%	79
Household Obligations	147	63.6%	37	16.0%	47	20.3%	231
Other	19	48.7%	8	20.5%	12	30.8%	39
Recreation/Entertainment	258	64.7%	61	15.3%	80	20.1%	399
Services	19	70.4%	1	3.7%	7	25.9%	27
Shopping	27	55.1%	10	20.4%	12	24.5%	49
Social	83	67.5%	8	6.5%	32	26.0%	123
Work/School	296	71.5%	21	5.1%	97	23.4%	414
Total	1297	67.7%	233	12.2%	387	20.2%	1917

Most conflicts are resolved by shifting activities within the same day (68%). 12% are resolved by moving the activity to another day and the remaining 20% are resolved by deleting the activity altogether. Skipped activities do not imply that the activity never gets done. In fact, such an activity could possibly be done by the same person in a different week (which we do not observe in a one week survey), by another household or non-household member, or could be replaced by another type of activity that meets the same goal. A skipped activity, here, is defined as an activity that is not immediately replaced by another activity of the same type at the same location by the same person within the same week. The resolution of a conflict is related to the conflict class, as shown in Table 10.6. If a complete overlap (i.e. class 3) occurs, then the activity is moved to another day or skipped outright over 60% of the time. However, if the "competing activity" falls within the original activity, then the activity is shortened or moved within the same day over 80% of the time.

Some differences are also noticeable between resolution strategies for different activity types, as shown in Table 10.7. Activities that are most likely to be skipped include drop-off/pickup activities, shopping, social and other activities. These activities are largely discretionary activities, with the exception, perhaps, of the drop-off and pick up of children. For these activities, it would be likely that responsibility has been transferred to another person, or another strategy found to serve the children. With the exception of social activities, these discretionary activities are also highly likely to be moved to another day and somewhat less likely to be shifted around in the same day. Those activities that were more likely to have timing and/or duration changed within the same day included work/school, services and basic needs. Work/school and basic needs are more pre-planned in nature and tend to form the basic, routine skeleton for the schedule. Thus, fine-tuning, as more spontaneous activities are added to the schedule, would reasonably cause these changes.

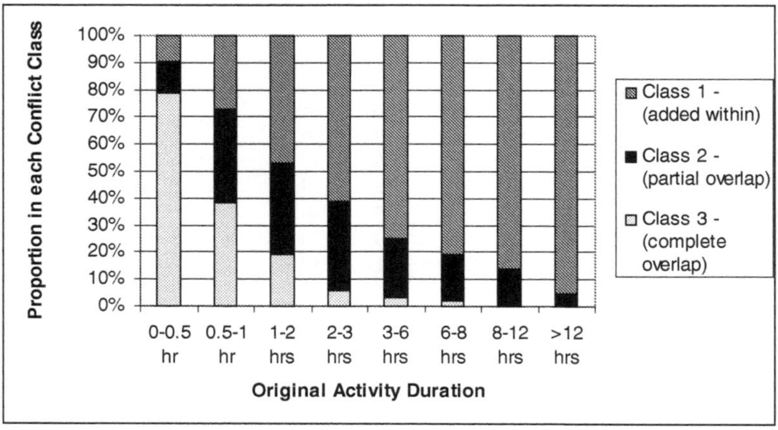

Figure 10.2

The Proportion of Conflict Classes by Original Activity Duration

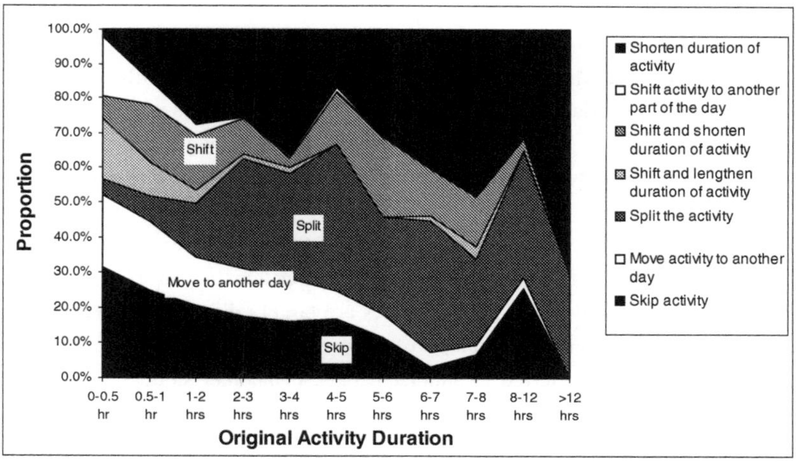

Figure 10.3

Conflict Resolution Strategies by Original Activity Duration

Clearly, duration of the activity must play a role in how activities are rescheduled. First, the duration of an activity affects the kinds of conflicts that occur with the activity. Figure 10.2 shows how short duration "original" activities are far more likely to be involved in conflicts with other activities that overlap them completely (i.e. Class 3 conflicts). As duration increases, the incidence of partial overlaps increases (i.e. Class 2 conflicts), and for "original" activities that are longer than 2 hours in duration, the majority of conflicts are with competing activities that fall entirely within the original activity (i.e. Class 1 conflicts).

The conflict resolution strategy is also related to the duration of the activity, as shown in Figure 10.3. As duration increases, opportunities to shorten the duration of or split the activity increase notably. Opportunities to shift the activity within the same day tend to decrease with increasing activity duration, especially when the activity's duration is not simultaneously shortened. The way that a conflict is resolved appears to be related, at least, to the kind of conflict, and characteristics of the activity, including its duration and the activity type. Yet each of these attributes are related to each other, which makes it difficult to sort out the true causal factors behind the choice of a conflict resolution strategy without a multivariate analysis. Furthermore, there are clearly other influences on the kinds of strategies chosen including the characteristics of the schedule (where are there "gaps" in the schedule that might accommodate a shifted activity?), and characteristics of the person (are certain persons more prone to fill their schedule, while others are more likely to reject opportunities that arise?). These influences are potential subjects for further research, and are not discussed in this paper. While analysis continues to sort out potential influences on rescheduling behaviour, it is certain is that no single attribute of the activity, the conflict, the schedule or the person can, by itself, explain the resolution strategy chosen.

Implications for the TASHA Scheduling Process Model

The prototype version of the Travel and Activity Scheduler for Household Agents (TASHA) assumes a set of rules for activity rescheduling in response to scheduling conflicts (see Miller and Roorda, 2003 for details). Some observations can be made about the appropriateness of these assumptions, in light of the empirical analysis provided in this paper. First, precedence rankings for activity types assumed in TASHA can be compared to the optimal precedence rankings found in the CHASE data. A comparison is shown in Figure 10.4. First, it is to be noted that the activity classifications used in TASHA are limited by what is available in the travel survey data upon which it is based, therefore, the activity classifications are not as precise as those found in CHASE, in particular for the "other" category.

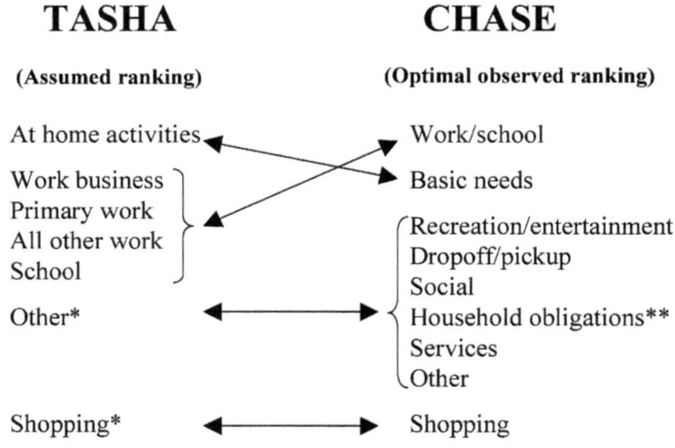

* In TASHA, a further differentiation is assumed between individual and joint shopping and other activities, with joint activities being assigned a higher precedence
** In CHASE, some household obligations done at home better correspond to "At home activities" in TASHA

Figure 10.4

Precedence Rankings Assumed in TASHA vs. those Observed in CHASE

Second, the currently assumed precedence rankings in TASHA correspond reasonably well to those found in the CHASE survey data. The clear exception is the number one ranking of at-home activities (the default activity in TASHA generated activity schedules, which in CHASE are found to have lower ranking than work/school. Scheduling strategies found in CHASE indicate that the TASHA model does not allow for the full range of observed rescheduling responses. Conflicts in TASHA are resolved by a) shortening the duration of the activity, b) shifting the activity to another part of the day, c) shifting and shortening the activity or d) splitting the activity. Once they have been added to the schedule, TASHA *does not* allow activities to be shifted and lengthened, moved to another day, or skipped or shifted to another person's schedule. These omissions indicate a need to extend the time horizon of the TASHA model from 24 hours to one week and to improve the representation of household interactions in the model to provide the appropriate structure to allow for these rescheduling responses.

CONCLUSIONS

Several conclusions can be made through the analysis of CHASE data from the first wave of the Toronto Area Panel Survey:

- It is feasible to observe conflict resolution strategies using revealed response data from the CHASE survey instrument.
- About one third of all activity modifications are made because of an activity conflict. Once an activity is entered into the schedule there is, overall, at least a 3.6% chance that it is subsequently modified or deleted because of a conflicting activity.
- Precedence, the degree to which an activity is routine or pre-planned, can be simply approximated using broad activity groups, yielding the following "optimal" precedence ranking: work/school, basic needs, recreation/ entertainment, drop-off/pickup, social, household obligations, services, other, shopping. This ranking is violated 27.7% of the time in the CHASE data.
- The precedence ranking is difficult to improve significantly by cross-classifying the data by sex or income, and the rankings themselves do not change very much across groups. Therefore, a single activity precedence ranking for all individuals may be an appropriate simplification.
- Assessment of the strategies used to resolve the conflict (once the displaced activity is chosen) shows that most conflicts (68%) are resolved by moving the activity within the same day. 12% are moved to another day and 20% are skipped, moved to another day outside the survey week or done by another person.
- Systematic differences in conflict resolution strategy can be found for different kinds of conflicts, for different activity types and for different activity durations. No single attribute of the activity, the conflict, the schedule or the person can, by itself, explain the resolution strategy chosen.
- Activity precedence rankings in TASHA are similar to those found in CHASE with the exception of in-home activities.
- A significant proportion of conflicts are resolved by moving activities to another day (12%), skipping activities (or shifting them to another week or another person) (20%), and shifting and lengthening activities (6%). TASHA's prototype version does not allow for these responses.

The following suggestions are provided for further research based on these results.

- An improved measure of activity precedence could be developed that is a function of activity type and other key attributes that are elements of precedence (e.g. level of commitment to other people, the degree of pre-planning, etc.). Such an improved measure could be used to develop better rules for predicting the outcome of scheduling conflicts.
- There are some attributes of activities that have an influence on the activity's precedence that cannot be observed, hence uncertainty will always exist in our measure of precedence. This

uncertainty could be incorporated into the measure of precedence by means of an error term, such that the rule base for activity scheduling/rescheduling becomes more stochastic in nature.

- This analysis of conflict resolution strategies has focussed on the influence of attributes of the activity and the nature of the scheduling conflict. Other influences include the characteristics of the schedule ("gaps" in the schedule that might accommodate a shifted activity?), and characteristics of the person (are certain persons more prone to fill their schedule, while others are more likely to reject opportunities that arise?). These should be further explored.
- The time horizon of the TASHA model should be extended from 24 hours to one week and the representation of household interactions in the model should be improved to allow for a full range of rescheduling responses.

ACKNOWLEDGEMENTS

The authors would like to express thanks to Juan Carrasco and Marek Litwin, who provided insights that were helpful in writing this paper, and to all those that participated in the Toronto Activity Panel Survey. Gratitude is also expressed for financial support from the Social Sciences and Humanities Research Council of Canada.

REFERENCES

Arentze, T.A. and H.J.P. Timmermans (2000). *Albatross: A Learning-Based Transportation Oriented Simulation System.* EIRASS, Eindhoven.

Arentze, T.A. and H.J.P. Timmermans (2004). A learning-based transportation oriented simulation system, *Transpn. Res. B*, **38**, 613-633.

Axhausen, K. (1998). Can we ever obtain the data we would like to have? In: *Theoretical Foundations of Travel Choice Modeling* (T. Gärling, T. Laitila and K. Westin, eds.), pp. 305-323. Pergamon, Oxford, UK.

Axhausen, K. W. and T. Gärling (1992). Activity-based approaches to travel analysis: conceptual frameworks, models, and research problems. *Transpn. Rev.*, **12**, 323-341.

Bhat, C.R. and K.T. Lawton (2000). *Passenger Travel Demand Forecasting.* Transportation Research Board, National Research Council (USA). Millennium Paper.

Doherty, S.T. and E.J. Miller (2000). A computerized household activity scheduling survey. *Transpn.,* **27**, 75-97.

Doherty, S.T., E. Nemeth, M.J. Roorda and E.J. Miller (2004). Design and assessment of the Toronto area computerized household activity scheduling survey. *Proc. 83rd Ann. Meet. Transpn. Res. B.*, Washington, D.C. (CD-Rom).

Ettema, D.F., A.W.J. Borgers and H.J.P. Timmermans (1993). Simulation model of activity scheduling behaviour. *Transpn. Res. Rec.*, **1413**, 1-11.

Ettema, D.F., A.W.J. Borgers and H.J.P. Timmermans (1994). Using interactive computer experiments for identifying activity scheduling heuristics. *Proc. 7th Int. Conf. on Trav. Beh.*, Santiago, Chile.

Gärling, T. (1994). Behavioural assumptions overlooked in travel choice modelling. *Proc. 7th Int. Conf. on Trav. Beh.*, Santiago, Chile.

Gärling, T.A., K. Brannas, J. Garvill, R.G. Golledge, S. Gopal. E. Holm and E. Lindberg (1989). Household activity scheduling. In: *Transport Policy, Management and Technology Towards 2001*, Vol. 4, pp. 235-248. Western Periodicals, Ventura, CA.

Jones, P.M., M.C. Dix, M.I. Clarke and I.G. Heggie (1983). *Understanding Travel Behaviour*. Gower, Aldershot.

Jones, P., F.S. Koppelman and J-P. Orfeuil (1990). Activity analysis: state of the art and future directions. In: *Developments in Dynamic and Activity Based Approaches to Travel Analysis* (P. Jones, ed.), pp. 34-55. Avebury, Aldershot.

Kawakami, S. and T. Isobe (1990). Development of a one-day travel-activity scheduling model for workers. In: *Developments in Dynamic and Activity-Based Approaches to Travel Analysis* (P. Jones, ed.), pp. 184-205. Avebury, Aldershot.

Lee-Gosselin, M. (1996). Scope and potential of interactive stated response data collection methods. *Proc. Household Trav. Surv.: New Concepts and Research Needs,* Irvine, California, pp. 115-133. Transportation Research Board Conference Proceedings 10, Washington D.C.

Miller, E.J. and M.J. Roorda (2003). Prototype model of household activity-travel scheduling. *Transpn. Res. Rec.,* **1831**, 114-121.

Pas, E. (1985). State of the art and research opportunities in travel demand: another perspective. *Transpn. Res. A*, **19**, 460-464.

Pendyala, R.M., R. Kitamura and D.V.G.P Reddy (1995). A rule-based activity-travel scheduling algorithm integrating neural networks of behavioural adaptation. Paper presented at the EIRASS Conference on Activity Based Approaches, Eindhoven, The Netherlands (copy obtainable from the authors).

Pendyala, R.M., R. Kitamura and D.V.G.P. Reddy (1998). Application of an activity-based travel demand model incorporating a rule-based algorithm. *Env. & Plan. B,* **25,** 753-772.

Recker, W.W., M.G. McNally and G.S. Root (1986a). A model of complex travel behavior: part I - theoretical development. *Transpn. Res. A*, **20**, 307-318.

Recker, W.W., M.G. McNally and G.S. Root (1986b). A model of complex travel behavior: part II - an operational model. *Transp. Res. A*, **20**, 319-330.

Roorda, M.J., S.T. Doherty and E.J. Miller (2005). Operationalizing household activity scheduling models: addressing assumptions and the use of new sources of behavioural data. In: *Behavioural Foundations of Integrated Land-Use and Transportation Models: Assumptions and New Conceptual Frameworks* (M. Lee-Gosselin and S.T. Doherty, eds.). Elsevier, New York, to appear.

Roorda, M.J., and E.J. Miller (2004). Toronto activity panel survey: a multiple instrument panel survey. *Proc. 7th Int. Conf. on Trav. Surv. Meth.*, Costa Rica.

11

SEQUENTIAL AND SIMULTANEOUS CHOICE STRUCTURES FOR MODELING INTRA-HOUSEHOLD INTERACTIONS IN REGIONAL TRAVEL MODELS

Peter Vovsha, PB Consult, Parsons Brinckerhoff Inc, New York, NY, US

John Gliebe, PB Consult, Parsons Brinckerhoff Inc, Portland, OR, US

Eric Petersen, PB Consult, Parsons Brinckerhoff Inc, Chicago, IL, US

Frank Koppelman, Northwestern University, Evanston, IL, US

INTRODUCTION

Intra-household interactions constitute an important aspect in modelling activity and travel-related decisions. Recognition of this importance has recently produced a growing body of research on various aspects of modelling intra-household interactions and group decision making mechanisms as well as first attempts to incorporate intra-household interactions in regional travel demand models. The previously published research works were mostly focused on time allocation aspect and less on generation of activity episodes, trips, and travel tours. In particular, the works of Townsend (1987), Golob and McNally (1997), Fujii *et al.* (1999), Borgers *et al.* (2002), Gliebe and Koppelman (2002), Goulias (2002), Meka *et al.* (2002), Zhang *et al.* (2002, 2004), Zhang and Fujiwara (2004) and Ettema *et al.* (2004) give examples of models for time allocation between various type of activities and household members. Though these works provide valuable insights into the intra-household decision-making mechanism they are not directly compatible with the structure of most travel demand models that are based on discrete units of travel and discrete choice modelling techniques.

Progress in Activity-Based Analysis edited by H. Timmermans

Most of the approaches including Townsend (1987), Golob and McNally (1997), Simma and Axhausen (2001), Borgers *et al.* (2002), Gliebe and Koppelman (2002, 2004), Scott and Kanaroglou (2002), Ettema *et al.* (2004), and Srinivasan and Bhat (2004) were limited to household heads only and did not consider explicitly the other household members as active agents in the intra-household decision making. This is another limitation that has to be lifted in order to integrate intra-household interactions in the framework of regional travel demand models. To date, no comprehensive approach has been proposed that would address interactions between all household members, include all types of individual, joint, and allocated activities, and also represent an operational framework that could be incorporated in regional travel demand models. The current paper presents an attempt to build a general and operational framework for incorporation of intra-household interactions in the regional travel demand model.

Classification of Intra-Household Interactions

As shown in Figure 11.1, the operational structure of intra-household interactions distinguishes between two principal mechanisms: activity coordination and resource allocation.

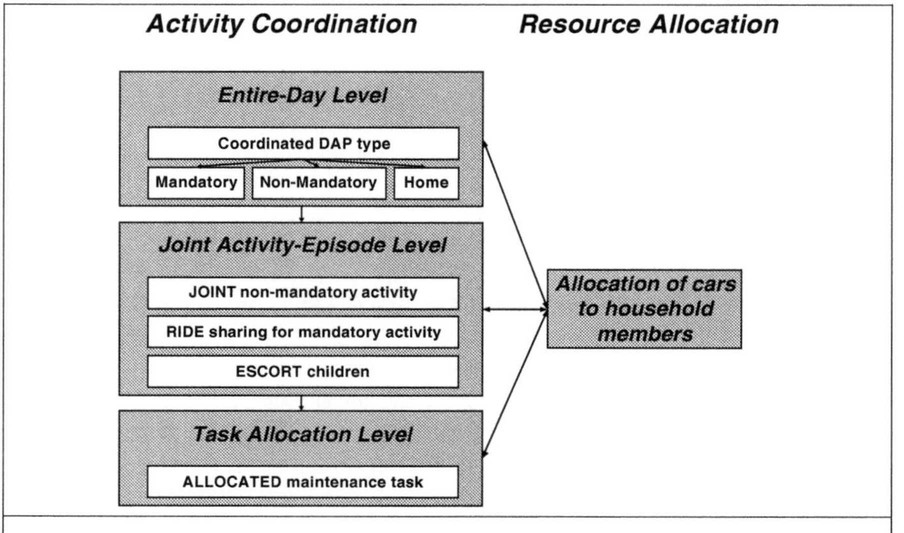

Figure 11.1
Operational Classification of Intra-Household Interactions

The activity coordination mechanism reflects the way household members interact in order to undertake various joint activities and/or travel arrangements as well as allocate household maintenance tasks to household members. It is based on the general behavioural phenomenon that joint participation in activities has an added "group-wise worth" that cannot be reduced to a simple sum of individual utilities for each participant. It also represents various compromises made by some household members in order to serve the other household members representing "altruistic" behaviour that cannot be explained by the individual utility maximization. Activity coordination is the focus of the current paper.

Resource allocation represents another facet of intra-household interactions. Even if activity agenda of all household members on a given day includes only individual activities, they have to interact in order to share constrained resources between them. In the context of travel demand modelling, the most important allocated (and frequently constrained) resource is household cars. First attempts to incorporate intra-household allocation of cars as a part of a travel demand model have been made by Wen and Koppelman (1999, 2000) and Miller *et al.* (2003). This aspect of intra-household interactions is beyond the scope of the current paper.

Activity coordination mechanism can be stratified by three following principal layers of intra-household interactions:

1. *Coordinated principal activity pattern (DAP) types at the entire-day level.* We consider three principal DAP types: (1) mandatory (work, university or school activities, which might include additional out-of-home non-mandatory activities); (2) non-mandatory travel (only non-mandatory activities at least one of which is out of home); and (3) staying at home or absence from town for the entire day. Statistical evidence shows strong coordination between household members at this principal level, resulting in such decisions as staying home for child care; coordinated work commutes; and household members taking time off together for major shopping, family events, or vacations.
2. *Episodic joint activity and travel.* Even if household members have chosen different pattern types (for example, one mandatory and the other non-mandatory) they may participate in shared activity and/or joint travel episodes. We propose a classification of typical joint activity and travel types that supports the development of operational choice models. In particular, we distinguish fully joint travel tours for shared activities from partially joint tours, in which household members share transportation without participation in the same activity.
3. *Intra-household allocation of maintenance activities.* Many of the routine household maintenance activities (shopping, banking, visiting post office, etc) are implemented and scheduled individually; however, generation of such an activity and its allocation to a

particular household member is a function of a household decision-making process. Thus, these activities require an intra-household interaction mechanism to be properly understood and modelled.

It is also assumed that a general hierarchy of intra-household decision making follows these three layers from top to bottom. It means that entire-day level decisions come first. Then, conditional upon the chosen daily pattern types for each household member, the decisions regarding joint activities and travel are made. Finally, maintenance activities are allocated to persons conditional upon the chosen daily patterns and participation in joint activities. These assumptions give a schematic and simplified view on the extremely complicated real-world variety of travel behaviour of the members of a household and numerous interactions between them. This view, however, has two important features: (i) the proposed structure gives a good coverage for most frequent cases of intra-household interactions observed in the household travel surveys, (ii) the proposed structure serves as a constructive framework for derivation of operational choice models that can be estimated based on available surveys and applied in a framework of a regional travel demand model.

Further classification of episodic joint activities is subject to the purpose of travel demand modelling. At this stage we do not model explicitly in-home activities. The following categories of out-of-home episodic joint activity and travel are distinguished:

1. *Joint travel generated by the shared activity.* This category is almost exclusively bound to non-mandatory activities (shopping, eating out, other maintenance, and discretionary activities) as well as almost exclusively implies a fully joint tour structure. Essentially, in the modelling procedure a single set of activity attributes (location, schedule, duration, and travel mode) should be considered for all members of the travel party.
2. *Joint travel to synchronized mandatory activities.* The activities are essentially individual but associated travel arrangements are joint that may require a certain time-space coordination from different household members. This category has a significant share of drop-offs and pick-ups of school children made by workers on the way to and from work. Additionally, a significant percentage of school children travel together to and from school generating fully-joint tours and joint half-tours. Also carpooling of workers for commuting to work is observed, though this type has a comparatively low percentage.
3. *Escorting* that is the purely "altruistic" purpose of driving some other household member without participation in the activity. Statistical analysis has shown that majority of escorting is associated with serving children who cannot drive alone and, in the case of preschool children, cannot even ride transit alone.

The proposed classification leads to a sequence of five models: 1-coordinated DAP, 2-joint travel for shared non-mandatory activity, 3-joint travel (ride-sharing) for mandatory activities, 4-escorting children, and 5-allocation of maintenance tasks.

COORDINATION OF DAILY ACTVITY PATTERN TYPES

Classification of DAP types can be done in many different ways. DAP definition normally includes a list of activities undertaken by the person in the course of entire day with some predetermined hierarchy of the activity types. DAP may also include activity sequencing and/or scheduling attributes, as well as travel related characteristics. In particular, the definition adopted for most tour-based model systems uses travel tours as basic units. Figure 11.2 below shows the structural dimensions along which DAPs are classified in the current research. DAP is classified by three main types:

Mandatory pattern (M) that includes at least one of the three mandatory activities – work, university, or school. This constitutes either a workday or a university/school day, and may include additional non-mandatory activities such as separate home-based tours or intermediate stops on the mandatory tours.

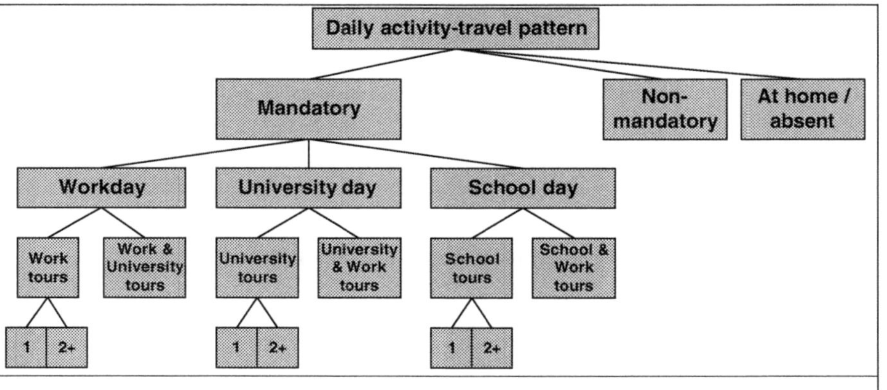

Figure 11.2
Classification of Daily Activity Patterns

Non-mandatory pattern (NM) that includes only maintenance and discretionary tours. By virtue of the tour primary purpose definition, maintenance and discretionary tours cannot include travel for mandatory activities.

At-home pattern (H) that includes only in-home activities. At the current stage of model development, at-home patterns are not distinguished by any specific activity (work at home, take care of child, being sick, etc). Cases with complete absence from town (business travel) were also combined with this category.

The M type is further classified by purpose and frequency of mandatory tours. The nature of mandatory activities – they are usually associated with both long duration and long commuting times – limits significantly the number of mandatory tours that can be implemented in the course of a day. The vast majority of observed cases include only one or two tours, where two-tour combinations include either two tours to the same primary activity or a combination of work and university/school activities. In contrast to the M type, the NM type includes a wider variety of tour frequencies and purposes that is difficult to cover by one choice framework. Thus, the associated details are modelled later in the model stream. Statistical analysis presented in Vovsha *et al.* (2004a) and Bradley and Vovsha (2005) has shown that there is an extremely strong correlation between DAP types of different household members, especially for joint NM and H types. It means that joint staying at home or having a non-mandatory travel day has additional utility beyond a person utility associated with these patterns when implemented alone. For this reason, DAP for different household members cannot be modelled independently. In the most general way the DAP type choice model can be represented by a matrix view in the Table 11.1 below. Each household member $m \in M$ has a row while available alternatives are represents by columns. The choice is associated with a value of 1 in the corresponding cell. The row totals are all equal to 1, while the column totals are not controlled and can take any value between 0 and $\overline{M}$ assuming that every alternative is available to every person.

Table 11.1
Coordinated DAP Model – A Matrix View

HH Members ($m \in M$)	**DAP Alternatives** ($i \in I \cup J$)				
	Individual Mandatory ($i \in I$)			**Potentially Joint** ($i \in J$)	
	W1	**W2**	**...**	**NM**	**H**
1st person	1				
2nd person					1
3rd person					1
...					

Alternative DAP types are broken into two groups. The first group $i \in I$ contains patterns with mandatory activities that assumed individual in a sense that choice of this pattern by one household member does not directly affect choice of the mandatory pattern by the other household member. The second group $i \in J$ contains two patterns – NM and H – that have a potential to be joint if several household members choose the same pattern.

The total number of possible matrices is equal to $\left(\bar{I}+\bar{J}\right)^{\overline{M}}$ where $\bar{I},\bar{J},\overline{M}$ denote the number of elements in arrays I,J,M consequently. This gives a maximum choice set for a simultaneous model. However, there are several important considerations that significantly reduce a dimensionality of the simultaneous model. First of all, most of the individual mandatory DAP types are only partially available for the corresponding person types. Secondly, and even more importantly, intra-household coordination of DAP types is relevant only for the NM and H patterns. Thus, simultaneous modelling of DAP types for all household members is essential only for trinary choice (mandatory, NM, H) while sub-choice of the mandatory pattern can be modelled for each person separately. These considerations result in the following tree representation (Figure 11.3):

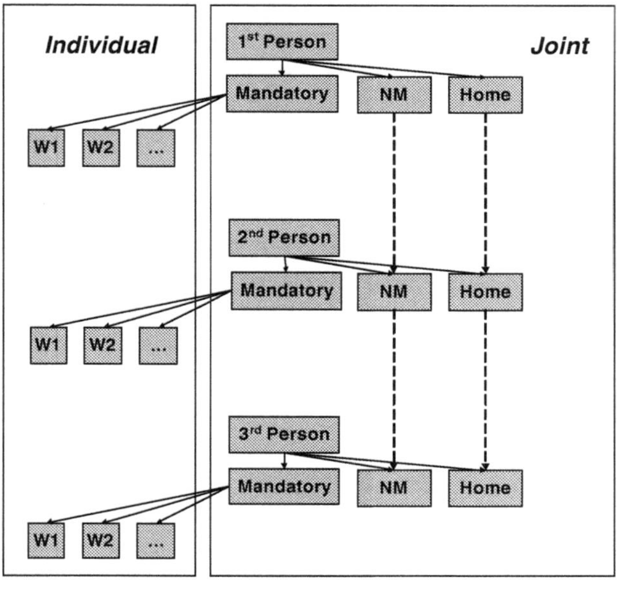

Figure 11.3
DAP Type Choice Structure

The following alternative choice constructs correspond to the tree representation in the Figure 11.3:

- Sequential processing of persons according to the intra-household hierarchy.
- Simultaneous modelling of potentially joint alternatives for all household members with subsequent modelling of individual alternatives.
- Parallel choice structure that considers combinations of main trinary choices at the upper level and individual sub-choices simultaneously in one choice structure.

Sequential processing of persons according to the intra-household hierarchy assumes that choices made by the persons modelled first are used as variables explaining choices of the subsequently modelled persons. Choice model for each person includes all individual and joint alternatives available for the person type. Linkage across person is implemented by means of using Boolean variables for potentially joint choices (indicators of either NM or H patterns) in the utility functions for the corresponding alternatives of the subsequently modelled persons. This was the preferred model structure for several regional travel models in US mostly for its simplicity in estimation and application (Vovsha *et al.*, 2004a). However, this approach does not have a full integrity in capturing intra-household interactions and relies on the ordering of persons in the household.

Simultaneous modelling of potentially joint alternatives for all household members assumes that for each person only a trinary choice (M, NM, H) is considered (Bradley and Vovsha, 2005). Sub-choice of the mandatory alternative is done by a separate choice model conditional upon the choice of mandatory alternative in the trinary choice. Compared to the sequential model, this structure is much more powerful for capturing intra-household interactions in the most integrative way. Even for a household of six persons the simultaneous combination of trinary models results in a total of $3^6 = 729$ alternatives that is a manageable number in estimation and application. For a limited number of households of size greater than six, the model is applied for the first six household members by priority while the rest of the household members is processed sequentially conditional upon the choices made by the first six members. However, this structure has also a drawback comparing to the sequential estimation. Higher integrity of intra-household interactions comes at the expense of disjoining the upper-level trinary choice (M, NM, H) from the lower-level choice of the mandatory sub-alternative.

Parallel choice structure considers combinations of main trinary choices at the upper level and individual sub-choices of mandatory alternatives simultaneously in one choice structure. Different from just mechanical combination of all alternatives for all household members that would result in infeasible choice structure, the parallel choice model requires only a limited principal combinations to be listed explicitly (as in the simultaneous approach described above) as upper-level nests (Gliebe, 2004; Gliebe and Koppelman, 2004, 2005). These nests correspond to the combination of activities where joint participation is essential. The structure of these nests captures different levels of intra-household interaction. Under each nest, the correspondent individual choices of mandatory alternatives are considered for each person individually. This greatly reduces the dimensionality of the model and makes the whole structure manageable in estimation and application. An example of a simple parallel choice structure for a case of a two-person household is shown in the Figure 11.4.

EPISODIC JOINT NON-MANDATORY ACTIVITIES

Episodic joint non-mandatory activities are associated with fully-joint travel tours. Each fully-joint tour is considered as a unit of modelling with a group-wise decision making regarding the primary destination, mode, frequency and location of stops etc. Formally, modelling joint activities involves two linked stages:

1. Generation stage attributed to the entire-household level that is done by means of a frequency-choice model that considers a number of joint tours $j_k = 0,1,2...\bar{J}_k$ as alternatives where $k \in K$ denotes segmentation by purpose / activity type.
2. Participation stage at which decision is made for each household member $m \in M$ and tour $j_k = 0,1,2...\bar{J}_k$ whether to participate or not in the joint tour.

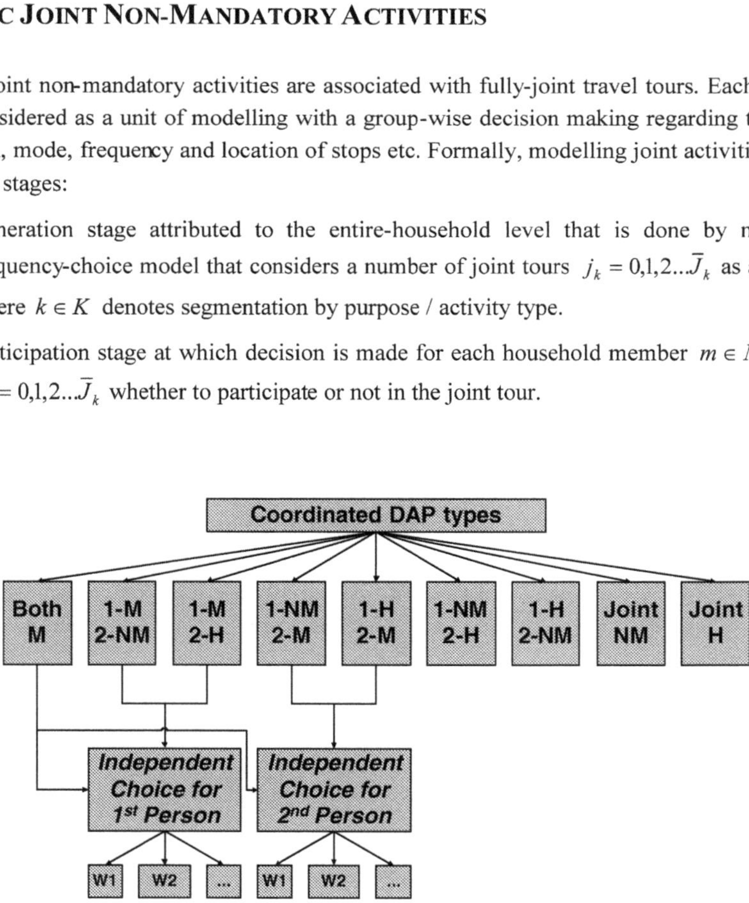

Figure 11.4
Parallel Choice Structure Applied for DAP

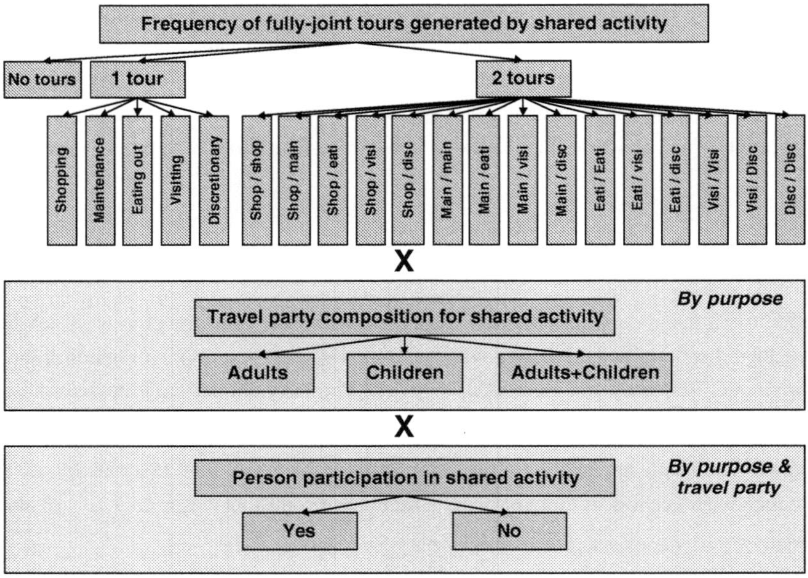

Figure 11.5
Example of Model Structure for Joint Non-Mandatory Activity

The number of travel purposes is limited to 4-5 (shopping, maintenance, discretionary, eating-out, visiting relatives and friends) and the observed maximum total number of fully joint tours implemented by a household during a regular workday is limited to 2-3. A simultaneous frequency-choice model can be formulated that would cover all possible frequencies and purpose combinations. A model adopted by Vovsha *et al.* (2004) included 5 purposes and maximum of 2 joint tours that resulted in 21 alternatives – see Figure 11.5.

The participation stage in the most general way can be viewed as a matrix where each household member has a row while joint tours are represented by column – see Table 11.2 below. Choice to participate is associated with a value of 1 in the corresponding cell. Different from the DAP type choice matrix in the Table 11.1 above, the row totals in the joint participation matrix are not controlled and can take any value from 0 to $\bar{J}$ since each person may choose not to participate in joint tours or participate in some of them, or participate in all of them. The column totals are constrained to be greater or equal to 2 since joint activity requires at least two members to participate. Household members having 1 in a particular column constitute the travel party for the tour that is a subset of household members $m \in M_j \subset M$.

Table 11.2
Joint Participation Model – A Matrix View

HH Members ($m \in M$)	Joint Tours ($j \in J$)		
	1st Tour (Purpose k_1)	2nd Tour (Purpose k_2)	3rd Tour (Purpose k_3)
1st person	1	1	
2nd person	1	1	1
3rd person		1	1
…			

Each person participation matrix constitutes a distinct alternative. Even if the number of tours is limited (say, 3) and number of household members is limited (say, 6) the resulting number of participation matrices would be $\left(C_{\overline{M}}^{2}\right)^{7} = \left(C_{6}^{2}\right)^{3} = \left(\frac{6*5}{1*2}\right)^{3} = 3{,}375$. If the frequency choice is combined with the participation choice in one simultaneous structure, the total number of alternatives would be even greater, though not significantly since the matrices for frequencies of 1 and 2 would be much simpler. Choice models with several thousands of alternatives are not infeasible if a parsimonious component-wise utility structure can be applied. However, for first practical implementations of the model, the following various decomposition schemes with sequential processing along some dimensions were applied:

- Sequential modelling of generation and participation stages, that results in two choice models applied in succession:
    - *Tour frequency* choice model that is applied for the entire household and yields probability of having a certain set of joint tours by purpose
    - *Person participation* choice model that yields probability of having a certain participation matrix conditional upon the chosen set of joint tours; the participation choice model in itself quite complicated and can be decomposed by sequential processing of tours (i.e. columns of the Table 11.2). Further on, the participation model for each tour can be decomposed into two models – see Vovsha *et al.* (2003):
        - *Party-composition* choice model that predicts a principal party composition in terms of the participating person types (adults, children, mixed).
        - *Person participation* choice model that assigns member of the travel party for each tour conditional upon the chosen party composition; this model in turn can be decomposed by persons into a sequence of binary choice models (to participate or not).

- Simultaneous modelling of generation and participation stages as one choice structure

Sequential modelling of generation and participation stages assumes that the tour-frequency choice model is applied first for the entire household. This choice model yields probability of a certain frequency combination of tours by purpose $P(j_1, j_2,...j_{\overline{K}})$. The utility function is normally formulated in such a way that advantage can be taken on the combinatorial structure of choice alternatives each of them essentially represents a combination of elemental tours. In particular, the following utility structure was applied by Vovsha *et al.* (2003):

$$V(j_1, j_2,...j_{\overline{K}}) = \alpha_{j_1+j_2+...j_{\overline{K}}} + \sum_{k \in K} j_k \times V_k , \qquad (1)$$

where,

$\alpha_{j_1+j_2+...+j_{\overline{K}}}$	=	constant that depends on the total number of tours only,
V_k	=	elemental utility of one tour specific to purpose.

In this structure it is assumed that the elemental purpose-specific utilities capture impact of all explanatory variables and if there are several tours in the alternative their utilities are linearly combined. For example, if there are two tours for shopping purpose the shopping tour utility component will be doubled. Constants are specific to the total number of tours only (0, 1, 2...) and allows for capturing a saturation effect for multi-tour alternatives.

The participation choice model yields probability of a certain participation matrix to be chosen. The participation matrix can be described in terms of Boolean variables $\{\omega_{jm} = 0,1\}$ or alternatively in terms of travel parties formed for each tour $M_j = \{m \in M | \omega_{jm} = 1\}$. If we consider tours sequentially, then the core choice model is essentially formulated for each tour separately and we can drop index j from the participation variables, i.e. consider only one column in the Table 11.2. Thus the participation problem is reduced to defining a probability $P(\widetilde{M})$ of a subset of the household members $\widetilde{M} = \{m \in M | \omega_m = 1\}$ to be chosen as a travel party. Even this reduced task is not simple for the model formulation and estimation because of variety of person types and possible combinations of them. One of the constructive ways to decompose this choice further is to define first a principal party composition $g \in G$ in terms of participating person types and then model person participation conditional upon the chosen party composition.

Possible party compositions_ $g \in G$ are defined in a mutually exclusive and collectively exhaustive way by person types. For example, in the model reported by Vovsha *et al.* (2003), three party compositions were defined – adults, children, and mixed. Every household member m can participate in only a subset of relevant parties G_m. There are in a common case several possible parties that can be formed within the same composition type $\widetilde{M} \subset M_g$. For example, if there are 3 adult household members there are 4 different adult parties that can be formed (1st and 2nd member, 1st and 3rd, 2nd and 3rd, and all three). The probability of a party to be chosen $P(\widetilde{M})$ is:

$$P(\widetilde{M}) = P(g|G) \times P(\widetilde{M}|M_g), \tag{2}$$

where,

$P(g\|G)$	=	marginal party composition choice probability,
$P(\widetilde{M}\|M_g)$	=	conditional participation choice probability.

While the first choice sub-model for marginal party-composition probability is comparatively simple and has a predetermined set of alternatives (three in the MORPC case), the second model for conditional participation choice probability is more complicated and the number of alternatives for this model is a function of the household size in composition. For example, consider a number of alternatives for adult travel party. If the household has only one adult member then adult party is infeasible. If the household has 2 adult members, adult party is feasible but participation of both adults is mandatory, i.e. there is only one participation alternative. If the household has 3 adult members, there are 4 possible ways to form an adult party. If the household has 4 adult members, there are already 11 possible participation alternatives. In this situation, it is not straightforward to formulate a choice structure that would incorporate all possible household sizes. One of the ways to simplify this model is to decompose it further into a sequence of binary choice models for each relevant person. It means that the conditional participation probability is assumed to have the following form:

$$P(\widetilde{M}|M_g) = \prod_{m \in M_g} P(m|g), \tag{3}$$

where,

$P(m|g)$ is a probability for a person to participate in the given party composition.

The advantage of the binary participation choice model is its simplicity in terms of the number of alternatives. The choice utility can incorporate numerous person, household, and other variables, as well as the purpose of the tour / activity type. Application of the binary choice model in a micro-simulation fashion does not guarantee a feasible travel party of size 2 or large. In the case of infeasible size of 0 or 1 the micro-simulation procedure is restarted until a feasible solution is generated. The model is automatically sensitive to the household size in a sense that relatively large parties would generally be generated for large households.

Application experience of this sequential structure (tour frequency, party composition by tours, person participation by persons) has shown that is performs reasonable well in practical terms. However, there are several serious weaknesses of the sequential approach that should be understood:

- Person participation in any joint tour is modelled independently of the other tours; it is assumed that since number of tours is limited, there is only a little saturation effect. In reality, in a case that there several joint tours implemented by the household in the course of a day there can be strong limitations for participation of the same person in all of them including scheduling conflicts when these joint tours are implemented by different travel parties at the same time.
- Person participation in any joint tour is modelled independently of the other household members; it is assumed that every person has an inherent propensity to participate in joint activities of a certain type (purpose and party composition). In reality, there can be a strong clustering effect when participation of some household members may be strongly linked to participation of the other ones. Alternatively there can be a substitution effect, especially for mixed parties where one of the household adults takes children while the spouse may stay at home or undertake some other activity.
- Frequency of joint tours is modelled independently from person participation in them. Person participation is conditional upon the set of generated tours, however, there is no upward linkage that would make frequency of tour explicit function of potential person participation.

Consider the first two aspects that relate to the person participation under condition of a fixed set of tours. To ensure integrity across both tours and persons we have to consider the whole participation matrix and not to decompose it by columns (tours) or rows (persons). However it is not a simple task to find a choice model structure that would correspond to such multidimensional choice in terms of the observed utility components and correlations across unobserved components. A simple MNL model that would consider all possible participation matrices with additive utility that is compound of components specific to either tour or person can be equivalently decomposed into a sequence of independent participation models. It is important to substantiate possible meaningful rules for grouping alternatives (participation matrices) based on some partial similarities.

One of the advantages of the matrix view in the Table 11.2 above is that it suggests several aggregations of alternatives that can be effectively used for substantiation of possible nests. First of all, there are aggregations naturally associated with the marginal totals, i.e. total number of tours in which each person participates as well as total number of persons participating in each tour. Secondly, there are aggregations associated with clusters of persons and subsets of similar activities (tours for the same purpose). The following practical conclusions for substantiating the corresponding choice tree can be made:

- Nesting by persons with subsequent sub-nesting by a number of tours in which the person participates (person "workload") can be useful since it gives a reasonable dimension for similarities across different participation matrices. Two matrices can be considered similar if all or at least some persons participate in approximately the same number of tours in both matrices. Alternatively, matrices that represent significantly different person workloads should be treated as distinct alternatives.

- Nesting by activities with subsequent sub-nesting by a number of participants (party size) also can be useful since it gives another dimension for similarities across different participation matrices. Two matrices can be considered similar if all or at least some tours have approximately the same number of participants. Alternatively, matrices that represent significantly different party sizes for all tours should be treated as distinct alternatives.

- Intra-household clusters by person types can be used as additional nests above persons or as alternative nests instead of persons or even along with the person nests. This means that workload variations within the cluster are more probable, while changing workloads between clusters lead to principally different matrices. Travel party composition types used for the model decomposition in the Columbus and Atlanta models can be also associated with person clusters that do not necessarily have to be mutually exclusive.

- Tour groups by purpose can be used as additional nests above tours or as alternative nests instead of tours or even along with the tour nests. This means that participation variations between tours made for the same purpose are more probable, while changing parties between tours of different purposes lead to principally different matrices. This dimension, however, is relevant only for infrequent cases with 3 or more joint tours made by a household on the same day.

Combining these considerations we arrive at the following choice structure shown in the Figure 11.6 for a case of a 3-person household with 3 joint tours. Since there are 4 possible parties for each tour we have $4^3 = 64$ possible participation matrices at the lowest choice level.

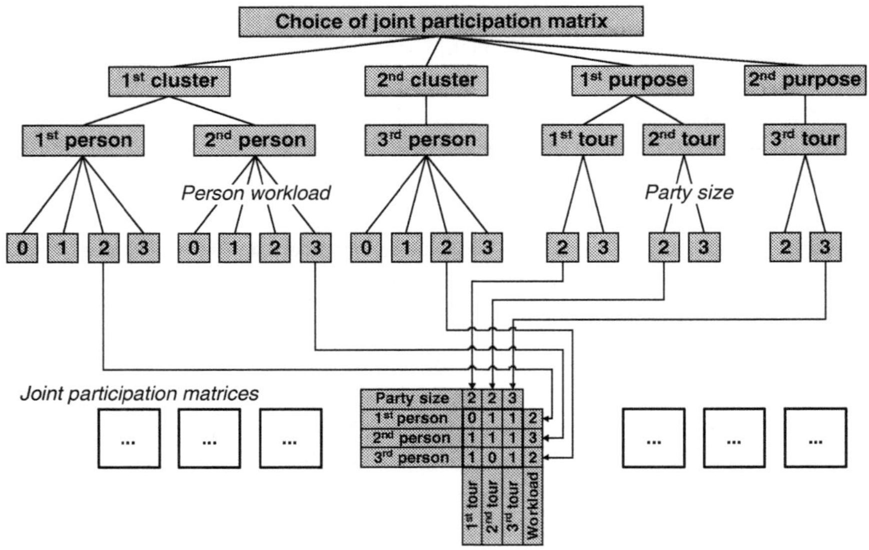

Figure 11.6
General Choice Structure for Joint Participation

Clusters that correspond to persons that are substitutable, for example two adults each of them can play the role of a driver of a mixed party with children, are treated as nests (i.e. unobserved similar components in the utility function). Person clusters that correspond to person types that are most frequently linked in the same party (for example non-working adult with preschool child) are treated through observed components of the entire-party utility function.

Since this structure is not a simple hierarchy where each lower-level alternative belongs to exactly one nest, it cannot be modelled by a simple nested logit model. However, it can be effectively handled by generalized nested structures of the GEV class. A structure of the utility function can be quite parsimonious because it is essentially combined of a limited number of pre-determined components. This greatly simplifies the model estimation and application. In particular, the following components should be taken into account:

V_{km} = elemental participation utility (person suitability for the activity)

V_{mn} = person workload utility to capture saturation effects at the person level

V_{kl} = party size utility to capture additional worth of joint participation

The participation matrix utility can be linearly combined of these components in the following way:

$$W\left(\left\{M_j\right\}_{j\in J}\right)=\sum_j \sum_{m\in M_j} V_{k(j),m} + \sum_m V_{m,n(m)} + \sum_j V_{k(j),l(j)} \,. \tag{4}$$

Each component corresponds to either the cell of the matrix or one of its margins (or aggregate margins) and they are all combined over (non-zero) cells. Further generalization of the model includes the frequency choice aspect. Combining frequency choice and participation choice in one choice structure essentially means that the variety of participation matrix to consider should include all possible matrices with variable (rather than fixed) number of columns. There are several ways to construct such a simultaneous model:

- Consider frequency of joint tours as one more upper level (above the choice of participation matrix) in the choice hierarchy shown in the Figure 11.6. This is the most straightforward way to combine frequency and participation choice model since when they are considered in a sequential fashion, frequency should naturally precede participation. This, however, means that a similar structure should be replicated under each frequency combination.
- Consider frequency as one more upper level only for tours, while person nest will continue to go to the root directly. This structure is simpler since additional nesting level and the corresponding multiplication of nests would relate to the tour side only.
- Consider a maximum observed number of tours for each purpose but allow for participation matrix to have empty columns with no persons assigned. This requires extension of the participation matrix rules: either no one or at least 2 persons have to participate.

All approaches mentioned above produce the same full set of participation matrices with variable number of columns as elemental choice alternatives. The differences relate to the way how the nests are structured and consequently how the correlation structure of the model is assumed.

Ride-Sharing for Mandatory Activities

Ride-sharing for mandatory activities relate to pure travel arrangement while the underlying activity for each participant is assumed individual with correspondingly individual choices of locations and durations. Thus, different from joint activities, the ride-sharing modelling technique does not require a generation model but rather a linking and synchronizing model. Ride-sharing arrangements may require limited adjustments of schedules of participants in order to synchronize their travel but it is assumed that location for each (mandatory) activity and the basic schedule are fixed for each household member and the household interaction are aimed at finding the best travel arrangements that would serve that individual locations and schedules rather than change them.

When modelling ride-sharing, it is assumed that for each household member it is known a number and purpose of mandatory tours I_m^k as well as location zone $z(i)$, preferred outbound time (departure from home) $\tau(i)$, and preferred inbound time (arrival back home) $\pi(i)$ for each tour. Ride-sharing can occur only for either outbound or inbound bunches of mandatory tours that share the same home end. Thus, the model can be essentially broken into two parts – outbound and inbound ride-sharing. In many cases, these two parts can be processed independently, especially when a worker and school child are involved with very different activity durations. However, for worker-worker and child-child compositions, two-way ride-sharing arrangements can be considered where ride-sharing decisions are not independent by directions.

The ride-sharing model considers partition of mandatory tours into ordered subsets of outbound and inbound half-tours $j = \{i_1, i_2 ...\}$. The length of the subset corresponds to the number of participants. One-tour length means travel alone; two-tour length means participation of two persons in a shared ride; three-tour length means participation of three persons in a shared ride, etc. Order of participants reflects their roles in the shared ride. The first tour corresponds to the driver; the second tour corresponds to the passenger with the longest ride (the last getting-off passenger for the outbound direction or the first getting-in passenger for the inbound direction), etc.

From the formal point of view modelling ride-sharing involves two subsequent stages:

1. Linkage and synchronization of outbound and inbound half-tours that is done by means of a *partition-choice* model that considers all possible partitions of mandatory half-tours into rides (alone and shared)
2. *Ordered participation choice* model that essentially considers a role of each participant (driver, passenger) and route along which activity locations of all ride participants are visited.

A tree choice structure example for a household with 2 workers (each having one work tour) and one child (having one school tour) is shown in Figure 11.7. The outbound and inbound sets of linking choices are not exclusive but rather should be combined. Thus, even for a comparatively small household of 3 persons, the linking- choice level includes 5*5=25 alternatives. For a large household the formal number of choice alternatives would grow up exponentially. However, in the real-world model estimation and application there are numerous considerations that allow for reduction of the ride-sharing choice model to a manageable size:

- Number of household members who actually have mandatory tours (workers and students) is limited in most of the households to 3-4 persons with predominantly one tour per person. Thus, for majority of cases the choice structure would be similar to shown in Figure 11.7.

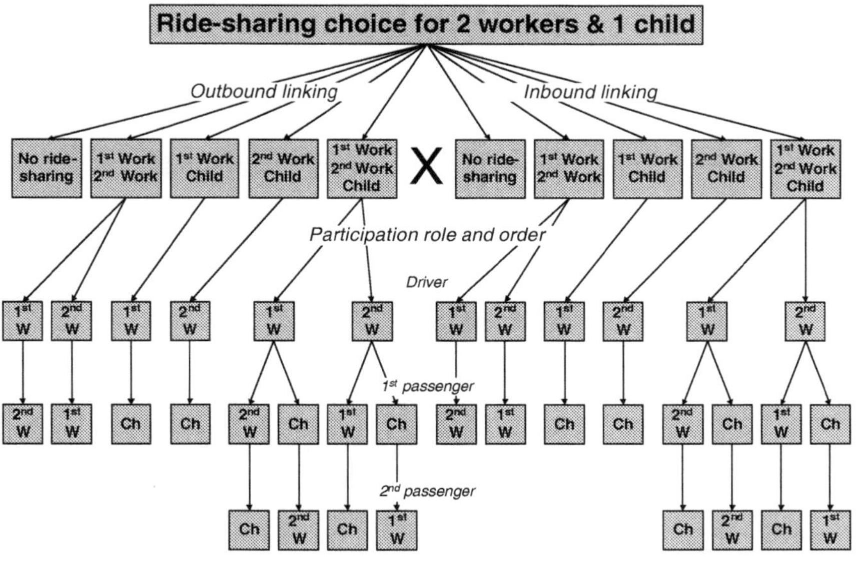

Figure 11.7

General Choice Structure for Ride-Sharing

- Many of the possible linkages can be rejected as impossible at the preliminary stage of synchronizing locations and departure/arrival times based on reasonable thresholds. These thresholds include maximum allowable differences in departure/arrival times (30 min performs quite well in practical terms) and maximum deviation from the shortest path to or from the location of activity for the driver (5 miles performs quite well). Application of these thresholds allows for cutting down a number of branches in the linking model significantly.
- A maximum size of travel party can be limited to 3 participants because larger travel parties for ride sharing to mandatory activities proved to be very infrequent even for large households.

In addition to a priori elimination of improbable alternatives, several constructive decompositions of the choice structure can be considered. The first one includes a natural breakdown into the linking and participation-role stages. Secondly, the linking stage itself can be implemented as a sequence of pair-wise choices rather than a single choice model. To illustrate the ways of decomposition of the linking model it is useful to put it into a matrix view (Table 11.3).

Table 11.3
Ride-Sharing Model – A Matrix View

Drivers' Half Tours	**Passengers' Half-Tours**				
	Outbound			**Inbound**	
	1st Worker	**2nd Worker**	**Child**	**1st Worker**	**...**
		Outbound:			
1st worker					
2nd worker		1	1		
		Inbound:			
1st worker				1	1
2nd worker					

Each ride-sharing matrix constitutes a distinct alternative for the choice model. Rows correspond to half-tours of potential drivers. Columns correspond to all half-tours. Each column can have not more than one assignment of 1. A cell value of 1 means that the corresponding passenger (column) travels with the driver (row). Assignment of 1 to the diagonal cell means being a driver of the ride. This is mandatory for all rows that have non-zero totals (i.e. valid rides must include the driver's half-tour). A column can consist of zeros only. It means that the corresponding passenger cannot find a ride and travel alone or by other modes (transit or non-motorized). A row can consist of zeros only. It means that the corresponding potential driver is either taken as a passenger by the other driver or travels alone by other modes. A minimal positive total for each row is equal to two, meaning that it must be a driver and at least one passenger for each ride.

The table cannot have cycling relations when two half-tours serve as drivers to each other. Inbound and outbound half-tours cannot interact, thus the appropriate parts of the table are blocked out. Also, if the same person has several mandatory tours on the same day they cannot be linked between themselves. These rules along with the threshold described below introduce many a priori zeros to the ride-sharing matrix, thus, reducing significantly a number of feasible alternatives.

Every non-diagonal cell is associated with potential pair of half-tours. A pair-wise measure of matching these half-tours in time and space can be formed as a linear combination of schedule discrepancies and route deviations associated with each direction of the ride (Vovsha and Petersen, 2005):

$$V_{i_1 i_2}^{out} = a \times [\tau(i_1) - \tau(i_2)]^2 + b \times [\min(L_{z_2} + L_{z_2 z_1} - L_{z_1}, L_{z_1} + L_{z_2 z_1} - L_{z_2})]^2$$
$$V_{i_1 i_2}^{inb} = a \times [\pi(i_1) - \pi(i_2)]^2 + b \times [\min(L_{z_2} + L_{z_2 z_1} - L_{z_1}, L_{z_1} + L_{z_2 z_1} - L_{z_2})]^2, \quad (5)$$

where,

L_{z_1} = distance from home to the destination of the first tour,

L_{z_2} = distance from home to the destination of the second tour,

$L_{z_1 z_2}$ = distance between the destinations of the first and second tour

The coefficients a and b are statistically estimated together with the other parameters. The minimum route deviation reflects the inconvenience of ride-sharing for the driver who has to visit the passenger's location first and then go to his/her own location (in the outbound case). If one of the persons cannot drive (like in a case of worker-child carpool) the corresponding route deviation term is set to a large number, thus ensuring that minimum would relate to the driver's deviation. Squaring discrepancies proved to work better than linear inclusion because the negative impact of large schedule discrepancies and route deviations on probability of ride-sharing is highly non-linear.

This pair-wise measure is instrumental for both the model decomposition and simultaneous formulations. One of the possible model decompositions is based on sequential processing of pairs ordered by the minimum discrepancy measure. Sequential processing of potential ride-sharing pairs requires development of a binary choice model that yields probability of sharing ride for two persons as a function of the matching measure as well as household, person, and other characteristics. For the inbound direction the utility function of ride-sharing should include indicator on outbound ride sharing with generally a strong positive impact. Then the following sequential procedure can be outlined (outbound as an example):

1. Order outbound half-tour pairs by the matching measure (and possibly person type considerations)
2. Take the highest-order unprocessed pair that has a matching measure within the threshold

   $$V_{i_1 i_2}^{out} \leq \overline{V}$$

3. Run a binary choice model to calculate probability of ride-sharing and simulate the choice outcome (1-share, 0-not)
4. If the choice outcome is 1 then engage the pair:
   a. If no one of the engaged persons has been previously engaged yet, then register a new shared ride
   b. If one of the engaged persons has been already engaged in a shared ride, then add a new participant to the previously registered shared ride and mark all person pairs in this ride as processed
   c. If both engaged persons have been already engaged in different shared rides then combine these rides, register a new ride that includes all participants from the

previous rides, discard the previous rides, mark all person pairs in a new combined ride as processed

5. If there are unprocessed pairs go to step 2; if no unprocessed pairs left then end

This procedure assumes a micro-simulation framework for the model application. Ordering of half-tour pairs by the matching measure can be enhanced using person-based priority rules. For example, processing of worker-child pairs can be done prior to processing worker-worker pairs.

The simultaneous choice approach that relates to the upper-level nests in the Figure 11.7 is also practically manageable taking into account that numerous branches of the tree can be pruned a priori. However, in some cases for large households the number of alternatives (ride-sharing matrices) can reach thousands. As for the most cases where choice alternatives are compound of elements, the utility function can be effectively combined from a limited number of components that greatly simplifies the choice model estimation and application even with thousands of alternatives. The following utility components can be used to construct the ride-sharing matrix utility:

$V_{i_1 i_2}^{out}, V_{i_1 i_2}^{inb}$	=	pair-wise half-tour ride-sharing utilities
V_n^{out}, V_n^{inb}	=	half-tour utilities specific to the party size of the ride
V_i^{both}	=	utility of having a ride in both directions

The ride-sharing matrix utility takes the following form:

$$V(\{j^{out}, j^{inb}\}) = \sum_{j^{out}} \left(V_{n(j)} + \sum_{i_1, i_2 \in I_j} V_{i_1 i_2} \right) + \sum_{j^{inb}} \left(V_{n(j)} + \sum_{i_1, i_2 \in I_j} V_{i_1 i_2} \right) + \sum_{i \in I^{both}} V_i^{both} \quad (6)$$

The person participation role model considers sequences of person within the ride in such a way that the first person plays the driver role, the second person corresponds to the passenger with the longest route, and so forth. The last person is the first passenger dropped off on the outbound half-tour or the last person picked-up on the inbound half-tour. The last person does not experience any route deviation. The order of persons from the driver to the shortest-leg passenger corresponds to the magnitude of potential deviations from the shortest route. The number of alternatives is equal to $n!$ and also some of them should be excluded since not every person can be a driver. For example, for a rare case of a ride-sharing of 4 persons and also assumi ng that they all can be drivers, we have 4!=24 alternative orders that all can be treated simultaneously in one choice structure. The utility for the participation role distribution is combined of person utilities associated with route deviation and additional driver-role component that is associated with person type and characteristics in the following way (example for a 3-person ride):

$$V_{i_1 i_2 i_3} = V_{i_1}^{driv} + V_{i_1 i_2 i_3}^{dev} + V_{i_2 i_3}^{dev} \tag{7}$$

ESCORTING CHILDREN

Escorting is a joint travel arrangement that is characterized by distinctive-in-kind roles of participants. There is always an escorting adult driver (in vast majority of the observed cases a single adult person, otherwise it becomes a joint tour for shared activities) and one or several escorted children. The important characteristic that distinguishes escorting from all other joint activity and travel arrangements is that only the escorted persons have a purposed activity to participate while the driver does not participate in any activity and implement a pure chauffeuring function. A dominant share of escorting involves children as passengers. Escorting of adult household members is observed rarely and mostly in households with low car ownership. Thus, we assume escorting children from now on in this paper. From the perspective of escorted person, he/she has a mandatory or non-mandatory tour to implement. The escorting service may cover the whole tour (two-way escorting) or only one of the half-tours (one-way escorting). From the perspective of the chauffeur, his/her tour may cover only one half-tour of the escorted child with no waiting at the activity location or both half-tours of the escorted child with waiting while the child is involved in his/her activity. Thus, for each tour of a child that demands escorting there are five possible alternatives:

- No escort
- Escort in outbound direction only (from home to activity)
- Escort in inbound direction only (from activity back home)
- Escort in both direction by means of two separate tours of the same driver or by different drivers without waiting
- Escort in both direction by means of a single tour of the same driver with waiting

To formalize escorting as a choice model we use the following notation:

$i \in I$	=	child tours that demand escorting,
$m \in M_a$	=	household adults that can play a chauffeur role,
$j \in J_m$	=	escorting tours made by each chauffeur.

Table 11.4
Escorting Tour Construction Model – A Matrix View

Chauffeurs ($m \in M_a$)	**Escort tours** ($j \in J_m$)	**Child tours demanding escort $i \in I$**					
		1st child				**2nd child**	
		1st tour		**2nd tour**		**1st tour**	
		Out	**Inb**	**Out**	**Inb**	**Out**	**Inb**
1st chauffer	1st escort	1				1	
	2nd escort			1	1		
2nd chauffer	1st escort		1				

The set of children's tours $i \in I$ with all pertinent characteristics of the person $m(i)$, tour purpose / activity type $k(i)$, departure-from-home time $\tau(i)$ for outbound half-tour, arrival-back-home time $\pi(i)$ for inbound half-tour, and location $z(i)$ is assumed known and fixed. The set of adult chauffeurs $m \in M_a$ with all pertinent characteristics of the person and availability to serve child tours I_m^{out}, I_m^{inb} within the time window left after scheduling the chauffeur's mandatory and joint activities (they are considered of higher scheduling priority) is also assumed known and fixed.

Modelling of escorting can be formalized as finding a set of escorting tours for each chauffeur $j \in J_m$ where each tour covers a subset of outbound I_j^{out} and inbound I_j^{inb} children's tours. The subsets I_j^{out} and I_j^{inb} are mutually exclusive across escorting tours $j \in J_m$ for either outbound or inbound subsets. Also not every child half-tour that demand escorting can be satisfied in a general case. Thus, $\bigcup_j I_j^{out} \subseteq I$ as well as $\bigcup_j I_j^{inb} \subseteq I$. The escorting tour construction problem can be presented in the following matrix view (Table 11.4).

Each escorting tour construction matrix constitutes a distinct alternative. The rules for construction of a feasible matrix are as follows:

- Every row corresponds to escorting tour of the chauffer. Cell values of 1 correspond to served child half-tours. Escorting tours for each chauffer are listed in a chronological order. The first escort tour can take any outbound or inbound child half-tours that fall into the available time window of the chauffeur I_m^{out}, I_m^{inb}. Each subsequent escorting tour of the same chauffeur has a narrower window available since the previous tour(s) block out additional time windows making available sets essentially tour-specific and dependent on the escorting matrix $I_{jm}^{out}, I_{jm}^{inb}$. The minimal row total allowed is 1 (at least one half-tour should be served; otherwise the escorting tour does not make sense). The maximum row total is controlled by the availability rules but it never can be greater than number of child tours in the set $\bar{I}$ multiplied by 2.

- Every column corresponds to a child half-tour (either outbound or inbound). The column total can be either 1 (get escort) or 0 (not).
- The allocation of 1's in each row should meet the internal tour feasibility conditions:
  - The bundle of outbound half tours of children I_j^{out} served by the tour should have close departure-from-home times and locations. A threshold on the discrepancy measure described for ride-sharing above proved to be useful for bundling outbound half-tours for escorting as well.
  - The bundle of inbound half tours of children I_j^{inb} served by the tour should have close arrival-back-home times and locations. The same threshold on the discrepancy measure is applied for bundling inbound tours as well.
  - All outbound half tours I_j^{out} start earlier than inbound half-tours I_j^{inb} served by the same escorting tour. Thus, only disjoint-in-time sets I_j^{out} and I_j^{inb} can be "bridged" by one escorting tour.

One of the problems with forming alternative escorting matrices is that the set of escorting tours $j \in J$ itself is variable and should be considered as a part of the choice model. One of the possible views on the forming alternative escorting matrices that also gives some insights into the possible decomposition of the choice structure is that it can be broken into two subsequent stages:

- Escorting tour set formation $j \in J$ by partitioning sets of children's half-tours I^{out}, I^{inb} into non-overlapping and non-exhaustive subsets I_j^{out}, I_j^{inb}. This problem is similar to the partitioning problem for the ride-sharing mechanism described above.

- Escorting tour allocation to chauffeurs $j \in J_m$. This problem is similar to the task allocation mechanism described in the section that follows.

However, the escorting tour formation cannot be done effectively without considering chauffeurs' availability constraints. More exactly, if "bundling" outbound and inbound child half-tours can be effectively done based on the half-tours themselves, "bridging" outbound and inbound half-tours cannot be done without imposing the chauffeur availability constraints. Still, if we consider all possible tour formations and then allocation to chauffeurs, for most households, the task of listing all resulting combinations is not insurmountable.

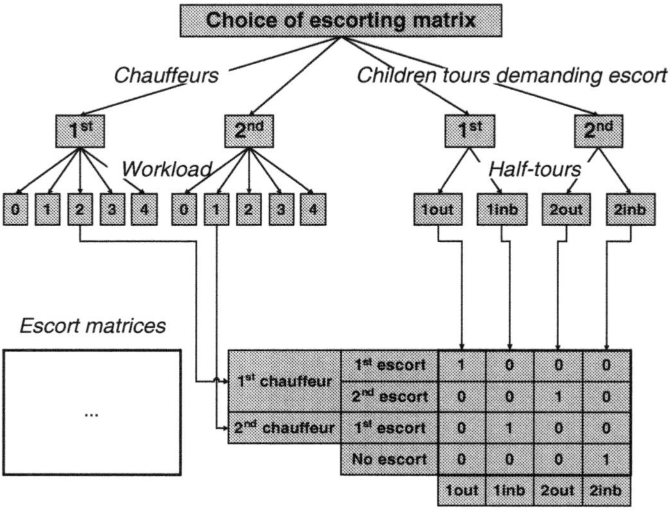

Figure 11.8
Escorting Choice Tree – General Case

For example, in a household with 2 potential chauffeurs and 2 child tours (i.e. 4 half-tours) to handle, we will have 52 possible tour formation sets (including various options of serving only some of the 4 half-tours) and then from 2 to 16 allocation-to-chauffeurs alternatives for each of tour formation sets which results in approximately 500 escorting matrix alternatives. The corresponding choice tree is depicted in the Figure 11.8.

However, after application of the tour-feasibility rules, many of the tour-formation sets fail at ether bundling or bridging stage. Further on at the chauffer availability check many of the matrices fail because at least one of the tours prove to be outside the available time window of the assigned chauffer. These two checks normally reduce the choice set size significantly. However, for large households the number of potential escorting matrices in the original list can come to thousands and thus decompositions of the choice structure is welcome even for computational efficiency. One of the possible decompositions assumes that the household chauffeurs are ordered starting from the least individual person types. Non-workers are considered the first-choice chauffeurs, followed by retired persons, then by part-time workers, then by full-time workers, then by university students, and finally by driving-age school children. Then, the choice model is developed for a single person and includes only residual chauffeuring alternatives left after the choices actually made by the previously modelled chauffeurs. Figure 11.9 shows as an example a 1 child tour to serve.

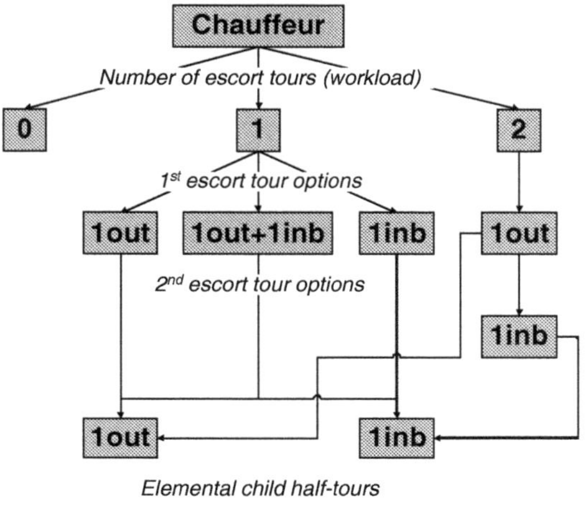

Figure 11.9
Escorting Choice Tree for a Single Chauffeur

Another possible decomposition is based on the ordering of child tours demanding escort rather than chauffeurs. Figure 11.10 shows an example of 2 chauffeurs available. More details are provided in Vovsha and Petersen (2005). In this case, household children are ordered first by age (from youngest to oldest) and then tours for each child are ordered chronologically. The choice model considers one child tour at a time to define probability of coverage by escorting for outbound and inbound half-tours and assignment of the chauffeur. Bundling of children's half-tours is done by inclusion of the chauffeur's previously assigned tours as the lower-level sub-choices.

In all cases of the model formulation only a limited number of the following components of the utility function should be considered:

V_i^{out}, V_i^{inb}	=	escorting utility for each child half tour (no escort has zero utility),
ΔV_i	=	additional child utility of escorting in both directions,
$V_{im}^{out}, V_{im}^{inb}$	=	chauffeur suitability and availability for each child half-tour,
$V_{mn}(T_m)$	=	chauffeur workload saturation effect,
$V_{jm}\left(I_{jm}^{out}, I_{jm}^{inb}\right)$	=	chauffeur tour disutility associated with bundling and bridging.

The entire-household escorting matrix utility can be expressed as a sum of these components in the following way:

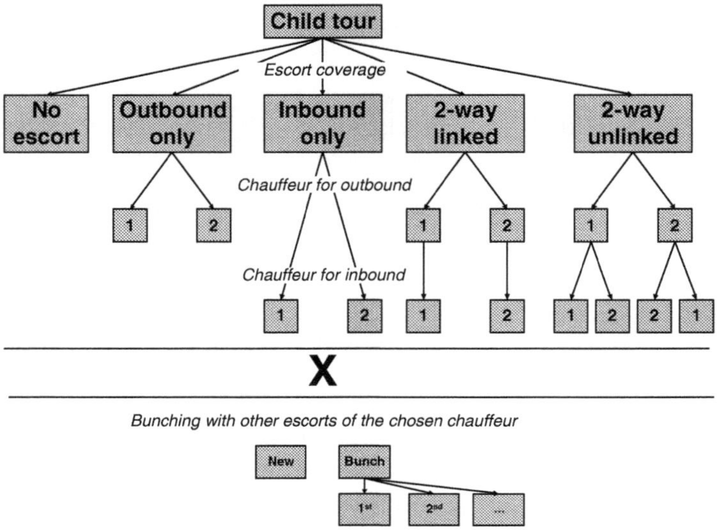

Figure 11.10
Escorting Choice Tree for a Single Child Tour

$$U = \sum_{i \in I} \left(V_i^{out} + V_i^{inb} + \Delta V_i^{both} \right) + \sum_{m \in M_a} \left(\sum_{i \in I_m^{out}} V_{im}^{out} + \sum_{i \in I_m^{inb}} V_{im}^{out} + V_{m,n(m)} + \sum_j V_{jm} \right) \qquad (8)$$

The first three components correspond to different aspects of the utility of the served children while the last three components relate to associated disutility aspects of the chauffeurs.

ALLOCATION OF MAINTENANCE TASKS

The essence of the current model is to generate these activities at the household level and then allocate to the household members. Since, travel details are dependent on the characteristic of the person and his/her individual decisions this model is better formulated in terms of maintenance tasks rather than tours. From the formal point of view, modelling allocated activities involves two linked stages:

1. Generation stage attributed to the entire-household level that is done by means of a frequency choice model that considers a number of maintenance tasks by type $\bar{j}_k = 0,1,2 \ldots \bar{J}_k$ as alternatives.

Table 11.5
Task Allocation Model – A Matrix View

HH Members ($m \in M$)	Maintenance Tasks ($j \in J$)		
	1st Task (Purpose k_1)	2nd Task (Purpose k_2)	3rd Task (Purpose k_3)
1st person	1	1	
2nd person			1
3rd person			

2. Task allocation stage at which decision is made for each task $j_k = 1,2...\bar{j}_k$ to which of the household members $m \in M$ this task is assigned for implementation.

There is an appealing analogy between the maintenance task allocation problem and the joint non-mandatory activity problem described above. The frequency choice model has the same structure. The task allocation model has replaced the joint participation model. However, the task allocation model is similar to the joint participation model with the only principal difference that a single person is assigned to each task instead of a party.

Taking into account that the number of allocated maintenance activity types is limited to 2-3 (major shopping, grocery and incidental shopping, banking, other maintenance) and observed maximum number of maintenance tasks implemented by a household on a regular workday is limited to 3-4, a single simultaneous frequency choice model can be formulated that covers all possible frequencies and activity types as alternatives. The allocation stage in a general way can be viewed as a matrix where each household member has a row while maintenance tasks are represented by columns (Table 11.5). Allocation is associated with a value of 1 in the corresponding cell.

Similar to the joint participation choice matrix in the Table 11.2, the row totals are not controlled and can take any value from 0 to $\bar{J}$ since each person may not to take maintenance tasks, or take some of them, or take all of them. The row totals correspond to person workloads. Different from the joint participation choice matrix, the column totals are constrained to be equal to 1 since each task is allocated to a single person. Each task allocation matrix constitutes a distinct alternative. Total number of possible task allocation matrices is $\overline{M}^{\bar{J}}$ if we assume that every person can be assigned to any task. It should be noted that for a realistic situation of a household with 4 members and 5 tasks, it will result in only 625 possible matrices that is quite manageable for simultaneous choice, taking also into account that a parsimonious component-wise utility structure can be applied. However, for larger households where the number of allocation matrices can exceed 10,000, decomposition of the choice structure should be applied.

For example, the task allocation matrix can be modelled by a sequence of two modelling steps, preceded by a frequency choice model (Vovsha *et al.*, 2004b). The task allocation choice model that is applied for each task independently and returns a choice of a person the most suitable for the task as a function of the activity type and person characteristics (person type, residual time window left after mandatory activities, the number of joint and escorting tours in which the person participates, etc). Next the resulting fractional matrix of allocation probabilities $P(m|j)$ obtained at the previous stage is discretionized. The entire matrix discretizing is applied with fixed margins $P_m = \sum_j P(m|j)$ and $1 = \sum_m P(m|j)$ that corresponds to rows and columns of the matrix consequently. The reason why matrix discretizing is applied instead of a simple Monte-Carlo pick for each row is that independent pick for each task may result in illogical allocations with one person overloaded while the others will not have a task.

The same way as we considered simultaneous formulations for the joint non-mandatory activity and escorting, it can be done for the task allocation model. The simultaneous structure is more complicated analytically that any of the particular models applied in sequence, but it has an important advantage because all choices are modelled in a fully consistent way.

Consider first the task allocation matrix under condition of a fixed set of tasks. This choice model should properly take into account complicated structures of differential similarities across various matrices. The matrix view in the Table 11.5 above suggests several aggregations that can be effectively used in the choice structure. Similar to the joint participation matrix, there are aggregations associated with the row marginal totals (person workloads in terms of the total number of tasks allocated) as well as clusters of persons and subsets of similar tasks (of the same purpose or activity type). The following practical conclusions for substantiating the corresponding choice tree can be made:

- Nesting by persons with subsequent sub-nesting by a number of tasks taken by the person can be useful since it gives a reasonable dimension for similarities across different task allocation matrices. Two matrices can be considered similar if all persons take approximately the same number of tasks in both matrices.

- Intra-household clusters by person types can be used as additional nests above persons. This means that workload variations within the cluster are more probable, while changing workloads between clusters lead to principally different matrices.

Considering these nesting principles we arrive at the structure shown in the Figure 11.11 for a case of a 3-person household with 3 maintenance tasks generated.

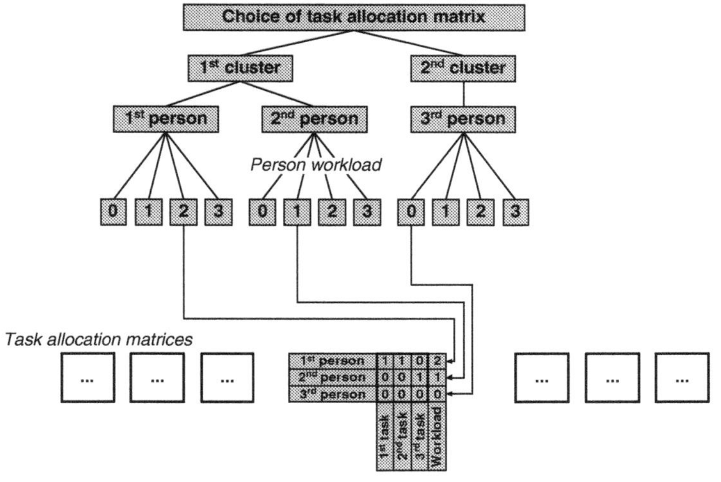

Figure 11.11
General Choice Structure for Maintenance Task Allocation

Since there are 3 person choices for each task we have $3^3 = 27$ possible allocation matrices at the lower choice level. Clusters correspond to persons that are substitutable and have similar daily activity patterns, for example, two workers while the third person can be a school child with a different propensity to implement maintenance tasks. Similar to the joint participation model, this structure is not a simple hierarchy and it cannot be modelled by a nested logit model. It requires generalized nested structures.

A structure of the participation matrix utility can be again effectively combined of a limited number of predetermined components in the following way:

$$U\left(\{m_j\}_{j\in J}\right) = \sum_j V_{k(j),m_j} + \sum_m V_{m,n(m)}, \qquad (9)$$

where,

V_{km} = elemental allocation utility (person suitability for the activity)

V_{mn} = person workload disutility to capture saturation effects at the person level

The further generalization of the model includes combining frequency choice and allocation choice in one choice structure. This essentially means that the variety of participation matrix should include all possible matrices with variable (rather than fixed) number of columns. There are two ways to construct such a model:

1. Consider frequency of joint tours as one more upper level (above the choice of participation matrix) in the hierarchy. This is the most straightforward way to combine frequency and allocation choices since when they are considered in a sequential fashion, frequency should naturally precede participation.
2. Consider a maximum observed number of tours for each person nest but allow for the allocation matrix to have empty columns with no persons assigned. This requires extension of the allocation matrix rules.

Both approaches produce the same full set of task allocation matrices as elemental choice alternatives. The differences relate to the way how the nests are structured and consequently how the correlation structure of the model is assumed.

CONCLUSIONS

The following conclusions can be made to summarize the current stage of the research:

- Incorporation of intra-household interactions constitutes an important aspect for further progress in modelling activity and travel-related decisions. The previously published research works were mostly focused on time allocation aspect and less on generation of activity episodes, trips, and travel tours that are necessary units for compatibility with regional travel demand models. Also, most of the approaches were limited to household heads only.
- The proposed approach distinguishes between three principal levels of intra-household interactions: 1) Coordinated principal daily pattern types, 2) Episodic joint activity and travel, 3) Intra-household allocation of maintenance activities. The proposed structure gives a good coverage for most frequent cases of intra-household interactions and serves as a constructive framework for derivation of operational choice models.
- Coordinated principal daily pattern types relate to the entire-day level of interactions. We consider three principal daily pattern types: (1) mandatory (work, university or school activities, which might include additional out-of-home non-mandatory activities); (2) non-mandatory travel (only non-mandatory activities at least one of which is out of home); and (3) staying at home or absence from town for the entire day. Statistical evidence shows strong coordination between household members at this principal level.

- Episodic joint activity and travel occur even if household members have chosen different pattern types since they may participate in shared activities and/or joint travel arrangements. We propose a classification of typical joint activity and travel types by the following categories that support the development of operational choice models:
  - Joint travel generated by the shared activity. This category is almost exclusively bound to non-mandatory activities (shopping, eating out, other maintenance, and discretionary activities) as well as almost exclusively implies a fully joint tour structure.
  - Joint travel to synchronized mandatory activities. This category has a significant share of drop-offs and pick-ups of school children made by workers on the way to and from work. Additionally, a significant percentage of school children travel together to and from school generating fully-joint tours and joint half-tours. Also carpooling of workers for commuting to work is observed, though this type has a comparatively low percentage.
  - Escorting that is a reported "altruistic" purpose of driving some other household member without participation in the activity. Statistical analysis has shown that majority of escorting is associated with serving children.
- Intra-household allocation of maintenance activities represents another important aspect of intra-household interactions. Many of the routine household maintenance activities (shopping, banking, visiting post office, etc) are implemented and scheduled individually; however, generation of such an activity and its allocation to a particular household member is a function of a household decision-making process.
  - It is shown that the choice structures can be reduced to a combination of a limited number of typical models that belong to the GEV class of logit-based models. These models together create an analytical framework for integrative modelling of the daily activity and travel of multiple household members, taking into account their interactions.
- The limited size of the paper and the authors' intention to discuss various types of intra-household interactions and the corresponding modelling structures in a comprehensive way made it impossible to include the results of statistical analysis and model estimation. The following list of sources where the estimation results were presented is suggested for an interested reader:
  - Parallel choice model (Gliebe, 2004; Gliebe and Koppelman, 2004, 2005)
  - Coordinated daily activity pattern type with sequential modelling of household members (Vovsha *et al.*, 2004a)

- Joint daily activity pattern type with simultaneous modelling of household members (Bradley and Vovsha, 2005)
- Generation and participation in joint non-mandatory activities (Vovsha *et al.*, 2003)
- Escorting children to school including ride-sharing and pure chauffeuring forms (Vovsha and Petersen, 2005)
- Generation and allocation of maintenance tasks (Vovsha *et al.*, 2004b)

REFERENCES

Borgers A., F. Hofman and H.J.P. Timmermans (2002). Conditional choice modeling of time allocation among spouses in transport settings. *Eur. J. Transpn. and Infr. Res.,* **2**, 5-17.

Bradley, M. and P. Vovsha (2005). A model for joint choice of daily activity pattern types of household members. *Proc. 84th TRB Conf.*, Washington, D.C. (CD-Rom).

Ettema, D., T. Schwanen and H.J.P. Timmermans (2004). The effect of locational factors on task and time allocation in households. *Proc. 83rd TRB Conf.*, Washington, D.C. (CD-Rom).

Fujii, S., R. Kitamura and K. Kishizawa (1999). Analysis of individuals' joint activity engagement using a model system of activity-travel behavior and time use. *Transpn. Res. Rec.*, **1676**, 11-19.

Gliebe, J. (2004). Models of Household Joint Decision Making in Activities and Travel. Ph.D. Dissertation, Northwestern University, Evanston, IL.

Gliebe, J. and F. Koppelman (2002). A model of joint activity participations between household members. *Transpn.*, **29**, 49-72.

Gliebe, J. and F. Koppelman (2005). An integrated model of individual and joint daily activity-travel patterns incorporating inter-agent choice set constraints. *Transpn. Res. B*, forthcoming.

Golob, T. and M.G. McNally (1997). A model of activity participation and travel interactions between household heads. *Transpn. Res. B*, **31**, 177-194.

Goulias, K. (2002). Multilevel analysis of daily time use and time allocation to activity types accounting for complex covariance structures using correlated random effects. *Transpn.*, **29**, 31-48

Meka, S., R. Pendyala R and M. Kumara (2002). A structural equations analysis of within-household activity and time allocation between two adults. *Proc. 81st TRB Conf.*, Washington, D.C. (CD-Rom).

Miller, E., M. Roorda and J. Carrasco (2003). A tour-based model of travel choice. *Proc.10th IATBR Conf.,* Lucerne, Switzerland (CD-Rom).

Scott, D. and P. Kanaroglou (2002). An activity-episode generation model that captures interaction between household heads: development and empirical analysis. *Transpn. Res. B*, **36**, 875-896.

Simma, A. and K.W. Axhausen (2001). Within-household allocation of travel – the case of Upper Austria. *Transpn. Res. Rec.*, **1752**, 69-75.

Srinivasan, S. and C. Bhat (2004). Modeling the generation and allocation of shopping activities in a household. *Proc. 83rd TRB Conf.*, Washington, D.C. (CD-Rom).

Townsend, T.A. (1987). The Effects of Household Characteristics on the Multi-Day Time Allocations and Travel/Activity Patterns of Households and their Members. Ph.D. Dissertation, Northwestern University, Evanston, IL.

Vovsha, P. and E. Petersen (2005). Escorting children to school: statistical analysis and applied approach. *Proc. 84th TRB Conf.*, Washington, D.C. (CD-Rom).

Vovsha, P., E. Petersen and R. Donnelly (2003). Explicit modeling of joint travel by household members: statistical evidence and applied approach. *Transpn. Res. Rec.*, **1831**, 1-10.

Vovsha, P., E. Petersen and R. Donnelly (2004a). Impact of intra-household interactions on individual daily activity-travel patterns. *Proc. 83rd TRB Conf.*, Washington, D.C. (CD-Rom).

Vovsha, P., E. Petersen and R. Donnelly (2004b). A model for allocation of maintenance activities to the household members. *Proc. 83rd TRB Conf.*, Washington, D.C. (CD-Rom).

Wen, C. and F. Koppelman (1999). Integrated model system of stop generation and tour formation for analysis of activity and travel patterns. *Transpn. Res. Rec.*, **1676**, 136-144.

Wen, C. and F. Koppelman (2000). A conceptual and methodological framework for the generation of activity-travel patterns. *Transpn.*, **27**, 5-23.

Zhang, J. and A. Fujiwara (2004). Representing heterogeneous intra-household interactions in the context of time allocation. *Proc. 83rd TRB Conf.*, Washington, D.C. (CD-Rom).

Zhang, J., H.J.P. Timmermans and A.W.J. Borgers (2002). A utility-maximizing model of household time use for independent, shared and allocated activities incorporating group decision mechanism. *Transpn. Res. Rec.*, **1807**, 1-8.

Zhang, J., H.J.P. Timmermans and A.W.J. Borgers (2004). Model structure kernel for household task and time allocation incorporating household interaction and inter-activity dependency. *Proc. 83rd TRB Conf.*, Washington, D.C. (CD-Rom).

12

AN EMPIRICAL COMPARISON OF ALTERNATIVE MODELS OF HOUSEHOLD TIME ALLOCATION

Junyi Zhang, Hiroshima University, Hiroshima, Japan

Akimasa Fujiwara, Hiroshima University, Hiroshima, Japan

Harry Timmermans, Eindhoven University of Technology, Eindhoven, The Netherlands

Aloys Borgers, Eindhoven University of Technology, Eindhoven, The Netherlands

INTRODUCTION

It is well known that household members interact in making decisions about the different activities that they perform and the related allocation of time. The importance and necessity of introducing group decision-making mechanisms into transportation should be evident. Joint activity participation, household resource allocation, task allocation and role specification are all key topics in activity-based analyses, and all involve the household as the decision-making unit.

However, to date, research in transportation on family/group decision-making issues is still scarce. Examining *joint activity participation*, Vovsha *et al*. (2003) pointed out that a significant portion of a region's travel is implemented by persons travelling together, either because they are sharing the same activity or because of cooperative arrangements where one of the travellers picks up or drops off the other on the way, and the travel decisions of the persons who travel together are naturally linked across such dimensions as choice of transport mode, destination, and time of day. Previously, Kostyniuk and Kitamura (1983) also argued that the relative importance of joint activity participation is evident in that joint activities tend to have a longer duration than non-work independent activities, and persons tend to stay out later and travel farther from home. Similarly,

Progress in Activity-Based Analysis edited by H. Timmermans

Chandrasekharan and Goulias (1999) analyzed solo versus joint trip making and showed that joint trip making is influenced by household size, household lifecycle, age, number of vehicles, and accessibility measures. Gliebe and Koppelman (2000) argued that joint activities arise from the same sources as independent activities, individual and household needs, but are motivated by different considerations, including efficiency resulting from time or money savings, altruism (e.g., helping another household member with a task, or providing social support) and companionship (e.g., preferences for joint versus individual leisure activities are primarily motivated by the need for companionship).

Car ownership is the most typical example of *resource allocation* within a household (especially in multi-adult, one car households). If more than one household member needs to use the same car, then group decision-making actions take place. The aspect has probably been addressed most in the sense that several activity-based models of transport demand do trace who is using the car, implying a constraint on the mode choice of other household members (e.g., Arentze and Timmermans, 2000).

As for *task allocation*, based on social and cultural norms and tradition different kinds of activities are often considered more suitable for particular household members. In their study about activity-travel behaviour, Hanson and Hanson (1981) showed that responsibilities for conducting some activities within a family (i.e. between men and women) is unequally distributed (child care is often a women's task), although there is tremendous variability in the allocation of specific tasks between the genders and ages from one culture to another (Rogers and Schlossman, 1990).

In addition to these decision problems related to the implementation of activity programs, the household is also the decision unit for long term decisions, involving the trade-off between residence, mode choice and job location. Examples are Timmermans, *et al.* (1992) and Badoe (2002).

Although increasingly more people are arguing in favour of a group decision model of activity-travel decisions, most of existing models either rely on individual choice models or do not consider the joint decision making process explicitly (Zhang *et al.*, 2002, 2004a and 2004b; Zhang and Fujiwara, 2004). It is also obvious that different contexts of activity-travel decisions usually involve different group decision-making mechanisms and consequently require different modelling approaches. It is not an easy task to cover various decision contexts in a single study. Therefore, this paper only focuses on household time allocation and attempts to empirically compare alternative models.

Conceptual Issues

Rules of Household Decision-Making

Household decision-making can be either consensual or accommodative (Davis, 1976). In the case of consensual decision-making, a household will continue to search for alternatives until one is found that satisfies the minimum level of expectations of all members. If decision-making under study is accommodative, there will be no way that one alternative can be satisfying to all. To arrive at the final decisions, the household members probably use various strategies, like bargaining, compromise, at different decision stages. Accordingly, individual household members might not maximize their own utility (Spiro, 1983). On the other hand, Kirchler (1995) argued that, for allocating its scarce resources and intensifying the emotional bonds between its members, a household will seek to minimize social and economic costs in decision situations by trying to make an optimal choice after passing through a commonly satisfying interaction process. Therefore, random utility maximization theory might be applicable to represent this kind of group decision-making mechanisms and will be applied to develop household time allocation models in this study.

Involvement and Relative Influence of Household Members

The involvement of household members differs according to decision types (Davis, 1976), and family life cycle (Cosenza and Davis, 1981). Concerning financial decisions (e.g., car and housing), the husbands' influence usually increases with technically complex items, whereas the probability for joint decision-making increases with increasing financial cost of the item. The spouses are observed to share money and asset management (i.e., saving and investment). For non-financial decisions, husbands and wives jointly decide about vacation, children's education, etc. Timmermans, *et al.* (1992) in their study of the choice of residential and job location involving dual earner households found that the influence of the man is stronger in the choice of job location, whereas on average women had more influence on the residential choice decision.

Intra-Household Interaction and Activity Dependency

Decisions made by different members within a household are not independent, suggesting the existence of *intra-household interaction*. The nature of such intra-household interaction is strongly influenced by the nature of the activity. A compulsory activity is by definition constrained to a particular household member and often also constrained by time, location and duration. This means that such activities likely receive a high priority and leave the household member performing such tasks with less flexibility. In turn, this may affect the allocation of other activities. Allocated activities will however also be influenced by role patterns within the household. Other mechanisms (e.g., turn-taking) may however prevail. This suggests a finer classification of activities might be desirable and this will be discussed later.

In addition to intra-household interaction, *activity dependency* will also influence the outcome of household decision-making process. Since each member has to perform different activities within his/her available time (e.g., 24 hours), the occurrence of one activity leads to the decrease of the probability of performing other activities.

Decision Constraints

Decisions on activity and travel behaviour are usually made under various constraints. These constraints include time and monetary constraints, capacity constraints, coupling constraints, authority constraints (see review by Ettema and Timmermans, 1997). Since this study focuses on the analysis of the relative influences of intra-household and intra-individual interactions, to simplify the discussion, only available time with respect to each household member is considered. Different members may have different time available. For example, some members may consider the whole day as a negotiation period while other members may take into account only a part of the whole day (e.g., the period except for work activity). However, since it is difficult to know how people perceive their available time, it is assumed here that each household member considers the whole period, which is a day in this study.

Classification of Activities

Every household has to perform a set of activities to survive or to give some meaning or pleasure. The utility of these activities will differ between individuals. Even though any number of activities can be modelled theoretically, proper classification helps meaningful interpretations of model estimation results. Here, activities are first classified into in-home activities and out-of-home activities. The out-of-home activities are further divided into independent, allocated and shared activities. An independent activity is an activity conducted by an individual household member, which is not a household task. Shared activities are those activities that require the presence of all or a subset of household members. An allocated activity is a household task that is assigned to a specific household member. The shared activities may be synchronized or non-synchronized activities. In the former case, household members carry out the shared activity together from the beginning to the end. In the latter case, household members share the activity partially. This study only deals with the synchronized activities.

This classification still assumes that decisions on activities within each category are homogeneous. This may however not be true in the sense that task allocation mechanisms may differ within a group of activities. In this study, therefore, a finer classification, involving more activity classes, will be used.

THEORETICAL COMPARISON OF MULTI-LINEAR AND ISO-ELASTIC HOUSEHOLD UTILITY FUNCTIONS

It is expected that different types of households may show different group decision-making mechanisms. For example, some households may attempt to maximize the utility of the weak (e.g., children) or the utility of the strong (e.g., wage-earners), while others might balance the utility of the household members. To represent such differing household decision-making mechanisms, the authors have proposed two representative types of household time allocation models (Zhang *et al.*, 2002, 2004a and 2004b; Zhang and Fujiwara, 2004). The first model adopts a multi-linear household utility function and the second uses an iso-elastic class of household utility function. The characteristics of these two types of household utility functions will be explained and theoretically compared below.

Characteristics of Multi-Linear Household Utility Function

The multi-linear household utility function is defined as follows:

$$HUF = H\left(u_1, u_2, \ldots, u_n\right) = \sum_{i=1}^{n} w_i u_i + \lambda \sum_{i=1}^{n} \sum_{i'>i} \left(w_i w_{i'} u_i u_{i'}\right) \quad (1)$$

$$w_i \geq 0 \text{ and } \sum_i w_i = 1 \quad (2)$$

where,

HUF denotes "Household Utility Function",
u_i is household member i's utility,
λ is parameter of intra-household interaction,

w_i is household member i's weight parameter, reflecting the relative influence of each member,

n is the number of household members.

One can see that the first term in the right side of equation (1) is the weighted average value of members' utilities. The weight w_i can be interpreted as a measure of a member's power or relative influence within the household. It also reflects the influence of the degree of involvement and/or the types of strategies adopted in the household decision-making process. The second term in the right side of equation (1) represents the influence of intra-household interaction. It is obvious that intra-household interaction is expressed in the form of Nash-type utility function without a reference point. The Nash-type utility function assumes that each household member identifies his/her most preferred outcome and the household then compromises by averaging along the resulting negotiation frontier (Curry *et al.*, 1991). Accordingly, the interaction parameter λ reflects household members' concern for achieving equality of utilities. A positive value of the interaction parameter λ means that the existence of intra-household interaction brings about an increase in household utility. The larger the value of λ, the higher the household's collective desire to choose a time allocation such that the utilities of all household members are approximately equal. On the other hand, a negative value of λ means that the existence of intra-household interaction reduces household utility and consequently suggests that the household does not prefer equality of members' utilities. The multi-linear household utility function finds its theoretical roots in "group decision theory" (e.g., Harsanyi, 1955). The multi-linear household utility function can include several special cases of utilitarianism-types of utility functions, as shown below.

Additive-Type Utility Function

$$HUF = \sum_{i=1}^{n} w_i u_i \quad (3)$$

In this case interactions between group members are not taken into account. This model can be arrived at when the household members first form their own separate utility function as a weighted average of part-worth utilities and then maximize the resulting mixture function.

Compromise-Type Utility Function

$$HUF = \sum_{i=1}^{n} (u_i / n) \quad (4)$$

In this case, household members have equal weights, called compromise weight by Curry and Menasco (1979).

Capitulation-Type Utility Function

$$HUF = \sum_{i=1}^{n} \overline{w}_i u_i \tag{5}$$

or

$$HUF = \sum_{i=1}^{n} w_i \overline{u}_i, \ \ \overline{u}_i = (n-1)^{-1} \sum_{i' \neq i} u_{i'} \tag{6}$$

where, $\overline{w}_i$ represents the average weight of other members relative to member i and is called capitulation weight, and $\overline{u}_i$ represents the average utility of other members relative to member i and is called capitulation utility. In this case, household interaction is taken into account by assuming that each household member uses other members' weights (utilities) as his or her own weight (utility) for joint decision-making.

Characteristics of Iso-Elastic Household Utility Function

The iso-elastic household utility function is defined as follows:

$$HUF = H(u_1, u_2, ..., u_n) = \frac{1}{1-\alpha} \sum_i w_i u_i^{1-\alpha}, \quad w_i \geq 0 \; and \; \sum_i w_i = 1 \tag{7}$$

where, α is a parameter indicating intra-household interaction and other notions are the same as in equation (1). The iso-elastic function is drawn from the research on social welfare function (Atkinson, 1983). The intra-household interaction parameter α describes how and to what extent the household positions its members (or considers the existence of its members) in the decision-making process and consequently makes its final decision-making. Therefore, different values of α and w_i, and the sign of α represent different household decision-making mechanisms. In other words, equation (6) can include various types of household utility functions as special cases. Several representative examples are given below.

Minimum Type Household Utility Function. If α is larger than one, increasing the utility of the weak member in a household leads to an increase in total household utility, where the weak member means his/her relative influence and/or utility is smaller. In case α reaches to positive infinity, equation (6) represents a "*minimum*" type household utility function. In other words, the household regards the utility of its weakest member as the household utility and maximizes it, as shown below.

$$HUF = \min\left(u_i \mid i = 1,2,...,n\right) \tag{8}$$

Nash-Type Household Utility Function. If α approximates one, equation (6) becomes a "*Nash-type*" household utility in the sense that each member first identifies his/her most preferred outcome of household decision-making and the household then compromises by averaging along the resulting negotiation frontier.

$$HUF = \prod_i \left(u_i\right)^{w_i} \tag{9}$$

Utilitarianism-Type Household Utility Function. If α is equal to 0, equation (6) ends in "*utilitarianism-type*" of household utility, which assumes that the household first averages its members' separate utilities and then maximizes the resulting mixture utility function, as shown in equations (3) ~ (5).

Autocratic Type of Household Utility Function. If α is negative, then equation (6) leads to "*autocratic type*" of household utility, which assumes that the household utility increases with the utilities of the strong household members.

Similarity and Dissimilarity

As discussed above, in representing intra-household interaction, the multi-linear and iso-elastic utility functions adopt different modelling strategies and overlay functionally.

Similarity. If $\lambda = 0$ and $\alpha = 0$, both household utility functions become additive types. In this case, the household utility is determined only by each member's relative influence (or power) within the household. Considering that power relationships within a household may be difficult to change , the additive type utility function suggests that even if decision situations change, the household might choose the same decision outcomes. The interaction term (the second term in the right side of equation (1)) for the multi-linear utility function represents the similar decision mechanism as shown in equation (9), i.e., the Nash-type function, which is a special case of iso-elastic utility function.

If $\lambda > 0$ and $\alpha < 1$, both types of functions suggest that household prefer the existence of intra-household interaction. In other words, the existence of intra-household interaction will lead to an increase in the household utility function. In this case, the relative influence parameter w_i will also work in the same way to bring about the change in household utility.

Dissimilarity. The two types of household utility functions integrate the above-mentioned common governing behavioural elements about household decision-making in a different manner. The multi-linear utility function is composed of an additive-type utility function and a Nash-type interaction term. In this sense, the multi-linear utility function uses the Nash-type interaction term to represent the households' consideration of equality during the joint decision-making process. In contrast, the iso-elastic function introduces the parameter α (see equation (6)), which is also called Atkinson's (1970) measure of aversion to inequality, to incorporate the influence of intra-household interaction. Because parameter α is the inverse of the elasticity of substitution along social indifference curves, it reflects households' preferences for trading off utility between their members. Due to the introduction of parameter α, the iso-elastic utility function can also represent "*minimum*" types of household decision-making mechanisms, which cannot be represented by the multi-linear utility function. Accordingly, the iso-elastic utility function seems more general and flexible in representing household decision-making mechanisms.

Summary of Household Time Allocation Models

The household time allocation model can be derived by maximizing the following Lagrange function, where T_i is household member i's available time and t_{ij} indicates the time of activity j.

$$L = H(u_1, u_2, \ldots, u_n) + \mu_i (T_i - \sum_j t_{ij}) \tag{10}$$

Specification of Members' Utility Functions. Before deriving the household time allocation model, one needs to specify each member's utility function. To incorporate the influence of *activity dependency*, which is usually ignored in conventional time allocation models, the multi-linear utility function is adopted to define each member's utility function as follows:

$$u_i = \sum_j r_{ij} u_{ij} + \sum_{j=1}^{J} \sum_{j'>j} \delta_i r_{ij} r_{ij'} u_{ij} u_{ij'} \tag{11}$$

where,

u_{ij} is household member i's utility for activity j,

δ_i is parameter of *activity dependency* for member i,

r_{ij} is household member i's weight (or relative interest) parameter for activity j, reflecting the relative importance of each activity for each member's utility, and

J is the number of activities for each member.

One can interpret the meanings of relative importance parameter r_{ij} and parameter δ_i of activity dependency, in the same way as for the relative influence parameter w_i and parameter λ of intra-household interaction. That is, r_{ij} refers to the importance of performing an activity for each member. The parameter δ_i reflects member *i*'s concern for achieving equality of utilities across different activities. Of course, δ_i can take any value along the real axis. A positive (or negative) value of δ_i means that the existence of activity dependency leads to an increase (or decrease) in each member's utility. Concerning the utility of each activity, the following logarithm function is adopted in order to derive an operational model.

$$u_{ij} = \rho_{ij} \, ln(t_{ij} + 1) \tag{12}$$

where,

t_{ij} is the time of individual *i* performing activity *j*, and

ρ_{ij} reflects member *i*'s heterogeneous evaluation of the allocated time for non-shared activity *j*.

Here, t_{ij} +1 is introduced to guarantee the computable logarithm function. To simplify the equation description, henceforth, it is assumed that t_{ij} refers to t_{ij} +1, and equation (12) is re-written as follows:

$$u_{ij} = \rho_{ij} \ln(t_{ij}) \tag{13}$$

One can see that, the utility (u_{ij}) for each activity is assumed to be non-negative and its marginal utility is monotonically decreasing. Kitamura (1984) examined the rationality of this function. Joh *et al.* (2002, 2003) suggested a more general and flexible utility function, but the use of this more complex function in models of household decision-making is left as a topic of future research.

Deriving Household Time Allocation Models. Different from other activities, the shared activity involves household members' joint activity participation. Since this study only deals with the synchronized activities, the involved household members share the activity time together from the beginning to the end. Therefore, $t_{is} = t_s$ holds for any involved member *i*. Reflecting this time constraint, Lagrange functions are shown below, with respect to the multi-linear and iso-elastic types of household utility functions, respectively, where the binary interaction term is only incorporated in the multi-linear function for the simplicity of discussion.

$$L = \sum_{i=1}^{n} w_i u_i + \lambda \sum_{i=1}^{n} \sum_{i'>i} (w_i w_{i'} u_i u_{i'}) + \sum_i \mu_i \left\{ T_i - \sum_j t_{ij} - \sum_s t_s \right\} \tag{14}$$

$$L = \frac{1}{1-\alpha} \sum_i w_i u_i^{1-\alpha} + \sum_i \mu_i \left\{ T_i - \sum_j t_{ij} - \sum_s t_s \right\} \tag{15}$$

Define

H in-home activity,
D_j jth out-of-home independent activity,
A_k kth out-of-home allocated activity, and
S_m mth out-of-home shared activity.

Then, maximizing equations (14) and (15) results in the following four types of household time allocation models:

(1) Time allocated to in-home activity

$$t_{iH} = T_i \cdot P(ns) \cdot P_i(H \mid ns) \tag{16}$$

(2) Time allocated to out-of-home independent activity

$$t_{iD_j} = T_i \cdot P(ns) \cdot P_i(D_j \mid ns) \tag{17}$$

(3) Time allocated to out-of-home allocated activity

$$t_{iA_k} = T_i \cdot P(ns) \cdot P_i(A_k \mid ns) \tag{18}$$

(4) Time allocated to out-of-home shared activity

$$t_{S_m} = T_i \cdot P(S_m) \tag{19}$$

where,

$$P(ns) = \frac{\sum_i (\psi_{iH} + \sum_j \psi_{iD_j} + \sum_k \psi_{iA_k})}{\sum_i (\psi_{iH} + \sum_j \psi_{iD_j} + \sum_k \psi_{iA_k} + \sum_m \psi_{iS_m})} \tag{20}$$

$$P(S_m) = \frac{\sum_i \psi_{iS_m}}{\sum_i (\psi_{iH} + \sum_j \psi_{iD_j} + \sum_k \psi_{iA_k} + \sum_m \psi_{iS_m})} \tag{21}$$

$$P_i(H \mid ns) = \frac{\psi_{iH}}{\psi_{iH} + \sum_j \psi_{iD_j} + \sum_k \psi_{iA_k}} \tag{22}$$

$$P_i(D_j \mid ns) = \frac{\psi_{iD_j}}{\psi_{iH} + \sum_j \psi_{iD_j} + \sum_k \psi_{iA_k}} \tag{23}$$

$$P_i(A_k \mid ns) = \frac{\psi_{iA_k}}{\psi_{iH} + \sum_j \psi_{iD_j} + \sum_k \psi_{iA_k}} \tag{24}$$

Here, '*ns*' represents the non-shared activity and the argument for each function is given as follows:

(i) Arguments of the multi-linear type household time allocation model

$$\psi_{iH} = mu_i mu_{iH} \rho_{iH} \tag{25}$$

$$\psi_{iD_j} = mu_i mu_{iD_j} \rho_{iD_j} \tag{26}$$

$$\psi_{iA_k} = mu_i mu_{iA_k} \rho_{iA_k} \tag{27}$$

$$\psi_{iS_m} = mu_i mu_{is} \rho_{is} + \sum_{i'} \left(\lambda w_i w_{i'} u_i mu_{i's} \rho_{i's} \right) \tag{28}$$

$$mu_i = w_i + \sum_{i'} \lambda w_i w_{i'} u_{i'} \tag{29}$$

$$mu_{ij} = r_{ij} + \sum_{j'} \delta r_{ij} r_{ij'} u_{ij'} \tag{30}$$

$$mu_{is} = r_{is} + \sum_{q \neq s} \delta r_{is} r_{iq} u_{iq} \tag{31}$$

(ii) Arguments of the iso-elastic type household time allocation model

$$\psi_{iH} = w_i u_i^{-\alpha} \Delta_{iH} \rho_{iH} \tag{32}$$

$$\psi_{iD_j} = w_i u_i^{-\alpha} \Delta_{iD_j} \rho_{iD_j} \tag{33}$$

$$\psi_{iA_k} = w_i u_i^{-\alpha} \Delta_{iA_k} \rho_{iA_k} \tag{34}$$

$$\psi_{iS_m} = w_i u_i^{-\alpha} \Delta_{iS_m} \rho_{iS_m} \tag{35}$$

$$\Delta_{ij} = r_{ij} + \delta_i \sum_{j'>j} r_{ij} r_{ij'} u_{ij'},\ j \in (H, D_j, A_k, S_m) \tag{36}$$

One can observe that both the derived models include a nested model structure for time allocation, in that the upper level treats the choice between non-shared activities and out-of-home shared activities, and the lower level treats the choice among non-shared activities. Furthermore, $P(S_m)$, $P(ns)$, $P_i(H \mid ns)$, $P_i(D_j \mid ns)$ and $P_i(A_k \mid ns)$ can be interpreted as both probabilities and proportional time shares that members perform the corresponding activities over a specified time, a zero-share meaning that no time is allocated to that activity, or the probability of performing that activity is zero. Accordingly, $P(S_m)$, $P(ns)$, $P_i(H \mid ns)$, $P_i(D_j \mid ns)$ and $P_i(A_k \mid ns)$ can also be used as task allocation probabilities.

Representing Heterogeneous Evaluation Structure about Activity Time

Decision makers and analysts alike do not know for sure the utility function of each activity. In addition, utility may change with the attributes of households and their members, as well as their mode choice behaviour, etc. To incorporate this kind of uncertainty and heterogeneity, equation (12) is re-written as follows, with respect to both the shared and non-shared activities.

$$u_{ij} = \exp\left(\left(\delta_j + \sum_k \beta_{jk} x_{ijk}\right) \ln\left(\sum_m \kappa_{jm} \tau_{ijm}\right) + \varepsilon_{ij}\right) \ln\left(t_{ij}\right) \tag{37}$$

$$u_{is} = \exp\left(\left(\delta_s + \sum_k \beta_{sk} x_{isk}\right) \ln\left(\sum_m \kappa_{sm} \tau_{ism}\right) + \varepsilon_{is}\right) \ln\left(t_s\right) \tag{38}$$

where,

x_{ijk} , β_{jk}	indicate the *k*th attribute of household member *i* affecting the non-shared activity *j*, and its parameter,
x_{isk}, β_{sk}	indicate the *k*th attribute of household member *i* affecting the shared activity *s*, and its parameter,
δ_j , δ_s	are the constant terms for the non-shared activities *j* and shared activity *s*,
τ_{ijm} , κ_{jm}	indicate the travel time of mode *m* spent for the non-shared activity *j* by household member *i*, and its parameter,
τ_{ism} , κ_{sm}	indicate the travel time of mode *m* spent for the shared activity *s* by household member *i*, and its parameter,
ε_{ij} , ε_{is}	are the error terms for activities *j* and *s*.

Equations (37) and (38) are adopted to reflect the heterogeneous influence of travel time on task and time allocation across households and/or their members. The logarithm function of travel time is used to ease model estimation. Because utility functions shown in equations (37) and (38) includes error terms, the estimations of household time allocation models should reflect the influences of these error terms. Equations (16) ~ (24) can be first log-transformed and then estimated using a seemingly unrelated regression (SUR) estimation procedure (Zeller, 1962).

Model Estimation and Discussion

Data for Model Estimation

To empirically compare the performance of the two representative household time allocation models, this study used a one-week activity diary survey administered in two small towns, Kakeya (population: 3,422, percentage elderly people: 33.3%) and Akagi (population: 4,036, percentage elderly people: 33.6%), in Shimane prefecture close to the Sea of Japan. These two towns were selected to investigate the spatial characteristics of time allocation patterns. To that end, one-week activity diary data from a total of 153 households were successfully collected. Each respondent was asked to report his/her activity and travel diary data from 6:00 am to 8:00 pm (i.e., the disposal time is assumed to be equal to 14 hours), reflecting the lifestyle of these two towns. The percentage of older people in these towns is substantial. When we compare the activity time between the elderly and others (Figure 12.1), it turns out that the out-of-home activity time for the elderly is 3.3 hours less than for the non-elderly. Figure 12.2 shows that more than 50% of the elderly have to rely on either "deliver/pick-up (37%)" provided by other household members, or a transit system (mainly bus: 10%).

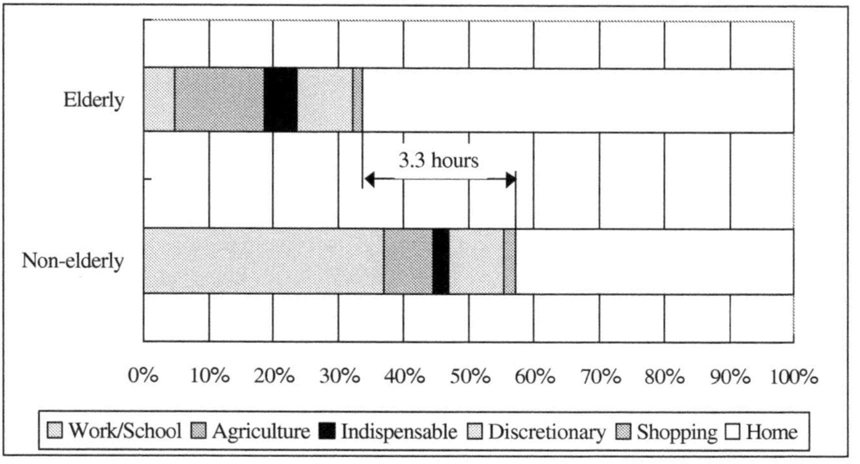

Figure 12.1
Time Allocation Pattern of the Elderly and Non-Elderly People

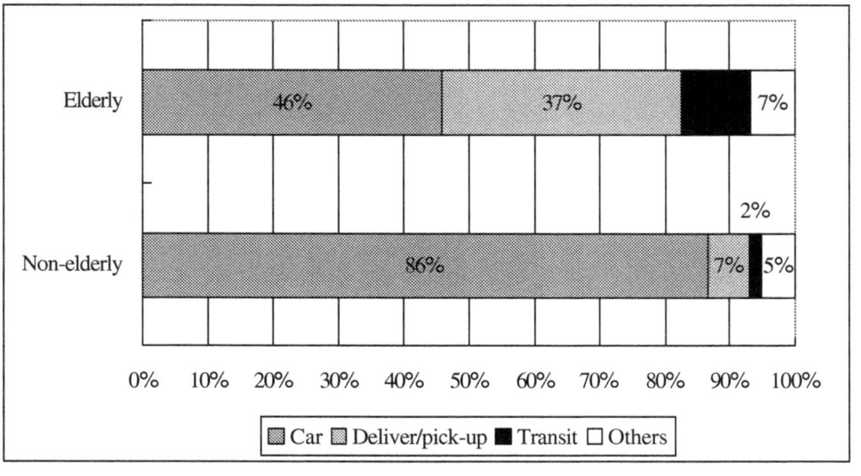

Figure 12.2
Modal Shares for the Elderly and Non-Elderly People

However, in February 2002, demand and supply adjustment regulations for the bus were abolished. As a result, private transport operators were allowed not only to freely open new services, but also to withdraw from unprofitable services. Since more than 80% of the female elderly people do not have a driver's license and are highly dependent on the bus services when going out, these elderly people will further have to rely on "deliver/pick-up" services after bus deregulation. However, household members who provide such "deliver/pick-up" services are often time-constrained.

Considering these matters, this study focuses on the joint decision-making about time allocation made by one household member who provides the "deliver/pick-up" service and another household member who receives "deliver/pick-up" service. The one-week activity diary data from 65 households, involving 454 person days were relevant and were therefore used for model estimation. The target activities include in-home activities, out of home independent activities (work, agriculture activity, indispensable activity and discretionary activity), allocated activities (shopping) and shared activities (the aforementioned activities performed together by household members).

Comparisons Based on Model Estimation Results

The following explanatory variables were used in model estimation: travel time by mode (car, deliver/pick-up, transit systems and others), age, and six dummy variables (living in Kakeya town: yes 1, no 0; holiday: yes 1, no 0; rainy day: yes 1, no 0; gender: male 1, female 0; housewife: yes 1, no 0 and driving license: yes 1, no 0). Model estimation results for the two household time allocation models are shown in Tables 12.1 and 12.2, respectively. Note that * = 5% level of significance, and ** represents 1% level of significance in these tables.

Comparison of Model Accuracy. The correlation between the observed and estimated activity time is shown in Table 12.3. Overall, the accuracy of the household time allocation model based on the multi-linear household utility function is better (correlation is 0.7749) than that of the model based on the iso-elastic utility function (the corresponding correlation is 0.7271). Both models show a higher correlation (0.8575 and 0.8257, respectively) for the non-shared activities performed by the members who receive "deliver/pick-up" services. Because most of these members are women, this study provides further evidence that the household time allocation model seems good at describing females' time allocation behaviour. Focusing on each activity, the multi-linear model is especially superior in representing out-of-home work activities for both members. In this case, its accuracy is about 10% higher than that of iso-elastic model. For other activities having a high correlation (e.g., 0.5), the difference between the two models is less than 5%. However, both models perform less well in describing time allocation for in-home activities and out-of-home agricultural activities of both members, and out-of-home dispensable and discretionary activities for the members who receive "deliver/pick-up" services. This may be an artefact of the data. The small difference in model accuracy in representing households' total activities suggests that any of the two household time allocation models can be used to properly represent household time allocation behaviour.

Table 12.1
Estimation Results of Multi-Linear Household Time Allocation Model

Variable	Household Member Providing "Deliver/Pick-up" Service			Household Member Receiving "Deliver/Pick-up" Service		
	Estimated Parameter		t-Statistic	Estimated Parameter		t-Statistic
Relative importance parameter for each activity.						
In-home activity	0.8990		-	0.9780		-
Out-of-home work activity	0.0190	**	16.187	0.0040	**	11.610
Out-of-home agriculture activity	0.0300	**	16.576	0.0070	**	11.715
Out-of-home indispensable activity	0.0080	**	16.414	0.0040	**	11.724
Out-of-home discretionary activity	0.0110	**	15.882	0.0040	**	11.143
Out-of-home shopping activity	0.0100	**	16.696	0.0030	**	11.747
Out-of-home shared activity	0.0230	**	10.086	0.0000		1.383
Activity dependency	0.0370	**	6.777	0.0920	**	5.308
Relative influence of each member	0.5380	*	2.157	0.4620		-
Intra-household interaction	0.2980		1.128	0.2980		1.128
Influence of travel time by mode						
Car	0.0810	**	12.359	0.0480	**	5.168
Deliver/pick-up	0.0790	**	5.961	0.0260	**	6.332
Transit systems	-0.0080		-0.979	0.0540	**	6.188
Others	0.0440	**	8.814	0.0180	**	6.665
Influence of characteristics of households and their members						
Constant term for agriculture activity	0.0910		0.276	12.2960	**	4.332
Constant term for other activities	7.3830	**	13.311			
Living at Kakeya town (Yes 1, No 0)	-0.3160	**	-2.890	1.8450	**	5.540
Holiday (Yes 1, No 0)	-0.4090	**	-3.918	0.6710	*	2.295
Rainy day (Yes 1, No 0)	-0.8160	**	-6.387	-1.0740	**	-3.068
Gender (Male 1, Female 0)	-0.1360		-0.907	4.1670	**	7.211
Age	-0.0460	**	-9.021	-0.0670	**	-6.390
Housewife (Yes 1, No 0)				-1.2747	**	-3.089
Driving license (Yes 1, No 0)				0.1140		0.204
Sample size	454 person*day					

Table 12.2
Estimation Results of Iso-Elastic Household Time Allocation Model

Variable	Household Member Providing "Deliver/Pick-up" Service			Household Member Receiving "Deliver/Pick-up" Service		
	Estimated Parameter		t-Statistic	Estimated Parameter		t-Statistic
Relative importance parameter for each activity.						
In-home activity	0.8797		-	0.9810		-
Out-of-home work activity	0.0083	**	18.803	0.0034	**	20.726
Out-of-home agriculture activity	0.0137	**	19.865	0.0064	**	20.651
Out-of-home indispensable activity	0.0032	**	19.359	0.0030	**	20.927
Out-of-home discretionary activity	0.0038	**	18.558	0.0037	**	19.311
Out-of-home shopping activity	0.0038	**	19.808	0.0026	**	20.918
Out-of-home shared activity	0.0874	**	13.475	0.00001		0.001
Activity dependency	0.0439	**	8.577	0.0010	**	2.927
Relative influence parameter for each member	0.8183	**	12.080	0.1817		-
Parameter of intra-household interaction	-1.3857	**	2.948	-1.3857	**	2.948
Influence of travel time by mode						
Car	0.0808	**	15.046	0.0620	**	8.950
Deliver/pick-up	0.0829	**	6.710	0.0213	**	10.989
Transit systems	0.0074		0.868	0.0508	**	9.786
Others	0.0382	**	10.728	0.0149	**	11.092
Influence of characteristics of households and their members						
Constant term for agriculture activity	0.0193		0.063	15.3975	**	5.737
Constant term for other activities	6.3525	**	13.572			
Living at Kakeya town (Yes 1, No 0)	-0.3295	**	3.436	1.8451	**	7.610
Holiday (Yes 1, No 0)	-0.3664	**	3.816	0.6713	**	3.442
Rainy day (Yes 1, No 0)	-0.3640	**	3.312	-0.7310	**	3.254
Gender (Male 1, Female 0)	-0.2704	*	2.065	4.1670	**	10.850
Age	-0.0217	**	5.045	-0.0625	**	8.188
Housewife (Yes 1, No 0)				-1.2747	**	4.858
Driving license (Yes 1, No 0)				0.1138		0.400
Sample size	454 person*day					

Table 12.3
Correlation between the Observed and Estimated Activity Time

Activity	Multi-Linear Model		Iso-Elastic Model	
	Household Member who Provides "Deliver/Pick-up" Service	**Household Member who Receives "Deliver/Pick-up" Service**	**Household Member who Provides "Deliver/Pick-up" Service**	**Household Member who Receives "Deliver/Pick-up" Service**
In-home activity	0.2958	0.2916	0.1856	0.2759
Out-of-home work activity	0.7307	0.6257	0.6627	0.5671
Out-of-home agriculture activity	0.1542	0.1421	0.1474	0.1403
Out-of-home dispensable activity	0.5327	0.3356	0.5300	0.2893
Out-of-home discretionary activity	0.5619	0.4642	0.5411	0.4258
Out-of-home shopping activity	0.7188	0.7517	0.7196	0.6492
Out-of-home shared activity	0.4528		0.4453	
Total non-shared activities	0.6847	0.8575	0.6445	0.8257
Household's total activities	0.7749		0.7271	

Table 12.4
Proportion of Zero Observations of Activity Time

Activity	**Household Member who Provides "Deliver/Pick-up" Service**	**Household Member who Receives "Deliver/Pick-up" Service**
In-home activity	1%	1%
Out-of-home work activity	59%	87%
Out-of-home agriculture activity	70%	76%
Out-of-home indispensable activity	83%	89%
Out-of-home discretionary activity	74%	80%
Out-of-home shopping activity	83%	89%
Out-of-home shared activity	62%	64%

Table 12.5
Relative Influence of Travel Time by Mode

Travel mode	Household Member who Provides "Deliver/Pick-up" Service		Household Member who Receives "Deliver/Pick-up" Service	
	Multi-Linear Type Model	Iso-Elastic Type Model	Multi-Linear Type Model	Iso-Elastic Type Model
Car	1.000	1.000	1.000	1.000
Deliver/pick-up	0.975	1.025	0.542	0.343
Transit system	-0.099	0.091	1.125	0.819
Others	0.543	0.472	0.375	0.240

Comparison of Household Decision-Making Patterns. Since the parameters of intra-household interactions are 0.2980 in the multi-linear model and –1.3857 (i.e., $1-\alpha = 2.3857$) in the iso-elastic model, both models show that the existence of intra-household interaction brings about an increase in households' utilities, and households first form their own separate utility function as a weighted average of part-worth utilities and then maximize the resulting mixture function incorporating the influence of intra-household interaction. However, the statistical insignificance of intra-household interaction parameters in the multi-linear model needs to be further examined. A reason might be the finer classification of activities, where some activities are not performed at all by a portion of respondents (Table 12.4). This suggests the necessity of applying a more efficient estimation method to incorporate the influence of zero observations. Based on current model estimation results, it might be concluded that the iso-elastic model seems suitable to explore the intra-household interaction.

Focusing on household members' relative influence in joint decision-making process, the multi-linear model estimates almost equal influences (0.5380 for the household members who provide "deliver/pick-up" services, and 0.4620 for the household members who receive the services). However, the iso-elastic model shows a very high influence (0.8183) by the members who provide "deliver/pick-up" services. The estimated parameters of intra-household interactions might result in such a large difference in relative influence. Concerning each member's time allocation, both models obtain positive and statistically significant values of activity dependency parameters. This implies that 1) activity dependency does takes place, and 2) people prefer the existence of activity dependency. Observing the relative importance parameters, the two models also show similar preference structures about activities under question. In other words, both models estimate that household members attach much more importance to in-home activity in the sense that each corresponding relative importance parameter is about 0.9.

Comparison of Heterogeneous Evaluation of Activity Time. First, the influence of travel time by mode was examined. Both models show that most of parameters for travel time by mode are statistically significant and positive. Since travel time is introduced in the form of a reverse function, this implies that an increase in travel time will reduce the time allocated to activities on average. This observation is intuitive and suggests the rationality of both models. In addition, both models show that travel time by transit system is not significant for the members who provide "deliver/pick-up" services. To understand the relative influence of travel time by mode, the ratio of travel time parameter relative to the corresponding parameter by car was calculated (Table 12.5). Both models show similar influence patterns. More specifically, both members show a higher preference for car travel time, and the household members who provide "deliver/pick-up" services regard "deliver/pick-up" travel time and the members who receive "deliver/pick-up" services regard transit system time, as nearly as important as car travel time.

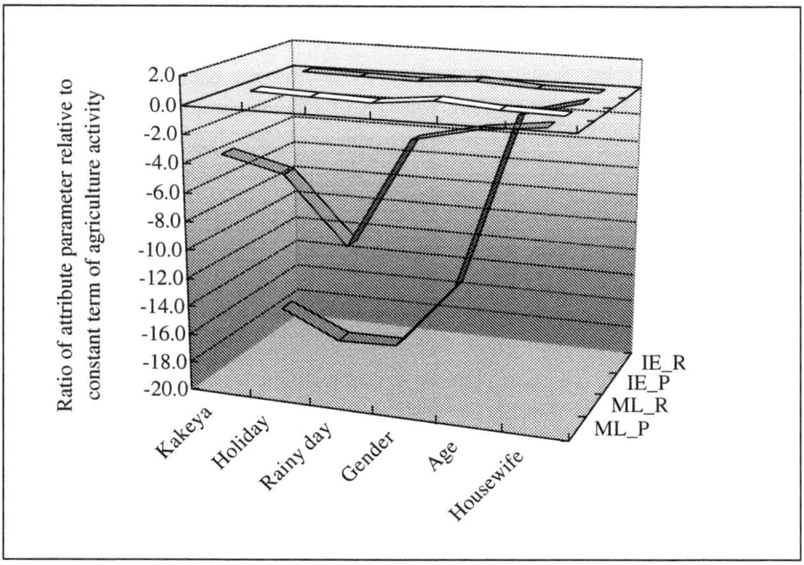

ML : Multi-linear type household time allocation
IE : Iso-elastic type household time allocation
P : Household member who provides "deliver/pick-up" service
R : Household member who receives "deliver/pick-up" service

Figure 12.3
Influence Patterns of Attributes of Households and Their Members

Next, the influence of attributes of household and their members was analysed. The relative influence of each attribute was examined by calculating the ratio of attribute parameter relative to the constant term of agriculture activity (Figure 12.3). It should be noted that any parameter could be used as a reference and the ratio of the driving license parameter is not shown in the figure because it is statistically insignificant. The two models show some differences in terms of the magnitudes of parameters of the following dummy variables: living in Kakeya town, holiday, and rainy day. Because these dummy variables are all related to the decision environment, this result suggests that the iso-elastic model is more sensitive to decision environments.

CONCLUSIONS

It has been widely recognized that travel patterns are influenced by the way in which households organize their activities in time and space and various types of activities are usually performed to satisfy the needs of both households and their members. Many households are multi-person households, and consequently household member interactions and exchanges within a household potentially might have a strong impact on household activity-travel patterns. Accordingly, decisions on activity-travel behaviour usually involve a group/household decision-making process. The focus on the decision maker as a group acting collectively suggests a reorientation of many existing travel choice theories and methodologies.

To represent household decision-making mechanism in the context of time allocation, the authors have proposed two types of household time allocation models based on the principle of random utility maximization. One assumes a multi-linear type of household utility function and another adopts the iso-elastic class of welfare function. In representing intra-household interaction, these two types of household utility functions overlap functionally. On the other hand, theoretical comparisons suggest that the iso-elastic utility function seems more general and flexible in representing household decision-making mechanisms.

To empirically compare these two representative household time allocation models, a one-week activity diary data, collected at a depopulated region in Japan, was used for model estimation. The estimation of both models on these data confirmed that both models have a relative high goodness-of-fit. Significant difference in model accuracy is not observed. Moreover, these two models all estimate that, on average, the households prefer the existence of intra-household interaction. This might be because a household is characterized by a high degree of intimacy among the members, strong, frequent and diverse interdependence that lasts over considerable periods of time in a widening array of settings. In the context of the random utility maximization principle, it is found that the household members first form their own separate utility function as a weighted average of part-worth utilities and then maximize the resulting function, and the household time allocation model seems good at describing female's time allocation behaviour.

In spite of the almost equal model accuracy and the similarity in observed household decision-making rules, the two models also show large differences in household members' relative influences. This implies that these differences might be behaviourally intrinsic and consequently, there may exist multiple solutions for household time allocation to achieve the maximum level of household utility. In other words, household members might follow different paths during the negotiation process to arrive at their most preferred outcomes. This seems intuitive and also suggests the necessity of introducing such mechanisms in the process of model development. However, considering the fact that there exist many zero observations in activity time, it seems that at this stage we cannot draw a definitive conclusion to decide which of the two models is the winner in representing household time allocation behaviour.

ACKNOWLEDGMENTS

Part of funding for the analysis was provided by the 21st Century COE (Centre of Excellence) Program "Social Capacity Development for Environmental Management and International Cooperation" at Graduate School for International Development and Cooperation, Hiroshima University, Japan. The authors also appreciate Ms. Yuko Oonishi's efforts in data preparation and model estimations.

REFERENCES

Arentze, T.A. and H.J.P. Timmermans (2000). *Albatross: A Learning Based Transportation Oriented Simulation System.* European Institute of Retailing and Services Studies, Eindhoven.

Atkinson, A.B. (1970). On the measurement of inequality. *J. Econ. Theory*, **2**, 244-263.

Atkinson, A.B. (1983). *The Economics of Inequality*. Clarendon Press, Oxford.

Badoe, D.A. (2002). Modelling work-trip mode choice decisions in two-worker households. *Transpn. Plan. & Techn.,* **25**, 49-73.

Chandrasekharan, B. and K.G. Goulias (1999). Exploratory longitudinal analysis of solo and joint trip making in the Puget Sound transportation panel. *Proc. 78th Ann. Meet. Transpn. Res. B.,* Washington, D.C. (CD-Rom).

Cosenza, R.M. and D.L. Davis (1981). Family vacation decision making over the family life cycle: A decision and influence structure analysis. *J. Trav. Res.*, **17**, 17-23.

Curry, D.J. and M.B. Menasco (1979). Some effects of differing information processing strategies on husband-wife joint decision. *J. Cons. Res.*, **6**, 192-203.

Curry, D.J., M.B. Menasco and J.W. van Ark (1991). Multiattribute dyadic choice: models and tests. *J. Mark. Res.*, **28**, 259-267.

Davis, H.L. (1976). Decision making within the household. *J. Cons. Res.*, **2**, 241-260.

Ettema, D.F. and H.J.P. Timmermans (1997). *Activity-Based Approaches to Travel Analysis.* Pergamon Press, Oxford.

Gliebe, J.P. and F.S. Koppelman (2000). A model of joint activity participation. *Proc. 9th IATBR Conf.*, Gold Coast, Queensland, Australia (Cd-Rom).

Hanson, S. and P. Hanson (1981). The impact of married women's employment on household travel patterns: a Swedish example. *Transpn.*, **10**, 165-183.

Harsanyi, J.C. (1955). Cardinal welfare, individualistic ethics, and interpersonal comparisons of utility. *J. of Pol. Econ.*, **63**, 309-321.

Joh, C.-H., T.A. Arentze and H.J.P. Timmermans (2002). Modelling individuals' activity-travel rescheduling heuristics: theory and numerical experiments. *Transpn. Res. Rec.*, **1807**, 16-25.

Joh, C.-H., T.A. Arentze and H.J.P. Timmermans (2003). Estimating non-linear utility functions of time use in the context of an activity schedule adaptation model. *Proc. 10th IATBR Conf.*, Lucerne, Switzerland (CD-Rom).

Kirchler, E. (1995). Studying economic decisions within private households: a critical review and design for a "couple experiences diary". *J. Ec. Psych.*, **16**, 393-419.

Kitamura, R. (1984). A model of daily time allocation to discretionary out-of-home activities and trips. *Transpn. Res. B*, **18**, 255-266.

Kostyniuk, L.P. and R. Kitamura (1983). An empirical investigation of household time space paths. In: *Recent Advances in Travel Demand Analysis* (S.M. Carpenter and P.M. Jones, eds.), pp. 266-289. Gower Publishing, Aldershot.

Rogers, B.L. and N.P. Schlossman (1990). *Intra-household Resource Allocation: Issues and Methods for Development Policy and Planning.* United Nations University Press, New York.

Spiro, R.L. (1983). Persuasion in family decision-making. *J. Cons. Res.*, **9**, 393-402.

Timmermans, H.J.P., A.W.J. Borgers, J. van Dijk and H. Oppewal (1992). Residential choice behaviour of dual-earner households: a decompositional joint choice model. *Env. & Plan. A*, **24**, 517-533.

Vovsha, P., E. Petersen and R. Donnelly (2003). Explicit modelling of joint travel by household members: statistical evidence and applied research. *Proc. 78th Ann. Meet. Transpn. Res. B.*, Washington, D.C. (CD-Rom).

Zeller, A. (1962). An efficient method of estimating seeming unrelated regressions and tests of aggregation bias. *J. Am. Stat. Assoc.*, **57**, 348-368.

Zhang, J. and A. Fujiwara (2004). Representing heterogeneous intra-household interaction in the context of time allocation. *Proc. 83rd Ann. Meet. Transpn. Res. B.*, Washington, D.C. (CD-Rom).

Zhang, J., H.J.P. Timmermans and A.W.J. Borgers (2002). A utility-maximizing model of household time use for independent, shared and allocated activities incorporating group decision mechanisms. *Transpn. Res. Rec.*, **1807**, 1-8.

Zhang, J., H.J.P. Timmermans and A.W.J. Borgers (2004a). A model of household task allocation and time use. *Transpn. Res. B*, **39**, 81-95.

Zhang, J., H.J.P. Timmermans and A.W.J. Borgers (2004b). Model structure kernel for household task and time allocation incorporating household interaction and inter-activity dependency. *Proc. 83rd Ann. Meet. Transpn. Res. B.*, Washington, D.C. (CD-Rom).

13

Incorporating Latent Determinants in Activity Time Allocation Modelling: Application to Value of Activity Time Estimation

K. P. Nepal, Tokyo Institute of Technology, Tokyo, Japan

D. Fukuda, Tokyo Institute of Technology, Tokyo, Japan

T. Yai, Tokyo Institute of Technology, Tokyo, Japan

Introduction

The role of time in the field of transportation planning has shifted as a subject of modelling and application from its relative obscurity a few decades ago to the central stage in recent years (Pas, 1998). This shift is due to the focus on activity-based transportation analysis. Activity-based models not only help us to analyze the links between the decisions about participation in different activities and the use of transportation systems and services, but also reveal the underlying motivational process of engagement in various activities and, hence, travel behaviour. The current themes of activity-based travel analysis are time allocation and duration, scheduling of activities in time and space, constraints on movements and activity choices, interactions between decisions and different individuals, household roles and structures and the processes of adaptation and change (Kalfs and Harvey, 2002).

There are a number of reasons for studying activity time allocation and valuation of activity duration, such as analyzing travel behaviour, examining transport policy options and evaluating

Progress in Activity-Based Analysis edited by H. Timmermans

transportation projects, to name only a few. The behavioural representation of such models is not easy as they require modelling of human behaviour. Transportation researchers have attempted to develop realistic and estimable models of activity time allocation using theoretical developments and experimental investigations for at least the last 20 years, and there has already been significant progress on some issues. It has been widely recognized that the basic social and psychological needs and/or desires motivate an individual's daily activity time allocation. However, even recent studies in activity time allocation modelling have not incorporated, at least explicitly, important but not directly observable motivational factors of time allocation. Also, despite the importance of calculating the benefits of saving time, models that estimate the value of activity time, broken down by activity types, have not been developed. Jara-Diaz and Guevara (2003) developed a combined work trip mode choice model and activity demand model from a common microeconomic framework to estimate the components of subjective value of travel time saving (SVTTS). However, their model can only estimate the value of time allocated to travel activity and work activity. Jara-Diaz (2003) has developed a theoretical framework for establishing all possible technical relations between the consumption of goods and assignment of time to each activity; however, the estimation of these models has not been explored.

Latent variable models (Loehlin, 1998; Bartholomew and Knott, 1999) are being used to model travel choices by integrating with discrete choice models (Ben-Akiva *et al.* 2002, Morikawa *et al.* 2002). They are also useful in activity time allocation modelling because of their abilities to handle measurement difficulty with respect to certain determinants of time allocation decisions. Latent variable models help us to estimate unobservable covariates indirectly through certain indicators and observable explanatory variables. They are particularly relevant when analyzing a decision maker's cognitive process, including attitudes, perceptions, intentions and other psychometric constructs. These latent determinants have significant role in an individual's activity choices and time allocation decisions. Examples include: social needs, desire for and priorities in relation to certain activities, inconveniences or difficulties encountered due to joint participations, time pressures etc. and, the necessity for a particular activity. These determinants are difficult to measure directly from observation, but they do have a significant influence on activity time allocation behaviour. A model that includes these latent factors constitutes a more realistic representation of an individual's socioeconomic environment and his or her perception of time use. This research has two purposes: one is to improve the existing modelling frameworks (Bhat and Misra, 1999; Prasetyo *et al.,* 2003; Jara-Diaz and Guavara, 2003) with respect to individual activity time allocation by incorporating latent determinants from the latent variable model; the other is to use this improved model to calculate the value of activity time by activity type.

This study deals with a micro-economic individual activity time allocation model based on the principle of random utility maximization. The model will contribute to a better understanding of the fundamental nature of activity time allocation decisions and the valuation of activity duration. In this study, we will examine individuals' decisions whether or not to participate in an activity and for how long. In addition, we will estimate individual's valuation of time allocated to various activities at the disaggregate level. We treat the travel modes and activity locations as fixed, so the total travel time and travel cost over the study period are exogenous to the model system and are not included in the individual's direct utility function. We pay particular attention to reformulating the existing basic modelling framework of activity time allocation. The mathematical formulation and derivation of an estimable extended econometric model with latent determinants will be clearly described. We conducted a small-scale empirical study, as a pilot investigation, to verify the significance of the model.

A General Micro-Economic Model of Activity Time Allocation

Micro-economic models are the basis of most existing research on modelling individual activity time allocation. They assume that individuals make choices concerning the allocation of time and money to different activities to maximize their total utility. Unlike structural equation models (Golob, 2001) or duration models (Mannering *et al.*, 1994; Ettema *et al.*, 1995; Bhat, 1996), they help us to estimate value of activity time by activity type, a quantity of particular relevance in transport economics. Becker (1965) laid down the foundations for explicitly incorporating time components in a utility function, although not directly but as the inputs of utility arguments (final commodities), and this study has had significant impact on economists and transportation researchers. Various other researches have built on Becker's foundation (see Kraan, 1996; Jara-Diaz, 1998, 2000; Meloni *et al.*, 2004 for detailed discussion). DeSerpa (1971) directly incorporated activity time allocation components, in addition to the good consumption components in the direct utility function, and theoretically proposed three distinct components of value of time in lower-bounded consumption-time constrained activities: value of time as a resource, value of activity time and value of saving time in the consumption-time constrained activity. Kitamura (1984) introduced random utility models of time allocations, under a utility maximizing principle, to formulate estimable models of activity time allocation. Yamamoto and Kitamura (1999) further extended this research, incorporating interactions between working and non-working days using the Tobit censored regression model (Tobin, 1958). Bhat and Misra (1999) employed the same concept to model weekly discretionary activity time allocations between in-home and out-home and between weekdays and weekends using log-linear regression model. Recently, Meloni *et al.* (2004) analyzed time allocations to discretionary in-home and out-home activities including trips using a nested-Tobit model. These models used only time allocation components in direct utility function and a time constraint. On the other hand, Kockelman (2001) developed a model for time and budget constrained activity demand analysis in micro-economic utility maximizing foundation using Roy's identity. Prasetyo *et al.* (2003) and Nepal *et al.* (2004) considered the effects of motivational factors of time allocation under time and money constraints.

Kitamura (1984) formulated the random utility of an activity as a multiplicative functional form of explanatory variables, a random disturbance term, and a natural logarithm of the time allocated to the activity. Since his model includes only activity time allocation components in the direct utility function with a time constraint only, the econometric specification in the form of the Tobit model is somewhat simplistic. Recently, therefore, other specifications for duration of activities have been formulated (e.g., Joh *et al*., 2002). Models using only a time constraint can no longer help us to estimate the value of activity time because income constraint is a must to calculate the marginal utility of income. In this study, we represent the random utility of an activity as the multiplicative functional form of random parameters and the natural logarithm of utility arguments over a study period. The random parameters are assumed to depend on a set of explanatory variables and a disturbance term in the exponential functional form in order to ensure these parameters to be strictly positive. As the activity time allocation model developed in this study is based at the individual level, the subscript for the individual is omitted for the convenience of notation.

Theoretical Development

The arguments of the direct utility function play a major role in the derivation of the activity time allocation model. We use DeSerpa's (1971) functional form of utility to derive a general model of activity time allocation. According to DeSerpa, the total direct utility of an individual depends on all activity time allocations and all goods consumed over the specified study period:

$$U = U(\mathbf{t},\mathbf{z}) = U(t_1, t_2, ..., t_j, ..., t_J, z_1, z_2, ..., z_k, ..., z_K) \tag{1}$$

where,

U is the direct utility function of an individual;
$\mathbf{t}$ is a J-dimensional vector of consumption time to all activities during the study period;
$\mathbf{z}$ is a K-dimensional vector of goods consumed during the study period;
t_j is the time allocated to activity j during the study period; and
z_k is the consumption of good k during the study period.

However, since separate consumption of goods and related prices are not generally included in the activity time use surveys, all goods are usually represented as composite goods (Z) with composite price (P). Hence, (1) can be re-written as:

$$U = U(\mathbf{t}, Z) = U(t_1, t_2, ..., t_j, ..., t_J, Z) \qquad (2)$$

The individual maximizes (2) under time constraint (3) and money budget constraint (4).

$$T - \sum_j t_j - T_t = 0 \qquad \Rightarrow \mu \qquad (3)$$

$$y - \sum_j r_j t_j - PZ - T_c \geq 0 \qquad \Rightarrow \lambda \qquad (4)$$

where,

r_j is the unit cost of participation to activity j other than cost of goods (positive: if individual pays, negative: if the individual is paid and $r_w = -w$);
w is the wage rate;
T_t is the total travel time during the study period;
T_c is the total travel cost during the study period;
y is an individual's fixed income apart from market labour;
μ is a Lagrange multiplier of time resource constraint; and
λ is a Lagrange multiplier of total money budget constraint.

Basically, there are two types of time use constraints: time resource, and consumption time (Bain, 1976). Time resource is the fixed time endowment for doing different activities, and, hence, a constraint. This time resource constraint requires that the amount of time allocated to specific activities add up to the total time available, as shown in (3). This relationship follows directly from the assumption that activities are performed one at a time, and that all available time is allocated to activities. Hence, the ratio μ/λ is seen as the value of extending time resource, and is known as the value of time as a resource (VOTR). Each activity can be performed only at the expense of time. The amount of time allocated to the consumption activity is partly a matter of choice, and partly a matter of necessity. The consumption time constraint applies only when it is binding. For example, individuals have to spend a minimum necessary travel time, but they generally prefer to spend as little travel time as possible. Hence, there is a value in lowering the minimum necessary travel time; this is known as the value of travel time savings (VTTS). However, since the proposed modelling framework assumes exogenous total daily travel time, only consumption times at the destinations are modelled. The consumption times at the destinations are, in general, non-binding because individuals are free to allocate their time to different activities except for a few exceptions of opening hours of supermarkets, movie show times at a theatre etc. When consumption time at the destinations is non-binding, the valuation of the consumption time to a particular activity is just the value of time as a commodity, and is known as the value of activity time (VOAT). The value of activity time is the marginal rate of substitution of time allocated to the particular activity for money and is usually estimated by taking the ratio of the marginal utility of time allocated to the activity and marginal utility of total money budget:

$$VOAT_j = \frac{\partial U / \partial t_j}{\lambda}, \qquad \forall j \tag{5}$$

The constrained utility maximization formulations (2), (3) and (4) can be solved by the Lagrange principle:

$$L = U(.) + \mu(T - \sum_j t_j - T_t) + \lambda(y - \sum_j r_j t_j - PZ - T_c) \tag{6}$$

From the first order conditions with respect to the decision variables t_j and Z, we can get the following relationships:

$$\frac{\partial U}{\partial t_j} = \mu + \lambda r_j \text{, and } \frac{\partial U}{\partial Z} = \lambda P \tag{7}$$

$$\frac{\partial U}{\partial t_j} = \mu + \frac{1}{P} \times \frac{\partial U}{\partial Z} \times r_j \tag{8}$$

If we want to focus on activity time allocation to leisure activities, and assume that work and income are exogenous, work activity will not be explicitly modelled as an activity. In general, individuals are not able to adjust the length of the work time within the given time resource period (day or week) according to their preferences. For work activity from (8),

$$\frac{\partial U}{\partial t_w} = \mu - \frac{1}{P} \times \frac{\partial U}{\partial Z} \times w \tag{9}$$

Substituting the value of μ from (9) in (8):

$$\frac{\partial U}{\partial t_j} = \frac{\partial U}{\partial t_w} + \frac{1}{P} \times \frac{\partial U}{\partial Z}(r_j + w), \quad \forall j \neq w \tag{10}$$

Econometric Specification

We need to make further assumptions to arrive at an estimable activity time allocation model from (10). We use the logarithmic Cobb-Douglas functional form of utility over its arguments. Kitamura (1984) has discussed the suitability of such a functional form of utility.

$$U = \sum_j A_j \ln(t_j) + B \ln(Z) \tag{11}$$

The functional form of the utility (11) reveals that for positive utility parameters, the increase in the values of arguments in utility increases the total utility but the marginal utility decreases, as shown in (12). The marginal utilities with respect to decision variables t_j and Z are:

$$\frac{\partial U}{\partial t_j} = \frac{A_j}{t_j}, \text{ and } \frac{\partial U}{\partial Z} = \frac{B}{Z} \tag{12}$$

where A_j is the random parameter associated with the activity *j*, and B is the parameter associated with the quantity of composite goods consumed (assumed as non-random). The functional form of A_j determines the shape of the utility function and varies over population. We assume that it depends on the activity's characteristics and on the individual's socioeconomic characteristics. To ensure that the parameter A_j is strictly positive, we used exponential functional forms of the parameters with deterministic and stochastic terms:

$$A_j = e^{\mathbf{x}_j \hat{\mathbf{a}}_j + \varepsilon_j} \tag{13}$$

where,

$\hat{\mathbf{a}}_j$ is a parameter vector associated with activity *j*;

$\mathbf{x}_j$ are explanatory variables associated with activity *j*; and

ε_j is a random disturbance term associated with activity j.

Since consumption of goods and related market prices are not generally observed in activity time use surveys, equation (4) allows us to substitute the expenditures on goods, G, during the study period in terms of other measurable variables using (14).

$$G = PZ = y - \sum_j r_j t_j - T_c \tag{14}$$

The econometric activity time allocation model (15) can be derived by substituting the functional form of utility and parameters from (12) and (13) into (10):

$$\frac{e^{\mathbf{x}_j \hat{\mathbf{a}}_j + \varepsilon_j}}{t_j} = \frac{A_w}{t_w} + \frac{B}{PZ}(r_j + w), \quad \forall j \neq w \tag{15}$$

We can take the natural logarithm to both side and arrange the terms:

$$\ln(t_j) = \mathbf{x}_j \hat{\mathbf{a}}_j - \ln\left(\frac{A_w}{t_w} + \frac{B}{G}(r_j + w)\right) + \varepsilon_j, \quad \forall j \neq w \tag{16}$$

We can include the activity participation cost for leisure activities to the market prices of goods or simply assume zero participation costs to all leisure activities. In so doing, (16) becomes:

$$\ln(t_j) = \mathbf{x}_j \hat{\mathbf{a}}_j - \ln\left(\frac{A_w}{t_w} + \frac{B}{G} \times w\right) + \varepsilon_j, \quad \forall j \neq w \tag{17}$$

Econometric Model of Activity Time Allocation with Latent Determinants

Diagrammatic Representation

The proposed theoretical framework for modelling activity time allocations that explicitly incorporates latent determinants consists of two model components: an activity time allocation model and a latent variable model. The ellipses in Figure 13.1 represent the latent variables, while rectangles represent the observed variables. Such type of representation is analogous with the path analysis (Loehlin, 1998). An individual's direct utility is unobserved and optimal time allocations to different activities are assumed as manifestations of the maximum utility. Similarly, latent variables are unobserved, and indicators are the manifestations for latent variables. The terms 'latent variables', 'latent constructs', 'latent factors', and 'latent determinants' are used interchangeably throughout this study.

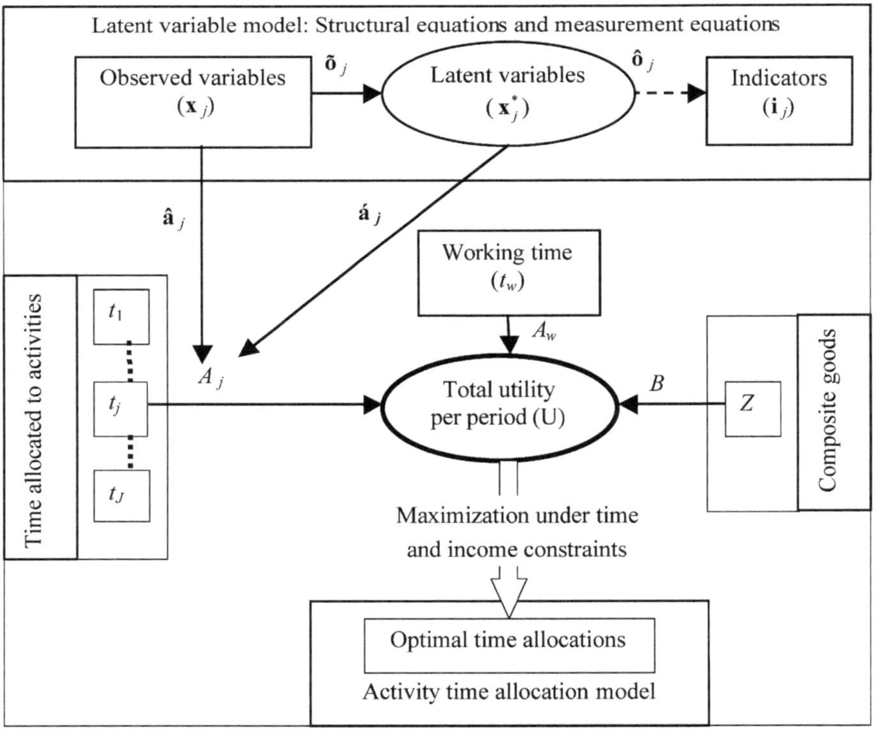

Figure 13.1
The Activity Time Allocation Model with Latent Determinants

Development of the Econometric Model

We extend the statistical activity time allocation model of leisure activities with exogenous work activity (17) in order to incorporate the latent determinants of activity time allocations. Rewriting equation (17) to distinguish explanatory variables for both observed variables, $\mathbf{x}_j$, and latent variables, $\mathbf{x}_j^*$, gives:

$$\ln(t_j) = \mathbf{x}_j^* \mathbf{\acute{a}}_j + \mathbf{x}_j \mathbf{\hat{a}}_j - \ln\left(\frac{A_w}{t_w} + \frac{B}{G} \times w\right) + \varepsilon_j, \quad \forall j \neq w \tag{18}$$

where $\mathbf{\acute{a}}_j$ are the parameters of the latent variables, and $\mathbf{\hat{a}}_j$ are the parameters of the observed explanatory variables. Equation (18) is the general expression of the activity time allocation model for activity j. However, it is sometimes more beneficial to model the systems of equations of all activities jointly:

$$\left.\begin{array}{l} \ln(t_1) = \mathbf{x}_1^* \mathbf{\acute{a}}_1 + \mathbf{x}_1 \mathbf{\hat{a}}_1 - \ln\left(\frac{A_w}{t_w} + \frac{B}{G} \times w\right) + \varepsilon_1, \\ \ldots \quad \ldots \quad \ldots \quad \ldots \quad \ldots \\ \ln(t_j) = \mathbf{x}_j^* \mathbf{\acute{a}}_j + \mathbf{x}_j \mathbf{\hat{a}}_j - \ln\left(\frac{A_w}{t_w} + \frac{B}{G} \times w\right) + \varepsilon_j \\ \ldots \quad \ldots \quad \ldots \quad \ldots \quad \ldots \\ \ln(t_J) = \mathbf{x}_J^* \mathbf{\acute{a}}_J + \mathbf{x}_J \mathbf{\hat{a}}_J - \ln\left(\frac{A_w}{t_w} + \frac{B}{G} \times w\right) + \varepsilon_J \end{array}\right\}, \quad j \neq w \tag{19}$$

Doing this requires more effort to specify the variance-covariance structures of the disturbances and constraints on the parameters across activities. The different model structures include covariance structures, random coefficient structures and unrestricted structures (see Greene, 1997 for different model structures and estimation approaches). For simplicity, we assume that the error terms are independently normally distributed across activities (cross-sectional heteroscedasticity):

$$\begin{array}{l} E(\varepsilon_j) = 0 \\ Var(\varepsilon_j) = \sigma_j^2 \\ Cov(\varepsilon_j, \varepsilon_{j'}) = 0, \, if \;\; j \neq j' \end{array} \tag{20}$$

Then, the variance-covariance matrix of error term can be stacked:

$$V=\begin{pmatrix} \sigma_1^2 & 0 & \ldots & 0 \\ \ldots & \ldots & \ldots & \ldots \\ 0 & \sigma_j^2 & \ldots & 0 \\ \ldots & \ldots & \ldots & \ldots \\ 0 & 0 & \ldots & \sigma_J^2 \end{pmatrix} \quad (21)$$

Since we assume the disturbances are independent across activities, we can derive the model for each activity and extend it to all activities when calculating the likelihood function. Extensions to other model structures are straightforward, but the complexity of estimation increases with the flexibility of model structures. Then, the density of time allocation to activity j from (18) can be written as:

$$f_1(t_j \mid \mathbf{x}_j^*, \mathbf{x}_j, t_w, G; \mathbf{\acute{a}}_j, \mathbf{\hat{a}}_j, A_w, B, \sigma_j) \quad (22)$$

Equation (22) cannot be estimated because it contains latent variables. Ben-Akiva *et al.* (2002) and Morikawa *et al.* (2002) discussed the general procedure of incorporating latent variables through structural equations and measurement equations in discrete choice models. The analogous procedure is described in the following paragraphs (equations 23 through 26) and heavily draws from Ben-Akiva *et al.* (2002) with a specific application to activity time allocation modelling.

Since the latent constructs are not directly observable, the activity time allocation model (22) should be integrated over the distribution of latent constructs to obtain the unconditional function. This requires the latent variable structural equation model:

$$\mathbf{x}_j^* = \mathbf{x}_j^*(\mathbf{x}_j; \mathbf{\tilde{o}}_j) + \mathbf{\grave{u}}_j \quad (23)$$

which describes the latent variables ($\mathbf{x}_j^*$) as a function of observable explanatory variables ($\mathbf{x}_j$), a set of parameters ($\mathbf{\tilde{o}}_j$), and a disturbance $\mathbf{\grave{u}}_j \sim D(\mathbf{Ø}_j)$, where D is the distribution function of $\mathbf{\grave{u}}_j$, and $\mathbf{Ø}_j$ is the parameter of the distribution. From this equation, we can obtain the density of latent variables $f_2(\mathbf{x}_j^* \mid \mathbf{x}_j; \mathbf{\tilde{o}}_j, \mathbf{Ø}_j)$. Then, we can express the desired function as:

$$\int_{\mathbf{x}_j^*} f_1(t_j \mid \mathbf{x}_j^*, \mathbf{x}_j, t_w, G; \mathbf{\acute{a}}_j, \mathbf{\hat{a}}_j, A_w, B, \sigma_j) \times f_2(\mathbf{x}_j^* \mid \mathbf{x}_j; \mathbf{\tilde{o}}_j, \mathbf{Ø}_j)\, d\mathbf{x}_j^* \quad (24)$$

It is difficult to estimate this model because we cannot directly observe $\mathbf{x}_j^*$, so we must have one latent variable measurement equation model:

$$\mathbf{i}_j = \mathbf{i}_j(\mathbf{x}_j^*; \hat{\mathbf{o}}_j) + \hat{\mathbf{\imath}}_j \tag{25}$$

which describes the observed indicators ($\mathbf{i}_j$) as a function of latent variables ($\mathbf{x}_j^*$), a set of parameters ($\hat{\mathbf{o}}_j$), and a disturbance $\hat{\mathbf{\imath}}_j \sim D(\grave{\mathbf{e}}_j)$, where D is the distribution function of $\hat{\mathbf{\imath}}_j$, and $\grave{\mathbf{e}}_j$ is the parameter of the distribution. From this equation, we can obtain the density of indicators $f_3(\mathbf{i}_j \mid \mathbf{x}_j^*; \hat{\mathbf{o}}_j, \grave{\mathbf{e}}_j)$. Then, we can express the interested function as:

$$\int_{\mathbf{x}_j} f_1(t_j \mid \mathbf{x}_j^*, \mathbf{x}_j, t_w, G; \acute{\mathbf{a}}_j, \hat{\mathbf{a}}_j, A_w, B, \sigma_j) \times f_2(\mathbf{x}_j^* \mid \mathbf{x}_j; \tilde{\mathbf{o}}_j, \boldsymbol{\O}_j) \times f_3(\mathbf{i}_j \mid \mathbf{x}_j^*; \hat{\mathbf{o}}_j, \grave{\mathbf{e}}_j)\, d\mathbf{x}_j^* \tag{26}$$

The function f_1 in equation (26) is the regression equation model of activity time allocation, and the functions f_2 and f_3 are the structural equation model and measurement equation model respectively from the latent variable model.

Estimation Approaches

Since equation (26) is computationally intense, requiring multidimensional integrals to be evaluated depending upon the number of latent variables, Walker (2001) discussed three approaches to simplify its estimation. The first approach is to include the indicators (responses) of the latent constructs directly as explanatory variables as used by Prastyo *et al.* (2003). Doing this does not require functions f_2 and f_3, and multidimensional integrals are no longer needed. The model vanishes to a simple regression equation or a Tobit censored regression equation (Tobin, 1958), depending on whether or not the dependent variable is censored. Both the second and third approaches, known as sequential and simultaneous estimation techniques, require the latent variable model. They differ only in the estimation procedures. In-depth discussions of these estimation procedures can be found in Ben-Akiva *et al.* (2002) and Morikawa *et al.* (2002) for integrated choice and latent variable models. The sequential estimation procedure, described below for activity time allocation modelling, is used for small-scale empirical analysis.

Sequential estimation uses a LISREL (Joreskog and Sorbom, 1999) model to estimate fitted latent variables from (23) and (25) separately and before the activity time allocation model (19). The sequentially estimated parameters are not consistent unless the distributions of the fitted latent variables are used in the activity time allocation model, as they contain measurement errors (Morikawa *et al.*, 2002). Assuming the error density $\varepsilon_j \sim N(0, \sigma_j^2)$, in other words, independent normal distribution across activities with mean zero and standard deviation σ_j, we can estimate the activity time allocation model for each activity j from the following likelihood function:

$$L_j = \prod_{n_j} \frac{1}{\sigma_j} \phi\left(\frac{\varepsilon_j}{\sigma_j}\right), \quad \forall j \neq w \tag{27}$$

where ε_j is the error term of activity j and can be obtained from (18), n_j represents all samples used to model activity j, and ϕ is the normal density function. However, with data from a limited period, we may not observe any of an individual's time allocations for some of the activities. We can overcome this problem by using a non-linear Tobit censored regression model for $t_j > 0$ and $t_j = 0$:

$$L_j = \prod_{n_{j1}} \frac{1}{\sigma_j} \phi\left(\frac{\varepsilon_j}{\sigma_j}\right) \times \prod_{n_{j0}} \Phi\left(\frac{\varepsilon_j}{\sigma_j}\right), \quad \forall j \neq w \tag{28}$$

where n_{j1} represents the samples with non-zero time allocation for activity j, n_{j0} represents the samples with zero time allocation for activity j, and Φ is the cumulative normal distribution. Since we assume the disturbances are independent across activities, we can estimate the joint model of all activities by combining the likelihood functions. The combined likelihood and log-likelihood of joint model are given in (29) and (30) respectively.

$$L = \prod_j L_j \tag{29}$$

$$LL = \sum_j \ln(L_j) \tag{30}$$

Value of Activity Time

After estimating the parameters of the integrated model (30), we can estimate the value of activity time of all leisure activities from the following relationship:

$$VOAT_j = \frac{\frac{\partial U}{\partial t_j}}{\lambda} = \frac{A_j}{t_j} \times \frac{1}{\lambda} = \frac{G}{t_j} \times \frac{A_j}{B}, \quad \forall j \neq w \tag{31}$$

Since the parameters are estimated at the individual level, the value of activity time is different for each activity and for each individual.

EMPIRICAL STUDY

Activity Survey and Socioeconomic Characteristics

We conducted a small-scale activity time use survey among individuals from Yokohama, Japan in December 2003. We visited 664 households, distributed pre-prepared questionnaires, and requested that respondents to mail them back. Of the 313 returned questionnaires (response rate 47%), 128 showed consistency for calibrations; they were from individuals who worked outside the home on weekdays. Although a sample of 128 is too small a number to draw sound conclusions, it is used here for demonstration purposes only. Prior to the survey, we summarized 15 distinct activities (Table 13.1). We requested that individuals marked the activity code, including the start and end time, if they had participated in that particular activity during the course of the day. We also collected the total daily travel costs, travel time and socioeconomic characteristics for each individual. In the data set we used for estimation, individuals were over 20 years of age; 46% were female and 54 % were male. We took age, gender, education, number of family members, number of years in the present home, and whether the wife worked outside as the major variables.

Unlike most existing time use surveys, individuals were also requested to give subjective ratings to four indicators on a 5-point semantic scale (5: very high, 4: high, 3: moderate, 2: low and 1: very low) for all activities on an average working day, except for work and travel, in order to avoid temporal and situation dependent indicators. In particular, we collected information about the following indicators in order to measure two latent variables: (i) difficulty with respect to participating in a particular activity due to lack of time (*TIME*); (ii) difficulty with respect to participating in a particular activity due to joint participation with other persons (*GROUP*); (iii) importance of a particular activity (*IMP*); and (iv) satisfaction obtained if participation in a particular activity was achieved (*SATIS*). The first two indicators are related to constraints or difficulties (*CONS*) and the latter two are related to the needs and desires of the individuals concerned (*NEED*). We decided to only collect four indicators for two latent variables in order to reduce the burdens on the respondents in this pilot investigation.

Table 13.1
Summary of Activities and Groups

Activity Groups	Specific Activities
Work activity	(1) work
Mandatory activity	(2) sleep (3) meal
Maintenance activity	(4) study (5) house keeping (6) shopping (7) medical treatment
Regular recreational	(8) hobby/amusement (9) sports
Irregular recreational	(10) conversations/relations (11) picnic/free walk (12) TV/radios (13) reading/entertainments (14) break/ rest
Travel	(15) travel

Table 13.2
Descriptive Statistics of Daily Time Allocations (Minutes)

Activity Group	Mean	Standard Deviation	Number of Samples with Non-Zero Time Allocation
Work	525	181	128
Mandatory activity	489	80	128
Maintenance Activity	168	138	115
Regular Recreational	43	41	87
Irregular Recreational	139	87	111
Travel	126	68	128

Instead of manually classifying each activity into different groups, we applied a factor analytic technique, based on the importance ratings of the individuals to 13 types of specific activities, other than work and travel. First, we perform an exploratory factor analysis using *LISREL* software for four factors. Second, we conducted a confirmatory factor analysis to group the activities, based on the information of factor structures, shown in the first column of Table 13.1. The descriptive statistics of the time allocation for different activity groups in the sample are summarized in Table 13.2.

Empirical Model Specifications

We specified simple structures of the latent variable model (structural equations and measurement equations) and the activity time allocation model. The structural equation model reads:

$$\begin{bmatrix} CONS \\ NEED \end{bmatrix} = \begin{bmatrix} \upsilon_{11} & \upsilon_{12} & \upsilon_{13} & \upsilon_{14} & \upsilon_{15} \\ \upsilon_{21} & \upsilon_{22} & \upsilon_{23} & \upsilon_{24} & \upsilon_{25} \end{bmatrix} \begin{bmatrix} AGE \\ GEN \\ EDU \\ NHM \\ INCO \end{bmatrix} + \begin{bmatrix} \omega_1 \\ \omega_2 \end{bmatrix} \tag{32}$$

where,

AGE is the age of the respondent (years);
GEN is the gender of the respondent (1 if female, 0 if male);
EDU is the education of the respondent (1 if more than high school, 0 if not);
NHM is the number of household members; and
INCO is the income of the respondent per year.

The measurement model was specified as:

$$\begin{bmatrix} GROUP \\ IMP \\ TIME \\ SATIS \end{bmatrix} = \begin{bmatrix} \tau_{11} & 0 \\ 0 & \tau_{22} \\ \tau_{31} & 0 \\ 0 & \tau_{42} \end{bmatrix} \begin{bmatrix} CONS \\ NEED \end{bmatrix} + \begin{bmatrix} 0.25 \\ 0.25 \\ \xi_3 \\ \xi_4 \end{bmatrix} \tag{33}$$

In equation (33), we fixed the variances of *GROUP* and *IMP* at 0.25 for identification after many trials of different combinations. In its simplest form, the activity time allocation model is specified below, where the parameters across activities are constrained to be equal, including identical error variances:

$$\left.\begin{aligned} \ln(t_1) &= \mathbf{x}_1^*\acute{\mathbf{a}}_1 + \mathbf{x}_1\hat{\mathbf{a}}_1 - \ln\left(\frac{A_w}{t_w} + \frac{B}{G} \times w\right) + \varepsilon \\ \ln(t_2) &= \mathbf{x}_2^*\acute{\mathbf{a}}_2 + \mathbf{x}_2\hat{\mathbf{a}}_2 - \ln\left(\frac{A_w}{t_w} + \frac{B}{G} \times w\right) + \varepsilon \\ \ln(t_3) &= \mathbf{x}_3^*\acute{\mathbf{a}}_3 + \mathbf{x}_3\hat{\mathbf{a}}_3 - \ln\left(\frac{A_w}{t_w} + \frac{B}{G} \times w\right) + \varepsilon \\ \ln(t_4) &= \mathbf{x}_4^*\acute{\mathbf{a}}_4 + \mathbf{x}_4\hat{\mathbf{a}}_4 - \ln\left(\frac{A_w}{t_w} + \frac{B}{G} \times w\right) + \varepsilon \end{aligned}\right\} \tag{34}$$

where,

$\mathbf{x}_j^* = \lfloor CONS_j, NEED_j \rfloor, \quad \forall j;$

$\mathbf{x} = [AGE, GEN, EDU, NHM, NHY, HW, INCO]$

NHY is the number of years in the present home; and

HW is the housewife (1 if the wife is not working outside the home, 0 otherwise).

Parameter Estimation and Empirical Results

We sequentially estimated the parameters of both the latent variable model and the activity time allocation model, first by the commercially available LISREL software and the later by a program in GAUSS 6.0 (Aptech Systems Inc., 2003).

We first estimated the parameters of the latent variable model (parameters of structural equations and measurement equations) from (32) and (33); they are shown in Tables 13.3 and 13.4 along with t-statistics. The goodness of fit statistics from LISREL model are shown in Table 13.5.

Table 13.3
Structural Equations Parameters

Latent Vars	Observed Variables	Mandatory Activities		Maintenance Activities		Regular Recreational		Irregular Recreational	
		Estimates	t-stat	Estimates	t-stat	Estimates	t-stat	Estimates	t-stat
	AGE	-0.0035	-0.43	-0.011	-0.83	0.011	1.34	0.012	1.34
	GEN	0.43	3.23	-0.21	-1.01	0.43	3.13	-0.28	-1.96
	EDU	0.42	3.29	-0.28	-1.39	0.25	1.98	-0.061	-0.46
CONS	NHM	-0.065	-0.76	-0.15	-1.11	0.10	1.17	-0.074	-0.80
	INCO	0.13	2.30	0.073	0.80	0.007	0.12	-0.15	-2.36
	Error var.	0.85		0.92		0.89		0.88	
	AGE	-0.012	-1.33	0.013	1.35	-0.01	-1.2	0.018	2.13
	GEN	0.35	2.46	0.42	2.53	0.12	0.84	-0.20	-1.49
	EDU	0.31	2.29	0.40	2.59	-0.11	-0.78	-0.20	-1.55
NEED	NHM	0.05	0.55	0.004	0.04	0.15	1.53	0.017	0.19
	INCO	0.20	3.19	-0.033	-0.48	0.17	2.56	-0.25	-3.87
	Error var.	0.81		0.87		0.84		0.78	

Table 13.4
Measurement Equations Parameters

Latent Variables	Observed Variables	Mandatory Activities		Maintenance Activities		Regular Recreational		Irregular Recreational	
		Estimates	t-stat	Estimates	t-stat	Estimates	t-stat	Estimates	t-stat
	CONS	1.11	12.91	0.33	4.68	1.05	12.71	0.83	11.32
GROUP	NEED	0	*	0	*	0	*	0	*
	Error var.	0.25	*	0.25	*	0.25	*	0.25	*
	CONS	0	*	0	*	0	*	0	*
IMP	NEED	0.71	10.14	0.37	5.98	0.57	8.78	0.57	8.78
	Error var.	0.25	*	0.25	*	0.25	*	0.25	*
	CONS	0.56	5.42	0.32	2.32	0.089	0.83	0.46	4.50
TIME	NEED	0	*	0	*	0	*	0	*
	Error var.	0.98	7.50	0.70	6.25	1.19	7.96	0.89	7.36
	CONS	0	*	0	*	0	*	0	*
SATIS	NEED	0.61	6.17	0.49	5.42	0.52	6.36	0.58	7.84
	Error var.	0.66	6.39	0.23	3.16	0.36	5.32	0.23	4.05

* Fixed parameters

Table 13.5
Summary of LISREL Goodness of Fit Statistics

Goodness-of-Fit Statistics	Mandatory Activities	Maintenance Activities	Regular Recreational	Irregular Recreational
Chi-Square	79.93	46.96	66.36	91.66
Degree of Freedom	14	14	14	14
P-Values	0.00000	0.00002	0.00000	0.00000
RMSEA	0.1930	0.136	0.172	0.209

Some parameters of structural equations and measurement equations are not significant, and goodness-of-fit statistics of the LISREL model are also poor. This might be because of a limited number of indicators for latent factors, a small number of data set, insufficient explanatory variables or a restricted model specification. We show both significant and non-significant parameters for illustration purposes. We calculated the latent variables *CONS* and *NEED* for each individual using the parameters of structural and measurement equation models. Table 13.6 summarizes two activity time allocation models with and without latent variables from (34). Note that all estimated parameters are common to all activities, including variances of the error terms. This model is too restrictive, but can easily be extended to estimate different sets of parameters for different activities, provided there are sufficient data to support the parameter estimation.

Table 13.6
Parameters of Activity Time Allocation Models without and with Latent Variables

Explanatory Variables		Activity Time Allocation Model without Latent Variables		Activity Time Allocation Model with Latent Variables	
		Estimates	t-stat	Estimates	t-stat
	CONS	---	---	-0.6223	-3.834
	NEED	---	---	1.1021	6.248
	AGE	0.0171	1.347	0.0035	0.295
	GEN	0.4197	1.204	0.4537	1.387
	EDU	0.1684	0.507	0.2937	0.943
	NHM	-0.0747	-0.575	-0.0950	-0.779
	NHY	-0.0193	-1.211	-0.0184	-1.227
	HW	-0.7309	-1.549	-0.5643	-1.273
	INCO	-0.2947	-3.491	-0.1724	-2.131
Parameter (A_w)		3.8000	0.493	45.15	0.916
Parameter (B)		1.9300	0.134	38.47	0.759
Error parameter	Standard dev.	3.0038	26.013	2.7955	26.052
Final Log-likelihood		-1130.24		-1096.96	
Number of samples		128		128	
AIC statistics		2280.48		2217.92	

The Akaike's Information Criterion (AIC) of these models indicates that the model with latent variables is better than the model without latent variables. There are two broad categories of variables included in the activity time allocation model: latent variables and individual socioeconomic characteristics. The signs and relative values of the estimated parameters are as expected. Negative parameter for the latent variable *CONS* shows that the time allocation to activities decreases if an individual is forced to participate in groups and he or she is under time pressure (lack of time). The positive parameter for *NEED* shows that the time allocation to activities increases if an individual feels that the activities are relatively important, and he or she gets more satisfaction from participating in them. The other socioeconomic variables can also be interpreted using a related equation (34). Note that some of the socioeconomic parameters are not significant. Due to the small data set, it is difficult to give a reason. The parameter for individual income is significant and negative because high-income people tend to work longer hours and have less time for leisure activities. Similarly, the gender dummy is positive because women are more likely to engage in leisure activities while the men work to earn more money. We estimated the value of activity time for each activity and for each individual using equation (31). The distributions of the value of activity time across individuals are diverse, ranging from nearly zero to very high values. Table 13.7 shows descriptive statistics for these distributions. The value of activity time for 'mandatory activities' is smaller and shows smaller variance than the value of time for 'maintenance activities' and 'recreational activities'.

Table 13.7
Values of Activity Time (Yen/minute)

Activity Type	Mean	Median	5 % Trimmed Mean	Standard Deviation	MAX	MIN
Mandatory activities	18.23	10.69	16.08	22.60	145.23	0.07
Maintenance activities	123.59	40.88	89.67	310.90	2933.42	1.04
Regular recreational	67.14	49.72	63.93	64.25	258.68	0.61
Irregular recreational	39.62	15.26	22.77	170.08	1792.74	0.07

Table 13.8
Values of Activity Time with Respect to Wage Rate (w)

Activity Type	Mean	Median	5 % Trimmed Mean	Standard Deviation	MAX	MIN
Mandatory activities	0.42	0.28	0.37	0.47	3.35	0.01
Maintenance activities	2.66	1.16	2.10	5.29	37.82	0.08
Regular recreational	1.72	0.83	1.39	3.07	16.23	0.09
Irregular recreational	0.83	0.33	0.63	2.07	20.65	0.01

We also computed the value of activity time in terms of the wage rate; the values are shown in Table 13.8. The result shows that the value of time for 'mandatory activities' is less than the wage rate, which is similar to current practice in transport economics. However, for activities like 'maintenance activities' and 'recreational activities', the value of activity time might be higher owing to the greater need for those activities.

SUMMARY AND CONCLUSIONS

This study has reformulated a general and basic microeconomic random utility model of individual activity time allocation, and has extended it to incorporate the latent determinants of activity participation decisions from a latent variable model. We have completed the mathematical derivation of the model step-by-step, and required no approximations and only a few assumptions to do so. The ultimate goal in developing this model was to improve the individual activity time allocation model by introducing latent variables. We have derived an estimable econometric activity time allocation model, incorporating latent determinants of time allocation from the latent variable model. The developed model is used to estimate the value of activity time of different activity types. We conducted a pilot empirical investigation using a small data set and a few indicators of the latent variables collected from individuals in Yokohama, Japan. The findings suggest that this model, which incorporates latent determinants of time allocation in a traditional activity time allocation model, is valuable not only for modelling activity time allocation, but also in calculating the value of activity time. We draw this conclusion from the highly significant parameters of the latent variables, the statistically improved model as indicated by model selection criteria, and the realistic estimates of the values of activity time for different activities.

There are a few major questions still to be answered. First, will the proposed integrated model work for a more detailed data set? Second, what kinds of indicators actually reveal the relevant latent variables and how can we design the questionnaires to obtain accurate responses with less burden for respondents? Third, can this model improve the accuracy of estimating the value of activity time, as compared with other methods? What are the practical difficulties to use the value of activity time, differentiated by activity types, when evaluating transportation projects? Further, more detailed empirical studies, applied to real-world projects, are needed in order to answer these important questions more clearly and confidently. Moreover, there are some limitations with respect to model specifications, such as uncorrelated disturbances and restricted model parameters across activities, that could be improved easily, provided the detailed data are available to estimate the parameters.

REFERENCES

Aptech Systems Inc. (2003). *GAUSS 6.0.* Maple Valley, USA.

Bain, J.H. (1976). Activity Choice Analysis, Time Allocation and Disaggregate Travel Demand Modeling. Master Thesis, Massachusetts Institute of Technology.

Bartholomew, D.J. and M. Knott (1999). *Latent Variable Models and Factor Analysis*. Second Edition, Oxford University Press, New York.

Becker, G.A. (1965). A theory of the allocation of time. *Econ. J.*, **75,** 493-517.

Ben-Akiva, M.E., J. Walker, A.T. Bernardino, D.A. Gopinath, T. Morikawa and A. Polydoropoulou (2002). Integration of choice and latent variable models. In: *In Perpetual Motion: Travel Behavior Research Opportunities and Application Challenges* (H.S. Mahmassani ed.), pp. 431-470. Elsevier, Oxford.

Bhat, C.R. (1996). A hazard-based duration model of shopping activity with non-parametric baseline specification and non-parametric control for unobserved heterogeneity. *Transpn. Res. B*, **30**, 189-209.

Bhat, C.R. and R. Misra (1999). Discretionary activity time allocation of individuals between in-home and out-of-home and between weekdays and weekends. *Transpn.,* **26**, 193-209.

DeSerpa, A. (1971). A theory of the economics of time. *Econ. J.*, **81,** 828-845.

Ettema, D.F., A.W.J. Borgers and H.J.P. Timmermans (1995). A competing risk hazard model of activity choice, timing, sequencing and duration. *Proc. 74th Ann. Meet. Transpn. Res. B.*, Washington, D.C. (CD-Rom).

Golob, T.F. (2001). Structural equation modeling for travel behavior research. Working Paper, Institute of Transportation Studies, University of California, Irvine.

Greene, W.H. (1997). *Econometric Analysis*. Prentice-Hall, New York.

Jara-Diaz, S.R. (1998). Time and income in travel choice: towards a microeconomic activity-based theoretical framework. In: *Theoretical Foundations of Travel Choice Modeling* (T. Gärling, T. Laitila, and K. Westin, eds.), pp. 51-73. Elsevier, Oxford.

Jara-Diaz, S.R. (2000). Allocation and valuation of travel time savings. In: *Handbook of Transport Modeling* (D.A. Hensher and K.J. Button, eds.), Vol. 1, pp. 303-319. Elsevier, Oxford.

Jara-Diaz, S.R. (2003). On the goods-activities technical relations in the time allocation theory. *Transpn.*, **30,** 245-260.

Jara-Diaz, S.R. and C.A. Guevara (2003). Behind the subjective value of travel time savings: the perception of work, leisure, and travel from joint mode choice activity model. *J. Transp. Econ. & Pol.*, **37,** 29-46.

Joh, C.-H, T.A. Arentze and H.J.P. Timmermans (2002). Modeling individual's activity-travel rescheduling heuristics. *Transpn. Res. Rec.*, **1807**, 16-26.

Joreskog, K. and D. Sorbom (1999). *LISREL 8.30 and PRELIS 2.30.* Scientific Software International Inc., Lincolnwood.

Kalfs, N. and A.S. Harvey (2002). Emerging developments in time use and mobility. In: *In Perpetual Motion: Travel Behavior Research Opportunities and Application Challenges* (H. S. Mahmassani, ed.), pp. 289-306. Elsevier, Oxford.

Kitamura, R. (1984). A model of daily time allocation to discretionary out-of-home activities and trips. *Transpn. Res. B,* **18,** 255-266.

Kockelman, K.M. (2001). A model for time- and budget-constrained activity demand analysis. *Transpn. Res. B*, **35,** 255-269.

Kraan, M. (1996). In search of limits to mobility growth with a model for the allocation of time and money. In: *Activity-Based Approaches to Travel Analysis* (D.F. Ettema and H.J.P. Timmermans, eds.), pp. 89-116. Pergamon, Oxford.

Loehlin, J.C. (1998). *Latent Variable Models: An Introduction to Factor, Path and Structural Analysis*. Third Edition, Lawrence Erlbaum Associates, New Jersey.

Mannering, F., E. Murakami and S.G. Kim (1994). Temporal stability of the travelers' activity choice and home-stay duration: Some empirical evidence. *Transpn.,* **21**, 371-392.

Meloni, I., L. Guala, and A. Loddo (2004). Time allocation to discretionary in-home, out-of-home and to trips. *Transpn.*, **31,** 69-96.

Morikawa, T., M.E. Ben-Akiva and D. McFadden (2002). Discrete choice models incorporating revealed preferences and psychometric data. *Econ. Models in Mark.*, **16**, 29-55.

Nepal, K.P., D. Fukuda and T. Yai (2004). A combined time allocation model of utility maximization and discrete choice model of marginal extension activity choice with the focus on individuals' psychological needs. *Proc. 10th WCTR Conf.*, Istanbul, Turkey.

Pas, E.I. (1998). Time in travel choice modeling: from relative obscurity to center stage. In: *Theoretical Foundations of Travel Choice Modeling* (T. Gärling, T. Laitila, and K. Westin, eds.), pp. 231-249. Elsevier, Oxford.

Prasetyo, I., D. Fukuda, H. Yoshino and T. Yai (2003). Analysis of travel time saving benefit by understanding individual needs and value of activity time: case study of Tokyo and Jakarta. *Transpn. Res. Rec.*, **1854,** 12-21.

Tobin, J. (1958). Estimations of relationships for limited dependent variables. *Econom.*, ***26***, 24-36.

Walker, J.L. (2001). Extended Discrete Choice Models: Integrated Framework, Flexible Error Structures, and Latent Variables. PhD Dissertation, Massachusetts Institute of Technology.

Yamamoto, T. and R. Kitamura (1999). An analysis of time allocation to in-home and out-of-home discretionary activities across working days and non-working days. *Transpn.*, **26**, 211-230.

14

AN ANALYSIS OF ACTIVITY TYPE CLASSIFICATION AND ISSUES RELATED TO THE *WITH WHOM* AND *FOR WHOM* QUESTIONS OF AN ACTIVITY DIARY

Konstadinos G. Goulias, University of California Santa Barbara, Santa Barbara, USA

Tae-Gyu Kim, City of El Paso, USA

INTRODUCTION

Activity-based models are evolving as possible replacements for trip-based transportation models for simulation and planning. This is mainly due to their more realistic representations of everyday life. At the heart of this realism is social behaviour in the form of time allocation to activities, which may be considered from three different perspectives: *motivational*, *microeconomic* (*household production)*, and *time-space constraints*. Chapin's research (1974) created the motivational foundation of activity-based approaches to travel demand analysis, focusing on the propensity and motivation of individuals to participate in activities (and travel), linking their behaviour to urban planning. During the same period, Becker developed his theory of time allocation from a household production viewpoint (Becker, 1976) applying micro-economic theory to a non-market resource and demonstrating the possibility of formulating time allocation models using economics reasoning (i.e., activity choice). In parallel, a third approach developed in time-space geography. Hägerstrand (1970) provided a conceptual base for *constraints* in human paths in time and space for a variety of planning horizons. These are *capability constraints* (e.g., physical limitations such as travel speed); *coupling constraints* (e.g., requirements to be with other persons at the same time and place); and *authority constraints* (e.g., restrictions due to institutional and regulatory contexts such as the opening and closing hours of stores).

Progress in Activity-Based Analysis edited by H. Timmermans

Cullen and Godson in two papers in the mid-1970s, as reviewed by Golledge and Stimpson (1997) and Arentze and Timmermans (2000), appear to be the first researchers bridging the gap between the motivational (Chapin) approach to activity participation and the constraints (Hägerstrand) approach by creating a model that depicts a routine and deliberated approach to activity analysis. Most subsequent contributions to the activity-based approach emerge in one way or another from these three initial frameworks with important operational improvements. Kitamura (1988), Bhat and Koppelman (1999), Arentze and Timmermans (2000), McNally (2000) and Pendyala and Bhat (2004) provide reviews from a variety of viewpoints.

Task allocation, coupling constraints, and roles within the household are three important considerations for building models of travel behaviour and they received considerable attention in the specialized literature (Townsend, 1987; Jones *et al.*, 1990; Arentze and Timmermans, 2000). At the core of all these approaches are the within-household interactions in activity participation (Gliebe and Koppelman, 2002; Zhang *et al.*, 2005; Pribyl and Goulias, 2004). However, modelling these interactions is not sufficient for building activity scheduling models because within household interactions alone fail to depict social behaviour in its entirety. In fact, individuals interact with other members of their wider family (e.g., older parents and a variety of other relatives) and are members of much larger social networks, including their work and school environments and a variety of formal and informal associations. As a result, individuals' motivations, commitments, and constraints are a direct function of their roles in a wider social world beyond the household.

Theories have been developed to reflect these more complex relationships (Goulias, 2003a, 2004). Understanding, defining the nature, and quantifying the implication of these wider roles and responsibilities requires different types of data than are usually collected in regional or national surveys. For example, qualitative research methods can provide insightful information (Goulias, 2003b; Clifton and Handy, 2003; Mehndiratta *et al.*, 2003; Burnett, 2004) and a variety of other techniques designed for more in-depth behavioural inquiry are excellent candidates (Pendyala and Bhat, 2004; Goulias *et al.*, 2004).

To aid our understanding we added strategic questions in a traditional activity survey and its diaries. We used two of these questions in a survey of the residents in Centre County (the area surrounding State College and the University Park campus of the Pennsylvania State University). The survey includes a two-day activity-diary in which respondents report every activity they pursue (each activity called an episode herein). For each activity episode, they also report *with whom* the activity was completed and *for whom*. In this first analysis of the ensuing database, we want to answer the following questions: (i) Do people spend most of their time alone or with others and in what proportions in a day?; (ii) Are the relative time allocation proportions different from day to day?; (iii) Do people pursue these activities for self-serving reasons or are they serving others?; (iv)

Are these relative time allocation proportions different from day to day?; (v) Can we identify relatively homogeneous groups of people with distinct character in their time allocation patterns?; and (vi) What are some key determinants of these different patterns? In the sections that follow, we describe the survey data, followed by data analysis and findings. The paper closes with a summary and conclusion.

DATA USED

The data used here were collected in Centre County, which is at the geographic centre of Pennsylvania, with an estimated 2001 population of 135,940 residents. Centre County is predominately rural although State College Borough and its adjacent areas have experienced a significant amount of urbanization. The University Park Campus (UP) of The Pennsylvania State University (PSU) with more than 41,000 students and 11,000 faculty and staff is also located in Centre County. The primary mode of travel in the county is private motor vehicle, although State College and the UP campus are well served by public transportation and the area immediately surrounding UP also experiences a high level of walking and bicycle traffic. The data used here are part of a larger project containing two components and named CentreSIM. First, a series of data collection projects using travel diaries started in 1992 to develop a Transportation Demand Management plan for the urbanized part of Centre County (called Centre Region).

In subsequent studies, data collection included an internet-based activity survey that was designed and implemented for PSU persons with e-mail accounts (Alam, 1998; Alam and Goulias, 1999). Second, a model development component started in early 1993 with the application of Geographic Information Systems in travel demand aiming at transportation operations applications (e.g., developer fee computation and assignment of taxation) and emergency management for potential local incidents. Within this second component, Alam (1998) also created a Geographic Information System (GIS) tool for assigning persons to locations within the UP campus on a building-by-building basis. Subsequently, the need emerged to build CentreSIM for the entire Centre County and a first demonstration of concept for a simple activity-based regional simulation model was provided (Kuhnau and Goulias, 2003). The same data were also used to estimate models for a synthetic schedule generator and household interactions predictor (Pribyl and Goulias, 2004).

The CentreSIM household and activity diary survey covers the entire Centre County and includes residents that work in Centre County and reside elsewhere. Each participating household provided voluntarily information about household composition and facilities available to the household members. In addition, each household member also reported personal information such as employment, driving ability, education and so forth. The survey also included a few questions about opinions and perceptions regarding the Centre County transportation system. Each person in the household provided a two-day complete record of the activities and the different transportation options selected. The sampling frame is a combination of several pools and includes a database of 46,448 household addresses in Centre County purchased from a commercial mailing list vendor in early October 2002, student address lists available through PSU, and a list of University Park Campus employees of Penn State who reside outside of Centre County. From this pool, 8,925 households were randomly selected for recruitment in the mail back household questionnaire (including a variety of demographic and social/economic questions). Of the responding households, 2,537 agreed to participate in the activity diary component. After data cleaning and verification, 1,471 persons (from 712 households) were selected for the analysis in this paper. Table 14.1 shows the number of persons, households, and relevant sociodemographic characteristics of the sample.

Table 14.1
A Selection of Sample Characteristics

Characteristics		**CentreSIM Sample**
Number of persons in the sample		1471
Number of households in the sample		712
Percent of males in the sample		48.2*
Persons per household (HHSIZE)		2.75
Children 1 to 4 years old per household (TOT1_4)		0.17
Children 5 to 12 years old per household (TOT5_12)		0.31
Children 13 to 15 years old per household (TOT13_15)		0.13
Children 16 to 18 years old per household (TOT16_18)		0.12
Cars per household		2.18
Number of employed (>=40 hours per week) persons in the sample		575 (39.8%)
Number of employed (<40 hours per week) persons in the sample		129 (8.9%)
College/University Students		148 (10.2%)
Total Combined Annual Household Income	<=$40,000	31.0%
	$40,001 to $70,000	32.5%
	=>$70,001	36.5%
Number of trips per day	Day 1	4.5
	Day 2	4.4
Number of activities (excluding trips) per day	Day 1	11.8
	Day 2	11.6
Number of in-home activities per day	Day 1	9.0
	Day 2	8.9
Total amount of time at home (min) per day	Day 1	983
	Day 2	1003
Total amount of time travelling (min.) per day	Day 1	92.5
	Day 2	85.0

*1.4% missing

Table 14.2
Codes for With Whom and for Whom Questions in the Activity Diary

With/For Whom	Code
ALONE-SOLO-SELF	0
Husband (Wife)	1 (2)
Son (Daughter)	3 (4)
Children – immediate family	5
Family members – immediate	6
Father (Mother)	7 (8)
Parents	9
Brother (Sister)	10 (11)
Siblings, unspecified	12
Grandson (Granddaughter)	13 (14)
Grandchildren, unspecified	15
Grandmother (Grandfather)	16 (17)
Grandparent(s), unspecified	18
Family other, not immediate	19
Friend's child/children	21
Friend, male (female)	22 (23)
Friends, multiple (unspecified)	24 (25)
Roommate, male (female)	26 (27)
Roommates, multiple (unspecified)	28 (29)
Co-worker (boss included), male	31
Co-worker (boss included), female	32
Co-workers (boss included), multiple	33
Co-worker (boss included), unspecified	34
Customer/client, male (female)	41 (42)
Customers/clients, multiple	43
Customer/client, unspecified	44

With/For Whom	Code
Agent/employee, male (female)	45 (46)
Agent/employees, multiple	47
Agent/employee, unspecified	48
Goods & Services Provider(s)*	49
Classmate, male (female)	51 (52)
Classmates, multiple (unspecified)	53 (54)
Teacher/professor/instructor, male	55
Teacher/professor/instructor, female	56
Teachers/professors/instructors, multiple (unspecified)	57 (58)
Neighbour, male (female)	61 (62)
Neighbours, multiple (unspecified)	63 (64)
Association† member, male (female)	65 (66)
Association† members, multiple	67
Association† member, unspecified	68
Other male (female)	71 (72)
Other relative, male (female)	73 (74)
Other relatives, multiple	75
Other relative, unspecified	76
Other child/children	77
Other, multiple	78
Other, not included above	79
Pet(s)	81
Farm animal(s), (Other animals)	82 (83)
Unknown (Missing)	998 (999)
Refused (Not Applicable or System Skip)	988 (888)

***Goods & Service Providers** include doctors, lawyers, store clerks, hairdressers, etc.

†**Associations** include churches, civic, fraternal and service organizations, etc

Table 14.3
Daily Average Time Allocation for the With Whom Question by Gender

	Women				Men			
	Average Minutes in Day 1 (Standard Deviation)	%	Average Minutes in Day 2 (Standard Deviation)	%	Average Minutes in Day 1 (Standard Deviation)	%	Average Minutes in Day 2 (Standard Deviation)	%
Total activity alone	537.5 (383.7)	38	534.4 (394.3)	37	485.4 (382.7)	34	500.0 (388.1)	35
Total activity with relatives	438.0 (376.7)	30	458.3 (387.1)	32	476.7 (373.3)	33	478.0 (373.4)	33
Total activity with others	293.5 (280.1)	20	273.6 (274.0)	19	262.4 (270.5)	18	249.4 (261.6)	17
Total activity with unknown/undeclared	83.3 (241.5)	6	91.7 (265.9)	6	115.2 (295.0)	8	122.9 (319.1)	9
Total travel time	86.5 (83.1)	6	81.1 (72.1)	6	99.3 (119.2)	7	88.8 (84.4)	6
Total	1439.0	100	1439.1	100	1439.0	100	1439.0	100

Activity types were reported using an open-ended format. In an attempt to reach a compromise between the large amount of possible categories offered by respondents and the categories used in the travel behaviour literature, we converted the activities offered by the respondents into a hierarchy of codes to reflect type and location of activity. To reflect the heavily home- and work-based character of activity participation and travel, the activity categories used for further data processing are activity categories that incorporate the location at which each activity was pursued. For example, all in-home activities were assigned a three-digit code that starts with the value 1 taking the format 1XX. All activities at work are coded as 2XX up to the value 250. All activities at school, day care and preschool are coded as 2XX with values starting at 250.

We also used a special taxonomy for picking up and dropping off people or goods and coded them as 3XX to parallel a popular travel category. Services (religious etc) are coded as 4XX, shopping as 5XX, social/entertainment/sports coded as 6XX, and some special codes were reserved for hobbies and visits, active sports and so forth. The same classification was used for multiple activities that were recorded as primary, secondary, and so on. The analysis reported here addresses the primary activity only. The two questions "for whom" an activity was pursued and "with whom" were also open ended. Based on the respondents' reporting, we developed the taxonomy in Table 14.2. The classification reflects the most important social networks of immediate family, relatives, work and work-related, learning and learning-related, and friendships. It also distinguishes between genders and social roles (e.g., agent/employee, service provider, classmate, and so forth). In this research, we offer the first analysis using the dataset. In doing so, we simplify the classification of activities and trips into alone (self), with family members (for family members), with others (for others), and unknown/undeclared in the data analysis.

Data Analysis

First, we review descriptive statistics for some of the key categories of activity participation and trip making, differentiating between men and women, consistent with other efforts in time use research that examine gender-roles. A pattern recognition exercise follows to identify homogeneous groups of people among the survey respondents.

Descriptive Statistics for Joint and Solo Activity Participation. Daily average time allocation to activities alone and with relatives comprise the lion's share in this sample (Table 14.3). For approximately 37 percent of their time women pursue activities alone. In each of the two days of the survey, they allocate 30 to 32 percent of their time with relatives. Men spend slightly less time alone and slightly more time with relatives. Overall, however, men and women exhibit similar patterns.

Table 14.4
Daily Average Episode Frequencies "With Whom" by Gender

	Women				Men			
	Total Episodes in Day 1 (Standard Deviation)	%	Total Episodes in Day 2 (Standard Deviation)	%	Total Episodes in Day 1 (Standard Deviation)	%	Total Episodes in Day 2 (Standard Deviation)	%
Activities alone	5.91 (4.24)	35	5.61 (4.12)	34	4.59 (3.57)	29	4.66 (3.64)	30
Activities with relatives	3.95 (3.56)	23	4.19 (3.70)	25	4.06 (3.18)	26	4.02 (3.15)	26
Activities with others	1.95 (2.09)	11	1.80 (1.98)	11	1.59 (1.79)	10	1.45 (1.64)	9
Unknown/Undeclared	0.62 (1.79)	4	0.63 (1.92)	4	0.83 (2.27)	5	0.82 (2.25)	5
Trips	4.54 (3.12)	27	4.47 (3.02)	27	4.53 (3.09)	29	4.38 (2.78)	29
Total	16.97	100	16.70	100	15.64	100	15.33	100

Table 14.5
Daily Average Number of Trips Among Persons Making at Least One Trip

	Women				Men			
	Trips in Day 1 (standard deviation)	**%**	**Trips in Day 2 (standard deviation)**	**%**	**Trips in Day 1 (standard deviation)**	**%**	**Trips in Day 2 (standard deviation)**	**%**
Alone	2.24 (2.39)	45	2.09 (2.31)	42	2.18 (2.38)	45	2.07 (2.24)	43
With Relatives	1.65 (2.14)	33	1.72 (2.20)	35	1.59 (2.07	33	1.65 (2.03)	35
With Others	0.93 (1.65)	19	0.93 (1.62)	19	0.75 (1.51)	15	0.7 (1.29)	15
With Unknown/Undeclared	0.20 (0.77)	4	0.20 (0.79)	4	0.33 (1.07)	7	0.34 (1.16)	7
Total	5.02	100	4.94	100	4.85	100	4.76	100
Sample Size	671		671		661		651	

Total activity with others is substantial for men and women and in similar amounts per day. The unknown and undeclared categories exhibit the same patterns as other data collection projects (i.e., men produce more missing data than women). We also find very small differences between the two days of the diary indicating consistent allocation to activities. It should also be noted that the missing information is not negligible and it is of similar magnitude as the total travel time per day. Total travel time is higher for men and it is higher for the first day of the activity diary for both men and women. These similarities between men and women and from one day to the next appear also in the standard deviations (Table 14.3) except for travel time for men in day 1. One important finding emerging from Table 14.3 is that although individuals participate in activities for longer periods of time with relatives (household members and other relatives), they also dedicate a substantial amount of time to activities with friends, co-workers, association members and a variety of other contacts including service providers – see the list in Table 14.2). Frequency of activity participation offers another side of time allocation (Table 14.4). In the data coding we included up to five persons for each activity, and this information was retained in five different variables one for each person mentioned. When we examine the first person mentioned in the diary for the activities pursued with others, 16.5 percent of the activities and travel are with husband and 17.7 percent with wife. Other family members claim another 9.8 percent and the response "mother" has another 6.4 percent

Co-workers also have a substantial share at 5.1 percent and a variety of friends spanning from 2.6 to 2.9 percent. The majority of combinations of persons in joint activities are dominated by household members and other relatives. However, there is also a substantial number of activities and time allocated with other personal job and education related contacts, friends, and acquaintances. Table 14.4 shows the daily average episode frequencies. A similar pattern emerges as Table 14.3 but this time relatives take twice as many activities per day as others. A comparison between Tables 14.3 and 14.4, however, shows that time spend with non-relatives is characterized by fewer longer episodes and time spend with relatives by more frequent and shorter episodes. Women also tend to have more activity episodes in a day than men and the day-to-day averages are extremely close. The number of trips, however, is remarkably similar between men and women and between the two days of the survey. Since men and women differ in their travel time, this indicates men tend to make longer trips (e.g., visit different destinations).

Table 14.5 provides additional information about the number of trips survey participants make alone and with others using data from persons that made at least one trip on the survey day. Men not only do not make more trips alone, but they also make slightly more trips with relatives. They also make fewer trips with others and they fail to record their with-whom answer more often than women. The standard deviations are very similar in both Day 1 and Day 2 and when comparing men to women. Day 2 shows consistently lower trip making because it contains more Sundays (13.5 percent of the days) than Day 1 (11.1 percent).

Descriptive Statistics for Self-Serving and Altruistic Activity Participation. As expected, the question "*for whom did you pursue each activity*" produced some interesting results. First, as Table 14.6 indicates, the majority of time in a day is dedicated to activities for one's self. These span from 66 to 69 percent of the average minutes in a day and they have relatively small standard deviations. Time dedicated for relatives and time dedicated for others show averages that are one order of magnitude different. The total amount of time dedicated for others in a day is higher than national averages. For an example of national averages, see the time use survey reports in http://www.bls.gov/news.release/atus.t01.htm, accessed November 2004). However, the relative split of time for household members and non-household members as well as the relative proportions of men's and women's time allocation are in agreement with time use surveys (an average of about 2 hours). Missing and undeclared time is of the same magnitude as for the with whom question and the men, again, are more likely to omit that information.

A somewhat different picture is offered by the analysis of episodes in each day of the survey. Almost half of the activities are dedicated to activities that are self-serving, with slightly higher daily frequencies for women than men. Similar daily rates are also for activities with others and the activities with unknown/undeclared persons. Larger differences are observed, however, in the frequency of episodes for relatives. Women dedicate on average almost one activity more per day than men for relatives and slightly more for others. In addition, there is higher variation in activity participation for relatives among women than among men.

Table 14.6

Daily Averages for the For Whom Question by Gender

	Women				Men			
	Average Minutes in Day 1 (Standard Deviation)	**%**	**Average Minutes in Day 2 (Standard Deviation)**	**%**	**Average Minutes in Day 1 (Standard Deviation)**	**%**	**Average Minutes in Day 2 (Standard Deviation)**	**%**
Total activity for self	943.0 (357.5)	66	963.2 (375.2)	67	972.8 (384.7)	68	992.1 (392.5)	69
Total activity for relatives	155.5 (212.8)	11	152.8 (217.0)	11	112.4 (178.6)	8	110.4 (185.6)	8
Total activity for others	155.4 (222.0)	11	136.2 (209.8)	9	138.9 (219.7)	10	130.8 (223.1)	9
Total activity for Unknown/Undeclared	98.0 (287.3)	7	105.7 (314.8)	7	114.8 (316.9)	8	116.5 (329.0)	8
Total travel time	86.5 (83.1)	6	81.1 (72.1)	6	99.3 (119.2)	7	88.8 (84.4)	6
Total	1439.0	100	1439.0	100	1438.6	100	1438.6	100

Table 14.7

Daily Average Episode Frequencies "for whom" by Gender

	Women				Men			
	Total Episodes in Day 1 (Standard Deviation)	**%**	**Total Episodes in Day 2 (Standard Deviation)**	**%**	**Total Episodes in Day 1 (Standard Deviation)**	**%**	**Total Episodes in Day 2 (Standard Deviation)**	**%**
For self	8.34 (4.23)	49	8.43 (4.43)	50	8.08 (4.27)	52	8.06 (4.27)	53
For relatives	2.26 (3.01)	13	2.20 (3.14)	13	1.38 (2.03)	9	1.40 (2.19)	9
For others	1.10 (1.72)	6	0.95 (1.53)	6	0.86 (1.35)	6	0.77 (1.33)	5
Unknown/Undeclared	0.73 (2.06)	4	0.65 (1.97)	4	0.75 (2.20)	5	0.73 (2.16)	5
Trips	4.54 (3.12)	27	4.47 (3.02)	27	4.53 (3.09)	29	4.38 (2.78)	29
Total	16.97	100	16.70	100	15.60	100	15.34	100

Table 14.8

Daily Average Number of Trips Among Persons Making at Least a Trip (for Whom)

	Women				Men			
	Trips in Day 1 (Standard Deviation)	**%**	**Trips in Day 2 (Standard Deviation)**	**%**	**Trips in Day 1 (Standard Deviation)**	**%**	**Trips in Day 2 (Standard Deviation)**	**%**
For Self	3.05 (2.40)	61	3.02 (2.43)	61	3.03 (2.43)	62	3.01 (2.39)	63
For Relatives	1.21 (1.91)	24	1.16 (1.93)	23	1.04 (1.79)	21	1.02 (1.69)	21
For Others	0.50 (1.33)	10	0.52 (1.27)	11	0.48 (1.45)	10	0.45 (1.13)	9
For Unknown/Undeclared	0.26 (0.95)	5	0.24 (0.86)	5	0.31 (1.01)	6	0.28 (0.95)	6
Total	5.02	100	4.94	100	4.86	100	4.76	100
Sample Size	671				661			

Table 14.8 contains a summary of the number of trips by each type. More than 60 percent of the trips are for self-serving reasons, whereas, slightly less than a quarter of the daily trips are for relatives. Trips made for others are on average approximately half a trip per day. From Table 14.8 also emerges that women are making more trips and more trips for relatives and others.

The study of differences and commonalities between men and women helps explain the persistence of gender-based social roles and behaviour. However, analyses of this type are limited in their ability to identify distinctly different behaviours. In addition, consideration of factors such as occupation and household context and lifecycle cycle stage offer better insight about behaviour than gender alone.

DAILY PATTERN IDENTIFICATION

Interaction among household members is examined in the spirit of past analyses by Goulias *et al.*, (2003) to develop patterns of activity participation and travel that go beyond the members of the household. In this exploratory analysis, we study daily activity participation of survey participants and their social network investment from two perspectives. First, we analyze the joint and solo activity and travel behaviour of our sample participants. Then we examine their stated self-serving and altruistic behaviour. All analyses include any missing data as a separate category to account for possible selection bias due to incomplete replies. To derive clusters of behaviour, we start from a one-cluster model and build in sequence models with more clusters, until estimation is no longer possible. This may happen if we observe lack of identification for one or more parameters, or if any increase in the number of clusters leads to insignificant improvements in the goodness of fit. The explanatory variables used here are at the person, household, and temporal levels. Differences among individuals in travelling together are determined by personal and household characteristics and, for this reason, we use a variable called "occupation" that also describes younger children at home, at a school, and so forth. This variable contains implicitly employment and age as explanatory variables. One of the most important variables for household task allocation is the number of children; therefore, we include the number of children by different age groups. In addition, instead of performing a separate analysis for weekdays and weekends, each model estimated below includes the day of the week as a categorical explanatory variable.

The technique selected to identify groups of homogeneous patterns of activity and travel behaviour in the CentreSIM survey data is *latent class cluster analysis*. This technique includes a *J*-category latent variable with each category representing a cluster; uses many "dependent" or clustering variables (named *criteria* variables herein); uses a mixture of multiple types of criteria variables (e.g., continuous, categorical, ordered, count); uses and tests the effect of covariates of many different types; is more flexible than many other clustering algorithms; is a model-based clustering approach and it provides probabilistic membership of observations in clusters; and provides convenient interpretable output.

In this chapter, we use notation and model formulation similar to Vermunt and Magidson (2002). Assume there is one latent variable, X representing the time allocation of a person during the two observation days in the CentreSIM Survey. Different categories of this variable X denote different types of activity-travel behaviour and the probability of belonging to each category of variable X represents the proportion of persons that choose that specific type of time allocation. Using observed data we would like to identify how many distinct groups we have and find the proportion of persons in each group. For each person in our sample we observe M measures (indicators) of activity and travel behaviour indicated by the symbol Y that can be used to infer membership in the categories of the latent variable X. A third set of variables, which are not included as criteria variables in the clusters, are used as explanatory variables and for this indicated with the symbol Z. The probability density of the Ys given a set of Z values is:

$$f(Y \mid Z) = \sum_{x} \pi(X \mid Z) f(Y \mid X, Z) \qquad (1)$$

where $\pi(X \mid Z)$ is the probability of belonging to a certain latent class given a set of covariate values. Lower case x in the Sum symbol denotes the categories of the variable X. If the Y variables belonging to different clusters (categories of variable X) are assumed mutually independent given the latent class and the covariates, we obtain:

$$f(Y \mid Z) = \sum_{x} \pi(X \mid Z) \prod_{m=1}^{M} f(Y_m \mid X, Z) \tag{2}$$

Since the scores on the latent variable given the covariates are assumed to come from a multinomial distribution, the probability of belonging to a given latent class can be calculated as follows:

$$\pi(X \mid Z) = \frac{e^{\eta_{X|Z}}}{\sum_{X} e^{\eta_{X|Z}}} \tag{3}$$

where the term η is a linear combination of the main effects of the latent variable (γ_{x_j}) and the covariate effects on the latent variable $(\gamma_{z_l x_j})$ defined as:

$$\eta_{X|Z} = \sum_{j=1}^{J} \gamma_{x_j} + \sum_{l=1}^{L} \sum_{j=1}^{J} \gamma_{z_l x_j} \tag{4}$$

One way to visualize this model is to consider a cross-classification table underlying the model in which latent and observed variables are included. This table has dimensions equal to the categories of all the variables when all variables are categorical. The cell values of this table are the entities we are trying to estimate using formulations as in Equation 4. As in many latent class models the likelihood function takes the familiar form shown below where θ denotes the unobserved parameters to be estimated.

$$LogL = \sum_{i} n_i \log f(Y_i \mid Z_i, \theta) \tag{5}$$

The parameters in equation 5 can be estimated by the Expectation Maximization (EM) algorithm, which produces maximum likelihood estimates under specific conditions. In the examples here, we use the Vermunt and Magidson (2002) method, which is a combination of EM with Newton-Raphson. Standard errors for the parameter estimates are computed using the Hessian matrix (matrix of the second order derivatives of the estimating equation). As the number of parameters to estimate increases, the degrees of freedom decrease rapidly, resulting in a variety of operational problems such as identification (inability to compute a parameter) or lack of convergence (subsequent estimation step parameters are not close enough). Most latent class models are also sensitive to local maxima of the likelihood function used in estimation, which can be circumvented by testing multiple models using different initial trial values for the parameters (see also Goulias, 1999). Estimation of models of this type is a hierarchical, iterative process in which we start with a one-cluster assumption and estimate a simple model. Then, experimentation proceeds by increasing the number of clusters until identification is no longer possible. For some parameters, the cluster sizes become too small to be meaningful, and/or the difference in goodness of fit between successive models is not significant. At this point, we select one or more models that appear to be a reasonable description of the observed data. We define alternate modelling options, such as correlations among criteria variables and variances within each cluster, and start another iterative cycle. This process continues until the addition of a more complex structure no longer yields a significant improvement (for nested models we can use a formal statistical likelihood-based step as a stop criterion).

Within these three steps, we also have two additional "mini-steps." For each model, we first develop starting values for the unknown parameters we are estimating that are drawn from a distribution of randomly selected moments. For a given set of starting values, we perform maximum likelihood iterations first using the EM algorithm until the values of subsequent iterations reach a predefined difference (or the total number of EM iterations reaches a maximum number). Then, the algorithm switches to a Newton-Raphson algorithm until a predetermined convergence criterion value is reached or the maximum number of iterations is reached. In this way, we can exploit advantages of both algorithms, i.e., the stability of EM when far away from the optimum and the speed of Newton-Raphson when close to the optimum (Vermunt and Magidson, 2002).

Statistical goodness-of-fit measures for latent class cluster models are the typical chi-square statistics that are used also in the cross-categorical data analysis (Agresti, 2002). The first measure is the likelihood ratio chi-square, G^2 or L^2. It has a chi-square distribution with degrees of freedom given by the number of "free" parameters (total number of different response patterns - the number of estimated model parameters – 1 if there are no covariates in the model). It represents the opposite of an R^2 in regression because it is the amount of unexplained associations among the criteria variables by the model. Therefore, higher values indicate models that do not fit the data well and lower values represent better fitting models. When two models are nested (i.e., they differ only in the number of estimated parameters), we could create the difference between the G^2 of these two models. This difference is chi-square distributed and can be used for hypotheses testing. A test of this type cannot be utilized between models that differ in the number of clusters because they are not nested. The L^2, the Bayes Information Criterion (BIC), Akaike Information Criterion (AIC) and the Consistent Akaike Information Criterion (CAIC) are computed to measure goodness of fit and to take into account model parsimony, penalizing models with many parameters. The lower the BIC, AIC or CAIC values, the better the model we estimate (McCutcheon, 2002). The approach followed in the analysis presented in this paper has some advantages over the more popular cluster analysis using the k-means and then using some kind of regression to identify the composition of each cluster (Goulias and Kim, 2001;Krizek, 2003).

Table 14.9
Solo and Joint Activity Clusters

Cluster	1	2	3	4	5	6	7	8
Cluster Size	0.3566	0.2273	0.1577	0.0666	0.0618	0.0588	0.046	0.0252
Criteria Variables								
Total daily time allocated to activities alone (DA_WSIG1)								
Mean	603.50	594.38	513.44	1.69	495.08	457.31	274.96	491.53
Total daily time allocated to activities with relatives (DA_WREL1)								
Mean	457.34	297.98	819.40	316.11	418.79	201.21	672.92	269.61
Total daily time allocated to activities with others (DA_WOTH1)								
Mean	308.31	447.46	28.06	242.95	187.17	361.60	150.83	243.42
Total daily time allocated to activities with unknown/undeclared (DA_WUNK1)								
Mean	0.04	0.12	0.19	816.22	277.08	235.27	172.41	177.70
Total daily time travelling alone (DT_WSIG1)								
Mean	49.84	33.51	0.04	0.55	28.98	49.40	22.60	190.17
Total daily time travelling with relatives (DT_WREL1)								
Mean	19.98	14.94	77.85	18.05	19.20	0.24	125.22	7.35
Total daily time travelling with others (DT_WOTH1)								
Mean	0.00	50.25	0.00	7.23	0.53	133.97	20.06	24.62
Total daily time travelling with unknown/undeclared (DT_WUNK1)								
Mean	0.00	0.00	0.00	36.19	12.18	0.00	0.00	34.61
Model Statistics	Cases	Params.	Log-likelihood		BIC	AIC	R-squared	
	1471	296	-57391.8		116943	115376	0.9448	

First, the latent class cluster method for identifying clusters is designed for combinations of continuous and discrete criteria variables, while the k-means method is defined for continuous variables only. Second, the method used here allows for probabilistic membership of each observation in each cluster. This provides flexibility in observation classification that the k-means does not. Third, post-processing of the cluster data using regression is not required because the method used allows the inclusion of covariates. There are other advantages of latent class methods in general and the specific implementation used here as illustrated in Vermunt and Magidson (2002).

Participation in Activities Alone and With Others. As in past research with solo and joint travelling (Chandrasekharan and Goulias, 1999) and using the same reasoning, we expand the domain of analysis to include activity participation with relatives or others (friends and/or co-workers). The criteria variables used for joint and solo patterns are:

total daily time allocated to activities alone (DA_WSIG1)
total daily time allocated to activities with relatives (DA_WREL1)
total daily time allocated to activities with others (DA_WOTH1)

total daily time allocated to activities with unknown/undeclared (DA_WUNK1)
total daily time travelling alone (DT_WSIG1)
total daily time travelling with relatives (DT_WREL1)
total daily time travelling with others (DA_WOTH1)
total daily time travelling with unknown/undeclared (DA_WUNK1)

The best model thus derived is an eight-cluster model. Table 14.9 describes the eight clusters in terms of the persons' behaviour within each cluster. The first largest cluster (35.7 percent) contains persons travelling mostly alone and spending most of their activity time alone, approximately ten hours per day. The second group of substantial size (22.7 percent of the sample) appears to travel the longest with others, about fifty minutes per day, and another 15 minutes with relatives. The persons in this group, however, also spend a good portion of their time pursuing activities alone and travelling alone. Both patterns appear to be weekday patterns with the first containing 83 percent weekdays and the second 87 percent. The third cluster, containing 15.8 percent of the sample, spends most of the daily time with relatives (approximately 820 minutes per day). As expected, this is accompanied by a substantial amount of time travelling with relatives. This is a cluster containing 54.5 percent of respondents that reported their behaviour on weekends. The fourth cluster spends very little time alone (activity and/or travel) and shows a large amount of time with unknown persons. The fourth cluster contains mostly persons that are unwilling/unable to provide accurate information, representing 6.6 percent of the sample. The other three clusters are smaller and have different combinations of time allocation. The different time allocation behaviour by a few clusters that contain large amounts of time with unknown data shows it is appropriate to include these cases in the analysis, detect possibly different behaviours, and reduce selectivity biases in the data analysis.

Table 14.10 provides further insights about these groups. First, the employed persons are more likely to belong to Cluster 1 and students (Kindergarten to 6th grade and Grade 7 to 12) have the highest probability of belonging to Cluster 2. Children at home and the disabled are more likely to belong to Cluster 3, which is also a cluster better aligned with Saturday and Sunday activity patterns. Children attending preschool or day-care have the highest probability of belonging to Cluster 1. This may be unexpected if one does not consider the substantial amount of time allocated with relatives in this cluster and the substantial presence of part time employees that may also be the companions of the pre-schoolers. As we move to older children the *preference,* shifts to Cluster 2 (recall it has less time with relatives than Clusters 1 and 3 but more time with others). College and University students show similar behaviour but they also have a large percentage in Cluster 1. Part-time and full-time workers have also a clearly higher preference for Cluster 1 with Cluster 2 by far the second most popular.

Table 14.10
Average Membership Probabilities for Solo and Joint Activity Clusters

Cluster	1	2	3	4	5	6	7	8	Total
Household role									
Child at home	0.126	0.277	0.460	0.098	0.020	0.000	0.000	0.020	1.000
Home duties	0.302	0.162	0.288	0.030	0.110	0.001	0.107	0.000	1.000
Looking for work	0.377	0.067	0.247	0.125	0.000	0.063	0.121	0.000	1.000
Retired	0.322	0.143	0.256	0.066	0.114	0.001	0.069	0.029	1.000
Disabled	0.323	0.200	0.343	0.000	0.000	0.000	0.067	0.067	1.000
Pre-school or day care	0.502	0.159	0.245	0.000	0.062	0.000	0.032	0.000	1.000
Grades K-6	0.129	0.513	0.282	0.023	0.012	0.000	0.029	0.012	1.000
Grades 7-12	0.143	0.468	0.143	0.047	0.031	0.103	0.066	0.000	1.000
College/University	0.292	0.361	0.020	0.096	0.047	0.127	0.016	0.041	1.000
Part-time (<40 hours/week)	0.440	0.161	0.154	0.069	0.033	0.067	0.042	0.035	1.000
Full-time (>=40 hours/week)	0.452	0.188	0.077	0.071	0.060	0.083	0.042	0.027	1.000
Missing	0.372	0.005	0.276	0.115	0.140	0.041	0.012	0.039	1.000
GENDER									
Female	0.350	0.255	0.162	0.057	0.059	0.060	0.041	0.017	1.000
Male	0.360	0.201	0.153	0.078	0.064	0.058	0.051	0.035	1.000
Missing	0.500	0.156	0.144	0.000	0.099	0.051	0.050	0.001	1.000
Day of the week									
Sunday	0.266	0.091	0.398	0.085	0.041	0.039	0.063	0.018	1.000
Monday	0.385	0.218	0.126	0.060	0.086	0.063	0.028	0.035	1.000
Tuesday	0.387	0.299	0.075	0.066	0.071	0.058	0.019	0.025	1.000
Wednesday	0.350	0.310	0.077	0.101	0.041	0.047	0.051	0.023	1.000
Thursday	0.437	0.218	0.088	0.051	0.075	0.058	0.038	0.037	1.000
Friday	0.396	0.241	0.120	0.062	0.040	0.074	0.048	0.020	1.000
Saturday	0.245	0.149	0.313	0.041	0.069	0.073	0.095	0.016	1.000

As shown in the descriptive analyses, men and women do not exhibit dramatically different behaviours, with only Cluster 2 having significantly more women than men at the 5 percent confidence level. The day of the week relationship with the clusters is very interesting with Cluster 3 as the most popular for Sundays and Saturdays. Clusters 1 and 2 are typical weekday time allocation patterns. However, the relatively large proportion of Clusters 1 and 2 on Sundays and Saturdays indicates that each of these clusters can represent scheduling preferences for any of the days of the week, and not all survey participants reserve weekends for family and/or friends. Their interactions are quite heterogeneous and depend on many other factors that are not included here. The findings here also raise concerns about the appropriateness in sample segmentation along the employment status of individuals, i.e., there are groups of employed persons that exhibit very different time allocation behaviours in a day.

Self-Serving and Altruistic Activity Participation and Travel. CentreSIM survey gave us the opportunity to study a few new ideas about activity and travel behaviour. One such idea is the "motivation" for allocating time during a day. In the first part of analysis, we found this sample (men and women) allocates a large amount of time in a day for self-serving tasks. The overall impression is that the sample even accounting for sleep time and other personal needs appears to be dedicating a substantial amount of time and activities/trips to self-serving reasons. This may be misleading due to averaging and a more detailed analysis through pattern recognition and sample heterogeneity analysis is required in this case too. The variables used to identify homogeneous groups are:

total daily time allocated to activities for self (DA_FSIG1)
total daily time allocated to activities for relatives (DA_FREL1)
total daily time allocated to activities for others (DA_FOTH1)
total daily time allocated to activities for unknown/undeclared (DA_FUNK1)
total daily time travelling for self (DT_FSIG1)
total daily time travelling for relatives (DT_FREL1)
total daily time travelling for others (DA_FOTH1)
total daily time travelling for unknown/undeclared (DA_FUNK1)

The latent class analysis for cluster identification reported on Table 14.11 shows the aggregate behaviour described above is the result of at least 10 distinct behavioural patterns. The first of approximately 37.8 percent of the sample contains persons that allocate more than 20 hours of time to activities for themselves and another 43 minutes of travelling for themselves. The second group of approximately 21 percent of the sample allocates a substantial amount of time to others but very little activity time and no travel time to relatives. The third cluster (15.5 percent of the sample) allocates a significant amount of time to relatives and others and they do the same for travel, particularly for relatives (60 minutes per day). The rest of the clusters have much larger amounts of time dedicated for unknown reasons/persons. This may be an indication of measurement error and confusion among the respondents in these groups. In a similar way as for the with whom cluster analyses, Table 14.12 provides additional description of the cluster composition. Children at home and disabled people are most likely to belong to Cluster 1 and as expected they will exhibit *selfish* behaviour, students (preschool through College) are also more likely to belong to Cluster 1. However, as we move to the older children, their probability of belonging to Cluster 1 decreases, while their probability of belonging to Cluster 2 increases and exhibits increasingly socially altruistic behaviour.

Table 14.11
Self-Serving and Altruistic Activity Clusters

	Cluster 1	**Cluster 2**	**Cluster 3**	**Cluster 4**	**Cluster 5**	**Cluster 6**	**Cluster 7**	**Cluster 8**	**Cluster 9**	**Cluster 10**
Cluster Size	0.3784	0.2095	0.1552	0.0437	0.0413	0.0383	0.0365	0.0337	0.0326	0.0308
Criteria Variables										
Total daily time allocated to activities for self (DA_FSIG1)										
Mean	1221.62	944.10	788.67	1021.69	33.09	575.91	666.66	746.16	1147.30	694.70
Total daily time allocated to activities for relatives (DA_FREL1)										
Mean	156.13	33.03	273.34	12.28	2.71	18.87	82.24	324.22	54.93	277.48
Total daily time allocated to activities for others (DA_FOTH1)										
Mean	0.13	367.89	267.70	0.44	15.92	220.56	323.03	1.15	157.74	106.67
Total daily time allocated to activities for unknown/undeclared (DA_FUNK1)										
Mean	0.05	0.16	0.12	235.81	1312.51	373.11	278.14	271.07	17.66	201.27
Total daily time travelling for self (DT_FSIG1)										
Mean	42.76	73.31	34.39	167.97	2.54	42.81	69.67	35.84	37.73	32.34
Total daily time travelling for relatives (DT_FREL1)										
Mean	17.73	0.00	60.34	0.41	2.35	0.09	7.53	60.00	9.08	50.28
Total daily time travelling for others (DT_FOTH1)										
Mean	*0.00*	*18.85*	*13.86*	*0.40*	*0.03*	*187.92*	*11.34*	*0.16*	*0.48*	*44.34*
Total daily time travelling for unknown/undeclared (DT_FUNK1)										
Mean	0.00	0.00	0.00	0.00	69.85	19.71	0.10	0.41	14.08	31.93
Model Statistics	Cases	Params.	Log-likelihood		BIC		AIC		R-squared	
	1471	376	-52787.1		108317		106326		0.966	

Part-time workers have the highest probability of belonging to Cluster 1, while full time workers have the highest probability of belonging to Cluster 2 and this again may be an indicator of the need to focus on self. Generally as household size and number of children five years or younger increase, the probability of belonging to Cluster 1 is also increasing, which is an indicator of possible distribution of task allocation that may be facilitated by the presence of more persons in the household and the gain of free time to be selfishly spent. As the number of children of age 16-18 increases, the probability of belonging to Cluster 3 also increases. Another way of analyzing the clusters derived here is to study the within-cluster composition. Table 14.13 shows the percent of persons within each cluster that belong to each of the occupation/role categories. For example, 25.1 percent of persons in Cluster 1 are full-time workers and 19.1 percent are retired. Full-time workers populate heavily Clusters 2, 3, 6 and 7. These are also the clusters with substantial amounts of time dedicated for others in a day and a substantial amount of travel for self and travelling for others. From among the most popular clusters, we cannot clearly differentiate between men and women.

Table 14.12
Average Membership Probabilities for Self-Serving and Altruistic Activity Clusters

CLUSTER	1	2	3	4	5	6	7	8	9	10	Total
OCCUPATION/ROLE											
Child at home	0.784	0.039	0.098	0.039	0.000	0.000	0.000	0.020	0.000	0.020	1.000
Home duties	0.431	0.025	0.290	0.019	0.000	0.015	0.031	0.114	0.030	0.046	1.000
Looking for work	0.500	0.187	0.063	0.000	0.063	0.060	0.000	0.000	0.063	0.066	1.000
Retired	0.440	0.135	0.109	0.059	0.048	0.021	0.031	0.060	0.040	0.057	1.000
Disabled	0.600	0.200	0.067	0.067	0.000	0.067	0.000	0.000	0.000	0.000	1.000
Pre-school or day care	0.596	0.000	0.121	0.060	0.000	0.032	0.000	0.031	0.066	0.094	1.000
Grades K-6	0.659	0.094	0.132	0.000	0.024	0.024	0.000	0.000	0.055	0.012	1.000
Grades 7-12	0.475	0.177	0.038	0.132	0.047	0.051	0.032	0.000	0.047	0.000	1.000
College/University	0.317	0.301	0.047	0.057	0.122	0.045	0.013	0.023	0.047	0.027	1.000
Part-time (<40 hours/week)	0.382	0.160	0.250	0.022	0.023	0.019	0.051	0.023	0.031	0.039	1.000
Full-time (>=40 hours/week)	0.243	0.304	0.201	0.034	0.035	0.054	0.057	0.031	0.023	0.019	1.000
Missing	0.500	0.115	0.115	0.077	0.039	0.039	0.000	0.039	0.000	0.077	1.000
GENDER											
Female	0.364	0.216	0.174	0.042	0.036	0.030	0.038	0.040	0.024	0.035	1.000
Male	0.392	0.203	0.134	0.047	0.048	0.047	0.036	0.028	0.039	0.027	1.000
Missing	0.418	0.215	0.213	0.000	0.000	0.055	0.002	0.000	0.098	0.000	1.000

Cluster 1 has more men and Clusters 2, and 3 have more women. Only Cluster 3 is significantly more popular for women than for men, showing that when we control for employment/occupation, gender is not a significant discriminator of activity and travel patterns of this type. Interestingly, this analysis shows that self-serving and altruistic behaviours are spread throughout the different occupation/role groups and the different age and gender groups. Respondents stated that many of their activities are pursued for multiple persons. Either their response would enumerate all the persons for whom the activity was pursued or they would make a written statement such for the *family*, for *company guests*, for *daughter's family,* for the *dogs,* and so forth.

Table 14.13
Within Cluster Membership Probabilities and Composition

CLUSTER	1	2	3	4	5	6	7	8	9	10
OCCUPATION/ROLE										
Child at home	0.072	0.007	0.022	0.031	0.000	0.000	0.000	0.021	0.000	0.022
Home duties	0.051	0.005	0.084	0.020	0.000	0.018	0.038	0.151	0.042	0.067
Looking for work	0.014	0.010	0.004	0.000	0.017	0.017	0.000	0.000	0.021	0.023
Retired	0.193	0.107	0.116	0.222	0.194	0.091	0.140	0.296	0.204	0.308
Disabled	0.016	0.010	0.004	0.016	0.000	0.018	0.000	0.000	0.000	0.000
Pre-school or day care	0.034	0.000	0.017	0.030	0.000	0.018	0.000	0.020	0.044	0.066
Grades K-6	0.099	0.026	0.049	0.000	0.033	0.036	0.000	0.000	0.096	0.022
Grades 7-12	0.073	0.049	0.014	0.175	0.066	0.077	0.051	0.000	0.084	0.000
College/ University	0.084	0.144	0.030	0.132	0.296	0.119	0.037	0.069	0.146	0.089
Part-time (<40 hours/week)	0.089	0.067	0.141	0.044	0.049	0.042	0.121	0.061	0.084	0.112
Full-time (>=40 hours/week)	0.251	0.567	0.505	0.299	0.329	0.546	0.614	0.361	0.279	0.246
Unknown	0.023	0.010	0.013	0.031	0.017	0.018	0.000	0.020	0.000	0.044
Total	1.000	1.000	1.000	1.000	1.000	1.000	1.000	1.000	1.000	1.000
GENDER										
Female	0.485	0.520	0.565	0.487	0.444	0.392	0.527	0.604	0.377	0.575
Male	0.500	0.466	0.417	0.513	0.556	0.588	0.473	0.396	0.582	0.425
Unknown	0.015	0.014	0.019	0.000	0.000	0.020	0.001	0.000	0.041	0.000
Total	1.000	1.000	1.000	1.000	1.000	1.000	1.000	1.000	1.000	1.000

A future task in this analysis includes differentiation among these multiple category answers to discern important differences and commonalities among respondents with multiple served recipients. Another key aspect here, however, is the location where altruistic and self-serving activities take place and location preferences that may be different for different genders. For example, 60 percent of the activities for ones self are at home. In contrast, 71 percent of the activities for daughters are outside the home and only 57 percent of activities for the son are outside the home. Examining the same proportion for the answer *mother*, we find 79.9 percent of the activities outside the home and for *father* 84.9 percent indicating a reversal of location preference. As we interpret the statistics in this analysis, it is worth noting the different data reporting styles by the respondents. The following two examples are indicative of joint activity reporting by a couple. Figure 14.1 shows perfect agreement in activity time and type, and also contains somewhat different ways to answer the *for whom* question. In this case, during coding a proper numerical value needs to be identified and be consistent throughout the survey database building.

Person	Date	Begin Time	End Time	Activity	With Whom	For Whom
Wife	January 30	11:00	11:10	Walked to bank	Husband	Self
		11:10	11:20	Banking	Husband and Bank Employee	Family
		11:20	11:25	Returned to Work	Husband	Self
		11:25	11:55	Went for Walk	Husband	Self
Husband		11:00	11:10	Walked with wife to Credit Union	Wife	Both of us
		11:10	11:20	Credit Union Transaction	Wife	Both of us
		11:20	11:55	Finished walk with wife	Wife	Both of us

Figure 14.1
First Example of a Portion of Husband-Wife Schedule

Person	Date	Begin Time	End Time	Activity	With Whom	For Whom
Wife	April 13	8:30	8:45	Go to Church	Husband and Daughter	Family
		8:45	10:30	Attended Church	Husband and Bank Employee	Family
		10:30	10:40	Went to Wal-Mart	Self	Father
		10:40	10:50	At Wal-Mart	Self	Father
		10:50	11:00	Went to Father's	Self	Father
		11:00	11:10	Return Home	Self	Self
Husband		9:00	9:10	Went to Church	Wife and Daughter	Family
		9:10	11:50	Attended Church	Wife and Daughter	Family
		11:50	12:00	Returned Home	Wife and Daughter	Family

Figure 14.2
Second Example of a Portion of Husband-Wife Schedule

In the second example (Figure 14.2), although the two persons report travelling together to church and could give the impression of possible miss-reporting of the departure and arrival time to church, it also shows the wife going elsewhere alone while the husband stays at the church site. The example is also indicative of the added activities women pursue helping others (in this example purchasing something for an older parent, deliver it, and then return home).

The example also demonstrates another important aspect of interactions that we need to study in travel behaviour to include a spatially distributed extended household spanning multiple generations (i.e., the grand parents, parents, and children). These three generations are very often defined as three different households that interact. As a result the within-household interactions alone miss important motivations for activity participation and travel.

SUMMARY AND CONCLUSIONS

In this chapter, using data from 1,471 residents of Centre County Pennsylvania, we examine time allocation to solo and joint time allocation, and for the first time in travel behaviour analysis, the stated altruistic and self-serving behaviour. Some of the findings here repeat findings in other regions. For example, workers (both full- and part-time) spend significant amounts of time travelling each day (1.5 hours) and very little shopping (about 8 minutes) each day. In addition, people without children in the home spend the most time per day in service activities and their counterparts with children spend a significant portion with the children. Both men and women allocate substantial amounts of time with family and others with women spending slightly more with family members. Both genders show little variability from one day to the next. Allocation of time with others, however, does not necessarily mean serving others. In fact, the proportion of time allocated for family and others is significantly lower than the proportion with family and others. Again, men and women have some similarities but women spend substantially more time with family members. When we examine the number of activities (episodes), we see a general agreement with the time dedicated to these activities. The gender and age of the recipient for whom the activity is pursued may influence the location of the activity. As one reviewer noted life cycle stage and activity purpose are key determinants of the activity location choice. However, the differences found among daughters and sons may also point out to other factors that require a more careful and detailed scrutiny in subsequent analysis.

The latent class cluster analysis reveals that a large amount of time in a day is dedicated to activities and travel alone and for reasons/purposes that are self-serving. For example, on average people in this sample spend 2/3 of their time for themselves. This happens even when we consider that a significant portion of the 24 hours is dedicated to sleep and rest. However, significant differences in all three behavioural aspects of the type of activity, solo versus joint participation, and self-serving versus altruistic are observed among the persons that work in different ways (part time and full time), among the different school age children, and persons that may appear to have reasons to stay home. The disabled and the retired, however, appear to be very active and diverse. The day of the week effects are very strong in this analysis with each day having its own "character" in terms of activity participation. A key finding here is that gender when controlling for other factors is not a major determinant (except for a small portion of behavioural patterns) indicating that the gender divide, when a more complete analysis is performed, disappears and the use of variables such as occupation should be preferred when available. Another important finding is that the segmentation of samples into workers versus non-workers may not be a sufficient and appropriate practice because as shown in the solo vs. joint and the self-serving vs. altruistic analyses workers and non-workers are found in substantial numbers in many different clusters. Therefore, workers and non-workers may have other characteristics in common such as caring for an older parent that determines their behaviour. To these large behavioural variations major contributor may be seasonality and weather that can be included in future analysis. The presence of many categories with large unknown values is a concern that also requires further scrutiny.

ACKNOWLEDGMENTS

We are grateful to the research personnel that worked at the Pennsylvania Transportation Institute on this project and they are: Michael Patten, Mark Hallinan, James Lee, Devani Perera, Brian Hoffheins, Brett Boyle, JinKi Eom, Joonhyo Kim, Ondrej Prybil, Michael Zekkos, Julie Whitt, and Avi Mukherjee. Funding for data collection was provided by the Pennsylvania Department of Transportation and the Mid-Atlantic Universities Transportation Center, a US DOT regional transportation centre. We are also grateful to the reviewers for the many comments, suggestions and corrections that improved earlier manuscripts. However, the contents of the chapter reflect the authors' viewpoints and they do not constitute a policy or official position of any State and/or Federal public agency.

REFERENCES

Agresti, A. (2002). *Categorical Data Analysis*. Wiley, New York, NY.

Alam, B.S. (1998). Dynamic Emergency Evacuation Management System Using GIS and Spatio-Temporal Models of Behavior. M.S. thesis. Department of Civil and Environmental Engineering. The Pennsylvania State University, University Park, PA.

Alam, B.S. and K.G. Goulias (1999). Dynamic emergency evacuation management system using GIS and spatio-temporal models of behavior. *Transpn. Res. Rec.*, **1660**, 92-99.

Arentze, T.A. and H.J.P. Timmermans (2000). Albatross: *A learning Based Transportation Oriented Simulation System*. European Institute of Retailing and Services Studies, Technical University of Eindhoven, Eindhoven.

Becker, G.S. (1976). *The Economic Approach to Human Behavior*. University of Chicago Press, Chicago, IL.

Bhat, C. and F.S. Koppelman (1999). Activity-based modeling of travel demand. In *Handbook of Transportation Science* (R.W. Hall, ed.), pp. 35-61. Kluwer, Boston, MA.

Burnett, P. (2004). Qualitative techniques for urban transportation. Paper presented TRANSTEC Conf., Athens, Greece.

Chandrasekharan, B. and K.G. Goulias (1999). Exploratory longitudinal analysis of solo and joint trip making in the Puget Sound transportation panel. *Transpn. Res. Rec.*, **1676**, 77-85.

Chapin, F.S. Jr. (1974). *Human Activity Patterns in the City*. Wiley, New York, NY.

Clifton, K.J. and S.L. Handy (2003). Qualitative methods in travel behaviour research. In: *Transport Survey Quality and Innovation* (P. Stopher and P. Jones, eds.), pp. 283-302. Pergamon, Oxford.

Gliebe, J.P. and F.S. Koppelman (2002). A model of joint activity participation. *Transpn.*, **29**, 49-72.

Golledge, R.G. and R.J. Stimson (1997). *Spatial Behavior: A Geographic Perspective*. The Guilford Press, New York, NY.

Goulias, K.G. (1999). Longitudinal analysis of activity and travel pattern dynamics using generalized mixed Markov latent class models. *Transpn. Res. B*, **33**, 535-557.

Goulias, K.G. (2003a). Transportation systems planning. In: *Transportation Systems Planning: Methods and Applications* (K.G. Goulias, ed.), pp. 1.1 – 1.45. CRC Press, Boca Raton, FL.

Goulias, K.G. (2003b). On the role of qualitative methods in travel surveys. In: *Transport Survey Quality and Innovation* (P. Stopher and P. Jones, eds.), pp. 319-329. Pergamon, Oxford.

Goulias, K.G. (2004). Transportation planning and modelling. In: *Handbook of Transportation Engineering* (M. Kutz, ed.), pp. 29.3-29.25. McGraw-Hill, New York, NY.

Goulias, K.G., M. Bradley, V. Noronha, R.G. Golledge and P. Vovsha (2004). Data Needs for Innovative Modeling. *Resource Paper Nat. Househ. Trav. Surv. Conf.: Understanding Our Nation's Travel*, Washington D.C., (http://www.trb.org/Conferences/NHTS).

Goulias, K.G., N. Kilgren, and T. Kim (2003). A decade of longitudinal travel behavior observation in the Puget Sound region: sample composition, summary statistics, and a selection of first order findings. *Proc. 10th IATRB Conf.*, Lucerne, Switzerland (CD-Rom).

Goulias, K.G. and T. Kim (2001). Multilevel analysis of activity and travel patterns accounting for person- and household-specific observed and unobserved effects simultaneously. *Transpn. Res. Rec.*, **1752**, 23-31.

Hägerstrand, T. (1970). What about people in regional science? *Pap. Reg. Sc. Assoc.*, **10**, 7-21.

Jones, P., F.S. Koppelman and J. Orfeuil (1990). Activity analysis: State-of-the-art and future directions. In: *Developments in Dynamic and Activity-Based Approaches to Travel Analysis* (P. Jones, ed.), pp. 34-55. Avebury, Aldershot.

Kitamura, R. (1988). An evaluation of activity-based travel analysis. *Transpn.*, **15**, 9-34.

Krizek, K.J. (2003). Planning, household travel, and household lifestyles. In: *Transportation Systems Planning: Methods and Applications* (K.G. Goulias, ed.), pp. 6.1-6.42. CRC Press, Boca Raton, FL.

Kuhnau, J. and K.G. Goulias (2003). Centre SIM: First-generation Model Design, Pragmatic Implementation, and Scenarios, In: *Transportation Systems Planning: Methods and Applications* (K.G. Goulias, ed.), pp. 16.1-16.14. CRC Press, Boca Raton, FL.

McCutcheon, A.L. (2002). Basic concepts and procedures in single- and multiple-group latent class analysis. In: *Applied Latent Class Analysis* (J.A. Hagenaars and A.L. McCutcheon, eds.), pp. 56-88. Cambridge University Press, Cambridge.

McNally, M.G. (2000). The Activity-based Approach. In: *Handbook of Transport Modelling* (D.A. Hensher and K.J. Button, eds.), pp. 113-128. Pergamon, Oxford.

Mehndiratta, S.R., R. Picado and C. Venter (2003). A qualitative survey technique to explore decision making behaviour in new contexts. In: *Transport Survey Quality and Innovation* (P. Stopher and P. Jones, eds.), pp. 303-317. Pergamon, Oxford.

Pendyala, R.M. and C. Bhat (2004). Emerging issues in travel behavior analysis. *Resource Paper Nat. Househ. Trav. Surv. Conf.: Understanding Our Nation's Travel*, Washington D.C., (http://www.trb.org/Conferences/NHTS).

Pribyl, O. and K.G. Goulias (2004). Simulation of daily activity patterns. *Proc. Conf. Progress in Activity-Based Analysis*, Maastricht, the Netherlands (CD-Rom), also this volume Chapter 3.

Townsend, T.A. (1987). The Effects of Household Characteristics on the Multi-Day Time Allocations and Travel - Activity Patterns of Households and their Members. Ph.D. Dissertation, Northwestern University, Evanston, IL.

Vermunt, J.K. and J. Magidson (2002). Latent class cluster analysis. In: *Applied Latent Class Analysis* (J.A. Hagenaars and A.L. McCutcheon, eds.), pp. 89-106. Cambridge University Press, Cambridge.

Zhang, J., H.J.P. Timmermans and A.W.J. Borgers (2005). A model of household task allocation and time use. *Transpn. Res. B*, **39**, 81-95.

15

Recent Developments in Activity Diary-Based Surveys and Analysis: Some Japanese Case Studies

Kazuo Nishii, University of Yamanashi, Yamanashi, Japan

Kuniaki Sasaki, University of Yamanashi, Yamanashi, Japan

Ryuichi Kitamura, Kyoto University, Kyoto, Japan

Katsunao Kondo, University of Marketing and Distribution Science, Kobe, Japan

Introduction

Activity-based approaches have a 25 year history in Japan (Kondo, 1974). Based on a time-geographic perspective (Hägerstrand, 1970) and early US work (Burns, 1970), the goal of this work was to improve travel demand analysis and better understand activity and travel behaviours. This shift in focus reflects significant changes in both problem formulation and methodology in transportation planning (Kitamura, 1988; Jones, 1990). It is caused not only by external factors such as technological innovation and increased variation in the social and economic environment, but also by internal factors such as diversification of individuals' values and lifestyles. In this chapter, we discuss four case studies based on activity diaries (AD) to provide examples of activity-based research in Japan focusing on the design and analysis of activity diaries (Kondo and Kitamura, 1987; Nishii and Kondo, 1992). It should be noted that the

Progress in Activity-Based Analysis edited by H. Timmermans

chapter is not meant to be a review of recent progress in Japan in this field of study. After the first generation in 1980s, the activity diary surveys were often designed as panel surveys to evaluate situational change of households (e.g., Nishii *et al.*, 1998; Takao *et al.*, 1998). One of other approaches concerns the use of person trip (PT) surveys. Fujii *et al.* (1997) and Fujii *et al.* (1998) were among the first to use PT surveys in activity analysis. They proposed new evaluation indicators of transportation policies based on activity analysis. This approach is currently continuing and developing as part of micro-simulation approaches and household-based analysis (e.g., Fujii *et al.*, 2000, Zhang *et al.*, 2004). In addition to these surveys, the use of new technology such as cellular phones and GPS is also increasing (e.g., Omori *et al.*, 1999; Asakura *et al.*, 2000; Osada *et al.*, 2004) to address policy issues that cannot be answered with traditional transport surveys.

ACTIVITY DIARY APPROACHES RELATED TO POLICY ISSUES

Jones *et al.* (1993) and Kitamura (1996) argued a decade ago that problem-solution-oriented analysis will be the mainstream of the next generation transportation demand forecasting methods. Various transport policies have to reflect future social needs and diversified values derived from such analysis. Such analysis emphasizes the derived nature of travel and needs to elucidate the decision structure underlying travel behaviours. In this regard, the activity-based approach has been developed as a useful tool for evaluating various transport policies (Hanson, 1979; Jones *et al.*, 1983).

Activity-based approaches have much in common with conventional trip-based approaches. They share the concern of estimating trip frequencies induced by facility improvement, modelling travel demand on non-workdays, and analyzing the effects of traffic management. However, the approach is particularly useful to address policy issues, mentioned in Table 15.1, which are concerned with the effects of social-economic change and various transport policies, based on the relationship between various activities in daily life and travel behaviours (Kostyniuk and Kitamura, 1983; Recker *et al.*, 1986). The activity-based approach needs data on (daily) activities. An activity diary (AD) can avoid some of the typical characteristics and limitations of a personal trip (PT) survey. For example, although a PT survey can derive characteristics of out-of-home activities by processing and transforming information on trip purposes, it is not easy to derive accurate information about discretionary activities such as leisure because the classification of trip purposes often lacks detail.

Table 15.1
Changes in Socio-Economic Circumstances and the Corresponding Issues in Activity-Based Analysis

Changes in Socio-Economic Circumstances	Corresponding Issues in Activity-Based Analysis
Aging society and low birth rate	Aged people's mobility and their activity time allocation and time use patterns
	Actual situation of various restrictions related to aged people's travelling and daily activities
	Aged people's latent demand for in-home activities caused by their restrictions
Variety of lifestyle patterns	Changes in work schedule (a five-day week, shortening of working hours, and work sharing) and their effect
	Women's participation in the labour market (increase of women's work rate and correcting differential wage) and its effect on household activity and travel patterns
	Activity patterns considering inter-dependencies among household members
	Effect of increase of leisure time on non-workdays
Progress of a highly intelligent society	Relationship between working hours (telecommuting, satellite office) and activity patterns
	Mobile telecommunication and activity patterns
Environmental policies	Effect of TDM policies on air quality (P&R, P&BR, road pricing)
	Improvement of railway services and its effect on commuters' activity patterns
Structural rebuilding and renewal in urban areas	Comprehensive transport policy and management aiming at sustainability and diversification of urban traffic
	Transportation system adaptable to a compact city

In addition, because PT surveys do not contain any information about in-home activities, it is quite difficult to estimate to what extent trip generation is likely to change as a function of changing socio-economics settings or transport policies in the sense that possible substitution cannot be taken into account. Activity diaries clearly have the advantage that detailed information about a long list of activities can be collected. In general, two types of activity diaries can be distinguished: the 24 hours AD and the consecutive multiple days diary (Ampt *et al.*, 1983). Table 15.2 lists some typical kinds of analysis that can be conducted with data about activities and with the diary at large. As shown in this table, the AD survey methods can serve a variety of purposes based on rich information on daily and weekly activities. They also enable us to challenge many issues on both surveying and modelling of activity patterns.

Table 15.2

General Purposes of Data Collection and Analytical Issues in AD Surveys

Aspects	Purposes for Data Collection	Analytical Issues
Activity data: Activity & travel linkages, trip chaining behaviours, and interdependency among household members	1)To easily acquire data about discretionary trips with a short distance as respondents have to answer the trips linked with their activities 2)To identify how the relationships among activities determine mode choice decisions in trip-chaining behaviour 3)To examine properties of trip production considering activity scheduling and the relationship between mandatory and discretionary activities 4)To explore properties of in-home activities and to examine the interdependencies among household members	a.) Issues on survey design & instrument; - Non-response bias - Reporting error - Sampling method, the role of incentives - Layout and design of questionnaire sheets - Respondent's fatigue and burden b.) Issues on modelling; - Activity linkages/ activity time allocation - Space time constraints - Household activity patterns
Diary data: Weekly activity patterns, weekdays / weekends activities, and dynamic characteristics	1) To examine a weekly activity pattern such as a daily shopping and eating-out 2) To compare activity patterns on weekdays and weekends 3) To examine heterogeneity underlying activity and travel behaviours based on the AD data from multiple points in time 4) To identify dynamic aspects like state dependency, irreversibility and irregularity by using panel AD data sets	- Interdependency among household members - Weekdays / weekends comparison of activity patterns - Variation of activity patterns across the week - Dynamic characteristics of activity patterns

To illustrate the potential of using activity diaries, the following section discusses four Japanese case studies. Each of these case studies was motivated by a specific problem. In addition, the case studies differ in terms of survey implementation and the analytic framework that was used, illustrating the spectrum of potential applications. Case 1 is an analysis of the impact of travel demand management policy on travel behaviour related to a park and bus ride social experiment (P&BR) in Kofu. A panel AD survey was conducted to measure dynamic change in activity patterns before and during the P&BR experiment. In this survey, the particular concern was how the activity pattern is influenced by the experiment.

The second example is an AD survey that was conducted as a supplementary survey for the fourth person trip survey in the Osaka-Kobe-Kyoto metropolitan area. This survey includes information about an individual's activity and travel linkages on three consecutive days and examines their time use patterns on non-workdays. The third case study is concerned with activity and travel behaviours of the users of a night express bus. The policy context here is that some Japanese bus companies have recently introduced a night express bus to expand their market share in reaction to their difficult financial situation. They expect that a night express bus service can promote an interchange of activities between a rural mid-sized city and metropolitan areas. It also produces an increase in efficiency in time use and activity patterns involving long distance movement. In addition, the bus service may influence diversification of activities caused by the activity patterns exceeding 24 hours.

Case 3 involves a two consecutive day diary related to movement between Kofu and Osaka. This survey aims to identify the characteristics of activity-travel patterns on the day travelling on the bus, on the second day, the day getting off the bus. The fourth case study is concerned with the relationship between telecommunication and activity patterns. It is clear that many kinds of innovative telecommunication tools have had effects on individual's daily activity and travel behaviour. The survey for this case study consists of two parts: an activity diary for university students to collect information on out-of-home and in-home activities and trips for an entire week, and a telecommunication record sheet about the use of mobile means of communication. The aim of this study was to analyze time use and allocation under space-time prism constraints. Rule-based methods were used for describing the relationships among telecommunication, activity and travel patterns.

THE CASE STUDIES

Case 1: Monitoring the P&BR Experiment in Kofu City

Kofu AD Survey. The P&BR social experiment, described in this case study, was conducted for two months in 1999, after an initial trial in 1997. The experiment aims at examining how workers change their transport mode from car to P&BR and at exploring the possibility of introducing a full-fledged P&BR system in Kofu city. The AD survey was conducted as a panel survey at two points in time. The first wave was before the experiment had started and the second wave during the experiment. Participants in this P&BR experiment were selected as the respondents of the first wave. They constitute the experimental group and their behaviours were monitored. In addition, a control group was formed, consisting of workers who live together with the participants in the experiment, and those who participated in the first P&BR experiment in 1997 but did not in 1999. The experimental group included 103 respondents, whereas the control group involved 131 respondents in wave 1. Respondents in the second wave involved only the participants of the experiment. Wave 2 consisted of 71 individuals from the original 103 respondents in wave 1. The social demographics of the experimental and control groups slightly differed. For example, males make up 83% of the experimental group, and only 55.3% of the control group. The share of office clerks in the experimental group is 60%, but only 37% in the control group. To improve the response rate and reduce the problem of non-response bias, the mail-out/hand-back method was employed to collect the data in wave 1. In wave 2, the mail-out/mail-back method was employed, since the respondents were already familiar with the survey. A gift coupon (1,000 yen, equivalent to 9 US dollars) was issued as an incentive to improve the response rate. The survey in wave 1 was conducted for three consecutive days. In wave 2, since the P&BR system was not available on weekends, the survey was conducted for three consecutive days excluding Saturday and Sunday. The AD survey sheets employed a pre-coded system for classifying activities in order to reduce respondents' burden. This system requested respondents to enter bars corresponding to the time spent on their activities in a column of pre-classified activities. This survey sheet was also used for the AD survey administered in the Kei-Han-Shin metropolitan area in 1999.

Analytical Framework and Some Results. The analytical framework of this case study is based on the concept of a time space path and time allocation of activity sequences in a day. A time space path represents characteristics of a chain of trips by focusing on the spatial dimension. The temporal dimension reveals time allocation and time use patterns underlying activity sequences. The first research question was why it is difficult for members of the control group to shift their transport mode from car to P&BR. It was assumed that the reason should be found in the characteristics of the activity and time use patterns. Figure 15.1 displays the distribution of in-home and out-of-home activity duration of both the experimental and control group at the Central Business District of Kofu City. The control group spent more time on in-home activities, including mandatory activities. In contrast, the experimental group spent more time on out-of-home activities, including work. The control group spent 25 minutes more on housekeeping compared to the experimental group. The average number of trips per day was 3.00 trips/day for the experimental group and 2.68 trips/day for the control group. The number of multi-stop trips was 8% higher in the control group. These results suggest that individuals in the control group prefer the car because they spend more time on in-home activities. Consequently, commuting time may be relatively more important to them. The higher number of multi-stop trips also points at a higher efficiency in organizing daily activities, favouring the car.

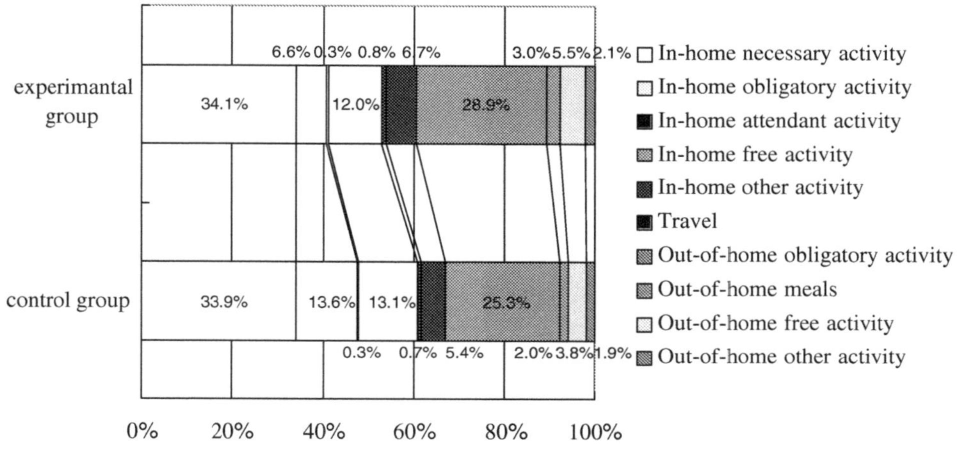

Figure 15.1
Share of Total Activity Time

The second research question was how the experimental group changed their activity patterns during the P&BR experiment. To answer this question, the analysis focused on the changes that can be derived from the AD panel data. When comparing trip-chaining behaviour of wave 1 (before the experiment) with wave 2 (during the experiment), it was found that, after shifting their commuting mode from car to P&BR, the percentage of respondents of the experimental group who conducted multi-stop trips in their daily trip chain increased by about 10%. The average number of activities per day increased by 0.9 and the average number of trips per day increased by 0.28. This tendency of making more stops in a trip chain was caused by the fact that they added activities around the P&BR terminal in the downtown area before going back home. Travel time increased by eight minutes on average. No significant difference was discovered in either the schedule of in-home or out-of-home activities. These findings suggest that the Kofu AD survey was very useful for evaluating the effects of the P&BR experiment on time use and time allocation patterns.

Case 2: AD Survey as a Supplement to a PT Survey in the Osaka-Kobe-Kyoto Area

Design and Implementation of the AD Survey. To investigate travel demand in the Osaka-Kobe-Kyoto metropolitan area, a PT survey was administrated three times: in 1970, 1980 and 1990. These PT surveys were used to design a master plan for the development of a comprehensive transportation network system in this area (Yoshida and Hasegawa, 2000). Dramatic changes in social and economic situations such as depopulation, low birth rate, aging, widespread use of information devices, and increased environmental consciousness required a new point of view for urban transportation planning. The AD survey that was administered as a supplement to the PT survey was regarded as a useful method for illuminating trip production mechanisms because the AD survey can explicitly deal with the derived nature of travel demand and provide information on individual decision-making. The activity diaries were used to analyze trends in trip production rates, to classify activity sequences based on time-use patterns, to examine substitution and complementary relationships between in-home and out-of-home activities, and to study the relationship between mobility levels and activity patterns. The diary data were especially used to analyze the causal relations between activity patterns and travel behaviour. A pilot survey was conducted in 1999 in preparation of the full AD survey in 2000. The survey was designed to measure individual characteristics (gender, age and office address) and facets of activities (timing, activity type, accompanying persons, destination and transport mode). A sheet, questioning about individual and household characteristics, was prepared and a set of activity diary sheets for three days was also prepared. One of two types of activity diary sheets was distributed to respondents: a fill-out sheet with open activity types and a pre-coded activity sheet. Respondents could freely enter their activities along the time axis on the former sheet, and could draw a line with an arrow based on the time axis and pre-coded activity groups on the latter sheet.

To collect data for workdays and non-workdays, the survey was administrated for three consecutive days in October and November 1999. Samples were selected at random with the household as one unit in large- and medium-sized cities, such as Kyoto City (Nakagyo-ku and Yamashina-ku), Osaka City (Kita-ku) and Otsu City. Diaries were delivered and retrieved personally. The average response rate was 54% in the pre-AD survey, 18% lower than the response rate of 72% for the pre-PT survey in 1999, which used the same mode of administration. The response rate for the AD survey in 2000 was 51% and the total sample included 1,798 individuals, whereas the response rate of the 2000 PT survey was 65.5% with a total of 430,000 individuals. These differences in response rates were probably caused by the fact that respondents were forced to complete troublesome and perhaps too many questions (all activities from morning to midnight were to be entered for three days). Table 15.3 shows the results of a comparison of the two types of survey sheets. The completion rate of the pre-coded sheet was approximately 10% lower, and had a higher number of blanks on the "travel mode" and "duration" facets. The latter may be caused by the fact that the answer space was rather small. With regard to trip production rates, no significant difference was discovered between the two types of survey sheets.

Table 15.3
Types of Questionnaire Sheet in the Pre-AD Survey

Type	Type E (Open End Type)	Type F (Pre-Coded Type)
How to entry a sheet	Respondents fill out the contents of their activities by drawing the time-interval line	Respondents fill out the line in the item corresponding to the contents of their activities.
The number of inquiries by phone	13 calls during the survey period	10 calls during the survey period
Response rate	58.8%	49.5%
The average time of missing activities	14.9 minutes	30.2 minutes
The percentage of unknown activity destinations	18.2%	16.8%
The percentage of missing transport mode data	7.4%	17.8%
Trip production rate	2.74 trips per day	2.78 trips per day

The AD Survey Results. First, trip production rates were analyzed. Table 15.4 shows the trip production rates on weekdays by gender and age comparing the AD survey with the PT survey. The weekday trip production rate according to the AD survey was 2.92 trips per person, 0.5 trips higher than the 2.49 trips per person per day derived from the PT survey. The table also shows that the trip production rate based on the AD survey is higher for the elderly (age 60-69). When focusing on the trip production rates by mode by day of week, it was found that the walking-trip production rate obtained from the PT survey was lower than that based on the AD survey for both workdays and non-workdays (PT survey: 0.47 trips on workdays, and 0.36 trips on Sunday; AD survey: 0.90 trips on workdays, and 0.74 trips on Sunday).

Second, activity time allocation patterns were analyzed. The results are reported in Figure 15.2, which portrays space-time prism heights and time allocation patterns of activity sequences by day of the week. The prism height is equal to the total available time, which is denoted as the difference between the 80 percentile value of the arrival time distribution of the last trip and the 20 percentile value of the departure time distribution of the first trip. Figure 15.2 indicates that the prism height on the weekend is about 12.5 hours and about 14 hours on weekdays. This difference in the prism height is caused by the tendency that individuals start with the first trip later on the weekend than they do on weekdays. The figure also indicates that the average duration of in-home activities on the weekend is 17.3 hours, which is 67 minutes longer than on weekdays.

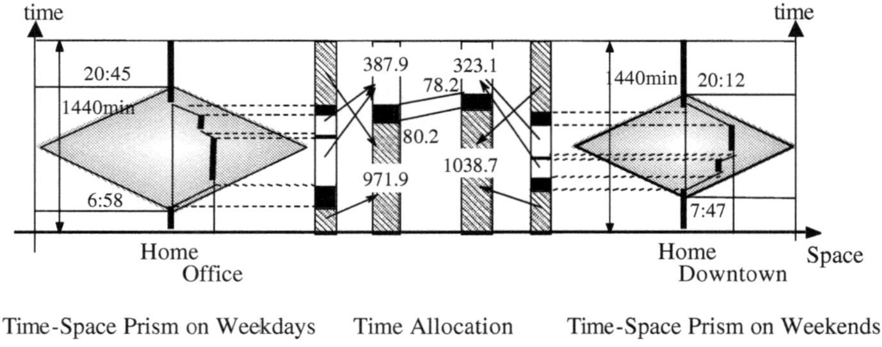

Figure 15.2
Time-Space Prism and Time Allocation by Day of the Week

On the other hand, the average duration of out-of-home activities on the weekend is 5.4 hours, which is 65 minutes shorter than on weekdays. Table 15.5 shows the participation rate by activity and the average duration of each activity by day of week. Participation rate is defined here as the ratio of the number of individuals, who conducted a particular activity on the survey day to the total number of individuals. When focusing on out-of-home activities, the participation rate for activities such as shopping for daily goods, personal business and miscellaneous tasks is 67.1% on the weekend and 82.9% on weekdays.

Table 15.4
Trip Production Rate on Weekdays by Age and Gender

Gender/ Age Group	Male		Female		All	
	AD Survey	PT Survey	AD Survey	PT Survey	AD Survey	PT Survey
15-19	2.85	2.29	2.66	2.35	2.76	2.32
20-29	2.66	2.43	2.64	2.45	2.65	2.44
30-39	2.73	2.83	3.23	3.14	3.01	2.99
40-49	2.90	2.88	3.67	2.94	3.30	2.91
50-59	2.78	2.79	3.29	2.59	3.06	2.69
60-69	3.54	2.42	3.06	2.10	3.30	2.26
Over 70	2.71	1.68	1.77	1.25	2.14	1.42
Average	2.88	2.55	2.96	2.43	2.92	2.49

The average duration of these activities is 171 minutes on the weekend and 310 minutes on weekdays. In contrast, the participation rate of discretionary activities such as recreation, leisure and entertainment is 42.9% on the weekend, while the average duration of these activities is 87.9 minutes. On weekdays, the participation rate for these activities is only 25.6% and the average duration only 31.2 minutes. In addition, it is noted there is hardly any difference in participation rates and average duration of mandatory and discretionary activities (for example, service to visitors in home and eating-out) between the weekend and weekdays.

Table 15.5
Participation Rate and Average Duration by Day of the Week

Activity	Classification	Weekends			Weekdays		
		Participation Rate (%)	Average Duration (Gross)	%	Participation Rate (%)	Average Duration (Gross)	%
In-home Activity	Unknown	5.0	4.8	0.3	4.5	4.6	0.3
	Essential: sleep, eat	99.6	577.5	37.2	99.0	529.6	33.7
	Mandatory: house-keeping	98.3	231.1	14.9	95.1	248.1	15.8
	Mandatory/ discretionary: visitors	8.3	9.8	0.6	8.5	7.8	0.5
	Discretionary: TV, game	93.8	332.5	21.4	92.6	289.4	18.4
	Others	3.3	1.4	0.1	5.4	5.3	0.3
Trip		87.5	76.5	4.9	90.9	82.2	5.2
Out-of-home Activity	Unknown	3.8	8.3	0.5	1.9	2.9	0.2
	Essential: sleep	0.4	0.3	0.0	1.4	4.6	0.3
	Mandatory: work	67.1	170.7	11.0	82.9	310.0	19.8
	Mandatory/ discretionary: eating-out	44.2	30.7	2.0	45.2	29.5	1.9
	Discretionary: leisure, sport	42.9	87.9	5.7	25.6	31.2	2.0
	Others	12.5	23.0	1.5	15.1	24.0	1.5

Case 3: AD Survey on Activity Patterns of Night Express Bus Users

Background and AD Survey. This case study was motivated by the fact that little is known about transport demand at night. Such information is however necessary to assess and precisely identify the demand structure for a night express bus service. A conventional one-day survey method is insufficient due to the lack of information. The survey tried to identify individual characteristics, characteristics of the trip by the night express bus on the survey day, user's awareness and needs for a night express bus, and activity diaries for two consecutive days involving the movement between Kofu and Osaka.

The night express bus service was launched on 30 September in 2000 and connects Kofu, a central city of Yamanashi Prefecture, and Osaka via Kyoto. It has two bus stops in Kofu, one stop in Kyoto, and several stops in Osaka. The frequency is one round trip per day, starting at 10 p.m. and terminating at 6:00 a.m. in Kyoto and at 7:00 a.m. in Osaka. The fee is 15,300 yen (about 140 US dollars) for a round ticket between Kofu and Osaka. The cost is relatively low compared to the Shin-Kan-Sen express train which amounts to about 12,000 yen (about 110 US dollars) for a one way ticket. The bus has 27 passenger seats, located in three lines and each seat is comfortably reclining. In case the number of reserved seats clearly exceeds the capacity, an extra bus is operated because passengers are not allowed to stand. The monthly average ratio of the number of passengers to the total number of seats of this night express bus has constantly been high (over 70%). The number of passengers riding the bus from Kofu to Osaka/Kyoto is slightly higher than the number of passengers going from Osaka/Kyoto to Kofu.

Table 15.6

AD Surveys on Activity Patterns of Night Express Bus Users

	First Survey	Second Survey
Period	2000.11.15 - 2000.12.6 Two months after the introduction	2002.11.2 - 2002.12.1 Two years after the introduction
Survey method	Hand-out and mail-back	Same as the first survey
Survey items	1) Individual characteristics, 2) Trips on the survey day, 3) Trip frequencies Kofu/Osaka before, 4) User awareness and needs, 5) Activity diaries at two consecutive days	1) - 5) items are same as the first survey
No. of distributed sheets	655 sheets (52.3%)	922 sheets (52.4%)
No. of samples/ Response rate	224 individuals (34.2%)	252 individuals (27.3%)

Table 15.6 proves a summary of the AD survey used in this project. The surveys were administered twice: the first survey in 2000, two months after the introduction of the bus service, and the second one in 2002. As shown in this table, the first survey involves all elements, described above. The format of the second survey was simplified and only involved trip characteristics, user awareness and needs and the activity diary. Information of trip characteristics was collected to analyze possible shifts in transport mode before and after the introduction of the night express bus service. Surveyors handed out a set of questionnaire sheets, an instruction sheet, and a self-addressed envelope to the arriving and departing passengers at the bus stop in Kofu. They were asked to mail back the completed questionnaires. The survey period was three weeks. All respondents received 1,000 yen (9 US dollars) as an incentive for cooperating in the survey.

Basic Results. The sample profile of the two AD surveys is as follows. The percentage of female respondents increased (58.0% in 2000; 64.4% in 2002). It was also found that people of widely different age used the bus service (26.7% in the 20s in 2000; 29.7% in 2002; dominant age group: 40s - 60s). The percentage of respondents in their 40s decreased by about 9%, while that in their 50s increased with 5%. The percentage of workers exceeded 60% in 2000 and decreased by 9% in 2002. The percentage of respondents who live in Yamanashi Prefecture was 63.8%, those who live in Osaka Prefecture was 19.6%, while 4.9% lived in Kyoto Prefecture. As for trip characteristics, the percent of the passengers departing from Kofu and arriving at Kyoto was 72.1% and 24.7% for Osaka. In contrast, the percentage arriving at Kofu and departing from Kyoto was 68.8%, while 29.4% departed from Osaka. The majority of night bus trips terminated in Kyoto. When comparing the distributions of trip purpose by point in time, it was found that the share of sightseeing and excursion activities was dominant and accounted for 47% in 2000 and 44% in 2002, followed by personal business activities. The share of social activities decreased by 7%, while business and home-coming activities tended to increase.

The following analysis is based on those users who arrived at their destination early in the morning, excluding those returning home. Table 15.7 shows the distribution of participation rates by activity and indicates that these users tended to spend time on eating and out-of-home activities after getting off the bus and gradually shift to sightseeing and other activities such as recreation and personal business. The mandatory activities tended to start around 9 o'clock in the morning. Users were classified into those whose dominant tour-purpose was sightseeing (Group 1) and those with mandatory activities (Group 2). As shown in Table 15.8, Group 1 is characterized by an average of 274 minutes travel time per day, an average duration of 153 minutes for shopping, eating-out, and other activities, and an average of 461 minutes for conducting in-door activities per day. The typical activity pattern of Group 1 has 3 stops for sightseeing and 1 stop for other activities.

Table 15.7
Average Participation Rate by Activity and Time Interval

Time Interval	Mandatory	Sightseeing	Eating out	Others	Trip
6:00~6:59	1.4%	4.0%	10.1%	14.2%	70.3%
7:00~7:59	4.5%	12.0%	18.0%	14.8%	50.7%
8:00~8:59	9.7%	32.3%	7.2%	15.6%	36.3%
9:00~9:59	18.7%	42.9%	3.0%	23.1%	12.3%
10:00~10:59	21.9%	45.9%	1.5%	14.6%	16.1%

Table 15.8
Average Activity Duration by Trip Purpose (in Minutes)

Tour Purpose	Sightseeing	Mandatory	Shopping	Eating-out	Others	Trip	In-door
Sightseeing	432.1	-	31.9	85.8	34.8	273.3	326.3
Mandatory	-	256.9	44.7	69.7	81.5	172.1	461.1

In contrast, Group 2 is characterized by an average number of 2.16 stops per day, an average duration of 196 minutes for out-of-home activities, and an average of 461 minutes per day for in-door activities.

Case 4: AD Survey Related to Telecommunication

Analytical Framework. The fourth case study concerns a survey on the relationship between telecommunication and activity patterns. Recent developments and innovations in information technology have brought about changes in individuals′ lifestyles. It is clear that many kinds of innovative telecommunication tools have had effects on individual's daily activity and travel behaviour. The rapid spread of mobile telecommunication has produced significant changes in relationships among communication, marketing, distribution and transportation. Especially, students are standing on the forefront where they can receive user-benefits from innovations in information technology. It can be safely assumed that workers and non-workers of the next generation will organize their daily life around mobile technology as they did when they were students. Identifying latent needs and expectations of mobile technology that students have becomes important for the marketing researcher and transportation planner alike.

In this case study, it is assumed that mobile telecommunication activities can be classified into three types. The first type concerns the activities to confirm planned activities; the second type concerns newly-added and accidental activities, and the third type concerns original demand for telecommunication. It is also assumed that a time space path is developed to satisfy temporal and spatial constraints. These paths can be revised by adding, modifying an deleting activity and/or changing attributes related to activity content, activity location and activity timing and duration.

Survey and Basic Results. The purpose of the survey among university students (Nishii *et al.*, 2003) was to identify relationships between activity patterns and mobile telecommunication. The total number of questionnaire sheets distributed was 258, and the response rate was 65.5%. The 153 sampled students were selected from five universities: UMDS, University of Tokyo, Kyoto University, Osaka City University, and Yamanashi University. Males constituted 80% of the sample. About 88% of the respondents were under-graduates. Almost 100% owned a cellular phone, while 22% owned a mobile PCs such as a laptop with Internet service. 22% of the respondents started using a cellular phone in 1998, 35% in 1999, 30% in 2000, and 13% in other years.

Table 15.9 shows the contents of the survey. The one-week activity diary collected information about in-home and out-of-home activities and travel, including activity type, start and end time, travel mode, and trip destination. One-week planned activities were recorded before respondents started completing the AD questionnaire sheet. The surveyed items also included the frequency of telecommunication, modes, and contents of telecommunication.

Characteristics of Telecommunication Activities. Students were divided into two groups. One group is composed of those who recorded telecommunication activities on Sunday, and the other group of those who were involved in telecommunication on Monday. Table 15.10 shows the average daily frequency of telecommunication activities for Sunday and Monday. The table indicates that the average number of telecommunications per day per person is 9.1 on Sunday and 12.6 on Monday. Standard deviations are large. The table also shows the average number of telecommunication activities by mode, classified into sending by phone, receiving by phone, sending by e-mail, receiving by e-mail, sending by web site. Results indicate that the proportion of telecommunication by e-mail is higher than that by phone.

Table 15.11 shows telecommunication contents by mode. The contents were classified into four categories: activities planned for the survey day, activities planned for other days, chatting, and urgent calls / messages. Students tend to prefer phone to e-mail for planned activities.

Table 15.9

Question Items and Contents of SCAT

Questioned Items	Contents
Weekly planned activities	Contents, starting time, and destination of the out-of-home activities planned in a week
Activity and travel patterns in a week (Activity Diary Survey)	The entire records of in-home activities, out-of-home activities, and trips in a week
Mobile telecommunications on a surveyed day	The records of all telecommunications on either Sunday or Monday: Contents, timing, partners, contents, and modes (receiving/sending by voice call and by SMS
Mobile telecommunications in relation to planned activities of the surveyed day	Contents of detail of telecommunications concerning planned activities on the surveyed day
Individual attributes and preferences	Individuals' attributes such as gender, age, residence type, and participation in part-time jobs, and their preferences for mobile telecommunication

Table 15.10

Frequencies of Telecommunication by Day of the Week

Mode		Sunday	Monday	Av. Freq. per Day
All		9.1	12.8	11.7
Voice call	Receiving	1.4	1.9	1.7
	Sending	1.1	1.4	1.3
Mailing (SMS)	Receiving	3.2	4.8	4.4
	Sending	2.2	3.5	3.1
Web site		1.2	1.2	1.2

Table 15.11

Distribution of Telecommunications by Content and Mode

Content	Planned Activities on the Surveyed Day	Planned Activities on other Days	Chattering	Urgent Calling or Sending
Receiving by phone	26.4%	20.4%	43.8%	9.4%
Sending by phone	27.2%	22.4%	36.8%	13.6%
Receiving by e-mail	16.5%	23.7%	55.5%	4.4%
Sending by e-mail	17.0%	21.6%	56.1%	5.3%

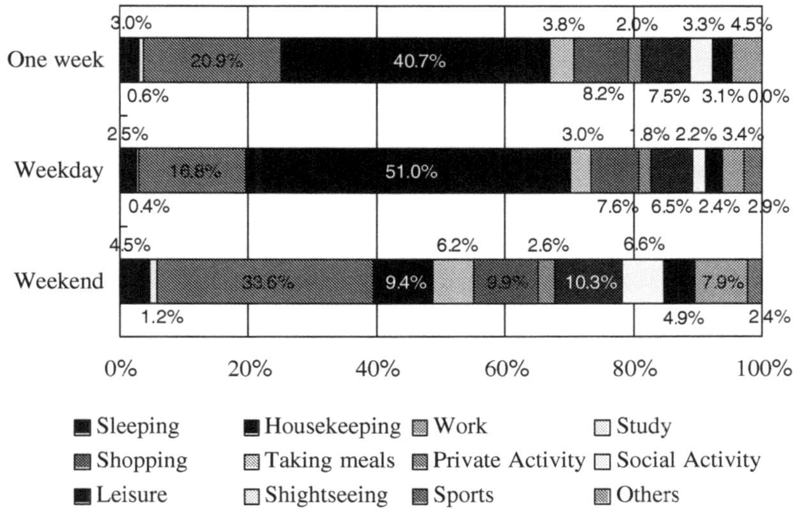

Figure 15.3
Average Time Use for Out-of-Home Activities

They also tend to use phones for activities, planned for the survey day, but e-mail for activities, planned for other days, suggesting that the choice of telecommunication mode is influenced by a sense of urgency. The majority of the receivers of both phone and e-mail are not their family members, but their friends and acquaintances (more than 84%).

Students' Activity Time Allocation Patterns. When focusing on the composition of activity contents by weekdays and weekends, it is found that 56% of the time in a week is spent on in-home activities, while 35% is spent on out-of-home activities and the remaining is allocated to travel time. On weekends, 8% more time is spent on in-home activities, while the amount of time spent on out-of-home activities on the weekends is lower.

Figure 15.3 shows the results for out-of-home activities in more detail. It shows that "studying" accounts for 51% on weekdays. On weekends, however, a variety of activities is conducted (7% of the available time is spent on leisure, 33% on work, 3% on personal business, and 15% on sightseeing and sport.

Discussion

The goal of this chapter has been to illustrate that activity diaries can be used in a variety of ways to provide answers to questions that are difficult if not impossible to answer by conventional trip-based surveys. Activity-travel diary provides more detail that is required to improve our predictions of transport demand. In addition, especially the data on time use for complete daily activity patterns allow different types of analyses, improving our understanding of travel decisions.

At the same time, however, activity diaries are more demanding and the costs may be higher. Especially, in case of a large-scale AD survey such as in Case 2, it thus may become very important to reduce the survey costs without reducing the level of accuracy. One of the promising solutions in this regard is to use new information technology such as PDA devices, web-based interactive questionnaires and so on, linked with GIS. Such new instruments can at least partially replace conventional activity-travel diaries. However, before these instruments should be used routinely, many issues should be examined in depth using controlled methodological research.

The focus of the present chapter has been on descriptive analysis. Research on how to use activity-travel diary data for travel demand forecasting needs further attention. New approaches, such as micro-simulation (e.g., Fujii and Kitamura, 1997) and rule-based decision tables (Arentze and Timmermans, 2000) look promising. Criteria for evaluating the effectiveness of such new models need to be clarified. In particular, three aspects need to be addressed. The first concerns the identification of the basic characteristics of activity patterns, based on their decision structures in multiple and complex time-space dimensions. The second concerns the development of models for explaining comprehensive activity-travel sequences. The third aspect that requires attention concerns the development of a quantitative method for evaluating the effects of transport policies on activity-travel sequences.

References

Ampt, S., J. Richardson and W. Brög (1983). *New Survey Methods in Transport*. VNU Science Press, Utrecht.

Arentze, T.A. and H.J.P. Timmermans (2000). Albatross: *A Learning-Based Transportation*

Oriented Simulation System. EIRASS, Eindhoven University of Technology, Eindhoven.

Asakura, Y., E. Hato, T. Daito and J. Tanabe (2000). Monitoring travel behavior using PHS based location data (in Japanese). *J. Infrastr. Plan. & Mngmt*, **597**, 95-104.

Burns, D., (1970). *Transportation, Temporal, and Spatial Components of Accessibility*. D.C. Hearth and Co., Lexington.

Fujii, S., A. Kikuchi and R. Kitamura (2000). A micro-simulation analysis of the effects of transportation control measures to reduce CO2 emissions: a case study in Kyoto City (in Japanese). *Transpn. Eng.*, **35**, 11-18.

Fujii, S. and R. Kitamura (1997). A micro-simulation model system of individuals' daily activity behaviour that incorporates spatial, temporal and coupling constraints (in Japanese). *Proceedings JSCE*, **14**, 643-652.

Fujii, S., R. Kitamura and K. Nagasawa (1998). An evaluation measure for policy and regional development based on an activity-based model (in Japanese). *J. Infrastr. Plan. & Mngmt*, **597**, 33-47.

Fujii, S., R. Kitamura and K. Seto (1997). Development of model system for individuals' daily activity-travel patterns that accounts for the utility of daily activities (in Japanese). *J. Infrastr. Plan. & Mngmt*, **562**, 83-96.

Hägerstrand, T. (1970). What about people in regional science? *Pap. Reg. Sc. Assoc.*, **70**, 37-52.

Hanson, S. (1979). Urban travel linkages: a review. In: *Behavioural Travel Modelling* (D. Hensher, ed.), pp. 81-100. Croom Helm, London.

Jones, P. (ed.) (1990). *Developments in Dynamic and Activity-Based Approaches to Travel, Analysis*. Avebury, Aldershot.

Jones, P., M. Dix, M. Clarke and I. Heggie (1983). *Understanding Travel Behaviour*. Gower, Aldershot.

Jones, P., F.S. Koppelman and J. Orfeuil (1993). Activity analysis: state of the art and future directions. In: *Developments in Dynamic and Activity-Based Approaches to Travel Analysis* (P. Jones, ed.), pp. 34-55. Avebury, Aldershot.

Kitamura, R. (1988). An evaluation of activity-based travel analysis. *Transpn.*, **15**, 9-34.

Kitamura, R. (1996). Challenges for travel demand forecasting: toward the development of next-generation model systems (in Japanese). *J. Infrastr. Plan. & Mngmt*, **530**, 17-30.

Kondo, K. (1974). Estimation of person trip patterns and modal split. *Proc. 6th Int. Symp. on Transpn and Traff. Flow Theory* (D.J. Buckley, ed.), pp. 715-742. Elsevier, Sydney.

Kondo, K. and R. Kitamura (1987). Time space constraints and the formation of trip chains. *Reg. Sc. & Urb. Econ.*, **17**, 49-65.

Kostyniuk, P. and R. Kitamura (1981). An empirical investigation of household time-space paths. *Resource Paper Int. Conf. on Trav. Demand An.: Activity-Based and Other New Approaches*,

Oxford.

Nishii, K. and K. Kondo (1992). Trip linkages of urban railway commuters under space-time constraints: some empirical observations. *Transpn. Res. B*, **26**, 33-44.

Nishii, K., K. Kondo and A. Ohta (1998). Empirical observations: time use patterns in Kofu shopping panel data (in Japanese). *Infrastr. Plan. Rev.*, **15**, pp.499-508.

Nishii, K., K. Sasaki, K. Kondo, and R. Kitamura (2003). New relationships between mobile communication tools and activity and travel patterns: empirical observations and decision structures. Paper presented at 10th EIRASS Conference, Portland, USA (copy obtainable from authors).

Ohmori, N., Y. Muromachi, N. Harata and K. Ohta (1999). Applicability of collecting travel behavior data using advanced communication systems (in Japanese). *City Plan. Rev.*, **34**, 169-174.

Osada, T., H. Koike and A. Morimoto (2004). Development of barrier-free map using GPS cellular phone with digital camera. *Proc. 83rd TRB Conf.*, Washington, D.C. (CD-Rom).

Recker, W.W., M.G. McNally and S. Root (1986). A model of complex travel behaviour. *Transpn. Res. A*, **20**, 307-318.

Takao, M., T. Morikawa, S. Kurauchi and K. Sasaki (1998). A time allocation model and its application to work time shift (in Japanese). *Proc. Infrastr. Plan.*, **21**, 783-786.

Yoshida, N. and T. Hasegawa (2000). Data collection and analysis for activity diary in Kei-han-shin metropolitan area (in Japanese). *Proc. Infrastr. Plan.*, **23**, 675-678.

Zhang, J., H.J.P. Timmermans and A.W.J. Borgers (2004). A model of household task allocation and time use. *Transpn. Res. B*, **39**, 81-95.

16

A Data Collection Strategy for Perceived and Observed Flexibility in the Spatio-Temporal Organisation of Household Activities and Associated Travel

Martin E.H. Lee-Gosselin, Université Laval, Quebec City, Canada

Introduction

Recent developments in activity-based modelling approaches, notably multi-agent micro-simulation, have led to greatly increased interest in ways of observing decision processes underlying the spatio-temporal organisation of activities. An appreciation of those decision processes requires observations at the individual and household levels, with due regard to interactions beyond the household, within social networks and with a wide range of other sources of activity opportunity and constraint. Considerable progress has been made using survey methods that record the evolution of planning and re-planning of activities over multi-day observation periods, such as the work of Doherty and colleagues using the computer-aided instrument known as CHASE (Doherty and Miller, 2000).

In this chapter, we first set the context for this work and then describe the overall strategy for this class of data collection in two in-depth panel surveys, one in the Toronto region and the other in Quebec City. These were undertaken to help specify the household travel behaviour sub-models of the ILUTE suite, itself under development since 1994 by a consortium of Canadian universities headed by Miller. The first waves of the two surveys were completed in 2003 on a total of 520 households in the two regions; two more waves should be completed by the end of 2005. The

Progress in Activity-Based Analysis edited by H. Timmermans

theoretical background of a new package known as OPFAST (Observations et Perceptions de la Flexibilité des Activités en termes SpatioTemporels / Observed and Perceived Flexibility of Activities in Space and Time) is briefly covered, and then the parts of the instrument package are described, in some detail for an in-depth interview component. Like CHASE, OPFAST records both anticipated as well as executed activities, in-home as well as out-of-home, over a seven-day period. Unlike CHASE, it uses paper instruments for all adults in the household and these are faxed daily to the survey laboratory using a machine that is lent to the household and installed during a start-up home interview. Interviewers then prepare a second, 1.5-2.0 hour home interview that takes place soon after the observation period has ended. This records retrospectively some, but not all, of the detail on activity planning attributes that CHASE achieves through computer-aided interrogation. On the other hand, OPFAST goes beyond CHASE in that it uses a simple gaming technique to record each adult's perceptions about the temporal and spatial flexibility of every recorded activity during the seven days. In addition, jointly-planned activities, household decision dynamics, the use of telecommunications, satisfaction with access to activities, and views of the future are explored in a simultaneous discussion with all participants in the household. The chapter concludes with an illustrative example of first-order results from the first wave in Quebec City, an overview of the methods for the second and third waves of the panel surveys, and some preliminary methodological conclusions.

DATA NEEDS FOR ACTIVITY-BASED MODELLING

The development of CHASE and its variants has been part of a greatly increased interest in a number of classes of disaggregate data. Discussions of activity and travel data needs in the mid to late 1990s, such as in Axhausen (1998) or Miller *et al.* (1998), considered the potential for improved microdata on *behavioural outcomes* as necessary and desirable, but not sufficient. It was recommended that the focus of data collection be broadened to activity generation/participation, activity scheduling, cognitive processes, dynamics over horizons from within-day to long term, and the policy sensitivity of a wide range of measures that affect activities (Lee-Gosselin and Polak, 1997). Recent activity-based research has started to respond, looking at ways of collecting data on *decision processes* underlying observed activities and associated travel, and their organisation in both space and time. Moreover, an appreciation of those decision processes requires observations at both the *individual* and *household* levels, with due regard to interactions beyond the household, within social networks and with a wide range of other sources of activity opportunity and constraint.

Such data aspirations are not easily satisfied, but two types of strategy have emerged. In the first, new ways of using data collected on behavioural outcomes are allowing inferences to be made about activity scheduling. For example, Arentze and Timmermans' (2000) Albatross makes substantial use of inferences about the timing, sequencing and grouping of activities from a specially-designed activity diary. Useful inferences about activity scheduling have also been made from traditional household travel survey data, for example in Toronto, Canada by Miller and Roorda (2003). More recently, surveys involving tracking technologies such as Global Position Systems (GPS), digital mobile phones and other location- and time-aware devices are making it possible to observe executed travel and dwell patterns with unprecedented detail, precision and accuracy. There is much current work on intelligent post-processing of the prodigious quantities of data produced, in order to minimise the need for respondent intervention or prompted recall, and the associated problems are not trivial. Nevertheless, the implications of using microdata thus obtained to make inferences about activity scheduling are potentially of great significance to modelling (see Doherty and Papinski, 2004), especially as some technologies lend themselves to observation over multi-week periods or even longer. Rarely, this is also possible using manual diaries, such as the six-week example reported by Axhausen *et al.* (2002).

In the second type of strategy, decision-making about the temporal and spatial organisation of activities is itself the focus of observation, in combination with the observation of outcomes, an approach that has had a longer history with some Stated Response data collection methods. Inspired by Jones *et al.* (1983), Gärling *et al.* (1986), Axhausen (1998) and others, much of the attention has been given to observing scheduling processes, incorporating the notions of a "repertoire" of possible activities on which a respondent draws to fill their schedule, the progressive refinement of schedules involving combinations of routinised, individually planned and impulsive activities, different levels of advanced planning, coordination and negotiation within households and social networks, and the simultaneous planning of multiple activities. This type of strategy favours multi-day, relatively in-depth data collection, and inevitably involves small to medium sample sizes. An example is the focus of the remainder of this chapter, but first we would note that the observation of decision processes over very long horizons is also important. In this regard, retrospective techniques covering decades, inspired by biographic and ethnographic approaches, are starting to find their place in transport studies. Examples include car-use (Lee-Gosselin and Turrentine, 1997), car ownership (Mohammadian and Miller, 2003) and residential mobility (Séguin and Thériault, 2000). In the first example, a history of decisions was used to build customised scenarios for a Stated Response survey.

The Need and the Opportunity for Data Collection on Activity Decision Processes in Urban Canada

In general, strategies to *infer* or to *observe* decision processes underlying activity organisation are not mutually incompatible. On the contrary, where a comprehensive understanding is sought of activities and travel in entire urban regions, a combination of the two may offer more than the sum of the parts. Such is the case in Canada, where a consortium of Canadian universities headed by Eric Miller of the University of Toronto since 1994 has been building up Integrated Land Use, Transportation and Environmental impact (ILUTE) model suites to be applied to Canada's major conurbations. From this emerged a detailed appreciation of the data needs of the next generation of activity-based models, and in particular the need for improved behavioural data from urban actors – individuals, households, firms and public agencies. The consortium targeted as test-regions several conurbations where there was a regular programme of large-sample household travel surveys, data from which could be used in complement to smaller-sample research datasets.

In 1998, the Consortium formed a partnership with an enlarged team of researchers, now known as the PROCESSUS (PROCesses of behaviour underlying Equity and Sustainability in Systems of Urban access and their Simulation / PROcessus Comportementaux Essentiels aux SystèmeS d'accès Urbain durables et équitables et à leur Simulation) Network, to undertake fundamental research on the behavioural foundations of ILUTE models. The network received a major collaborative grant from Canada's federal social science council for the period 2000-2005. Part of this allowed an innovative longitudinal panel survey of day-to-day activity and travel in Greater Toronto and Québec City. In both regions, university research centres that are part of PROCESSUS collaborate closely with regional planning and transport agencies, notably in connection with travel surveys, data conditioning, analysis and the development of computing and geomatics support.

The choice of Toronto and Quebec City was motivated, in part, because both regions undertake their household travel surveys at five-year intervals and in the same years. The most recent surveys were completed in 2001. Moreover the methodology has been reasonably stable for the past two decades. The sample sizes are enormous by current North American standards -- more than 100,000 households in Greater Toronto and over 27,000 in Quebec City. Thus it was possible to adopt a "layered" approach, in which activity and travel behaviour benchmarks could be established, and limited inferences about activity scheduling could be made, on traditional data, while the research datasets could target decision behaviour in-depth. The comparison between Toronto and Quebec City is interesting, too. Greater Toronto has over 4 million in habitants and growing, while the Quebec City region has about 650,000 inhabitants, a number that has grown very little in the past decade. Congestion levels, and the pressure to further develop the road and public transport networks, are much higher in Toronto than in Quebec City. Quebec City has about three times per-capita length of urban motorway that is found in Toronto. This does not favour public transport in a region that is slightly below what conventional wisdom suggests to be the minimum population required to support modes with dedicated rights of way; and currently there is a lively debate about reinforcing an existing strategy of high levels of public transport service in selected corridors by the (re-)introduction of light rail.

The Toronto and Quebec City Panel Surveys: Design Considerations for a Joint Strategy

Given the high interest in the possibility of simulating the longer-term relationship between land use and transport in Toronto and Quebec City using both traditional and research databases, the study team decided to focus new data collection on the evolution of spatio-temporal patterns and the levers that affect those patterns. While they opted for longitudinal panel surveys, they wanted to use in-depth survey techniques that build a cumulative understanding of how households generate options and make decisions. The survey methods were thus "free" to change from wave to wave, embracing the inevitable conditioning of panel members in earlier waves. This would be in sharp contrast to those panel surveys that use identical methods in each wave – in effect a repeated cross-section on the same people. One consequence of the "cumulative experience" design principle was that panel refreshment would not be possible, making panel conservation a high operational priority. Observing decision-making about activities and travel in a panel format was, to the team's knowledge, unprecedented. The choice of a medium-sized sample – about 250-300 households per region, was pragmatic give the variety of household types and locations required, and the resources available. Similarly, three waves was the maximum that could be organised within the life of the grant. To reduce the number of factors, the team decided to hold the constant as possible the season of the year for the three instances of participation for a given household.

The choice of survey methods for observing decision processes was not lightly undertaken, and was the subject of two design workshops, one in Toronto and a second in London involving some of the team's international collaborators. These built on the experience of team member and Doherty and colleagues in both regions using CHASE (Doherty and Miller, 2000). CHASE makes innovative use of a self-administered "daytimer" type of programme on a laptop computer. This allows respondents to sketch-plan a week of activities, and then edit their schedule at the end of each day to fill gaps and change entries to reflect what actually took place. Similarly, the sketch schedule for the remaining days of the seven-day observation period can be updated. Very significantly, the data collected consists not only of the activity and travel outcomes, but also a chronological record of each addition, deletion and modification to the computerised schedule. Moreover, the CHASE programme systematically queries the respondent for attributes of schedule changes at the time they are entered.

It was agreed by the team that a number of methodological questions should be answered by this research if possible, in addition to the creation of the datasets needed to help specify the ILUTE submodels of household activity and travel patterns. Among the most important were: the feasibility of using relatively onerous computer-aided self-administered instruments (CASI) such as CHASE on a wide spectrum of households, the feasibility and perceived burden of computerised versus non-computerised methods to study decision processes – including in the non-computerised case the retrospective recording of the characteristics of decisions taken as much as 7 –12 days earlier -- and the data-entry overhead involved compared to the real-time data entry offered by CASI, the feasibility of interactive home-interview methods, similar to some Stated Response methods, to engage respondents in classifying the way they went about deciding to do the activities that were recorded during an observation period, the usefulness of a semi-structured sub-interview about linked activities, household decision dynamics, issues of satisfaction and anticipated changes during the life of the panel survey and beyond.

It became clear that CHASE be predominant in one region, and that a whole new protocol to be set up. An additional complication was that the Toronto Panel was to be in English and the Quebec City Panel in French. The team decided that the only practical solution was to implement CHASE for the entire first wave sample in Toronto, and to develop and implement a new protocol for the first wave sample in Quebec City. The essence of resulting non-computerised survey package for Quebec City is that, like CHASE, it traces the evolution of decisions about activities and travel throughout a seven-day period, but with less detail on each change in the schedule, and with a much more substantial follow-up interview seeking to understand holistically the temporal and spatial activity planning that was revealed during the observed week, and respondents' perceptions about this.

The price paid for this methodological agenda was that Toronto and Quebec were not identically phased: Toronto completed Wave 1 from March 2002 to April 2003, while Quebec City undertook a series of pilots leading to the launch of a stable Wave 1 method in January 2002, but continued data collection over two years, until December 2003. Both Toronto and Quebec completed the target of 270 households. In Quebec City, as noted above, a new package was developed: the first 20 households were "full pilots" and these were not retained in the final sample as the versions they had completed had been significantly superseded. Further methodological details of Toronto's first wave will be found in Roorda and Miller (2004). The Quebec City methodology, now known as OPFAST, is discussed below.

THE OPFAST SURVEY PACKAGE

Theoretical Background

The Québec Panel Survey shares with Toronto the parallel collection of data on both behavioural outcomes and underlying decisions, but as noted above OPFAST places greater emphasis than CHASE on the *perceptions* of respondents about the ways they organise their activities in time and space. It seeks to assess the room for manoeuvre, or flexibility, that they believed they had during the observed week. An exploration was undertaken (Ramadier *et al.*, 2005) of research traditions that have approached spatio-temporal behaviour in terms of the relationship that an individual builds with space-time as s/he perceives it. To summarise, some of these, such as those seeking to explain differences in perception by sociodemographic factors or lifestyle, have offered only limited perspectives. Ramedier *et al.* concluded that a more compelling alternative is:

> "…..a social psychological understanding of an individual's relation to space (which) permits a more accurate appreciation of the dynamics of behavioural change in a gi ven day or week (e.g. Gärling *et al.*, 1986). This recognises dynamic, decision-making processes that depend on their specific and limited spatio-temporal context. On the other hand, there are some broad behavioural rules that depend on a stable social and geographic context that itself helps define a lifestyle. Studies in environmental psychology suggest that attachment to places or spatial representations of cities, the most stable factors in space/individual relationships, strongly affect the definition of lifestyle. However, some theoretical perspectives (e.g. Dewey and Bentley, 1949) have long seen the individual and its environment as mutually defined, enabling more emphasis on the environmental situation. According to this perspective, the decision-making processes related to daily mobility are a product of the relationship between the individual, defined by his/her lifestyle, and the spatio-temporal constraints of the moment. It is thus important to study both temporal and spatial fixity as viewed by individuals, households and other personal networks. These and other theoretical perspectives from social psychology suggested that the relationship between the individual and space/time is a powerful way to reveal *both* behavioural rules and decision processes linked to spatial mobility. This relationship allows us to move beyond the mere description of spatial behaviour and attempt to understand the dynamic aspects of the observed phenomena."

A central idea that arose from this perspective was the simultaneous consideration of temporal and spatial fixity or flexibility for each executed activity. After Gärling *et al.* (1986) and others, a three-way distinction is often made between activities that are *routine*, those that require to be *pre-planned*, and those that are *impulsive or spontaneous*, but this is usually applied only to the temporal dimension. Crossing this trichotomy for both time and space yields a classification with nine possibilities, and this was incorporated into the survey design – based on the perceptions of the respondents.

Summary of the OPFAST Instruments and Procedures

The OPFAST package consists of four steps:

1. *A startup in-home interview (45-60 minutes) of the household, during which:*
   - Basic information is gathered on the structure of the household, household vehicles, residential mobility history, activity repertoire with travel modes etc.,
   - All members over 15 are instructed in the use of the paper instruments
   - A fax machine is lent to the household and installed

2. *Observation of one week of activity using paper instruments* (see Figure 16.1):
   - On a single Weekly Planning Sheet, organised into 7 columns with a time scale similar to desk diary, each household member over 15 is asked to record all known activities for the coming week during the startup interview; on the evening of each day, they are asked to add, delete and modify entries (by crossing out, erasing, etc.) for any of the days remaining
   - Respondents use seven daily Activity and Travel Logs to record actual behaviour, one day at a time
   - Once per day, the current state of the Weekly Planning Sheet, and that day's Activity and Travel Log, are faxed to the university; the date and hour of transmission is automatically recorded on each sheet
   - Support is provided by telephone based on material received

3. *In-depth home interview (1.5 – 2 hours) soon after the 7 day observation period, including:*

   - Validation of paper instruments
   - A sorting game that addresses the perceived spatial/temporal fixity of activities
   - Retrospective recording of details on planning horizons and interdependence
   - Holistic interpretation by each adult of the way they approached the spatial and temporal organisation of their activities
   - Semi-structured group interview about "projects" (sets of jointly-planned activities), activity negotiation, ways the recorded activity pattern could have been improved, use of telecommunications in planning and negotiation, expectations about the future, and unsatisfied demand

4. *Data entry routines:*
   - Entry screens and relational database in Microsoft Access™
   - Designed to align with CHASE data format, where possible
   - GIS support to the geocoding of activity locations
   - Usable for direct entry by interviewer in startup interview

Households are drawn from published directories. Recruitment is effected by letter, with telephone follow-up. A form of quota sampling is used to achieve a representative range of households: the composition of the sample is cumulatively monitored on life-stage, residential location and vehicle ownership characteristics to identify any underrepresented categories for special recruitment efforts are required. For example, in the current application, intermediaries were used in the latter stages of Wave 1 to recruit disadvantaged subpopulations and elderly people living alone. In addition, some direct recruiting was allowed of households that were known to survey staff, but these could not be close relatives.

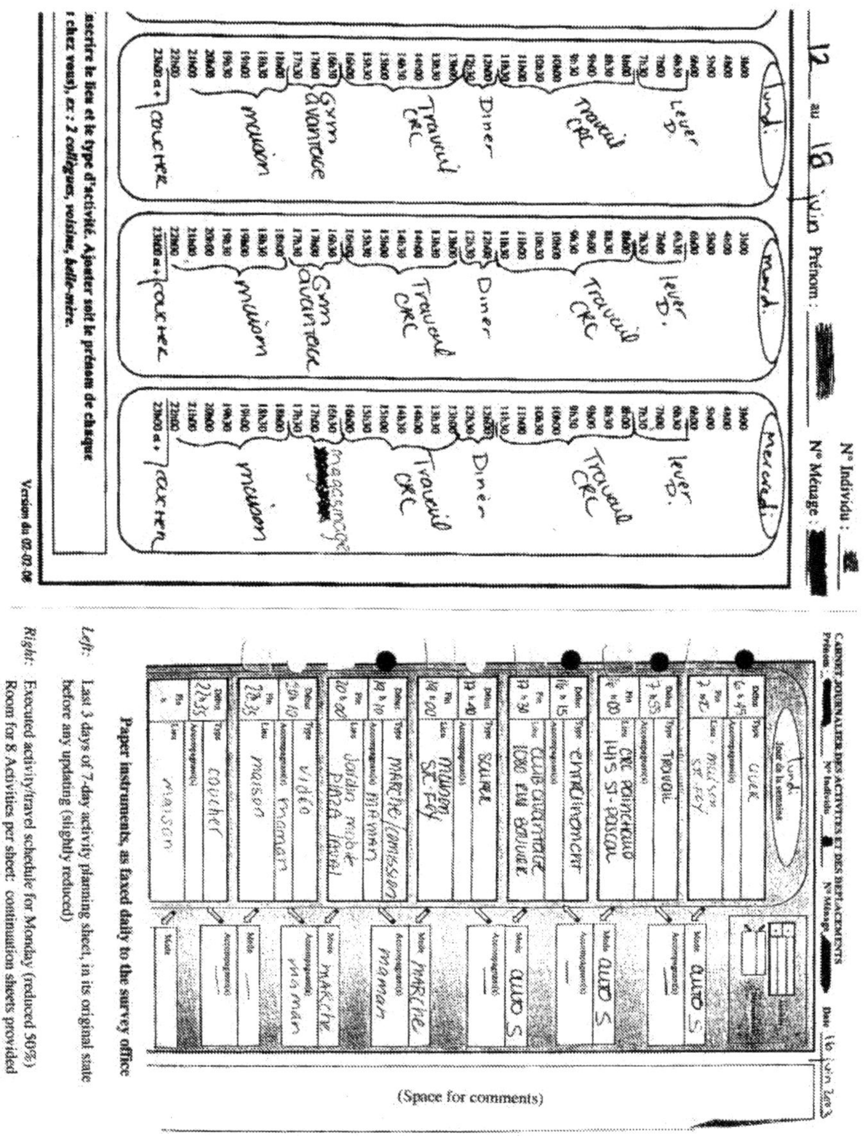

Figure 16.1
Detail of OPFAST Paper Instruments, Completed and Faxed Daily by Respondents

The self-completion parts of the OPFAST package were designed to be used individually by all household members 16 years old and above. It records the activities and travel of children under 16 only to the extent that they coincide with those of adults or require adults to provide transport. However, the second interview provides an opportunity to explore the influences of all children's activities, and children are invited to participate in this interview. This helps interpret associations between household structure and life stage, recorded in the first interview, and patterns of spatio-temporal flexibility. In addition, there were some promising trials of simplified diary instruments for children, but for the current study, these were not added to the OPFAST package.

It is a requirement of the Québec panel that, for a household to be eligible, all household members over 15 under the same roof and who are involved in each other's activities must participate: 55% of the sampled households yielded information on the interactions between two or more such adults. A choice of incentives was offered worth between Cdn$20 and Cdn$30 per household.

Discussion of the Second Interview

The comparison of the Weekly Planning Sheet and each day's Activity and Travel Log provides a first view of decision-making about activity organisation, but the in-depth interview that takes place soon after the seven-day observation period is the major source of observations of decision processes. Here there is an important structural difference between OPFAST and CHASE. While OPFAST intentionally observes many of the same attributes of decisions as CHASE, the use of paper instruments precludes the multi-level branching that is possible with CASI methods. Instead, decision attributes are queried retrospectively during the in-depth interview, but for parsimony are explored in detail only for activities that were pre-planned and impulsive on the temporal dimension. It is important to note that the sorting of each and every executed activity over the seven days into routine, pre-planned and impulsive categories in time and space is done not by the researcher but by each individual participant, using a simple sorting game (the placing of colour-coded labels on to the Activity and Travel Log). The game has arbitrary rules that further limit our interpretation of the results, notably that the distinction between pre-planned and impulsive is that, for the latter "one hour in advance, I did not know [*that*] (temporal dimension) [*where*] (spatial dimension), I was going to do the activity".

Having thus sorted all the activities, the respondent is in a good position, with guidance if needed from a trained interviewer, to reflect on issues such as planning horizons and the level of coordination with others in each instance. This retrospective questioning also uses some threshold branching, for example to identify impulsive activities that were decided fewer than five minutes in advance and hence to detect the method of communication if others were involved (by which, incidentally, OPFAST was made more sensitive to the role of mobile phones). Threshold branching is also used to record attributes of deletions and significant modifications that the interviewer selects in advance by comparing the Weekly Planning Sheet and corresponding Activity and Travel Logs, as well as the evolution of the Weekly Planning Sheet over the week (recalling that its current state is faxed once per day to the survey office).

The design of the sorting game also recognises the opportunity to seek holistic views of activity organisation after each participant has thought about their entire set of recorded activities during seven days, typically 50 – 80 of them (note that in-home activities are included, but simplifications are allowed as they are in CHASE, such as "morning preparations"). A series of questions are structured, such as looking for common elements among temporally impulsive activities. As well, these questions seek insights about both the temporal and spatial dimensions.

A further advantage of the sorting game and the interpretation of the results by respondents is that it frames the context for the concluding part of the in-depth interview. Up to this point each respondent works individually, although in the presence of any others, but the concluding questions comprise a semi-structured group interview. These questions cover five main areas: examples of "projects" – jointly planned activities (for the team's use of this concept, see Miller, 2005); the dynamics of household decision making, including identifying whose schedule is the least flexible and thus tends to constrain those of others, and the role of telecommunications; ways in which, with hindsight, the observed week could have been made to work better; a vision of how a similar week might look 3-5 years into the future; a discussion of activities that respondents would have wished to execute during the observed week, but that they could not "get to". In sum, this is a complex interview protocol that draws on a number of theoretical frameworks, and on developments in both Revealed Behaviour and Stated Response survey methods. In particular, it is designed to strengthen the understanding of spatio-temporal activity organisation holistically, building on respondents' own perceptions about personal and household activity. The OPFAST package uses a reflexive approach that facilitates the chances in the second and third waves to recognise and exploit the cumulative effects of participation in the earlier wave(s). As noted above, the view of the team is that panel "conditioning" is normal and substantial given the extent of participation required, and that the researcher and respondents are effectively engaged in a joint learning process.

RESPONSE TO THE FIRST WAVE IN QUEBEC CITY

In the first wave, 250 Québec City households participated. In these households, 381 adults executed 26,267 activities and 12,840 trips. Table 16.1 shows overall rounded percentage of activities that fall into the nine spatio-temporal classes.

Table 16.1
Perceived Spatio-Temporal Flexibility of Activities

		In Space			
		Routine	**Pre-Planned**	**Impulsive**	**Total**
In time	Routine	52%	1%	<1%	53%
	Pre-planned	14 %	8%	1%	23%
	Impulsive	18 %	2%	4 %	24%
	Total	84%	11%	5%	100%

It should be remembered that these results are affected by the instructions given to respondents for recording activities. In-home activities that are similar (such as various forms of relaxing, or morning preparations) are grouped together, whereas out-of-home activities are mostly counted individually. To lay the groundwork for deriving behavioural rules that would be useful to the specification of micro-simulation models, initial descriptive analyses are focussing on differences in spatio-temporal routinisation between different types of individual and household, and on the extent to which access to different services, and out-of-home activities requires advanced planning, negotiation and coordination with others.

DATA COLLECTION STRATEGY FOR THE SECOND AND THIRD WAVES

After the two different approaches to observing decision processes in the first waves of the Toronto and Quebec City Panel Surveys, the team adopted the same approach for Wave 2. This consists of a Computer-Aided Telephone Interview (CATI), developed by Matthew Roorda in Toronto, that fills in details of two days of executed activities that are recorded by the individual respondents on a mailed memory jogger. The starting day of the two-day period is randomised over all days of the week. In most cases, the memory jogger is mailed back and activities are entered into the CATI to reduce the duration of the interview. The CATI includes a limited Stated Adaptation component that selects one single activity and one pair of related activities for which hypothetical changes in constraints (for example a traffic delay) are posed. Respondents are questioned on how they would have coped with this situation, and this is used to build an understanding of how flexibility is used to deal with different classes of activity. Quebec's use of the CATI is augmented in two ways. First, respondents are again asked to apply the spatio-temporal flexibility classification to each activity on the memory jogger, which in all cases is returned to the survey office. Secondly, a supplementary household-level questionnaire is used to update responses to the initial sub-interview about: activity negotiation; ways the recorded activity pattern could have been improved; expectations about the future; and unsatisfied demand. On the critical issue of panel retention, much attention was given to keeping in touch with panel households, for example using newsletters and pre-wave motivational letters. Toronto achieved 84% retention in Wave 2. Quebec City, which started Wave 2 in May 2004, expects to achieve a similar retention rate; of particular note is that attrition through loss of contact between waves was limited to only three of their households.

The data collection strategy for Wave 3, to be completed in both Toronto and Quebec City during 2005, is again similar for the two regions. Both repeat their mail-out, mail-back and CATI instruments that were developed for Wave 2. Again, two days of actual activities and travel are observed. However, the Stated Adaptation component is replaced by a new component that focuses on activities that are routine in time. The mail-out, mail-back package contains, for the first time in the whole survey, a seven-day activity pattern form on which each of the over-15s are asked to identify activities and trips that they "normally do every week". Using different form designs, both Toronto and Quebec City seek respondent's perceptions about the degree of flexibility in each activity's location, accompanying people and travel mode. Quebec had recorded perceptions at each wave and it was shown feasible in pre-tests to build on this, using a fairly detailed log format: activities are recorded one per row, and both multiple-choice and open-ended responses are used. Toronto opted for a large (43 x 56 cm) "skeleton diary" on which a typical week is sketched and the perceptions of flexibility are indicated with coloured pens. Both regions entered data in advance of the CATI from both the two-day diary and the routine activities forms, and in Quebec from the supplementary household-level questionnaire. During the CATI, additional attributes are sought for each routine activity: Quebec's more detailed routine activity form means that this part of their CATI is shorter.

While it was desirable to use in Wave 3 an approach similar to that of Wave 2, which had a much lower respondent burden than Wave 1, it is hoped to return to a seven-day observation period for a subset of Wave 3 respondents. This presents formidable difficulties for retrieving details of each and every activity in a telephone interview, and so certain very recent techniques for GPS-aided surveys, developed in another PROCESSUS project, may be deployed. Meanwhile, through the continuous comparison of the experience with CHASE and OPFAST, insights will be again shared into the ongoing design process for both approaches.

Conclusion: Methodological Lessons

At the time of this paper, a detailed comparison of data from CHASE and OPFAST has not been completed, but we can draw some useful inferences from the experience of implementing the two total designs adopted, particularly in response to the issues raised in Section 4. These are, of course, only a small part of all the survey design issues the two teams have faced.

CHASE and OPFAST are very high burden survey packages, and it was not known whether they could be feasibly implemented on medium size samples comprising a wide spectrum of households. The largest previous CHASE samples, in Hamilton Ontario and Quebec City, each had about 40 households. In the panel surveys, it proved possible to recruit 270 households for each method. Based on comparisons of demographic variables with data published for the Toronto and Quebec City regions by federal and provincial agencies, the final samples were without obviously serious selection biases, despite that response rates from the directory listings were of the order of 10% in both cities. Without the use of intermediaries, however, it is clear that the Quebec City sample would have had a significantly larger proportion of higher socio-economic status households that is found in the population. Largely owing to the persistence of excellent field staff, the high response burden of Wave 1 was overcome. Although only Toronto did a follow-up survey of respondent satisfaction (Doherty *et al.*, 2004), and there is still much left work to do to scrutinise data quality, initial results are that there seem to be remarkably few instances of item non-response or aborted participations in either region. Further, neither method seems to have suffered badly from respondent fatigue towards the end of the observation period. However, a significant number of respondents warned that they would not be willing to undertake 7 days of either method in subsequent waves. Impressions from field staff during Wave 2 suggest that was more frequent with CHASE.

The wishes of respondents were respected by changing to a two-day observation period in Wave 2: the change in Toronto was greater in that paper and CATI methods replaced the CASI, whereas Quebec respondents were on more familiar ground with the paper instruments. It is interesting to note that the retention rate – around 85% between Waves 1 and 2 – was very similar in both regions. It is reasonable to assume that retention would have been significantly lower if the Wave 1 instruments had been largely re-used. Again, the reduced burden of Waves 2 and 3 was possible because the design of these panel surveys intentionally built upon the learning and conditioning of respondents in Wave 1, thereby precluding the use of panel refreshment. We believe the training of respondents during the second interview of Quebec City's Wave 1 was essential to the reliability of the data on perceived spatial and temporal flexibility collected in that region's subsequent waves.

On the feasibility of computerised versus non-computerised methods to study decision processes, it would be difficult to judge the relative merits without a side-by-side test of CHASE and OPFAST using the same region, language and field staff. In addition, as discussed above, these two packages have a common core but do not measure exactly the same variables. The reliance of OPFAST on an in-depth second interview provides a rich opportunity to recover from partial or ambiguous response, and there have been no significant concerns reported by field staff about the apparent accuracy and authenticity of the characteristics of decisions reported 7 to 12 days later. For a wide range of respondents, staff reported that OPFAST was understandable and comfortable. This was also true of the Stated Response methods used to engage respondents in classifying the way they went about deciding to do the activities that were recorded during an observation period: in most households this exercise, which is done individually and simultaneously, took about 20 minutes. Compared to CHASE, however, a high price was paid in terms of the data-entry overhead involved compared to the real-time data entry offered by a CASI, but this again is partly explained by the open-endedness of some many of the interview questions, which require careful post-coding by the interviewers. The same remark particularly applies to the holistic responses on the "why" of spatio-temporal flexibility, and to the sub-interview with OPFAST about linked activities, household decision dynamics, issues of satisfaction and anticipated changes. This latter is proving to be a very useful source of explanation for the perceived flexibility patterns, and therefore it was continued as a paper questionnaire in Quebec City's Waves 2 and 3.

Despite the post-processing costs for OPFAST, the per-household total costs of the two methods were fairly similar, after taking into account their different stages of development. With care, a non-profit organisation such as a university should be able to achieve Wave 1 usable data for about Can$330 to $360 per household (around €225 at contemporary exchange rates), and about double that for all three Waves combined, assuming similar lower-burden packages for Waves 2 and 3. It cannot be over-emphasised that the careful nurturing of the panel members is absolutely central to the success of the overall longitudinal panel strategy. Initial indications are that CHASE and OPFAST are both viable, albeit labour-intensive, options for initiating a panel survey of decision processes underlying household activity and travel patterns. Much data analysis will be required before issues of data validity and stability can be compared for the core variables, but it is already clear that, taken as whole packages, they are likely to be more useful in combination than in substitution. Future methodological research should allow this to be tested on common ground.

ACKNOWLEDGEMENTS

This research was supported by *Access to Activities and Services in Urban Canada*, a Major Collaborative Research Initiative of SSHRC 2000-2005, with additional support from the Ministère des Transports du Québec. Research on GIS and GPS-aided surveys was funded by the GEOIDE. We gratefully acknowledge the counsel and support of regional transit and planning agencies in the execution of the panel surveys and in the complementary use of their data. Much appreciated contributions to the data collection strategy were made by team members Eric Miller, Sean Doherty, Matthew Roorda, May Lynne Fong, Thierry Ramadier, Marius Thériault, Paul Villeneuve, Francis Joud, Alexandre Frenette, Talia McCray, Carole Leclerc, and Sharon Frenette and international collaborators John Polak, Kay Axhausen, Peter Jones and the students and employees who were on the "front line" in the in-depth interviews.

REFERENCES

Arentze, T. and H.J.P. Timmermans (2000). *Albatross: A Learning-Based Transportation Oriented Simulation System.* EIRASS, Eindhoven.

Axhausen, K.W. (1998). Can we ever obtain the data we would like to have?. In: *Theoretical Foundations of Travel Choice Modelling* (T. Gårling, T. Laitila and K. Westin (eds.), pp. 305-324. Pergamon Press, Oxford.

Axhausen, K.W, A. Zimmersman, S. Schönfelder, G. Rindfüser and T. Haupt (2002). Observing the rhythms of daily life: a six-week travel diary. *Transpn.*, **29**, 95-124.

Dewey, J. and A.F. Bentley (1949). *Knowing and the Known.* Beacon, Boston.

Doherty, S.T. and E.J. Miller (2000). A computerized household activity scheduling survey. *Transpn.*, **27**, 75-97.

Doherty, S.T., E. Nemeth, M.J. Roorda and E.J. Miller (2004). Design and assessment of the toronto area computerized household activity scheduling survey. *Transpn. Res. Rec.,* in press.

Doherty, S.T. and D. Papinski (2004). Is it possible to automatically trace activity scheduling decisions?. *Proc. Conf. Progr. Activity-Based An.*, EIRASS, Maastricht, The Netherlands (Cd-Rom).

Gärling, T.., J. Saïsä, A. Böök and E. Lindberg (1986). The spatiotemporal sequencing of everyday activities in the large-scale environment. *J. Env. Psych.*, **6**, 261-280.

Jones, P.M., M.C. Ddix, M.I. Clarke and I.G. Heggie (1983). *Understanding Travel Behaviour.* Gower, Aldershot.

Lee-Gosselin, M.E.H. and J.W. Polak (1997). Summary of workshop on data needs. Activity-Based Travel Forecasting Conference, New Orleans, dot/usdoe/epa, pp. 173-182.

Lee-Gosselin, M.E.H. and T.S. Turrentine (1997). Investigating past and future choice-sets of car users, with special reference to perceptions of safety. *Proc. 8th IATBR Conf.*, Austin, Texas.

Miller, E.J. (2005). Propositions for modelling household decision-making. In: *Integrated Land-Use and Transportation Models: Behavioural Foundations* (M. Lee-Gosselin and S.T. Doherty (eds.). Elsevier, New York. forthcoming.

Miller, E.J., D.S. Kriger and J.D. Hunt (1998). Integrated Urban Models for Simulation of Transit and Land-Use Policies. TCRP Project H-12, Final Report, University of Toronto and Delcan Corporation, Toronto.

Miller, E.J. and M.J. Roorda (2003). Prototype model of household activity-travel scheduling. *Transpn. Res. Rec.*, **1831**, 114-121.

Mohammadian, A. and E.J. Miller (2003). Dynamic modelling of household automobile transactions. *Proc. 82nd TRB Conf.*, Washington, D.C. (CD-Rom).

Ramadier, T., M.E.H. Lee-Gosselin, F. Frenette and F. Joud (2005). Conceptual perspectives for measuring spatio-temporal behaviour in urban areas. In: *Integrated Land-Use and*

Transportation Models: Behavioural Foundations (M. Lee-Gosselin and S.T. Doherty (eds.). Elsevier, New York. forthcoming.

Roorda, M.J. and E.J. Miller (2004). Toronto activity panel survey: demonstrating the benefits of a multiple instrument panel survey. *Proc. 7th Int. Conf. Trav. Surv. Meth.*, Costa Rica (CD-Rom).

Séguin, A.-M. and M. Thériault (2000). Residential trajectories of professional workers in the metropolitan area of Quebec city. *Congrès Annuel de l'Association Canadienne de Sciences Régionales*, Toronto (Cd-Rom).

17

THE DESIGN AND IMPLEMENTATION OF AN ACTIVITY SCHEDULING SURVEY USING THE INTERNET

Tomás Ruiz, Technical University of Valencia, Valencia, Spain

INTRODUCTION

Travel behaviour researchers are increasingly recognizing the need for in-depth research into the activity scheduling process (Pas, 1982; Jones *et al.*, 1983; Axhausen and Gärling, 1992). Numerous modelling efforts have been made to better understand the scheduling process (Gärling *et al.*, 1994; Ettema *et al.*, 1994, 1995; RDC, 1995; Arentze and Timmermans, 2000; Doherty *et al.*, 2001; Joh *et al.*, 2003; Miller, 2004). Yet, few attempts of collecting data on the underlying activity scheduling process can be found in the literature. Hayes-Roth and Hayes-Roth (1979) pioneered this stream of research. They used 30 "think-aloud protocols" from five subjects to investigate the kinds of behaviour exhibited when they were posed with a series of errands to perform in a simulated urban environment. Using the same technique, Chen (2001) conducted another experiment investigating the process of activity scheduling and rescheduling. Subjects were randomly assigned a day to fill out an activity diary. Before the day began, subjects were asked to record activities they had planned for the assigned diary day. On the night following the diary day, telephone interviews were conducted to obtain from subjects the executed activities of the day. Ettema *et al.* (1994) developed Magic, a computer programme for self completion of activity (re-)scheduling tasks, and used this tool in an interactive computer experiment to identify the types of steps taken by subjects to perform a scheduling task. Later, several similar but extended computer programmes were developed. Doherty and Miller (2000) developed CHASE, a computer-aided self-interview of

Progress in Activity-Based Analysis edited by H. Timmermans

activity scheduling for households, allowing users to record their scheduling decisions over a multi-day period. Lee *et al.* (2000) and Lee and McNally (2001) developed iCHASE and REACT! and utilized the Internet for sending data from a respondent's home to a central database. Rindsfüser *et al.* (2003) developed EX-ACT, a computerized hand-held survey instrument to gather information as closely as possible to real decisions points.

In this study, data collection methods based only on the Internet have been used for collecting the required data. Internet surveys are currently in vogue largely because of several assumptions about how they stack up against more-traditional survey modes (see Dillman, 2000; Couper *et al.*, 2001; Thompson *et al.*, 2003). Their potential advantages are that they are less time consuming, are much cheaper to conduct, are easier to execute, have the ability to include visual aids and animations, have the ability to automate skip patterns and randomise questions, and can collect information about response behaviour. In the transport field, Denike (1998) carried out the first known large scale Internet survey on travel behaviour and likely response of different market segments to the initiation of improved transit service among teaching staff, students and other staff at University of British Columbia. Internet travel diaries were developed by Adler and Rimmer (1999) primarily to test the concept of respondent-interactive geo-coding. Another Internet travel survey application was developed to obtain traveller responses to an Advanced Traveller Information Systems (ATIS) (Kraan *et al.*, 2000). The Internet has been used to complement the data collected by Global Positioning Systems devices (Lee *et al.*, 2001; Stopher *et al.*, 2004). The Resource Systems Group Inc. (2002) has designed web-based survey templates for household travel diary surveys, travel origin/destination surveys, travel mode choice surveys and transit customer satisfaction surveys. Arentze and Timmermans (2002) used stated adaptation and stated choice games experiments administered through the Internet to examine how individuals adjust their activity-travel patterns. An Internet-based origin-destination travel survey was developed by Abdel-Aty (2003). This study involved mail-out and roadside distribution of the survey instrument but offered by mail back an Internet retrieval options to respondents. Bricka and Zmud (2003) also utilized an Internet retrieval tool option in a household travel survey to capture those highly mobile respondents that might otherwise be excluded in a conventional telephone household survey. Finally, Hojman *et al.*, (2004) carried out an Internet-based survey to elicit people's preferences for improved road safety.

The literature reviewed outlines no attempt of gathering information on the activity scheduling process using data collection methods based only on the Internet. The research described in this chapter tries to fill this gap. The remainder of the chapter is organized as follows. First, the objectives of the research project and the target population are defined. Then, the survey design and the implementation procedures are described. This is followed by a description of the main results, and finally the implications of these findings are discussed and some conclusions are drawn.

STUDY OBJECTIVES AND POPULATION OF INTEREST

To date, investigations on how the activity scheduling process occurs in reality have been focused on observing a continuous multi-day period limited to a week. In this survey, it was decided to expand the research period to more than a week. Considering the amount of data required, however, respondents were asked to complete an activity-travel diary for only one to four non-consecutive days spread over a four-week period. Therefore, the first objective of this study was to examine the differences in the scheduling process with regard to the time horizons. The interest was in the sequence of activity attribute decisions and the order in which they are planned or in other words, the "planned time horizons": decisions made early that day or the day before, planned decisions made a few days before, or pre-planned decisions (a few weeks before). The second objective of this study was to explore the possibilities of using data collection tools based exclusively on the Internet for gathering the required information: a personal Activity Agenda, the process of scheduling activities and travel, and the decisions rules adopted throughout that process. The population of interest was university students and staff of the "Avenida de los Naranjos" campus of the Technical University of Valencia, Spain (UPV), which at the moment of the survey (November and December 2003) consisted of 2,058 faculty members, 1,221 other staff and 26,082 students.

SURVEY METHODOLOGY

Survey Design

The survey was developed along the following lines. First, on November 13, 2003, an electronic mail was sent to all students and staff members at the "Avenida de los Naranjos" campus of the UPV. The content of the e-mail indicated the main objective of the Internet survey, an invitation to collaborate in the survey and a hyperlink to the website. A lottery prize consisting of three tickets with a value of 300 € each for acquiring office material was offered as an incentive to participants. After that, the participants would follow the instructions: first, they were requested to complete an on-line questionnaire with their demographic and socio-economic characteristics; second, they were asked to make a plan about their activities and associated travel for personally assigned survey days to complete an activity diary for each day. In the case investigators detected participants who had not completed any of the on-line questionnaires, they were notified with up to three e-mails reminding them to do so. The survey administrator sent out e-mails to alert interviewees' potential mistakes they might have made on a specific day. Finally, after each survey day, participants received an e-mail in which they were invited to enter the website and modify/add/delete the planned activities and travel if necessary to describe their executed activity-travel schedule.

Instrument Design

A multi-screen website was designed using the HTML language interface and ASP programming. When entering the website, the respondent was given a brief introduction and instructions on how to complete the survey. If the respondent agreed to participate in the survey, he/she was asked to define and confirm a key code and to provide his/her e-mail address. The respondent was then instructed to complete a preliminary screen to describe himself/herself (Figure17.1). An example of a completed daily Activity Agenda was shown to the respondent (Figure 17.2). Next, the future survey days marked on a multi-day Activity Calendar appeared (Figure 17.3). Respondents were randomly given between one to four non-consecutive survey days spread across a four-week period. Respondents were asked to build up a planned activity schedule for each designated day in the Activity Calendar by selecting the day and adding activities to the right time slot in the Agenda. The added activities were those that the respondent has already thought about conducting.

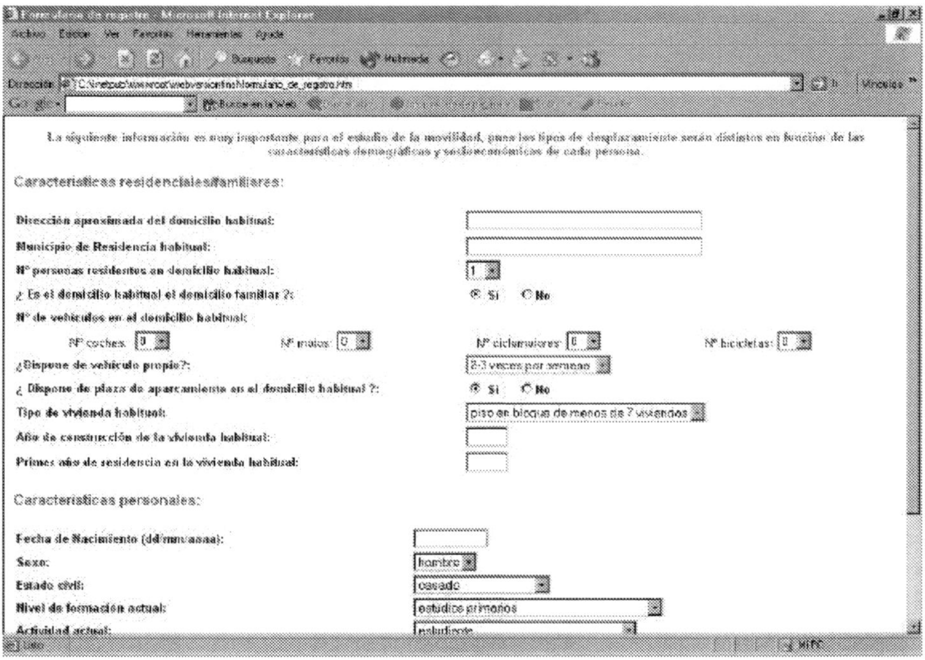

Figure 17.1
Demographic and Socio-Economic Screen

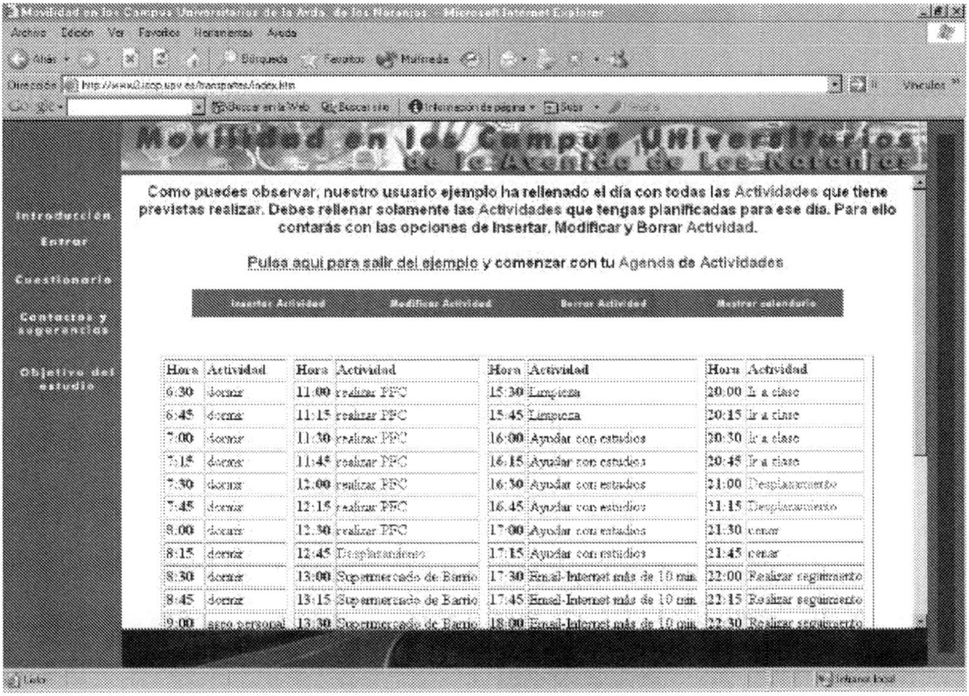

Figure 17.2
Example of a Completed Daily Activity Agenda Screen

Table 17.1
Total Response

	Corrected Population	Responses with Demographic Data	Valid Responses: Planned or Executed	Valid Responses: Planned and Executed	Valid Responses: Different Executed Schedule
Teaching and Researching Staff	2,058	579	131	49	24
Other staff	990	278	65	24	12
Students	26,082	3,528	707	256	102
Total	29,130	4,385	903	329	138

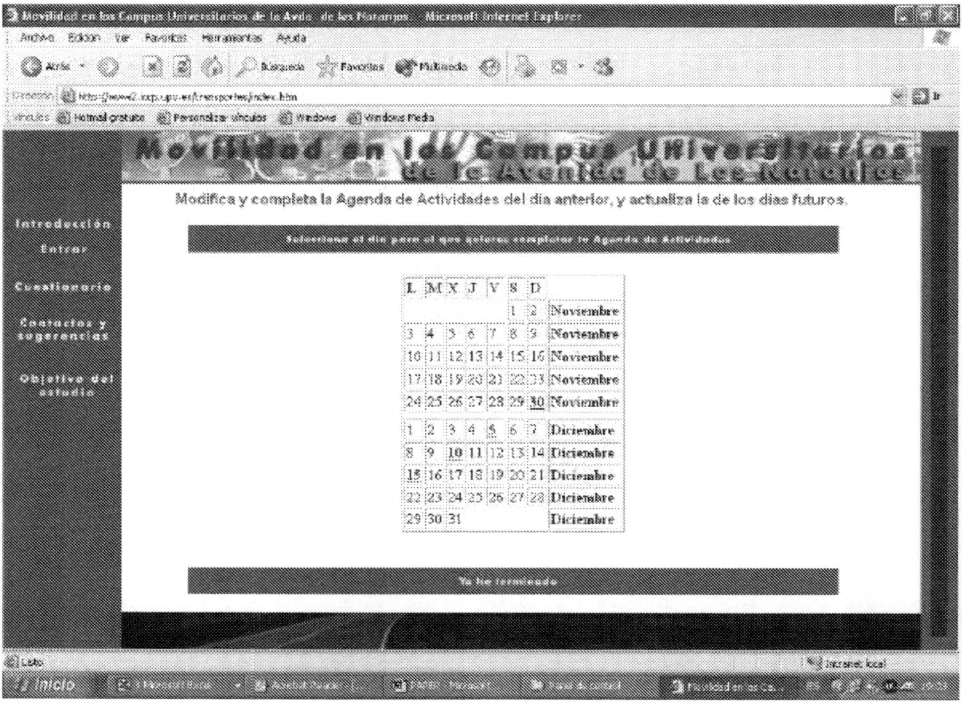

Figure 17.3
Activity Calendar Screen

The day after each survey day, respondents received a reminder, requesting them to update the multi-day Activity Agenda, reflecting their actual behaviour. Respondents were asked to include all activities that lasted longer than 10 minutes. A respondent's daily agenda was displayed for each survey day as a series of rows depicting the time from 6.30 to midnight at 15 minutes intervals. The daily agenda included the options Add, Modify, Delete and Show Agenda. Each selected option led respondents to a different dialog screen (Figure 17.2). For example, the Add dialog screen prompted the respondent to define the main activity type and the secondary activity type which was carried out at the same time in terms of location, starting time, activity duration, daily frequency, accompanying people, and, if travelled, transport mode, start time, travel time and a complementary activity which the respondent was conducting during the trip (Figure 17.4). The Modify/Delete dialog screens showed the list of all activities and travel, and by selecting an activity or a travel episode, respondents were allowed to change any of their attributes or delete the activity and associated travel.

SURVEY RESULTS

Before discussing some scheduling characteristics, we will first report response rates and evidence of fatigue effects in the following subsections.

Response Rates

Overall, 4385 university members visited the web page and provided the data about their demographic and socio-economic characteristics (Table 17.1). This meant a response rate of 14 percent of the target population, corrected for those e-mail addresses that remained inactive since they were created. 903 university members completed a total of 1,595 daily planned or executed activity surveys. Only 329 participants accessed the website twice for providing information about both their planned activity schedules and their executed activity schedules, involving 581 survey days. Of these, 138 respondents reported differences between planned and executed schedules, involving 195 surveys days.

Fatigue Effects

The number of respondents who provided valid data about their planned or executed activity schedules was largely the same for the number of survey days assigned in the Activity Calendar: 240, 223, 229 and 221 respondents for respectively a one-day, a two-day, a three-day and a four-day Activity Calendar. However, the number of survey days obtained decreased if one considers subsequent designated days in the Activity Calendar (Figure 17.5). This result suggests the existence of fatigue effects. On the other hand, the respondents who completed a two-day or three-day Activity Calendar and gave data for the entire Calendar (two and three survey days per respondent, respectively) were more than those who partially completed the Calendar (Figure 17.6). Few respondents who completed the four-day Activity Calendar gave data for only two survey days. The majority of the latter decided to fill only one survey day or to complete the entirely survey. As a result, most of the respondents who filled out Activity Calendars with many days tended to complete the majority of them. Finally, the fatigue effects were also detected in reporting activities and travel episodes while respondents completed longer than one-day Activity Calendars (Table 17.2). The number of executed activities and trips that each respondent provided per day decreased from 16.9 for one-day Activity Calendars to 15.9 for three-day Activity Calendars, although this decrease can also be explained by differences in the number of planned and the number of executed activities.

Scheduling Behaviour and Activity-Travel Patterns

Overall, respondents who had an executed schedule that was different from the planned schedule reported a total of 2,666 add-decisions to define their planned activity schedules and only 593 to describe their executed activity schedules. For the latter, they also reported 905 modify- and 109 delete-decisions to schedule 2,358 activities and 793 trips. On average, this represented 16.8 add-, 4.6 modify- and 0.6 delete-decisions made to schedule 12.1 activities and 4.1 trips per respondent per day. The earliest activity schedule was planned 17 days in advance, and the latest the day before the survey day. Depending on this time horizon, the characteristics of the scheduling process were different (Table 17.3).

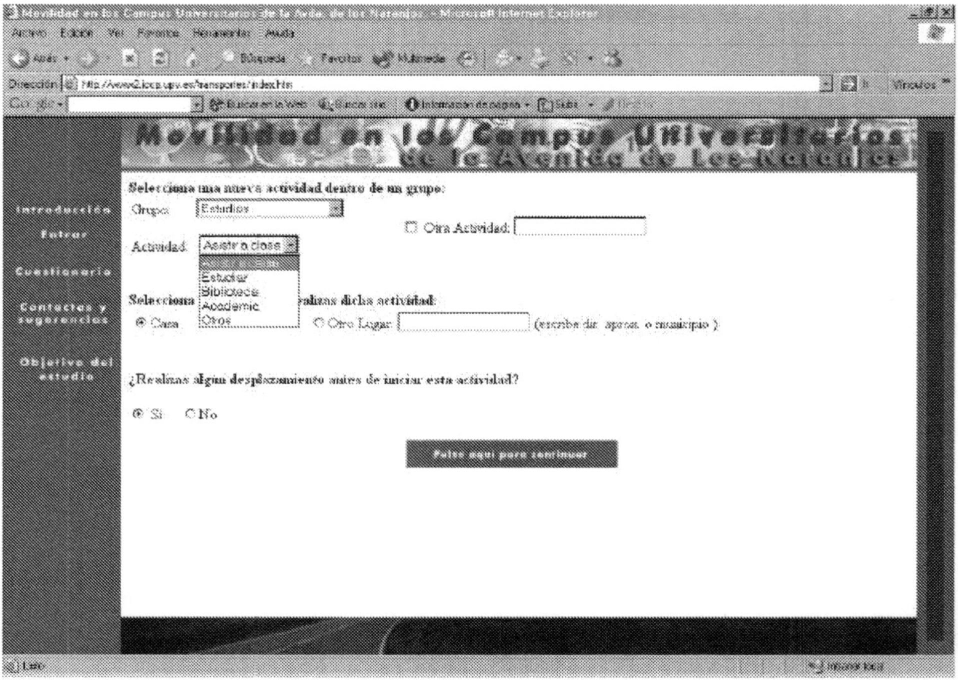

Figure 17.4
Add Dialog Screen

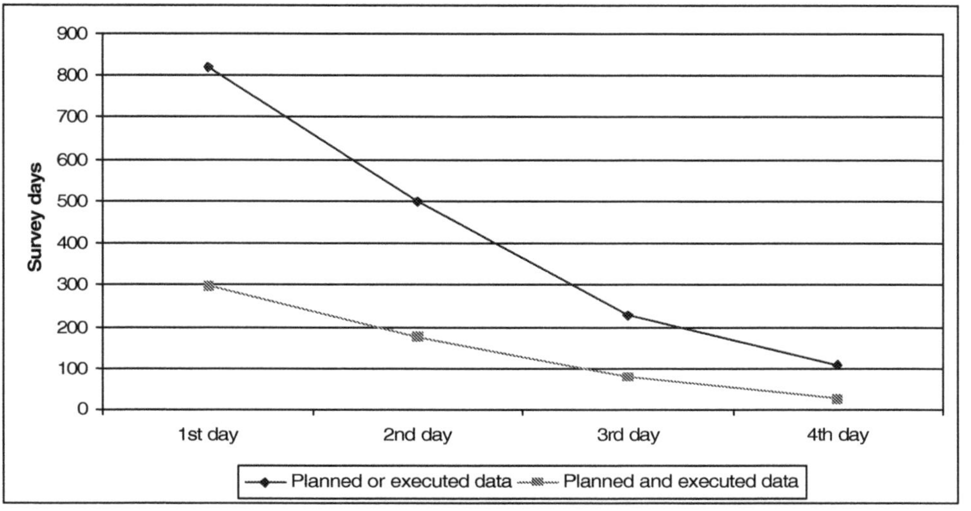

Figure 17.5
Fatigue Effects of Respondents: Survey Days Collected Considering Subsequent Designated Days in the Activity Calendar

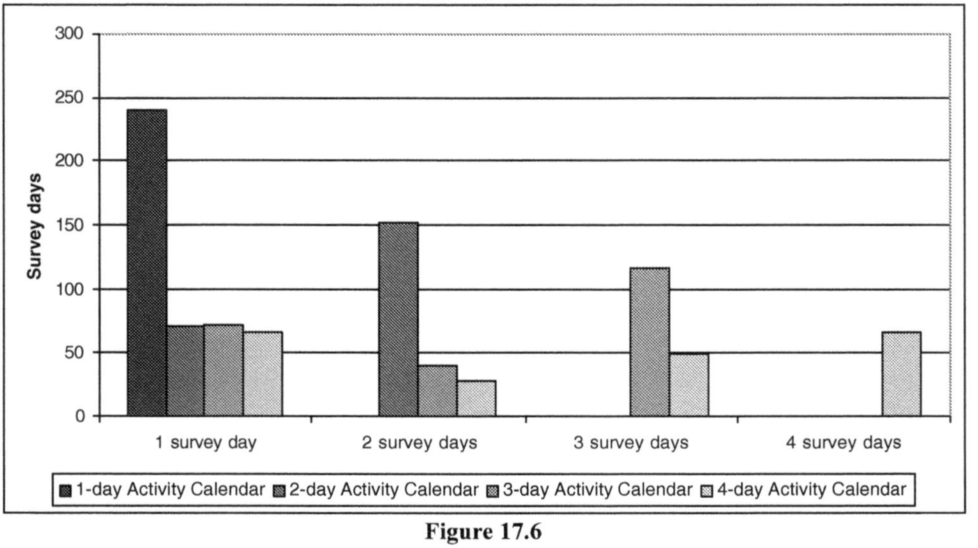

Figure 17.6
Fatigue Effects of Respondents: type of Activity Calendar Assigned by the Number of Survey Days Completed

Table 17.2
Activities and Trips per Respondent per Day for Planned and Executed Schedules per Type of Activity Calendar Completed

	Planned	Executed		Planned	Executed
1-day Activity Calendar	15.1	16.9	3-day Activity Calendar	13.2	15.9
2-day Activity Calendar	13	16.2	4-day Activity Calendar	13.9	16.1

Table 17.3
Scheduling Steps per Respondent per Day in Different Time Horizons for those whose Executed Schedule differs from Previously Planned Schedule

	Plan Time horizon					
	1 Day in Advance		2-6 Days in Advance		7-17 Days in Advance	
	Planned	Executed	Planned	Executed	Planned	Executed
Activities	10.3	12.2	9.3	11.8	10.4	12.2
In-home	7.2	8.4	7.1	8.4	8.3	8.9
Out-of-home	3.1	3.8	2.2	3.4	2.1	3.3
Trips	3.8	4.2	3.3	3.8	3.6	4.1
Deletions		0.5		0.5		0.6
Modifications		5.3		4.3		3.9
Additions		2.7		3.5		3.1
Observations	88		52		55	

Deletion was the less frequent scheduling decision to define the final schedule. Respondents only made between 0.5 and 0.6 delete-decisions per day. On the other hand, respondents made between 3.9 and 5.3 modify-decisions per day to define the final schedule. Respondents also made between 2.7 and 3.5 add-decisions to describe their executed schedules. The number of delete-decisions slightly increased with an increasing time horizon. On the other hand, the number of modify-decisions decreased with an increasing time horizon. Add decisions to define the final schedule were more likely to occur for activity schedules planned two or more days before the survey day than for those planned the day before. The total number of attributes modified/added per survey-day when a modify-decision was made to define the final schedule decreased with an increasing time horizon (Table 17.4). The same relation was observed for the number of modify-decisions per respondent per day. The number of attributes modified/added per modify-decision slightly increased as the time horizon also increased. The most common modification was a change of start time of an activity, representing more than one third of all modifications. Changes in activity duration and with-whom were recorded less frequently.

Figure 17.7 and Figure 17.8 represents the average number of daily activities and trips scheduled by respondents by day of the week and activity/mode type. The data analysed was the final schedules defined by respondents whose previously planned schedules were different. In-home activities were observed in the final scheduled more often on Friday to Sunday, whereas the average number of out-of-home activities as expected was higher on weekdays. No obvious trends exist over the course of the week for different types of activities, except for the decrease of work/school activities from Friday to Sunday and the increase of leisure activities especially on Sunday. Slight increases in housekeeping and social related activities are present on Sunday.

The trip pattern over the course of the week for different mode types is shown in Figure 17.8. Public transport trips followed a decreasing trend over the week while car non-driver trips increased over the weekend. Car driver trips increased from Monday to Thursday and then diminished from Friday to Sunday.

Table 17.4

Modified/Added Attributes in Modify-Decisions per Respondent per Day x 10^{-2} in Different Time Horizons for those whose Executed Schedule differs from the Planned Schedule

	Plan Tine Horizon					
	1 Day in Advance		2-6 Days in Advance		7-17 Days in Advance	
	Total	%	Total	%	Total	%
Starting time	4.2	48.2	3.5	45.0	3.2	46.6
Duration	2.6	29.7	2.3	29.5	0.6	9.0
With-whom	1.6	18.2	1.7	22.5	1.3	18.8
Type of activity	0.3	3.4	0.2	2.3	0.3	3.7
Location	0.0	0.5	0.1	0.8	0.1	2.1
Sub-total	8.7	100.0	7.7	100.0	6.9	100.0
Travel time	0.9	39.1	0.6	36.3	0.7	38.3
Travel mode	0.6	29.2	0.5	30.0	0.5	26.6
Activity while travelling	0.6	26.0	0.4	25.0	0.4	25.5
Trip o/d	0.1	5.7	0.1	8.8	0.2	9.6
Sub-total	2.2	100.0	1.5	100.0	1.7	100.0
Total	10.8		9.2		8.6	
Total per Modify-decision	2.0		2.1		2.2	

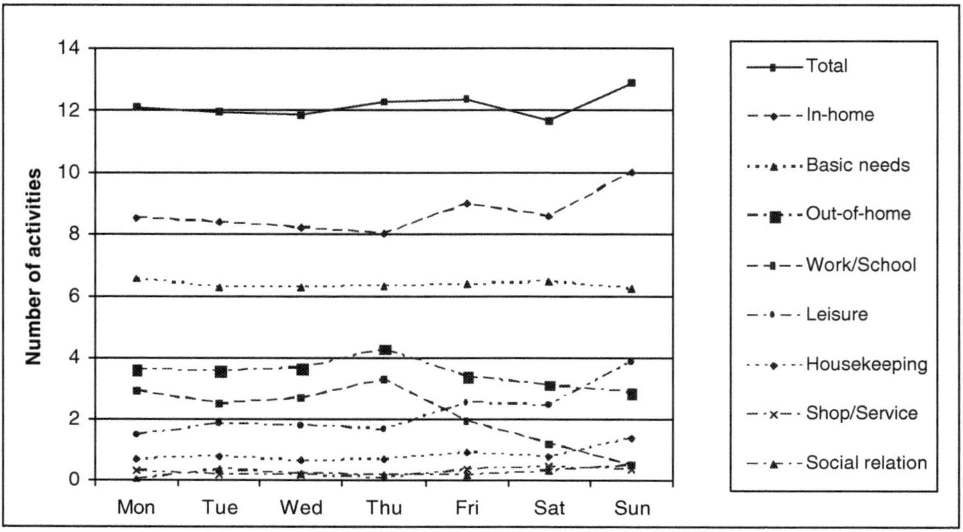

Figure 17.7
Average Number of Observed Daily Activities by Day of the Week and Activity Type

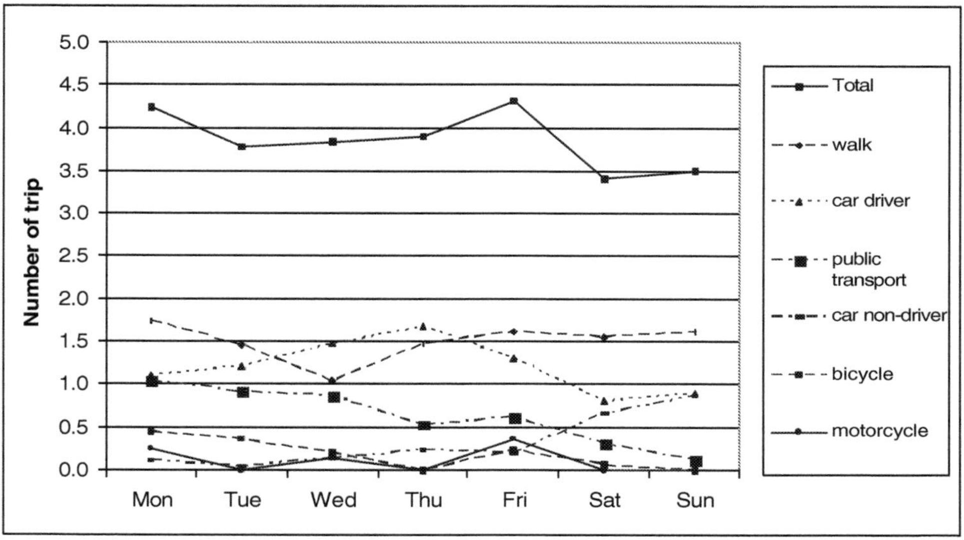

Figure 17.8
Average Number of Observed Daily Trips by Day of the Week and Mode Type

DISCUSSION AND CONCLUSIONS

This research project involved the design and implementation of an activity scheduling survey using data collection methods based only on the Internet. Information on both planned activity-travel schedules of participants amounting up to 17 days in advance and executed schedules was collected. A preliminary analysis of the data indicates that the characteristics of the scheduling process were different. Both the number of add and delete decisions for describing executed schedules were more likely to occur when respondents had planned those schedules two or more days before the survey day. This suggests that the probability of changing activity schedules is a function of time horizon. On the other hand, the number of modify-decisions decreased with an increasing time horizon. The number of attributes modified/added per modify-decision slightly increased as the time horizon also increased.

Despite the burdensome on-line questionnaire, it was possible to collect data on the activity-travel scheduling process using data collection methods based only on the Internet. 138 respondents made a plan for their schedules for 195 days and they also reported the details of the actually executed schedules for those survey days. Taking into account the well-known advantages of Internet-based surveys, especially associated to costs, speed and flexibility, these are sufficient reasons to be optimistic about the use of the Internet for gathering this complex data, although one cannot forget the characteristic sampling bias associated to Internet surveys. In this case, a special effort is needed to recruit a representative sample of respondents.

On the other hand, the activity-travel scheduling data collected in this research project were very limited. Only two observations of the scheduling process were made. Therefore, it would be appropriate to extend the number of observations asking respondents to review their planned agenda several times before the survey day. Activity schedules would then likely undergo numerous transformations and adjustments before implemented.

There is also a need for improving the website design to further facilitate and encourage participation. First of all, the number of steps that respondents have to take while entering data should be reduced. Secondly, a much clearer explanation of the reasons why these kinds of data is being asked is also desirable. Finally, incompatibilities in hardware or software should be reduced through intensive checking of different computers configurations. Additionally, there are some enhancements that are costless. For example, initial contact to potential respondents by e-mails should be distributed over time (or a large team of people should be involved) to make available more resources for the non-response follow-up execution and to allow assistance to the respondents. The non-response follow-up can be partially automated.

Acknowledgements

The author is grateful to Sean Doherty for providing valuable comments on the early stages of the website design. Thanks also go to Sole Gómez, Javi Piqueres, Alex Moya, Dani Porres and Adolfo Lozano for their help in the website design, implementation and survey execution, to Chang-Hyeon Joh for his valuable comments on an earlier draft of this chapter, and to two anonymous referees for their helpful observations. The author would like to acknowledge the financial support provided through a research grant from the "Agencia Valenciana de Ciencia y Tecnología"

References

Abdel-Aty, M.A. (2003). Hybrid distribution and response techniques for an origin-destination travel survey. *ITE Journal*, **73**, 22-27.

Adler, T. and L. Rimmer (1999). Computer-Based Intelligent Travel Survey System: CASI/Internet Travel Diaries with Interactive Geocoding. Final Project Report SBIR Contract No. DTRS57-98-C-00068, White River Junction, VT, United States.

Arentze, T.A. and H.J.P. Timmermans (2000). Albatross: *A Learning-Based Transportation Oriented Simulation System.* EIRASS, Eindhoven.

Arentze, T.A. and H.J.P. Timmermans (2002). Response Rates, Sampling Bias and Reliability of Internet-Based Travel Surveys: Experiences with a Stated Adaptation Experiment. Research Report, EIRASS, Eindhoven, The Netherlands.

Axhausen, K.W. and T. Gärling (1992). Activity-based approaches to travel analysis: conceptual frameworks, models, and research problems. *Transpn. Rev.*, **12**, 323-341.

Bricka, S. and J. Zmud (2003). Impact of Internet retrieval for reducing non-response in household travel surveys. *Proc. 82nd Ann. Meet. Transpn. Res. B.*, Washington, D.C. (CD-Rom).

Chen, Q. (2001). An Exploration of Activity Scheduling and Rescheduling Processes. Ph.D. Dissertation, Department of Civil and Environmental Engineering, University of California-Davis, USA

Couper, M., M. Traugott and M. Lamias (2001). Web surveys and administration. *Publ. Opin. Quart.*, **65**, 230-253.

Dillman, D.A. (2000). *Mail and Internet Surveys, The Tailored Design Method.* John Wiley & Sons, New York

Denike, K. (1998). Travel Patterns to and from UBC and Response of Market Segments Collected via an Internet Web-based Scenario. University of British Columbia, Vancouver, Canada, www.trek.ubc.ca/research/pdf/denike.pdf .

Doherty, S.T. and E.J. Miller (2000). A computerized household activity scheduling survey. *Transpn.*, **27**, 75-97.

Doherty, S.T., E.J. Miller, K.W. Axhausen and T. Gärling (2001). A conceptual model of the weekly household activity-travel scheduling process. In: *Travel Behaviour: Patterns, Implications and Modelling* (E. Stern, I. Salomon and P. Bovy, eds.), pp. 148-165. Elgar Publishing Ltd., Cheltenham, U.K.

Ettema, D.F., A.W.J. Borgers and H.J.P. Timmermans (1994). Using interactive computer experiments for identifying activity scheduling heuristics. *Proc. 7th IATBR Conf.*, Santiago, Chile.

Ettema, D.F., A.W.J. Borgers and H.J.P. Timmermans (1995). SMASH (Simulation model of activity scheduling heuristics): empirical test and simulation issues. Paper presented at the *Activity Based Approaches Conference*, Eindhoven, The Netherlands.

Gärling, T., M.-P. Kwan and R.G. Golledge (1994). Computational-process modelling of household activity scheduling. *Transpn. Res. B*, **28**, 355-364.

Hayes-Roth, B. and F. Hayes-Roth (1979). A cognitive model of planning. *Cogn. Sc.*, **3**, 275-310.

Hojman, P., J. de D. Ortúzar and L. Rizzi (2004). Internet-based surveys to elicit the value of risk reductions. *Proc. 7th Int. Conf. on Trav. Surv. Meth.*, Costa Rica (CD-Rom).

Joh, C.-H., T.A. Arentze and H.J.P. Timmermans (2002). Understanding activity scheduling and rescheduling behaviour: theory and numerical illustration. *GeoJournal*, **53**, pp. 359-371.

Jones, P.M., M.C. Dix, M.I. Clark and I.G. Heggie (1983). *Understanding Travel Behaviour*. Gower, Aldershot.

Kraan, M., H.S. Mahmassani and N. Huynh (2000). Traveller responses to advanced traveller information systems for shopping trips. *Transpn. Res. Rec.*, **1725**, 116-123.

Lee, M.S., S.T. Doherty, R. Sabetiashraf and M.G. McNally (2000). iCHASE: an Internet computerized household activity scheduling elicitor survey. *Proc. 79th Ann. Meet. Transpn. Res. B.*, Washington, D.C. (CD-Rom).

Lee, M.S., J.E. Marca, C.R. Rindt, A. Koos and M.G. McNally (2001). Evaluation of a shared-use electric vehicle program: integrating a web-based survey with in-vehicle tracking. Paper presented at the *17th Pac. Rim Reg. Sci. Assoc. Meeting*, Portland, USA.

Lee, M. and M.G. McNally (2001). Experiments with a computerized self-administrative activity survey. *Transpn. Res. Rec.*, **1748**, 125-131.

Miller, E.J. (2004). An integrated framework for modelling short- and long-run household decision-making. *Proc. EIRASS Conf. on Progress in Activity-Based Analysis*. Maastricht, the Netherlands, also in this volume.

Pas, E.I. (1982). Analytically derived classifications of daily travel-activity behaviour: description, evaluation and interpretation. *Transpn. Res. Rec.*, **879**, 9-15.

RDC, Inc. (1995). Activity-Based Modeling System for Travel Demand Forecasting. Final Report prepared for Metropolitan Washington Council of Governments, Washington, D.C.

Resource System Group Inc. (2002). Documentation for SBIR Phase II Final Report: Computer-Based Intelligent Travel Survey System, DTRS57-00-C-10030. Prepared for the FHWA. http://www.fhwa.dot.gov/ohim/trb/sbir/sbir.htm

Rindsfüser, G., H. Mühlhans, S.T. Doherty and K.J. Beckmann (2003). Tracing the planning and execution of activities and their attributes: design and application of a hand-held scheduling process survey. *Proc. 10th Int. Conf. on Travel Beh. Res.*, Lucerne, Switzerland (CD-Rom).

Stopher, P.R., A. Collins and P. Bullock (2004). GPS surveys and the Internet. Paper presented at the *27th Austr. Transp. Res. Forum*. Adelaide, Australia.

Thompson, L.F., E.A. Surface, D.L. Martin and M.G. Sanders (2003). From paper to pixels: moving personnel surveys to the web. *Pers. Psych.*, **56**, 197-227.

18

ACTIVITY-BASED TRAVEL FORECASTING MODELS IN THE UNITED STATES: PROGRESS SINCE 1995 AND PROSPECTS FOR THE FUTURE

Peter Vovsha, PB Consult, Parsons Brinckerhoff Inc, New York, NY, US

Mark Bradley, Bradley Research & Consulting, Santa Barbara, CA, US

John L. Bowman, Transportation Systems & Decision Sciences, Brookline, MA, US

INTRODUCTION

The focus of our paper is the use of activity-based models in practice for urban and regional planning in the United States. While our focus is on projects that have been carried out for immediate application in the U.S., we recognize that much other research is being done in other research settings, both inside and outside the U.S. that will help determine the types of activity-based models that are used in practice in the future.

We have chosen to focus our topic because regional planning in the US is at a critical stage where the adoption of activity-based models is accelerating, with the potential for much more acceleration in the future. This paper provides an opportunity to step back and look at what types of modeling developments have been successfully implemented since the previous EIRASS conference in 1995. This section provides an overview of developments before 1995. In the next section, we discuss the conceptual areas of applied models that we feel need the most work, and describe five specific features that have been implemented in one or more of four important model systems developed since 1995. The section that follows describes the authors' latest model system design for two new

Progress in Activity-Based Analysis edited by H. Timmermans

development projects. The next section also identifies important topics for further enhancement of the design. In the section that follows, we discuss factors that remain as hindrances to the acceptance of activity-based models for planning by government agencies in the U.S. This is followed by a brief concluding section.

Transportation and land use planning in the US is done primarily at the regional level, typically for an area that includes a large city and the surrounding suburbs and satellite cities. Each county or city within a region may also do its own planning, but it will typically use a version of the same planning model used by the regional metropolitan planning organization (MPO). The large majority of MPOs use the aggregate 4-step modeling approach first introduced in the 1960s.

Although a great deal of marginal refinement to the 4-step approach has been done since the 1960s, the basic drawbacks of the approach still remain. These include: (i) Person-trips as the unit of analysis: The models do not capture the interactions between trips made in the same trip chain, the interactions between trip chains made during the same day, or the interactions between the trips made by people in the same household; (ii) Spatial aggregation: All trip origins and destinations within a given zone are modeled as if they are located at the same point in space; (iii) Temporal aggregation: Typically, only two or three periods of the day are considered (e.g. AM peak hour, PM peak hour, off-peak), and the proportion of trips made in each period is treated as constant and not sensitive to changes in traffic congestion or other factors; (iv) Demographic aggregation: All households within a given zone are treated as identical, or, at best, segmented along a few dimensions such as income, household size and car ownership.

By the time of the previous EIRASS conference on activity-based methods in 1995, a number of tour-based models had already been implemented in order to replace trips with tours as the primary unit of analysis. Early applications were carried out in both the US and Europe: San Francisco Bay Area (Ruiter and Ben-Akiva, 1978), The Netherlands (Daly *et al.*, 1983; Gunn *et al.,* 1989), Boise, Idaho (Shiftan, 1999), Stockholm (Algers *et al.*, 1995), New Hampshire (Rossi and Shiftan, 1997), and Italy (Cascetta and Biggiero, 1997). These early tour-based applications were successful in eliminating the most glaring weakness of the 4-step approach, the poor ability to deal with non-home-based trips made in the middle of home-based trip chains. They did not, however, deal with any of the other weaknesses listed above. They did not consider interactions between tours made at different times in the same day, or by different people in the same household. They did not offer any large improvements to deal with the issues of spatial, demographic and temporal aggregation. All three types of aggregation can cause significant aggregation bias due to the fact that logit models and gravity models are non-linear—i.e., the logit probability at the average value is not necessarily equal to the average of the logit probabilities across all individual values. This fact is often overlooked.

KEY CONCEPTS AND FEATURES OF APPLIED MODELS SINCE 1995

Since 1995, a number of activity-based travel demand model systems have been implemented in the United States that address the issues mentioned above. These include model systems developed for the following metropolitan areas (the public agency involved in each case is noted in parentheses): Portland, Oregon (Bowman *et al.*, 1999; Bradley *et al.*, 1998; Bradley *et al.* 1999), San Francisco County (Bradley *et al.*, 2001; Jonnalaggada *et al.*, 2001), New York City (Vovsha *et al.*, 2002; Peterson *et al.*, 2002), and Columbus, Ohio (Vovsha *et al.*, 2004a, 2004b; Vovsha and Bradley, 2004). The improvements that were begun by the tour-based approaches before 1995, and continued by these more recent projects, fit into two major conceptual categories: disaggregation and integration. Disaggregation involves reducing the size of spatial, temporal and demographic categories used in the models, enabling them to more realistically represent behavior, which in some important situations is very sensitive to minor differences in time, space and household characteristics. Integration is conceptually more complex than disaggregation. The remainder of this subsection discusses the concept of integration in more detail, defining the subcategories of intra-person integrity and intra-household integrity.

"Intra-person integrity" is achieved by realistically representing outcomes of the individual daily activity and travel pattern. All modeled activity episodes, their durations, locations, and travel tours associated with visiting out-of-home activities are consistent and feasible within the person time-space constraints. Furthermore, the models accurately represent important correlations among these activity and travel outcomes, and of each outcome with the factors affecting it. The factors include urban and transportation system attributes, as well as the characteristics of the person and household.

A high degree of intra-person integrity is difficult to achieve, because the outcomes being modeled are so complex that it is necessary to decompose the outcome into pieces. Integrity is then sought by effectively linking multiple sub-models. Intra-person integrity was in the core of the original concept of the daily activity pattern choice model (Bowman and Ben-Akiva, 1999; 2001; Bhat and Singh, 2000). The major breakthrough that makes this approach operational is the integrative formulation of the daily pattern, defined in terms of a number and structure of travel tours. This provides the necessary input to tie together a subsequent set of conditional travel models, since the great variety of observed individual daily activity patterns and structural complexity of the choice model in combination with a huge number of possible activity location alternatives make it impossible to model all dimensions in one choice model.

"Intra-household integrity" is achieved by expanding the integration to encompass all members of a household. The models accurately represent the coordination of daily patterns among different

household members. This coordination includes, among other things, participation in joint activities and joint travel, as well as intra-household mechanisms for allocation of maintenance activities and household cars. Numerous ways exist to improve intra-household integrity:

- Use household composition variables (for example, the presence of children of particular age categories) as explanatory variables in activity pattern or tour generation models for workers and other individuals. This approach can be classified as implicit because the decision unit is still the individual.
- Model jointly—or at least in a coordinated manner—the individual daily activity pattern types (or related activity-travel characteristics) for several household members. Most frequently, time allocation units are used for modeling and the Structural Equation System is employed (Golob and McNally, 1997; Fujii *et al.*, 1999; Simma and Axhausen, 2001; Meka *et al.*, 2002). The next section of this paper describes another approach, used by the authors, that uses a linked set of discrete choice models (Vovsha *et al.*, 2004a)
- Model explicitly the joint participation of household members in activities and travel. This component has been modeled in terms of either episode generation or time allocation between individual and joint activities (Gliebe and Koppelman, 2002; Scott and Kanaroglou, 2002; Ettema *et al.*, 2004). Explicit modeling of joint tours has been incorporated into the Columbus model (Vovsha *et al.*, 2003b).
- Model explicitly the within-household allocation of maintenance activities to household members (Borgers *et al.*, 2002; Srinivasan and Bhat, 2004). This component has been incorporated into the Columbus modeling system (Vovsha *et al.*, 2004b).
- Explicitly allocate cars to household members, accounting for actual availability of a car for a particular person's travel tour (Arentze and Timmermans, 2000, 2004; Wen and Koppelman, 1999, 2000; Miller *et al.*, 2003).

Various hierarchical structures have been used to decompose the complex activity-travel pattern and schedule into manageable pieces. The choice of component models, their specific implementation, and their order in the hierarchy all bear heavily on the degree to which integrity is achieved. The linkages between the models are also very important, and they naturally fall into two categories, those seeking downward integrity and those seeking upward integrity.

Downward integrity means that all lower-level decisions in the choice model hierarchy are properly conditional upon the upper-level decisions and take into account a gradually narrowed scope of lower-level choice alternatives as the upper-level choices progress. Downward integrity is not an automatic property of hierarchical cascades of choice models, especially if different activity dimensions such as number of tours/activities, their location, and timing are considered. For example, the first activity-based models for Portland, San Francisco, and New York had independent-by-tours mode, destination, and TOD choice models that could produce conflicting choices for different tours made by the same person. Downward integrity is ensured by properly sequencing the models, tracking the important variables from choice to choice that accurately describe the feasible scope left for each subsequent choice, and preventing conflicting choices for the same individual. A time-use approach provides an operational framework for downward integrity because time serves as an ultimate and constrained resource for any type of activity. Activities and tours in a priority hierarchy can be scheduled using residual-time-window variables—determined by higher priority tours—to explain and limit the generation and scheduling of lower priority tours.

Upward integrity means that when modeling upper-level choices the composite measure of quality of the lower-level choices available for each upper-level alternative is properly taken into account. Attention to upward integrity is important because decomposing the models into a hierarchy prevents some factors that affect upper level models from being directly usable in those models. For example, travel conditions can affect whether, or how frequently, a person goes to work. This outcome is on an upper level in an activity-based model hierarchy, but the exact travel conditions are attributes of the mode, time and route choices, which are on lower levels of the model hierarchy. Bowman and Ben-Akiva (2001) first achieved a significant degree of upward integrity in a daily activity pattern model system through the use of expected maximum utility variables derived as utility logsums from the model system's conditional tour models. A lack of upward integrity can result in model systems where the upper level activity models are insensitive to changes in the travel environment. It can also assign inappropriate probabilities to outcomes, or allow infeasible outcomes such as a person-day pattern with a work tour from 7:00AM to 10:00 PM and three additional non-work tours in the same day. Each of the next five subsections describes a specific feature that has been implemented in one or more of the above-mentioned activity-based model systems (Portland, San Francisco, New York and Columbus) to help achieve one or both of the objectives of disaggregation and integration. Table 18.1 identifies the five features and, for each, which model incorporates it.

Consistent Generation of All Tours and Trips Made During a Person-Day

The full day activity schedule approach (Bowman, 1998; Bowman and Ben-Akiva, 1999) was the first operational discrete choice framework for simultaneously modeling the key aspects of an individual's day-long activity pattern: the purpose and location type of the primary activity of the day (subsistence, maintenance or discretionary; in-home -out-of-home); the number of intermediate stops made on the way to and from the primary activity (for out-of-home patterns only); the number of work-based tours made during the day (for work patterns only), and the number and purpose of additional home-based tours made during the day. The first application of this approach was for the Portland model system, developed for the purpose at looking at the response to peak-hour congestion pricing, a type of policy that their 4-step model was not fully responsive to because trip generation and time-of-day distributions were not sensitive to travel times or costs.

Shortly afterwards, this same approach was adopted for the San Francisco County model system. Because of relatively limited data and budget to create this system, the full day pattern approach used for the Portland models was simplified in the following ways: (i) maintenance and discretionary tours were grouped as "Other", (ii) instead of using detailed person-based mode/destination/time-of-day choice logsums in the day pattern generation model, as had been done in Portland, more simple zone-based accessibility measures were used to approximate the logsums, greatly reducing the run time for the entire system. (A variation on this same approach is also used in the New York, Columbus and Atlanta models), and (iii) the location of work activities was treated as a long term choice, implying that the generation of the activities in the person-day was directly conditional on both the home and work locations, rather than just the home location. The New York and Columbus models use more of a "cascading" model approach – first generating mandatory tours, then maintenance tours, then discretionary tours, then intermediate stops on all tours. The residual time window remaining after higher priority tours and activities are generated and scheduled can be used in the generation of subsequent tours and activities. The advantage of this approach is that it is simpler and more flexible, particularly when dealing with interactions between household members, as discussed below. A disadvantage is that it does not directly capture substitution between trip chaining versus making multiple tours, as in Portland and San Francisco.

Table 18.1
Key Features of Activity-Based Model Systems Developed for Selected U.S. Metropolitan Planning Organizations since 1995

	Feature	Portland	San Francisco	New York	Columbus
1	Consistent generation of all tours and trips made during a person-day?	Yes	Yes	Yes	Yes
2	A full population stochastic micro-simulation framework?	No in 1st version, Yes in later versions	Yes	Yes	Yes
3	Greater spatial detail than the TAZ level for land use and walk/transit access?	No in 1st version, Yes in later versions	No	No	No
4	Explicit modeling of interactions between activity patterns of household members?	No	No	No	Yes
5	Greater temporal detail for activity and travel scheduling?	Somewhat (5 time periods)	Somewhat (5 time periods)	Somewhat (4 time periods)	Yes (1 hour periods)

Shift to a Microsimulation (MCSM) Model Application Framework

Microsimulation of travel choices is not a new concept. Some trip-based and tour-based models before 1995 were applied to each individual or household within a sample representing the population of the region. The method, usually referred to as 'sample enumeration', rather than microsimulation, involves calculating the probabilities of all alternatives for each individual, and summing the probabilities for each alternative across all members of the sample, to yield aggregate shares for each alternative. The fundamental difference between sample enumeration and Monte Carlo microsimulation (MCSM) is that, instead of summing the probabilities, MCSM uses them to predict a single outcome for each individual by making Monte Carlo draws. MCSM has important advantages over sample enumeration. The advantages occur because MCSM relieves the modeler from having to keep track of multidimensional matrices of choice probabilities, which for a complex model system can be huge. As a result, it becomes feasible to include more components of choice, to distinguish more segments of the population, to disaggregate the spatial and temporal aspects of the models, and to use the household as the decision unit for some components of the outcome while using the individual as the decision unit for other components. All of these features should lead to improved realism and accuracy of aggregate model predictions derived from the disaggregate MCSM model outputs.

On the other hand, MCSM, and the more complex models that it allows, can impose greater demands for intra-person and intra-household integrity, if the accuracy of the behavioural models depend on the proper accounting of spatio-temporal, resource and household constraints, or if the demand models are to be coupled with traffic simulation models that similarly depend on individual itineraries that obey the same constraints.

The first implementation of the Portland model used sample enumeration, simulating a specified fraction of the full population (typically about 10%), keeping track of all choice probabilities, and expanding the results to match the full population. MCSM of the full population was first implemented in Portland in order to provide forecasts of activity sets for the early development of the TRANSIMS model system. Although simulating about 10 times as many individuals, the model run time was actually reduced because it was no longer necessary to carry multiple cascades of probabilities from one model to another. For tour-based models, this issue is especially important because the locations of intermediate stops are conditional on the locations of both the tour origin and the tour primary destination. In a sample enumeration framework, this requires applying a stop location choice model for every possible combination of tour origins and destinations, while in an MCSM framework it is only applied for a single O-D pair. Similarly, an MCSM was adopted in New York largely because the aggregate 4-step approach proved infeasible. With so many zones, modes and population segments to consider, the sheer size and number of aggregate O-D matrices that would need to be calculated was impractical. MCSM, however, proved to be feasible and practical, and provided the freedom to implement a more advanced activity-based model approach. MCSM was also adopted for the other model systems listed—San Francisco and Columbus. Application of an MCSM activity-based model system requires the input of a synthetic population representing the true population for the base year and forecast years. The four regions have used slightly different methods and variables to generate their synthetic populations from the National Census PUMS 5% sample. Each region has used its own set of control variables for sampling, based primarily on which variables are available as forecasts from land use models or other regional planning agencies. But in all cases, the control variables have included some combination of household income, household size, number of workers in household, age of the head of household, and household type in terms of presence of children and senior citizens.

Greater Spatial Detail for Land Use and Walk and Transit Accessibility

With modern GIS systems, data on land use and the location of residences and business is typically available at a much finer level than is used for transportation analysis zones (TAZ). Although shifting to finer spatial detail is not strictly part of "activity-based modelling", it has been made possible by the introduction of the MCSM approach. Because residences and trips are simulated one at a time, there is no need to store huge O-D matrices that include every possible location. Any inputs and outputs that still require storage as O-D matrices, such as travel times and costs for car and transit and output trip tables for assignment, can still be used at the TAZ-to-TAZ level. The MCSM framework is flexible enough to use two different levels of geographic detail for different types of data.

The second version of the Portland model system used 9,400 link faces to locate individual trip origins and destinations, rather than the 1,250 TAZ in the Portland zone system. This change was made to accommodate requirements of TRANSIMS as it existed at that time. In current work being carried out for the Atlanta Regional Commission (ARC), data on land use characteristics, pedestrian facilities, and transit accessibility are used at the level of 200 meter square grid cells. In both studies, it was found that using a greater level of spatial detail allows the modeller to use much more detailed estimates of walk access and egress times for transit, as well as non-motorized travel times for short trips. These changes greatly improved the estimation of certain mode choice model parameters related to walk and transit modes. Greater spatial detail is also beneficial for modelling destination choice and location choice. The model is less dominated by "size variables" that measure the number of opportunities in an aggregate zone, and thus captures more of the variables that drive location choice—the land use at a specific location, the accessibility to reach that specific location by various modes, and the accessibility to other locations in the walkable surroundings (measured using land use density variables). Two complicating issues arise when using location choice models with many thousands of spatial alternatives in the MCSM framework. One issue is that to be computationally feasible the model can be applied to only a sample of locations for any single location choice. Destination sampling is typically used in model estimation, but less commonly in model application. A second issue is that the random stochastic procedures used both in MCSM simulation and destination sampling introduce simulation error into the forecasts—the results change somewhat when the random number sequence is changed. Controlled tests performed by Castiglione *et al.* (2003) using the San Francisco model system showed that the percent variation between any two runs increases as the level of spatial disaggregation in viewing the results changes. Thus, when moving from zones to even smaller areas, we cannot expect the predicted number of trips to or from any particular location to be stable across runs. When aggregating the results to larger areas, however, the predictions become much more stable. This is a general, often misunderstood principal behind the detailed microsimulation approach. Capturing more aspects of behaviour at a disaggregate level will improve the aggregate forecasting ability of the model, but we do not claim that the model has the ability to accurately forecast at the finest levels of disaggregation (as that would require the use of a crystal ball).

Explicit Interactions among Activity Patterns of Household Members

Works of Golob and McNally (1997), Fujii *et al.* (1999), Simma and Axhausen (2001), Borgers *et al.* (2002), Gliebe and Koppelman (2002), Meka *et al.* (2002), Scott and Kanaroglou (2002), Zhang *et al.* (2002, 2004), Ettema *et al.* (2004), Srinivasan and Bhat (2004), and Zhang and Fujiwara (2004) give examples of models for time allocation and activity episode generation among various types of activities and household members, providing valuable insights into the intra-household decision-making mechanism and its importance for realistic modelling activities and travel. Among the recent applied model systems, the Columbus system represents a major advance by capturing intra-household interactions in three separate ways (Vovsha *et al.*, 2003b, 2004a, 2004b): (i) the type of activity pattern of each individual is directly conditional on the type of activity pattern made by other household members. So, if a child stays home all day because of illness, this also increases the chance that at least one parent will stay home also; (ii) home-based tours that are made by more than one person from the household are generated for the household, rather than generating them separately for each individual (and viewing them as independent individual tours). This is especially important for the accurate modelling of HOV demand, because, as it turns out, a large percentage of all HOV demand is generated by intra-household joint tours, and (iii) maintenance activities are generated for the household and then allocated to individuals, rather than generating them separately for each individual.

Greater Temporal Detail for Activity and Travel Scheduling

The Portland and San Francisco model systems both introduced models of time of day choice in the form of a joint model of the time a person leaves the home to begin a tour and the time they return home to end a tour. In both systems, the day is broken down into 5 separate periods: Early (before AM peak), AM peak, Midday, PM peak and Late (after PM peak). The AM and PM peak periods were defined to be periods of up to 3 hours, specific to the traffic patterns in each region. The New York model used a similar approach, but with only 4 periods—combining the Early and Late periods into a single Off-peak period. While those model systems provided a great improvement over most existing trip-based and tour-based model systems that had no time-of-day choice model, their time-of-day choice models can still be viewed as their weakest area. The reasons are that (i) most departure time changes due to traffic congestion, pricing, etc. tend to involve shifts within the greater 3 hour peak, e.g. from the "peak of the peak" to one of the shoulder periods. These shifts are not captured when the day is only broken into 4 or 5 periods, and (ii) using such long periods does not allow one to model shifts in activity scheduling or the interrelationships between activity scheduling and activity generation in a very meaningful way.

The Columbus model system provides two major advances over the other model systems discussed above: (i) the day is broken down into 1 hour time periods, and (ii) tours for various purposes are generated and scheduled in a consistent way. Work and school tours are generated first. After those tours are scheduled, the amount of time remaining is used to model the generation of remaining non-mandatory tours. Each time a tour is scheduled, the hours of the day used by the tour are made unavailable for subsequently modelled tours. This time-of-day choice model is essentially a continuous duration model (Vovsha and Bradley, 2004) transformed into a discrete choice form. The enhanced temporal resolution opens a way to explicitly control the person time window left after scheduling of each tour and use this residual time window as an important explanatory variable for generation and scheduling of the subsequently modelled tours.

CURRENT MODEL SYSTEM DESIGNS

This section discusses the latest design issues in subsequent development efforts for regional models in Atlanta and Denver. These latest designs were based on systematic analysis of the earlier designs and ways to achieve all five features mentioned above as well as a better integrity and balance between model system components. In the model system design, there always has been a question, what is the better behavioural unit that represents a decision maker for trip (or tour) generation stage – household or person. Conventional travel demand models are mostly household-based while the contemporary activity-based models tend to be person-based. The choice of the decision-making unit is less crucial if simple statistical models are applied that link person/household characteristics to the number of generated trips/tours. The use of MCSM with activity-based models enables more detailed segmentation by household and person types, and is much more sensitive to the choice of the decision-making unit. Additionally, since ensuring proper integration is one of the main challenges, it is important to find a right balance and linkage between household and person dimensions. Fortunately, MCSM allows this to be resolved by using the household for some choice dimensions and person for other dimensions. MCSM also makes it easier to incorporate explicit intra-household interactions of various types.

The latest integrated model system design places every activity into one of four household generation/participation categories: (i) *Individual*: tours for individual activities are generated and scheduled at the person level (with possible dependence on household variables, but without direct coordination of choices). The frequency of these activities is modelled for each person either as a part of the daily activity/travel pattern (as designed for Atlanta and Denver), or by means of a frequency choice model; (ii) *Coordinated*: coordinated activities are implemented by individuals within the household, but a mechanism is in force within the household to coordinate their generation and scheduling with those of other household members; (iii) *Allocated*: some activities are generated at the entire-household level because they reflect the collective household needs. However, they are assigned to individuals for scheduling and implementation. and (iv) *Joint*: some activities are generated at the entire-household level and implemented by several household members travelling together (and frequently sharing the same activity). For these types of tours, a tour-frequency model is used for the household, followed by a person participation model that is applied for each generated tour. Activities are also grouped into three main purpose categories: *mandatory*_(including going to work, university, or school), *maintenance* (including shopping, banking, visiting doctor, etc), and *discretionary* (including social and recreational activities, eating out, etc). Table 18.2 summarizes the main assumptions made regarding the possible combinations of activity generation/participation and purpose categories. Only five out of the twelve possible combinations are allowed, which greatly simplifies the modelling system, while preserving behavioural realism and covering most of the observed cases.

Table 18.2
Modelled Activity-Travel Purpose and Generation/Participation Categories

Purpose	Generation/Participation			
	Individual	Coordinated	Allocated	Joint
Mandatory		X		
Maintenance			X	X
Discretionary	X			X

Travel for mandatory activities is assumed to be coordinated. Participation, location, and scheduling of the activities are coordinated to some degree among household members, but the resulting mandatory tours are implemented individually. This allows for the possibility that household members may tend to coordinate schedules so that they go to work (or school), or stay home, on the same days, and coordinate their work hours. Maintenance activities may be either allocated or joint. It is assumed that the maintenance function is inherently household-based, even if it is implemented individually or related to a need of a particular household member, like visiting doctor. Even in these cases, maintenance activities are characterized by a significant degree of intra-household coordination, substitution, and possibly sharing. Discretionary activities may be either individual or joint. It is assumed that these activities are not allocated to household members since they do not directly relate to household needs. Thus, these activities are either planned and implemented together by several household members, or are planned and implemented individually. It is assumed that all else being equal, there is a predetermined structure of priorities in the activity generation and scheduling procedure along both dimensions. Mandatory activities take precedence over maintenance activities, while maintenance activities take precedence over discretionary activities. Joint activities are considered superior to allocated activities, while allocated activities are in turn considered superior to individual activities. Combination of these two priority principles yields the following order of generation and scheduling activities that serves as the main modelling skeleton for the model system design: coordinated mandatory activities, joint maintenance activities, joint discretionary activities, allocated maintenance activities, and individual discretionary activities. Stages 2, 3 and 4 are partially combined, with the generation of joint and allocated tours done simultaneously, and the person-participation in those tours done sequentially in the noted stages.

These assumptions form the basis of a practical model system hierarchy (see Figure 18.1) attaining a new level of intra-person and intra-household integrity. Linkages account for the correlations observed among household members for the most basic aspects of the daily patterns (go to work or school, stay at home, or have a day-off for a major out-of-home non-mandatory activity). To do this, the individual daily pattern is decomposed into two main parts. The first part—pattern type (level 1 of Figure 18.1)—is modelled for all household members taking into account long term choices (level 0) and intra-household interactions. Then the pattern details (level 3) are modelled for each person, given the pattern types of all household members. Joint and allocated activity tours are modelled explicitly. Activity and tour generation occurs in joint household models. Then, participation in the joint tours is modelled for each person, and allocated tours are assigned to persons. Essentially, in this structure, important aspects of the individual daily pattern emerge as the result of the numerous intra-household participation and allocation mechanisms (level 2), and the individual incorporates them into a complete pattern (in levels 3, 4 and 5), along with their mandatory activity, if applicable (from level 1) and their individual discretionary activities.

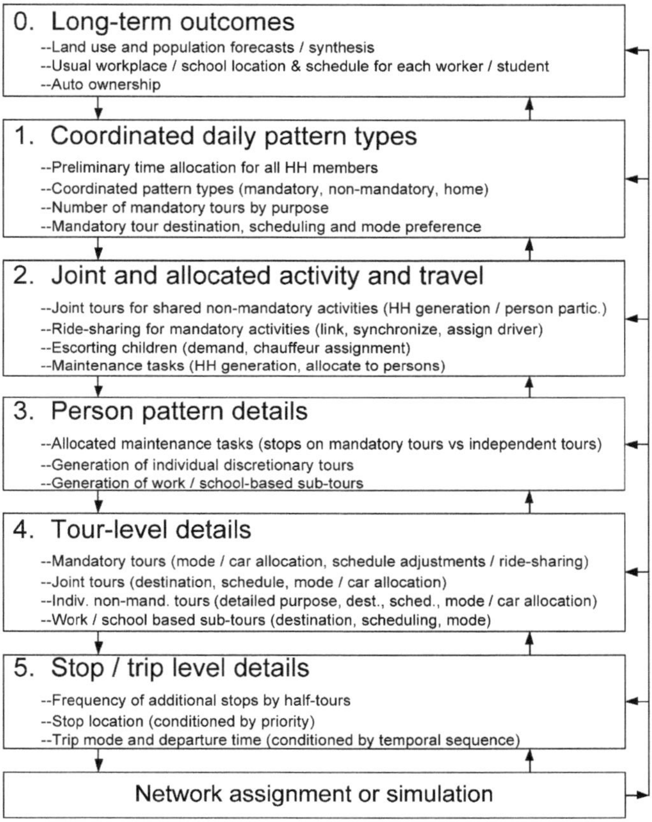

Figure 18.1

Household Activity and Travel Model Hierarchy for Model Systems in Development

Downward integrity is achieved in this structure by adhering to constraints at lower levels that are imposed by outcomes of higher levels. Enhanced resolution of the time-of-day (TOD) choice models allows for explicit tracking of time-use attributes (time windows available for implementing activities and travel tours) for each person at each stage of the tour generation and scheduling procedure. In particular, it has proven to be beneficial to model time-of-day choice for mandatory activities (that normally take the biggest share of the daily time budget) first and then condition the further generation and scheduling of non-mandatory activities on the size of the residual time windows left after the mandatory activities have been scheduled. In addition to the intra-person integrity achieved through the use of coordination in level 1 and household models in level 2, the individual models of levels 3-5 are implemented in a prescribed person-priority sequence so that, within each level, the models for lower priority persons are conditioned by the outcomes of higher priority persons. Design issues have been identified that require further research as model development proceeds. Several of the most important issues are discussed below.

Upward Integrity. Discrete choice modellers have traditionally used the log-sum (expected maximum utility over the lower-level choices) technique to "inform" the upper-level choices about what can happen down the hierarchy. This technique can be used in the MCSM framework as well, but it is extremely intensive computationally when it comes to calculation of tour mode choice log-sums for destination choice (takes more than 60% of running time of the model system) and it is not feasible to use full logsums (across all destinations, modes and TOD periods for all tours and stops) in the daily activity pattern model. Possible simpler approaches include the use of substitute variables, such as composite travel impedance variables, or partial logsums based on key dimensions of the conditional models. Another possible solution to improve upward integrity would be to exploit the overall iterative framework of the model application and use generated lower-level outcomes from the previous iteration as variables in the upper-level choices at the next iteration. A summary measure of the simulated outcome is its utility, which could be used instead of using the expected maximum utility arising from all possible outcomes. This approach can be interpreted as "learning process". The time-use framework might also be effectively used in this iterative procedure. Instead of feeding-back computationally intensive but actually quite abstract log-sums contracted over multiple choice dimensions, a simple variable—representing total travel time spent by the individual to realize the activity pattern in time and space—could be fed back and considered at the next iteration for a choice of the new daily pattern. To make the upper-level choice sensitive to the total expected travel time a continuous time allocation model (with travel budget as input variable) could be applied first. Then the daily pattern type and the subsequent chain of choices could be made conditional upon the expected time allocation. With this technically simple approach, the whole model chain would be sensitive to network improvements since these improvements are finally expressed in time savings.

Intra-Person Integrity. Further research is also needed to better understand the interrelationship between activity generation and scheduling stages, and their positioning in the model system hierarchy. Similarly, relationships should be further explored between such dimensions as activity locations/durations and tour configuration, in terms of a distribution of activity episodes by tours. Also, possible substitution between in-home and out-of-home (travel) activities can be considered as a part of the downward vertical integrity issue.

Intra-Household Integrity. Some ideas for refining this approach further include explicit modelling and linking of activities to pick-up or drop-off household members with the activity schedule of the person who is picked up or dropped off, and predicting the activity pattern type of all household members simultaneously, obviating the need to assume a fixed hierarchy of interdependence across person types.

Temporal Disaggregation. For the Atlanta model system we are testing further enhancements, introducing various types of "time pressure" variables to ensure that the activity scheduling and activity generation models are as consistent as possible, i.e., those that participate in more activities will tend to participate in each activity for a shorter duration, and vice versa. We may also test reducing the temporal resolution from 1 hour to, say, 30 min. Since the discrete choice time-of-day/duration models mimic continuous duration models in using mostly pseudo-continuous independent variables, one can change the duration of the periods without substantially changing the specification of the model.

Population Synthesis for MCSM. The different population synthesis approaches have never been evaluated and compared to determine which control variables are most necessary and which non-controlled variables are synthesized reliably enough to include in the models that produce forecasts (Bowman, 2004). We are currently carrying out such tests and comparisons for the Atlanta region, including an attempt to backcast to match 1990 Census distributions.

Spatial Disaggregation. Two basic approaches have been tried for spatial disaggregation. Portland uses the link face as the spatial unit. In Atlanta, a decision has been made to use grid cells of 200 meters square. An attraction of the grid cell approach is that the land use data becomes independent of the definition of the networks and the zone system. One can adjust the networks and zone system over time without having to redefine the land use variables each time. Furthermore, each time the zone system and networks are made more detailed, the model system will already be capable of locating the trip ends in the more detailed system. These are not the only approaches possible for disaggregation. Another proposal has been to use hexagonal or other geometric cells that accommodate the curvature of the earth better than square grid cells.

Acceptance of Activity-Based Travel Demand Models in Practice

Along with the current successes of the new-generation models, and the general sense that this approach represents a major breakthrough in travel demand modelling, it is also important to recognize the problematic side of these models, especially in how practitioners, planners and final decision makers may view them. It should also be noted that, to date, the amount of money and effort spent to develop and maintain conventional 4-step models in the US is, in aggregate, larger by an order of magnitude than the amount spent on the development of activity-based models. Most

MPOs are not yet investing in activity-based development, and of those that are, most invest small sums compared to the amount they continue to spend on trip-based models.

For modellers, the clear and strong advantages of the new generation of models are their behavioural realism and their ability to come closer to an understanding and modelling of individual and household behaviour. Planners will not appreciate these advantages, however, unless they see how it permits the travel demand models they use to better address their needs. Transportation planning decisions are generally based on aggregate forecasts of demand for and performance of transport facilities. In order to see the relevance and importance of micro-simulating the decisions of individual travellers, practitioners need to first understand how this new approach leads to more realistic and more policy responsive forecasts – at the aggregate level.

Modellers and researchers should accommodate the necessary pragmatism of practitioners in their assessment of travel demand models. To find a common language between the two communities and move activity-based models into practice, the advantages of the new models should be clearly translated into terms of realism of aggregate travel forecasts acceptable in transportation planning community rather than formulated in terms of "realism in understanding and modelling individual travel behaviour", as is common in the transportation research community. The key requirement to convince practitioners to adopt the new models is a demonstration of the "tangible" advantages of the new models over conventional ones in a practical context of particular types of projects or policy issues.

To summarize the current state of the acceptance in practice we will structure the subsequent discussion into four inter-related topics: (i) Objective theoretical advantages of activity/tour-based models that need to be better explained to practitioners, (ii) "Tangible" practical advantages of activity/tour-based models that need to be communicated more actively, (iii) Concerns that stem from misunderstanding and mistrust of model complexity by practitioners, and (iv) Valid concerns that need to be addressed in future research

Theoretical Advantages that Need to be Better Explained

One of the reasons why acceptance of the new generation of models is still very partial is that many practitioners believe that the conventional trip-based 4-step models are quite good and in general produce reasonable results. It is important to realistically and critically re-evaluate conventional models and help practitioners understand their limitations. It should be noted that "dethronement" of the 4-step approach in many respects can be done based on its internal deficiencies even before any comparison of the outcomes to those produced by new models. In particular, two major deficiencies of the trip-based models should be demonstrated: numerous internal inconsistencies across different model outcomes, and inability to replicate the base year statistics without strong mechanical adjustments of the model parameters. Internal inconsistencies of the 4-step model include unavoidable and uncontrolled discrepancies between amount of home-based and non-home-based trips produced by and attracted to each zone, imbalanced mode shares for outbound and inbound trips to or from the same zone, and other numerous conflicting outputs. It is important to demonstrate that in many cases these discrepancies are comparable in magnitude with the marginal advantages and disadvantages of the compared transportation alternatives.

The 4-step modelling paradigm in reality proved to be inseparable from the "culture" of mechanical static adjustments that relate to almost all model components. It includes adjustment of trip generation rates to match the VMT targets, K-factors introduced into trip distribution models, adjustment of mode-specific constants to match aggregate modal shares, direct adjustments of trip tables to match traffic or transit counts, etc. It should be explained to practitioners that these adjustments only help pass a static validation of the model but may well be irrelevant in the future. If the new models are able to replicate behaviour without making so many post-modelling adjustments, this fact needs to be communicated as an indicator of their greater predictive validity.

It has been relatively easy to explain the advantages of the tour-based modelling technique in terms of consistency of mode, destination, and timing choices for all linked trips. It is more difficult to explain how the tour-based technique actually works, in part because the set of dimensions for tour modelling includes seven components (primary destination, entire-tour time of day, entire-tour mode, stop frequency, stop location, trip time of day, and trip mode), while for trip modelling, only three components (destination, time of day, and mode) are considered. However, actual visualized examples that appeal to the practical intuition are valuable. In general, practitioners respond with interest and understanding to examples of how conventional models that treat each trip separately can produce conflicting and illogical results.

Tangible Practical Advantages that Need to be Communicated More Actively

One of the primary advantages of activity-based models is a full incorporation of the time-of-day dimension as an integral part of the model system. The conventional model structure is inherently incapable of comprehensive treatment of time-of-day choice. Actually, the placement of the trip distribution by time-of-day periods in the 4-step framework has never been well established theoretically and different modellers have followed different simplified conventions regarding time-of-day choice. In some cases 4-step models have been developed for different time-of-day periods with complete segmentation from the trip generation stage. This has an advantage of using time specific level-of-service variables in trip distribution and mode choice models. However, the problematic side of this approach is that it is very difficult to make the time-of-day choice component of trip generation reasonably sensitive to network improvements and policies. A more conventional approach is to apply trip generation, trip distribution, and mode choice models in a daily fashion, while having time-of-day choice as the last model before assignment. This approach, however, is characterized by inherent problems in defining day-representative level-of-service variables for trip distribution and mode choice models that normally results in serious and unreasonable simplifications.

In 4-step models, the time-of-day distribution model normally takes a form of flat peak factors, or in the best case is sensitive to the level-of-service variables for each particular trip. Also, in all cases, a simplified trip-based modelling framework does not take into account trip timing in combination with the duration of the underlying activity. Application of models of this type may result in naï ve prediction of massive shifts of trip in some period (say trips to work in the AM peak period) as the result of growing congestion or policy measures without subsequent analysis of inevitable shift of the return trips in the PM peak period and impacts on these shifts on the other trips made during the day. The following are practical examples of important projects and policy measures that can be effectively handled by the new-generation models but cannot be handled by conventional models: (i) differential by time-of-day toll strategies and parking policies. A trip-based model will predict a modal or time shift within each period independently; thus, for example, reducing a toll in the AM period would not make a difference for the PM period. With an activity-based model, changing the AM toll would result in demand changes for both AM and PM periods; (ii) shorter workday, or changing opening and closing hours for offices or shops. A trip-based model would not be sensitive for most of these policies or in the base case would predict time shifts for trips only in the period directly affected by the policy. An activity-based model incorporating activity duration as a part of the tour time-of-day choice would be able to predict numerous derived effects like consistent change of departure and arrival hours and rescheduling of the whole daily pattern with the subsequent implications for congestion in the transportation network.

A constructive discussion normally arises around the common over-sensitivity of the conventional models (a long-standing criticism) that may be well attributed to ignoring linkages across trips within the same tour. In this regard, the argument that an activity-based model has the tendency to exhibit a reasonable conservatism, compared to a conventional model, is generally well accepted. A favourable response is also shown to the incorporation of intra-household interactions in the model. High-Occupancy-Vehicle (HOV) facilities and differential-by-occupancy toll facilities are commonly a major focus of transport planning in US. Thus, the explicit modelling of joint travel—that is believed to make forecasts for such projects more realistic—may be presented as a clear advantage of the activity-based models. There is a distinct discrepancy between the conventional planning approach, focused on inter-household work HOV travel, and the reality that nearly 75% of HOV travel is intra-household and carried out for non-mandatory purposes (Vovsha *et al.*, 2003). Conventional models treat HOV as "mode" making very crude assumptions regarding its availability to each individual traveller. As a result, forecasts for HOV facilities attributable exclusively to the level-of-service variables as well as sensitivity to various toll strategies are often significantly over-predicted since they do not consider the intra-household constraints on carpooling. The new generation models can successfully capture intra-household HOV travel and, in doing so, may reorient the discussion of HOV travel and facilities in a more productive direction.

Another important practical advantage of activity-based models to conventional models is a better sensitivity to structural demographic changes that can produce a significant difference for long-term forecasts as well as for short-term policies that are targeting particular population slices. Conventional models applied in a fractional-probability fashion are very limited in terms of the population segmentation, especially at the trip distribution and mode choice stages. They normally include only 3-4 income groups and 3-4 car ownership groups. The trip generation stage can also incorporate 5-6 household size categories and 2-3 number-of-workers categories. Activity-based models applied with MCSM are virtually unlimited in the number of population segments. In particular, they incorporate person type attributes (worker status, age, gender) and household composition types (presence of workers in combination with children of different age groups) that can have significant impacts on travel behaviour. The recent sensitivity analysis implemented with the New York model has shown that changing proportion between full-time and part-time workers in favour of full-time workers can add up to 10% of traffic and transit ridership to the CBD area since full-time workers not only implement work trips more frequently but also have longer distances and travel to a different spatial cluster of jobs compared to part-time workers. Also, workers with children and (especially) preschool children are characterized by significantly shorter trip distances for maintenance and discretionary purposes that can result in about 5% of the daily VMT corrections if properly accounted. Conventional models cannot incorporate these types of effects.

Concerns that Stem from Misunderstanding and Mistrust of Model Complexity

Some practitioners have voiced a scepticism about the complexity of the model cascade, seeing in it more of an opportunity to introduce new errors, as well as the possibility of "compounding of errors", rather than yielding additional accuracy. As a part of the response to this concern, it is important to demonstrate the real magnitude of hidden aggregation biases pertinent to conventional models, and to explain how these biases can be eliminated in the new model framework, using real numerical examples. It is important to confront the widely spread belief that "simpler is better" or "less complex is more robust". As mentioned above, non-linear models can produce erroneous results when the input variables are aggregated. From this point of view, any plausible assumption about the distribution of the input variables will work better than the average value. Examples of frequently applied aggregations in the conventional models that can produce huge aggregation biases include the following list:

- Using average zonal walk distance to transit in a mode choice model, because transit share is extremely sensitive to walk accessibility, with a very steep sensitivity within the range from 0 to 1 mile (a distance that is contained in many traffic zones).
- Using average zonal parking cost in a mode choice model, because the average value falls on the steep part of the choice curve while the typical bi-value (low-high) distribution would result in a relatively low response.
- Using a single child category without segmenting by the age of the child in the trip generation and other models because the more realistic three different groups of children – preschool, school children of pre-driving age, and school children of driving age– have distinctly different travel behaviour, as well as different impacts on the travel behaviour of the household adult members.
- Using a single home-based-work trip purpose without segmentation by income group, because workers of different income groups are characterized by very different commute distance and spatial structure of jobs. Mixing different income groups in one aggregate trip distribution model produces unrealistic structure of spatial interaction, an outcome that is typically adjusted away in 4-step models by the introduction of K-factors.

The variability of MCSM (repeated runs do not match exactly) is still perceived by many as a drawback that complicates the comparison and unambiguous ranking of transportation alternatives. It is important to introduce into the planning culture an acceptance of handling the probabilistic outcomes of the travel demand models (maximum and minimum values along with averages), and to provide guidance on how to constructively exploit variability of MCSM in order to support the decision-making procedures. Many practitioners point out that the newer models may not have obvious advantages over conventional ones in terms of replication of traffic counts or other observed statistics for the base year. Moreover, in many respects it is easier to adjust a conventional travel demand model to fit traffic counts, because aggregate adjustments can be naturally incorporated into the aggregate model structure. It is important to understand that the replication of base-year observed data does not imply that a model can produce accurate travel forecasts for future and changed conditions. These two properties of the model are not necessarily parallel. Static validation and adjustments have very weak relationship to the dynamic validation. The main reason of the fully-disaggregate modelling of individuals is <u>not</u> that we hope to predict exactly the behaviour of each and every person. Rather, the aim is to ensure realistic aggregate sensitivity of the model to changing transportation and land-use environment, which cannot be adequately achieved by modelling directly at the aggregate level. Gradual transition from conventional models with parallel development and comparison is possible. It has been recognized that it would be beneficial to develop a conventional model and a new activity-based model in parallel, for the same region (based on the same surveys and other data sources) in order to compare them in various applications. Indeed, one of the most important things to be done to prove a practical superiority of the new models is to demonstrate their performance against 4-step models for real projects.

This type of comparison is planned in the framework of the Atlanta and Denver model improvement projects, pending the availability of needed funds, where the existing conventional models are being maintained and enhanced for several years, along with the parallel development of new activity-based models. Contrary to the prevailing opinion that switching to a new-generation model would require the agency to "throw out" the existing conventional model and "jump" into a multi-year development process with a great deal of uncertainty, the new and conventional models can co-exist for a certain period of time with gradual replacement of the conventional model components by the new ones. In particular, the following two-stage transition can be considered:

1. Replacement of the trip-generation and time-of-day models with a daily activity pattern model.
2. Replacement of the trip distribution and mode choice models with tour-based models of mode and destination choice, as well as the corresponding adjustment of the network processing procedures.

Many practitioners think that activity-based models are characterized by specific estimation requirements and cannot be supported by travel surveys not specifically designed for the activity-based approach. Actually, only two specific new aspects have been added to the survey format – explicit recording of joint activities and more systematic reporting of in-home activities. The core structure of the household travel surveys is equally suitable for estimation of conventional and activity-based models. Also the traditional size and scope of household interview surveys (5,000-10,000 households surveyed during one or two days) proved quite sufficient to support the model structures described above.

Concerns that Need to be Addressed in Future Research

Although many of the concerns and scepticism involved in moving the new generation of travel demand models into practice can be addressed by better explanation and practical demonstration of the advantages of the new models, there are a number of fundamental issues that relate to some theoretically unresolved problems. The following issues, in our view, can be classified as valid concerns that need to be addressed by the research community in order to accelerate the widespread application of activity-based models in practice. First, the complexity of activity-based models and the larger number of interacting model components makes it difficult to trace the sensitivity of the model to input factors in an analytical sense. In our experience with sensitivity tests with model systems for San Francisco, New York and Columbus, in many cases it was difficult to distinguish between program "bugs", Monte-Carlo variability, and valid model system responses until numerous tests were implemented. This greater complexity may be an inevitable price to pay for behavioural realism, since travel behaviour cannot be described exhaustively by a small number of analytical formulas. However, more work can be done in order to better understand and describe the output of the activity-based model system framework from the analytical point of view, including estimation of the Monte-Carlo "clouds" for statistics of interest.

Second, one of the important theoretical achievements associated with aggregate models is a closed and elegant theory of the network equilibrium in combination with logit-based (or entropy-based) demand models. This theory guarantees a unique stationary point for the equilibrium state as well as provides effective analytical methods for finding this equilibrium even for large and over-congested networks. So far, no attempts have been made to extend the theory of the network equilibrium to the activity-based models. The major theoretical problems associated with this extension relate to analytical complexity of the model chain, details of how the choice models are applied in micro-simulation, and Monte-Carlo variability. However, a closer look at these complications may show that none of them is essentially "fatal".

Third, the purpose of a realistic description of travel behaviour and the corresponding intricate structure of decision-making have led many researches to the understanding that the analytical framework of the activity-based models should be extended to incorporate various non-compensatory decision rules and mechanisms. The micro-simulation framework opens a way to explicitly model interactions between participating agents (persons, households, firm) on the individual basis and "skim" aggregate behaviour patterns without the explicit analytical formulation of the closed choice models. This concept proves very attractive and has also produced numerous (currently academic) attempts to formulate simulation models with numerous heuristic components and rules (Cascetta and Biggiero, 1997; Arentze and Timmermans, 2000; Peterson *et al.*, 2002; Miller *et al*, 2003). Though this way may eventually be a new breakthrough into more flexible modelling paradigms it is important in our view to preserve a reasonable level of theoretical foundation (comparable to the theory of random-utility choice) before these types of constructs can be seriously considered for practical application. In particular such theoretical attributes as clear formulation of behavioural assumptions, analytical properties of the resulting model structures, and ways to statistically estimate the model parameters have to be addressed.

CONCLUSIONS

Several conclusions regarding the experience to date with activity-based modelling in the US can be drawn. There is a growing interest and an increasing number of applications of travel demand models of the new generation. These important properties of the new models include an activity-based conceptual platform, the focus on the tour as the base unit for the modelling of travel, and the use of a micro-simulation technique that operates on households and persons at the fully-disaggregate level. The new generation of models brings much stronger behavioural realism to the travel demand forecasting process, by ensuring an internal logical consistency among the various activity/travel components for each household, person, and tour. The new generation of models is characterized by crucial changes in the structure of these models, compared to conventional models. Although these new modelling structures are evolving rapidly, and are following somewhat different specific paths of developments, it is already possible to summarize the basic structural features of the new generation of models. Among them is the consistent generation of all tours and activity episodes made during a person-day, grater level of spatial and temporal resolution, and explicit incorporation of intra-household interactions, a significant and important new component that has been entirely missing in the conventional travel demand models. The key conceptual lines of the ongoing development include further disaggregation in typological, spatial, and temporal terms and better integrity in modelling activity and travel decisions for each person and entire household. The analytical structure of the new generation of activity-based tour-based models in application is fundamentally different from the conventional aggregate models. Instead of fractional-probability calculations at the level of origin-destination pairs of zones, the model is applied at the level of individual households, persons, and tours with no explicit restrictions on the number of variables or population / travel segments. The first and recent experiences of development and application of the new generation models in Portland, San Francisco, New York, and Columbus has revealed some challenging issues that should be addressed in future research. These include a better linkage between the activity generation/scheduling stage and travel simulation stage, exploration of the variability of micro-simulation, formulation of the global network equilibrium conditions, and others. It is important to effectively promote the development and application activity-based models and demonstrate their clear advantages to practical planners in a meaningful way. For the larger transportation planning community, the most compelling aspects of activity-based models may be their conceptual consistency, added policy responsiveness, and their inherent realistic conservatism. Probably the most constructive way to further moving activity-based models into practice is a parallel implementation of the new model systems and conventional 4-step model system in the same metropolitan area with the same database. This would allow for implementation of various cross-comparisons and sensitivity tests that are necessary to clearly demonstrate superiority and practical usefulness of the new generation of travel demand models to a wide community of transportation planners.

REFERENCES

Algers, S., A. Daly, P. Kjellman and S. Widlert (1995). Stockholm model system (SIMS): application. *Proc. 7th World Conf. Transp. Res.,* Vol.2, pp. 345-361. WCTR, Sydney, Australia.

Arentze, T.A. and H.J.P Timmermans (2000). *Albatross: A Learning-Based Transportation Oriented Simulation System.* EIRASS, Eindhoven.

Arentze, T.A. and H.J.P. Timmermans (2004). A learning-based transportation oriented simulation system. *Transpn. Res. B*, **38**, 613-633.

Bhat C. and S. Singh (2000). A comprehensive daily activity-travel generation model system for workers. *Transpn. Res. A*, **34**, 1-22.

Borgers A., F. Hofman and H.J.P. Timmermans (2002). Conditional choice modeling of time allocation among spouses in transport settings. *Eur. J. Transpn. & Infr. Res.*, **2**, 5-17.

Bowman, J.L. (1998). The Day Activity Schedule Approach to Travel Demand Analysis. Ph.D. Dissertation, Department of Civil and Environmental Engineering, Massachusetts Institute of Technology.

Bowman, J.L. (2004). A comparison of population synthesizers used in microsimulation models of activity and travel demand. Working paper, www.JBowman.net.

Bowman, J.L. and M. Ben-Akiva (1999). The day activity schedule approach to travel demand analysis. *Proc. 78th Ann. Meet. Transpn. Res. B.,* Washington, D.C. (CD-Rom).

Bowman, J.L. and M. Ben-Akiva (2001). Activity-based disaggregate travel demand model system with activity schedules. *Transpn. Res. A*, **35**, 1-28.

Bowman J.L., M. Bradley, Y. Shiftan, K. Lawton and M. Ben-Akiva (1999). Demonstration of an activity based model system for Portland. In: *Selected Proc. 8th World Conf. on Transpn. Res.* (H. Meersman, E. Van de Voorde and W. Winkelmans, eds.), pp. 171-184. Pergamon, Oxford.

Bradley, M., J.L. Bowman and K. Lawton (1999). A comparison of sample enumeration and stochastic microsimulation for application of tour-based and activity-based travel demand models. *Proc. Eur. Transpn. Conf.*, Cambridge, UK (CD-Rom).

Bradley, M., J.L. Bowman, Y. Shiftan, K. Lawton and M.E. Ben-Akiva (1998). A System of Activity-Based Models for Portland, Oregon. Report prepared for the Federal Highway Administration Travel Model Improvement Program. Washington, D.C.

Bradley, M., M. Outwater, N. Jonnalagadda and E. Ruiter (2001). Estimation of an activity-based micro-simulation model for San Francisco. *Proc. 80th Ann. Meet. Transpn. Res. B.,* Washington, D.C. (CD-Rom).

Cascetta, E. and. J. Biggiero (1997). Integrated models for simulating the Italian passenger transport system. In: *Proc. 8th IFAC/IFIP/IFORS Symp. on Transpn. Systems in Chania, Greece*, (M. Papageorgiou and A. Pouliezos, eds.). Vol. 1, pp. 315-321. Pergamon, Oxford.

Castiglione, J., J. Freedman and M. Bradley (2003). Systematic investigation of variability due to random simulation error in an activity-based microsimulation forecasting model. *Transpn. Res. Rec.*, **1831**, 76-88.

Daly, A.J., H.H.P. van Zwam and J.Van der Valk (1983). Application of disaggregate models for a regional transport study in The Netherlands. In: *Proc. 3rd World Conf. on Transpn. Res.*, Vol.1, pp. 363-573. VNU Publisher, Hamburg.

Ettema, D.F., T. Schwanen and H.J.P. Timmermans (2004). The effect of locational factors on task and time allocation in households. *Proc. 83rd Ann. Meet. Transpn. Res. B.*, Washington, D.C. (CD-Rom).

Fujii, S., R. Kitamura and K. Kishizawa (1999). Analysis of individuals' joint activity engagement using a model system of activity-travel behavior and time use. *Transpn. Res. Rec.*, **1676**, 11-19.

Gliebe, J. and F. Koppelman (2002). A model of joint activity participations between household members. *J. Transpn.*, **29**, 49-72.

Golob, T. and M.G. McNally (1997). A model of activity participation and travel interactions between household heads. *Transpn. Res. B*, **31**, 177-194.

Gunn, H.F., A.I.J.M Van der Hoorn and A.J. Daly (1989). Long range country-wide travel demand forecasts from models of individual choice. In: *Proc. 5th Int. Conf. on Trav. Beh. Res.*, pp. 119-137. Avebury, Aldershot.

Jonnalagadda, N., J. Freedman, W. Davidson and J. Hunt (2001). Development of a micro-simulation activity-based model for San Francisco – destination and mode choice models. *Proc. 80th Ann. Meet. Transpn. Res. B.*, Washington, D.C. (CD-Rom).

Meka, S., R. Pendyala and M. Kumara (2002). A structural equations analysis of within-household activity and time allocation between two adults. *Proc. 81st Ann. Meet. Transpn. Res. B.*, Washington, D.C. (CD-Rom).

Miller, E., M. Roorda and J. Carrasco (2003). A tour-based model of travel choice. *Proc.10th Int.Conf. on Travel Beh. Res.*, Lucerne, Switzerland (CD-Rom).

PB Consult, (2004). Progress Report for the Year 2003: Modeling task 2 – Activity/Tour-Based Models. Regional Transportation Plan Major Update Project for the Atlanta Regional Commission, Atlanta Regional Commission, Atlanta, Georgia, USA.

Petersen, E., P. Vovsha and R. Donnelly (2002). Managing competition in micro-simulation. *Proc. 81st Ann. Meet. Transpn. Res. B.*, Washington, D.C. (CD-Rom).

Rossi, T. and Y. Shiftan (1997). Tour based travel demand modeling in the US. In: *Proc. 8th IFAC/IFIP/IFORS Symp. on Transpn. Systems in Chania, Greece*, (M. Papageorgiou and A. Pouliezos, eds.). Vol. 1, pp. 409-414. Pergamon, Oxford.

Ruiter, E.R. and M. Ben-Akiva (1978). Disaggregate travel demand models for the San Francisco bay area. *Transpn. Res. Rec.*, **673**, 121-128.

Scott, D. and P. Kanaroglou (2002). An activity-episode generation model that captures interaction between household heads: development and empirical analysis. *Transpn. Res. B*, **36**, 875-896.

Shiftan, Y. (1999). A practical approach to model trip chaining. *Transpn. Res. Rec.*, **1645**, 17-23.

Simma, A. and K.W. Axhausen (2001). Within-household allocation of travel – the case of Upper Austria. *Transpn. Res. Rec.*, **1752**, 69-75.

Srinivasan, S. and C. Bhat (2004). Modeling the generation and allocation of shopping activities in a household. *Proc. 83rd Ann. Meet. Transpn. Res. B.,* Washington, D.C. (CD-Rom).

Vovsha, P. and M. Bradley (2004). A hybrid discrete choice departure time and duration model for scheduling travel tours. *Proc. 83rd Ann. Meet. Transpn. Res. B.,* Washington, D.C. (CD-Rom).

Vovsha, P., E. Petersen and R. Donnelly (2002). Micro-simulation in travel demand modeling: lessons learned from the New York Best Practice Model. *Transpn. Res. Rec.*, **1805**, 68-77.

Vovsha, P., E. Petersen and R. Donnelly (2003a). Experience, tendencies, and new dimensions in the application of activity-based demand modeling systems for Metropolitan Planning Organizations in the United States. *Proc. 10th Int. Conf. on Travel Behavior Res.*, Lucerne, Switzerland (CD-Rom).

Vovsha, P., E. Petersen and R. Donnelly (2003b). Explicit modeling of joint travel by household members: statistical evidence and applied approach. *Transpn. Res. Rec.*, **1831**, 1-10.

Vovsha, P., E. Petersen and R. Donnelly (2004a). Impact of intra-household interactions on individual daily activity-travel patterns. *Proc. 83rd Ann. Meet. Transpn. Res. B.,* Washington, D.C. (CD-Rom).

Vovsha, P., E. Petersen and R. Donnelly (2004b). A model for allocation of maintenance activities to the household members. *Proc. 83rd Ann. Meet. Transpn. Res. B.,* Washington, D.C. (CD-Rom).

Wen, C. and F. Koppelman (1999). Integrated model system of stop generation and tour formation for analysis of activity and travel patterns. *Transpn. Res. Rec.*, **1676**, 136-144.

Wen, C. and F. Koppelman (2000). A conceptual and methodological framework for the generation of activity-travel patterns. *J. of Transpn.*, **27**, 5-23.

Zhang, J. and A. Fujiwara (2004). Representing heterogeneous intra-household interactions in the context of time allocation. *Proc. 83rd Ann. Meet. Transpn. Res. B.,* Washington, D.C. (CD-Rom).

Zhang, J., H.J.P. Timmermans and A.W.J. Borgers (2002). A utility-maximizing model of household time use for independent, shared and allocated activities incorporating group decision mechanism. *Transpn. Res. Rec.*, **1807**, 1-8.

Zhang, J., H.J.P. Timmermans and A.W.J. Borgers (2004). Model structure kernel for household task and time allocation incorporating household interaction and inter-activity dependency. *Proc. 83rd Ann. Meet. Transpn. Res. B.,* Washington, D.C. (CD-Rom).

19

TWO APPLICATIONS OF GIS-BASED ACTIVITY-TRAVEL SIMULATORS

Nobuaki Ohmori, The University of Tokyo, Tokyo, Japan

Noboru Harata, The University of Tokyo, Tokyo, Japan

Katsutoshi Ohta, Toyo University, Japan

INTRODUCTION

Understanding mechanisms underlying individual travel behaviour in urban space is essential for urban transportation planning and policy. Conventionally, travel is considered a demand, derived from the desire to engage in activities at certain locations. Hence, understanding the relationships between travel behaviour and daily activity engagement is effective in estimating individual and household responses to policy measures and to changes in environmental constraints. In addition, public policy should provide individuals with a greater set of options to choose from and distribute these options among the population in more equitable ways (Burns, 1979). It is therefore important to evaluate the potential of engaging in activities considering the wide variety of constraints that individuals face. Based on this viewpoint, extensive research on the feasibility of engaging in activities given spatio-temporal constraints has been conducted. This stream of research has been referred to as space-time accessibility (e.g., Lenntorp, 1978; Burns, 1979; Kwan, 1998; Miller, 1999). Especially in recent years, methodologies to analyze travel behaviour on the temporal dimension are necessary for evaluating the impact of Transportation Demand Management (TDM) policies and Intelligent Transport Systems (ITS) technologies. In addition, travellers can acquire information not only on transport systems but also on activity opportunities both pre-trip and en-route from various media such as TV, websites and car-navigation systems. Therefore, travellers

Progress in Activity-Based Analysis edited by H. Timmermans

can make decisions about activity-travel schedules more dynamically than before. There has been research on investigating scheduling behaviour using data collected by computer software such as "MAGIC" (Ettema *et al.*, 1994), "CHASE" (Doherty and Miller, 2000) and "React!" (Lee and McNally, 2001), which were developed to trace the process of activity scheduling. Zhou and Golledge (2004) developed a real-time tracking system of activity scheduling/schedule execution on the platform of Personal Digital Assistants (PDA).

Geographic Information Systems (GIS) are a very effective tool for the management, analysis and representation of spatial elements of the transportation network and individual travel patterns. In recent years, the development of software and progress in database preparation have contributed to a wider use of GIS in transportation (Thill, 2000; Dueker and Ton, 2001). GIS has also been used for studies on activity-travel patterns under spatio-temporal constraints (e.g., Segawa and Sadahiro, 1995; Kwan, 1998; Miller, 1999; Weber and Kwan, 2002; Kim and Kwan, 2003). In this context, the ability to visually represent travel patterns in GIS is considered very useful for understanding travel behaviour and responses to policy options and to changes in environmental constraints, not only for practitioners involved with transportation planning and policy, but also for students studying travel behaviour. Moreover, GIS can be used for travel surveys, showing the respondents information on travel patterns, the transportation network and urban environment more realistically. For example, "CHASE-GIS" (Kreitz and Doherty, 2002; Wermuth *et al.*, 2003) and "React!" use a GIS-based map for collecting activity locations. Spatio-temporal travel tracking data automatically collected with Global Positioning System (GPS) or Global System for Mobile Communication (GSM) can be also used to represent travel routes and speed profiles. As seen more recently, the three-dimensional visualization of activity-travel patterns using 3D-GIS (Kwan, 2003) can be more powerful to represent individual movement in time and space, which potentially improves our intuitive understanding of activity-travel behaviour.

This paper presents the development and applications of GIS-based activity-travel simulators. One application was developed especially for the purpose of instructing students in understanding the theory of space-time prisms/accessibility and travel behaviour under spatio-temporal constraints. It was used in a graduate course as an educational tool (SMAP for Education: SMAP-E) (Ohmori *et al.*, 2003). The other application was developed as a decision-support system for activity planning using interactive surveys to collect information about the activity scheduling process of tourists' leisure activities (SMAP for Leisure: SMAP-L). Both systems were programmed with MapBasic software by customizing MapInfo GIS. In both systems, time use during a day was represented on a timeline and the spatial component of travel pattern was represented on a map. In addition, the feasibility of alternative activity-travel patterns was tested, based on space-time prism constraints. A GIS database of the transportation network and activity opportunities with some attributes was used in these systems.

GIS-BASED ACTIVITY-TRAVEL SIMULATORS

More than twenty years ago, the Household Activity-Travel Simulator (HATS) developed at the Transport Studies Unit (TSU) at Oxford University was used very successfully in trying to better understand household travel decisions and the constraints within which those decisions are made (Jones *et al.*, 1983). On the HATS game board, spatial components of activity-travel patterns, the transport network and activity opportunities were represented on a map and its temporal components were represented on a timeline, using information from the activity diaries of all members of a household. When a policy measure was introduced, household members discussed and considered the changes in the constraints on the game board and new activity-travel patterns were simulated on this board. The gaming simulation approach was especially interested in inter-personal linkages and constraints, gaining more realistic responses than simply asking hypothetical questions. This tool had originally been developed for policy evaluation but was also used by practitioners and students to better understand the interrelationships between activity and travel behaviour (Jones, 1982). Following the TSU study, many other studies have adopted gaming simulation or interactive stated response survey methods (e.g., Phifer *et al.*, 1980; Burnett and Hanson, 1982; Achmed *et al.*, 1995; Lee-Gosselin and Turrentine, 1997).

It is very important to specify the choice set of feasible activity-travel patterns more realistically. The space-time prism constraint formalized by Hägerstrand (1970) is a very useful concept for specifying alternative feasible activity-travel patterns in space-time. Burns (1979) developed the concept of space-time accessibility. Lenntorp (1978) operationalized Hägerstrand's approach by developing the PESASP model that calculated the total number of possible space-time paths given a specific activity programme, the urban transport network and urban environment, and a set of constraints. CARLA, which was developed at TSU, is a similar model to generate alternative activity patterns, based on combinatorial algorithms (Jones *et al.*, 1983). Recent studies operationalized space-time accessibility using the concept of space-time prism on a real world transport network using GIS (Kwan, 1998; Miller, 1999; Weber and Kwan, 2002; Kim and Kwan, 2003). There has been some research to explore the possibility of developing integrated models of activity-travel patterns and GIS. GISICAS (Golledge *et al.*, 1994; Kwan, 1997) examined the integration of GIS and computational process model (CPM) for activity scheduling under space-time constraints. SMART (Stopher *et al.*, 1996) is a proposal for an activity-based travel model to operate in GIS that was assumed to integrate household activities, land use patterns, traffic flows and regional demographics.

The authors developed the GIS-based gaming simulation system, SMAP, integrating GIS and an activity-travel pattern generation model (Ohmori *et al.*, 2003). MapInfo GIS software was used as the system platform. MapBasic programming software was used for customizing MapInfo.

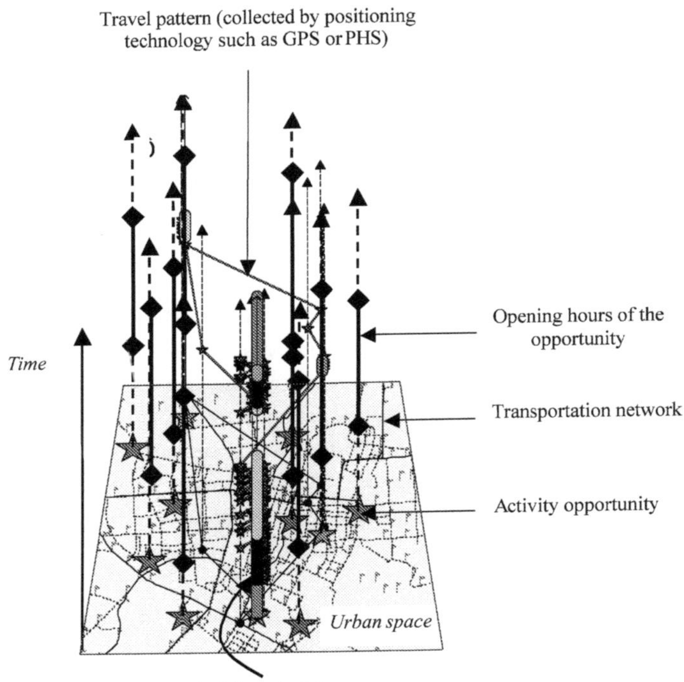

Figure 19.1
Concept of GIS Data Use in GIS-Based Activity-Travel Simulator (SMAP)

Users could operate the system by clicking the added menu bars in MapInfo. In HATS, transparent sheets were used and layered for drawing activity schedules, travel patterns, and the transportation network and opportunities (Jones, 1982). Each GIS layer in SMAP substitutes this function. Figure 19.1 shows the concept of GIS data use in SMAP for the spatio-temporal dimension. The activity-travel pattern generation model could enumerate feasible activity-travel patterns under spatio-temporal constraints of the individual's scheduled activities and opening hours of activity opportunities. When activity schedules of two individuals are used, the model can also in addition to personal transport modes (walk, bicycle, car-driver and public transport) explicitly deal with the availability of the car-passenger mode as a function of car-driver's schedule constraints. The series of system operations were logged and saved in a text file, which made it possible for the researcher to post-analyse the simulation results, including respondents' answers to the questions. The initial application of SMAP aimed at better understanding constraints that affect travel behaviour of elderly households and their responses to changes in the constraints in a local city in Japan. In particular, showing their one-week activity-travel patterns and the places they visited using the GIS was very useful for the respondents to reflect on their lifestyle.

Based on this original SMAP, two applications of GIS-based activity-travel simulator were developed. Figure 19.2 shows the basic structure of the GIS-based activity travel simulator. Input data consist of travel demand (activity schedule, individual/household characteristics, and travel tracking data), transportation supply (travel times between two locations calculated from transport network data) and activity opportunities (location and opening hours of activity opportunities). Using the above information, alternative activity-travel patterns are generated and/or the feasibility of the patterns is tested in the system based on the concept of space-time prism constraints. Then, spatial and temporal components of activity-travel patterns are represented using the GIS. By changing constraints, changes in alternative activity-travel patterns can be simulated. Table 19.1 shows a comparison of the three versions of SMAP. Characteristics of each SMAP and differences among the three will be explained in detail in the following section.

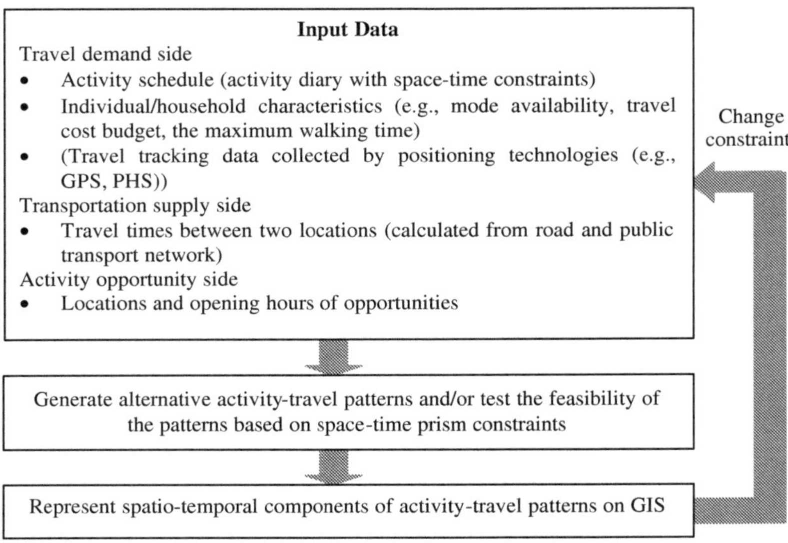

Log file of the operation process is stored in the computer.

Figure 19.2
Basic Structure of GIS-Based Activity-Travel Simulator (SMAP)

Table 19.1
Comparison of the Three Versions of SMAP

	SMAP	SMAP-E	SMAP-L
Main study purpose of the development and application	To understand the effects of spatio-temporal and inter-personal constraints on participation in out-of-home activities of elderly households	To instruct graduate students in understanding the theory of space-time prisms and activity-travel behaviour under spatio-temporal constraints	To understand the scheduling process of tourists' leisure activities
Method	Face-to-face interview	Self completion	Face-to-face interview
Input data			
(activity schedule)	One-week activity diary of all household members	One-week activity diary	Pre-planned activity schedule
(transport network)	Road network Bus network	Railway network	Road network
(activity opportunity)	Opportunities which the respondents visited during the reporting week	Opportunities which the respondents visited during the reporting week and railway stations	All the opportunities in a tourism information magazine
Study area	Local city (Akita city)	Mega city (Tokyo metropolitan area)	Local city (Awaji-shima Island)
Travel mode	Car driver Car passenger Bus with walk access/egress Taxi Bicycle Walk	Railway with walk access/egress	Car
Alternative activity-travel pattern in a prism	One out-of-home discretionary activity by different modes, times of day and days of the week, based on two individuals' prism constraints	One out-of-home discretionary activity by train at different times of day, based on an individual's prism constraint	Multiple out-of-home activities (trip chain) by car and different routes, based on a party's prism constraint

SIMULATOR OF SPACE-TIME PRISM AND ACCESSIBILITY: SMAP-E

SMAP-E is an improved version of SMAP. The improvement implies that system users can operate the model themselves. Users can simulate the potential path area (PPA), which consist of available opportunities within the available time, and the feasibility of engaging in an activity given the prism constraint. Using real activity schedules from the users' activity diary data allows users to conduct simulations of more realistic situations in an urban environment. The concept of space-time prism would be easy to understand if the travel speed is assumed to be constant across space (the shape of the PPA in this case is an ellipse and that of the potential path space (PPS) is a combination of two cones). However, in real urban space, the shape of the PPA is not that simple because it depends on the transportation network system. It may be difficult for (under)graduate students to imagine the actual PPA they face in daily life, even if they understand the space-time prism theory. For the above reasons, simulation exercises using real activity diary data is considered useful to better understand the theory of space-time prism and accessibility.

Development of SMAP-E

Since the original version of SMAP required several complicated procedures to operate, an interviewer operated it and respondents answered the questions by looking at a computer screen. In contrast, SMAP-E was developed so that system users could operate it themselves. Input data consist of three components: travel demand, transportation supply and activity opportunities. Travel demand side data (an activity-travel diary with space-time constraints of scheduled activities) are necessary for calculating the prisms. Detailed spatio-temporal travel tracking data, collected by positioning technologies such as Personal Handyphone System (PHS) (Ohmori *et al*, 2000; Asakura *et al*., 2001) are used for representing travel routes on the GIS map. Data on spatio-temporal components of the transportation network and activity opportunities are prepared. Although the only available mode is the train (with walk access/egress time) in this study, it is enough for the users, who reside in the Tokyo metropolitan area and travel mainly walking or by train, to simulate their activity-travel patterns. The attributes of the railway network include link length and the average travel speed. Minimum travel times between two stations are calculated using the railway network. Instead of preparing road network data, access and egress times are calculated as 1.3 times the straightline between the opportunity and the railway stations divided by the walking speed. Since the timetable of each railway route is not considered in the study, service hours are simply set such that all routes have the same hours. Opportunity data consist of location points and opening hours of all the activity opportunities that the system user visited during the diary survey week.

SMAP-E was designed to calculate and visualise the spatial volume of the space-time prism (PPA) and the feasibility of engaging in activities given the prism. A special C program that can be run from a menu bar of the customized MapInfo was developed. Activities are classified into three types: (a) activities fixed in space and time, (b) activities fixed in space but flexible in timing, and (c) discretionary activities (see Ohmori *et al*., 1999; Ohmori *et al*., 2003). The end time of an Activity (a) is the earliest start time of a prism, while the start time of the next Activity (a) is the latest end time of the prism. The introduction of Activity (b) enables a more realistic modelling of the PPA. That is, it allows one to model the timing of out-of-home discretionary activities adjusted to both opening hours of opportunities and activity scheduling constraints of other household members (Ohmori *et al*., 2003).

Application to the Graduate Course

SMAP-E has been applied to the course "Environmental Information Exercise in Spatial Planning and Policy" for graduate students at the University of Tokyo in the years 2001, 2002 and 2003. The objective of this exercise was to complement the lecture "Environmental Information System in Spatial Planning and Policy" and to help the students to better understand interrelationships between environmental information and individual activity patterns under space-time constraints. The numbers of students who participated in the exercise were 17 in 2001 year, 11 in 2002 year and 13 in 2003 year. About half of the students had never experienced any GIS software until the exercise.

In the first week, the students and the instructors travelled together around Tokyo by car equipped with GPS and PHS for practicing travel tracking data collection. From the second week to the third week, the students recorded their one-week activity diary in a diary survey sheet and carried the PHS device for the purpose of automatically collecting data on spatio-temporal movement. The off-line PHS system PEAMON (Asakura *et al.*, 2001) was used in the 2001 and 2003 exercises, while the on-line PHS system operated by NTT DoCoMo, Inc. (NTT DoCoMo website) was used in the 2002 exercise. Positional data were collected from morning to midnight at a specific time interval during the reporting week. In the 2002 exercise, several maps showing the participant's own travel trajectory of the day were sent to them by e-mail, after 23:00 every other day. They were asked to correct their activity diary recorded in the diary sheet, if any, by checking their travel patterns in the map.

Before using SMAP-E, the students were assigned to submit their first report titled "My one-week activity-travel patterns" in the six week. This report was based on an analysis of only the travel demand side data, i.e., activity-travel diary and PHS tracking data. Then, the students participated in a lecture on activity-travel patterns under spatio-temporal constraints. Having basic knowledge about space-time prisms and accessibility, the students conducted the simulation exercise using SMAP-E and submitted a second report titled "Understanding activity-travel patterns using SMAP-E."

The one-week activity diary and PHS spatio-temporal tracking data of all users were stored on the server. The users were able to access these data through the intranet. To operate SMAP-E, first, a user selects her/his identification number and sets constraints such as the maximum walking time. Next, s/he selects a target day from seven days and sets fixed activity scheduling constraints. As shown in Figure 19.3, time use during the day is represented separately in terms of in-home, travel and out-of-home activities, and by activity type, at the left side timeline. The travel trajectory of the day is represented as a drawn poly-line on the right side map. The C program continuously calculates the PPA for all prisms. The PPA of a selected prism can be identified on the map and simulated after changing constraints. Actually, railway stations within the PPA are displayed and colour-coded by the available time at the station to identify available opportunities. Furthermore, the user can test the feasibility of engaging in a target activity at a specific site (one of her/his visited opportunities during the reporting week) within a prism as shown in Figure 19.4.

As mentioned before, the objective of the exercise was to help graduate students to understand the important elements directly affecting space-time accessibility. The students were assigned to conduct the following simulation exercises ((1) to (8) corresponding to Figure 19.5):

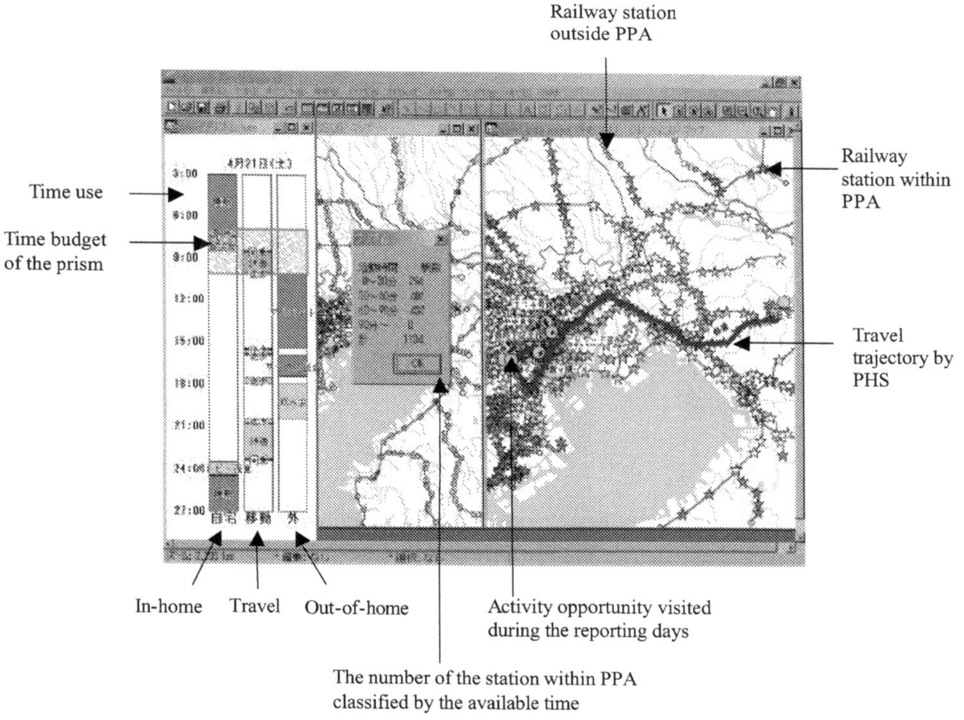

Figure 19.3
Representation of Activity-Travel Patterns and Available Opportunities Within a Space-Time Prism in SMAP-E

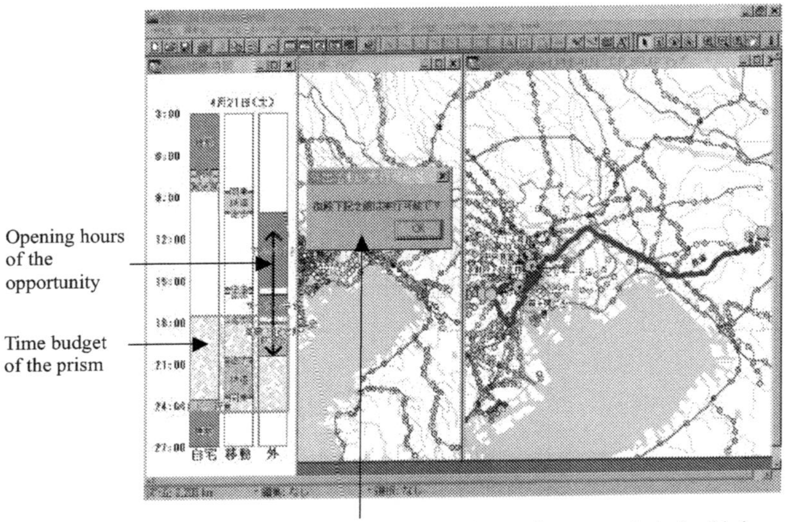

The message says, "The activity engagement at the opportunity is feasible in this prism."

Figure 19.4
Representation of the Feasibility of Engaging in the Target Activity on SMAP-E

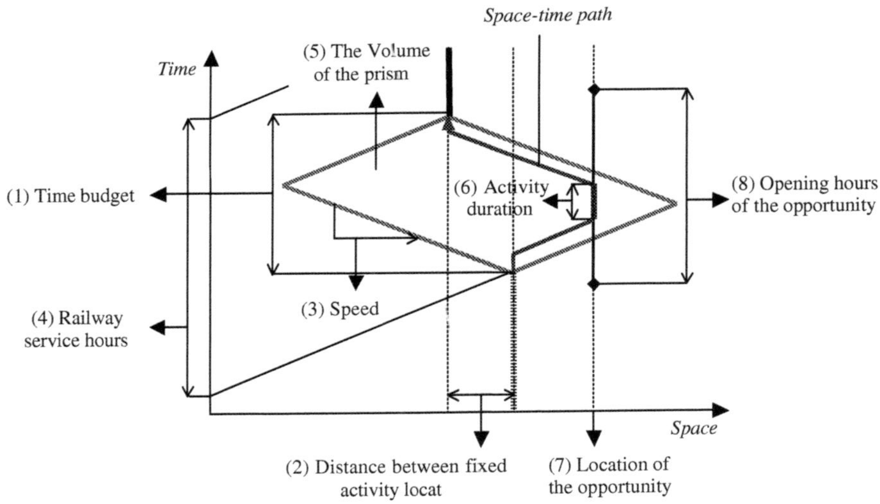

Figure 19.5
Operational Variables in the Simulation Exercises in SMAP-E

- Compare the volume of prisms (locations and the number of the railway stations in a prism, classified by the available time) at different times of the day, and simulate the volume of a prism before and after changing (1) time budget, (2) distance between fixed activity locations, (3) travel speed and (4) railway service hours.
- Simulate the feasibility of engaging in a discretionary activity in a prism before and after changing (5) the volume of a prism, (6) activity duration, (7) location of an opportunity and (8) opening hours of the opportunity.

Finally, they submitted the second report that consisted of a discussion of the simulation results and their opinion about the effectiveness and possible improvements of SMAP-E.

Analysis of the Students' Reports

In the first report, before the simulation exercises using SMAP-E, there were descriptions about their time use and travel patterns. For example, "Travel time occupied a large amount of a day", "Every day, I sleep for a very long time", "I found that my time use was not efficient", "My activity-travel patterns varied every other day", "I didn't go to any other places than the university, what a boring life!" and "It is the first time I identified the travel routes of my trips". These statements indicate that analysis of their own one-week activity-travel patterns by themselves should be effective for reflecting on their daily life. In the simulation exercise, on the other hand, the students were able to realize that spatio-temporal constraints affect their activity-travel patterns by examining not only their own activity schedule but also the influence of the transportation network and activity opportunities. The usefulness of a series of simulations using SMAP-E is discussed below.

Regarding the simulation of the volume of prisms, there were descriptions demonstrating that the students could understand the basic prism characteristics. For example, "A longer time budget made the prism larger" and "A longer distance between fixed activity locations with the same time budget made the PPA more elliptical depending on the transport network". Furthermore, the following descriptions demonstrate that they have a better understanding of the prism dynamics. For example, "Changing fixed activity locations changed the locations of potential discretionary activities" and "Suburbanization of home locations could involve the suburbanization of available discretionary activities". The visualization of space-time prisms on the GIS map could be helpful for the students' intuitive understanding of the prisms. There were also some discussions in the reports that constraints related to activity opportunities (i.e., opening hours and the minimum required activity duration) were also important in addition to as the activity schedule and the transport network. For example, "The longer opening hours of opportunities did not make the opportunity available given the very strong constraints of the activity schedule". SMAP-E has the potential of decision-support system to make daily activity-travel patterns more efficient or to assist in residential location choice decisions. Based on the students' use of SMAP-E, it can be concluded that the system could evaluate the effects of a wide variety of space-time related policy options on individual space-time accessibility. In particular, it was found to be a very effective system for simulating activity-travel patterns after changing major elements affecting space-time accessibility. However, further improvement of the system could enable users to conduct more realistic simulations.

SIMULATOR OF LEISURE ACTIVITY PLANNING: SMAP-L

Activity-travel scheduling is a necessary task when people travel and participate in activities. Especially for tourists visiting unfamiliar places, not only travel information (e.g., travel time, travel route, toll road fee, congestion point, accident point, parking place and parking fee) but also activity information (e.g., location, opening hours, entrance fees of opportunities) is useful for travelling and participating in activities efficiently. Usually, tourists can get information about leisure spots (activity opportunities) from a variety of media, such as information magazines, TV and websites, before leaving home for leisure travel. In recent years, when travelling by public transportation, travellers can obtain information about the alternative routes from websites. Car-navigation systems can also provide information on the shortest route, congestion and accident points in real-time. However, integrated information on transport systems and activity opportunities is typically not provided to travellers. Such kind of information should be helpful for leisure activity planning.

Development of SMAP-L

SMAP-L was also developed using MapInfo GIS software with MapBasic for its customization. The authors prepared a GIS database on the road network and activity opportunities in Awaji-shima Island as input to SMAP-L. The Digital Road Map (DRM), which is the standard road network data in Japan, was used for preparing database of travel times and routes between every pair of locations in Awaji-shima. Two different travel speeds were assigned to each link for respectively the highway and the expressway. The opportunity data were generated by the authors and consisted of all the activity opportunities (247 in total, including amusement parks, historic spots, restaurants, etc.) that appeared in the tourism information magazine "RURUBU Awaji-shima'02" with characteristics such as activity type, location, opening hours, entrance fee and recommended activity duration to spend at the site. Users of SMAP-L can simulate their activity-travel patterns by successively selecting destinations from the list of opportunities, and deciding on the activity duration at each destination and the money spent for travel (toll road and parking fee) and activities (entrance fee and food expenses). SMAP-L automatically searches travel times and routes (based on the minimum travel time) between two locations and represents the activity-travel pattern on a map and a timeline using GIS. SMAP-L can check time constraints such as the latest time to return home, the opening hours of the opportunities, and monetary budget constraints. In addition, SMAP-L can provide information about alternative opportunities closer to the preceding destination, alternative travel routes with good scenery, recommended activity duration at each opportunity and recommended lunch time. The above-mentioned functions of SMAP-L enable the researcher to collect data on individual dynamic decision-making processes under time and money constraints.

Figure 19.6 shows a snapshot of SMAP-L, while Figure 19.7 illustrates the concept of activity scheduling under spatio-temporal constraints in SMAP-L. As compared to SMAP-E, SMAP-L was used for the planning of trip-chain in a prism. SMAP and SMAP-E automatically produce alternative activity-travel patterns under space-time prism constraints, whereas the feasibility of the activity pattern in SMAP-L is tested after making the whole activity schedule.

Application to One-Day Leisure Activity Planning

SMAP-L was applied to one-day leisure activity planning in Awaji-shima Island to investigate its usefulness as a decision-support system and to examine travellers' activity scheduling behaviour. Awaji-shima Island is located at about 50 km from Osaka city and can be reached almost only by car. The main industry on the island is tourism. The authors administered a questionnaire survey in January 2003 among 33 respondents living in the Osaka metropolitan area. Among these respondents, 3 were university students, 17 were full-time workers whereas 13 were part-time job workers or housewives. The age of the respondents ranged from 20 to 60 years of age. There were 13 male and 20 female respondents. Respondents were asked about their experience with and attitude toward leisure travel, and asked to plan a one-day home-based leisure tour to Awaji-shima Island with family members or friends.

Only two of the respondents had never visited Awaji-shima Island whereas 26 had visited the island several times before the survey. Twenty-one of the respondents answered they liked planning leisure tours, but 5 among them were having trouble making the plan. Most of the respondents answered that they usually obtained information on leisure activities from magazines and websites. The information which respondents wanted when going-out for leisure activities was what type of activity could be conducted at the leisure spot, opening hours of the facility, travel times and routes from their home, weather condition and what kind of food they could have. The information that respondents wanted before starting travel was on leisure spots, whereas that during travel was on road congestion.

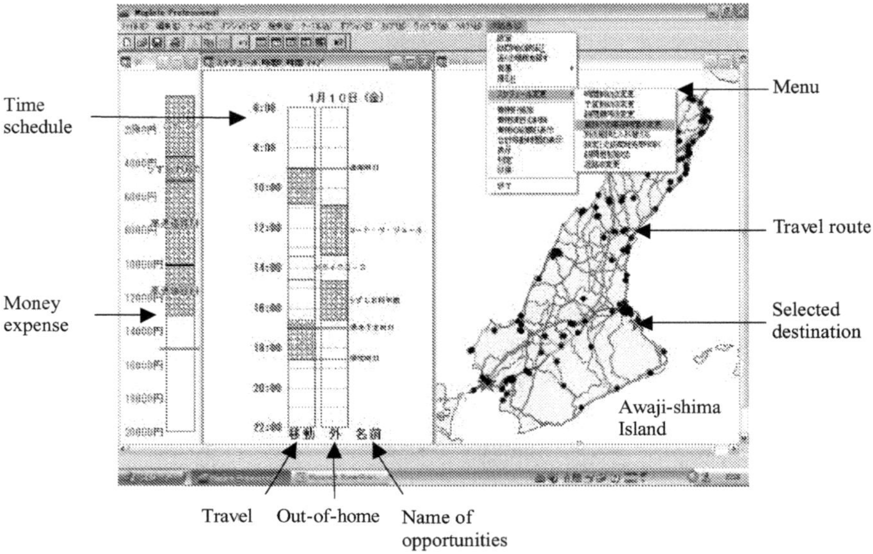

Figure 19.6
Snapshot of SMAP-L

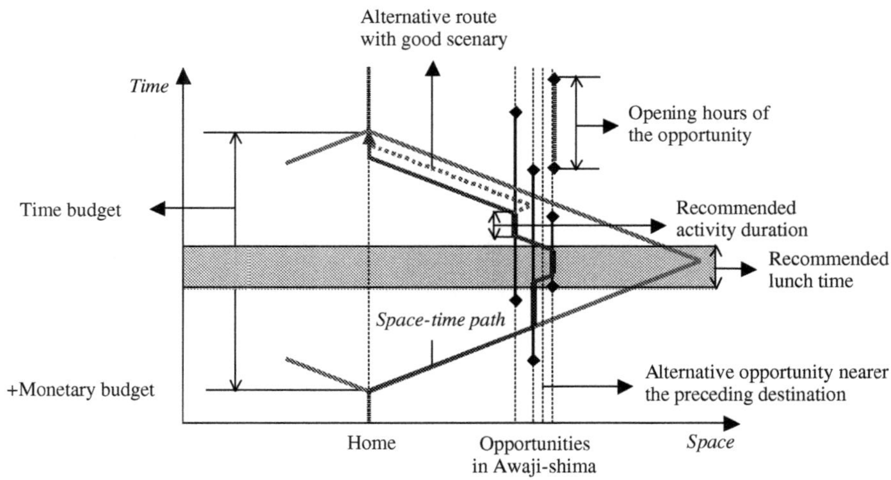

Figure 19.7
Activity Scheduling in SMAP-L

In planning a one-day leisure tour schedule to Awaji-shima Island, the respondents were asked to refer only to information in the "RURUBU Awaji-shima'02" magazine, which the authors had delivered to them in advance. Information in the magazine consisted of activity type, brief introduction, explanation of location, telephone number, opening hours and entrance fee of leisure spots and restaurants, and a map showing their location and road network. Eighty percent of respondents spent more than 20 minutes to complete the schedule. After that, an interviewer visited their home or office to conduct a face-to-face interview. The purpose of the interview was to ask the respondents in detail about their pre-planned activity schedule and to simulate the schedule using SMAP-L. It took about 60 minutes to finish an interview. The interviewer operated SMAP-L on a laptop PC. The respondents looked at the PC screen and answered a series of questions to simulate the pre-planned activity-travel schedule. First, departure and arrival time at home, monetary budget and the number of travellers were inputted as an initial setting. Then, respondents selected a destination from the list of leisure spots and decided on the activity duration at the destination. Next, travel time from her/his home to the first destination and the activity duration were represented on the timeline, and travel route was represented on the map in SMAP-L. The entrance fee at the destination, which was prepared as an attribute of the opportunity database was automatically added to the money expense indicator at the left side of the display. However, the toll road fee and other expenses such as lunch had to be manually added to the indicator by the interviewer. In the same way, until "going back home", the respondent made decisions about destinations, activity duration and money expenses successively, examining travel time, activity duration, and money expenses for travel and activities. Respondents who did not have or could not make a complete pre-planned schedule were asked to make a complete schedule with SMAP-L. The results of these sessions were stored on the laptop for post analysis. The interviews were also video-recorded.

Analysis of Activity Scheduling Processes

Analysis was carried out on the respondent's scheduling process using the data collected with SMAP-L. The analysis of the pre-planned activity schedules revealed substantial inter-personal differences in scheduling patterns as shown in Table 19.2. All the respondents planned the sequence of destinations they planned to visit in advance, but some of them did not plan or could not decide on some elements of the activity-travel schedule such as travel routes, location of having meals and departure/arrival time at home and each destination. Only 7 of the respondents took into account the maximum "monetary budget". In terms of combinations of each element, scheduling pattern 2, in which destinations, routes, meals and timing were pre-planned, was dominant. The number of respondents exhibiting scheduling pattern 4, in which only destinations were pre-planned, was not small (7 respondents).

Table 19.2
Scheduling Patterns and Elements of Pre-Planned Activity Schedules

Scheduling Pattern	Destination	Route	Meals	Timing	Money	Number of Respondents
1	X	X	X	X	X	5
2	X	X	X	X		10
3	X	X	X			8
4	X					7
5	X	X	X		X	2
6	X	X		X		1
Number of respondents	33	26	25	16	7	33

Scheduling patterns 1–4 represented 90% of the total number of patterns. It was found that 14 respondents made a plan to first visit the destination where they looked forward to visiting most, while only 3 of them made a plan to last visit there. This result may suggest that they expected the pre-planned schedule not to be completely feasible and considered it better to visit there earlier. In the face-to-face interview, when respondents were simulating their activity schedule with SMAP-L, the schedule would sometimes not be feasible because of the original time and monetary budget constraints. For example, the time arriving at home was later than the pre-planned time, or the time arriving at the restaurant was before the opening hours. In such cases, they had two alternative responses: "modify" the current schedule, or "ignore" the original pre-planned constraints.

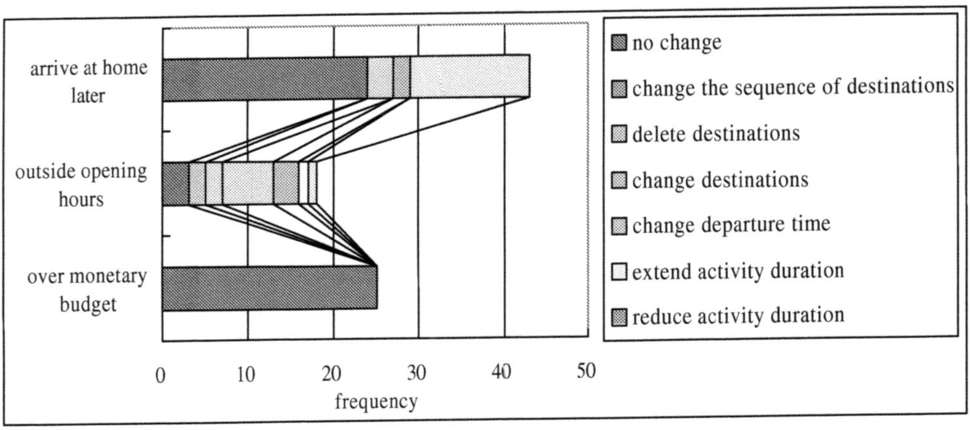

Figure 19.8
Modification of Activity Schedule

Figure 19.8 shows the responses. When arriving at home later than the latest time in which each respondent decided to return home, the most dominant response was "no change", whereas "reduce activity duration at destinations" was also chosen many times. The delay time from the pre-planned arrival time of most respondents who chose "no change" was within 1 hour. In contrast, when arriving at a destination "outside the opening hours", respondents modified the schedule by "changing destinations", or by changing time elements such as "change departure time from home", and "extend" or "reduce activity duration at the preceding opportunity". On the other hand, it appeared that even if the total money expense exceeded the limit of the pre-planned monetary budget, no one was worried about spending money for a one-day leisure tour. The results may suggest that tourists do not consider total time and monetary constraints, and flexibly and dynamically change the pre-planned schedule. However, further research needs to analyse revealed preference data of real leisure tours (e.g., Kemperman *et al.*, 2004).

CONCLUSIONS

Two applications of GIS-based activity-travel simulators were developed based on SMAP, a software tool integrating GIS and an activity-travel pattern generation model. One application was SMAP-E developed specifically for instructing students in their understanding of the theory of space-time prisms and accessibility. It was used in a graduate course as an educational tool. The participants simulated the volume of prisms and the feasibility of engaging in a discretionary activity, using their personal one-week activity diary data, before and after changing important variables concerning the activity schedule, the transport system and/or activity opportunities affecting space-time accessibility. An analysis of students' reports suggested that SMAP-E had contributed to improve their understanding of the theory of space-time prisms and human activity-travel patterns under spatio-temporal constraints. SMAP-E was also effective for participants to reflect on their daily life activities. The other application was SMAP-L, a decision-support system for activity planning using interactive surveys to collect information on the activity scheduling process of tourists. Face-to-face interviews were conducted with SMAP-L to simulate pre-planned activity scheduling of a one-day leisure tour. Experience suggest that SMAP-L is a very useful tool for supporting activity scheduling decisions, providing information on travel times, routes and opportunities, and examining time and monetary budget constraints. Data on the scheduling process collected with SMAP-L were used to analyze respondents' scheduling process in activity planning. The analyses revealed inter-personal differences in scheduling patterns and in response patterns when travellers found that the schedule was infeasible given time and monetary constraints.

GIS-based activity-travel simulators developed in this study were thus very useful both as an interactive survey tool and a decision-support system. There is a wide possibility of contributing to the further progress in activity-based analysis in the near future. First, GIS-based activity-travel simulators can contribute to the evaluation of space-time accessibility measures after introducing various space-time related policy options, such as congestion pricing and flexible working hours. Since the change of travel cost by the introduction of congestion pricing at peak periods affects travel behaviour, it would be effective for system users to investigate alternative activity-travel patterns (alternative routes, departure times, destinations and modes) with information on travel costs. In the case of flexible working hours, because the timing of work activity engagement depends on the worker's discretion, not only the volume and timing of prisms but also space-time accessibility can be changed. It would be meaningful to investigate the feasibility of engaging in discretionary activities at different working hours when deciding on the departure time to commute. Travel tracking data collected by GPS could be also used for representing detailed travel routes and speed profiles, and for estimating vehicle emissions.

Second, it is very effective to investigate individual travel patterns and encourage travellers to become involved in environmentally friendly travel patterns as demonstrated by the Travel Blending (Rose and Ampt, 2001) and Travel Feedback Programs (Taniguchi *et al.*, 2003). In the future, the GIS-based activity-travel simulator allows users to investigate alternative activity-travel patterns and to find more suitable patterns in various situations, which could encourage voluntary behavioural changes, without the help of an operator. To that effect, an object function to determine better alternatives should be a key element. For example, it is possible to introduce some object functions to minimize travel times, travel distance, fuel consumption, environmental damage, travel costs, etc., into the GIS-based activity-travel simulator. Finally, a system developed on a web-GIS for getting real-time information on travel and activities will be used as an enhanced car-navigation system with a real-time activity-travel scheduling function, incorporating variability and uncertainty in travel times and activity durations.

The GIS-based activity-travel simulator developed in this study can be classified as "constraints-based models" (Timmermans *et al.*, 2002). The models are valuable for demonstrating the potential impact of policies affecting the space-time environment on activity-travel patterns, but do have the limitation that they do not take into account that travellers will adjust/reschedule their activity patterns when faced with a changed space-time environment (Arentze and Timmermans, 2000). Incorporating explicit rescheduling models such as Aurora (Timmermans *et al.*, 2001; Joh *et al.*, 2002, 2003) or TASHA (Miller and Roorda, 2003) into the GIS-based activity-travel simulator could contribute to the further development of constraints-based models using GIS technologies. It would shift the relevance of the GIS simulator from a personal information system to a policy instrument which can be used to assess accessibility implications taking rescheduling options into account. Such a link is a challenge in its own right because these (re)scheduling models tend to be relatively complex.

ACKNOWLEDGEMENT

The work was funded by The Japan Research Center for Transport Policy and Foundation for Fusion of Science & Technology. Programming of SMAP-L mostly depended on Yoshiaki Shouji. Dr. Takuya Maruyama assisted in preparing the railway network data for the Tokyo Metropolitan Area. The Institute of Urban Transport Planning Co., Ltd. and NTT DoCoMo, Inc. assisted in using PHS location tracking systems. The authors appreciate three anonymous reviewers for useful comments.

REFERENCES

Achmed, A.K.F., S. Carapetis and M.A.P. Taylor (1995). Rural transport in Bangladesh: Impact of non-motorized transport on households' activity-travel patterns. *J. East Asia Soc. for Transpn. St.*, **1**, 893-909.

Arentze, T.A. and H.J.P. Timmermans (2000). *Albatross: A Learning-Based Transportation Oriented Simulation System*. European Institute of Retailing and Services Studies, The Hague.

Asakura, Y., A. Okamoto, A. Suzuki, Y.H. Lee and J. Tanabe (2001). Monitoring individual travel behaviour using PEAMON: a cellular phone based location positioning instrument combined with acceleration sensor. *Proc. 8th World Congress on ITS* (CD-Rom).

Burnett, P. and S. Hanson (1982). The analysis of travel as an example of complex human behaviour in spatially-constrained situations: definition and measurement issues. *Transpn. Res. A*, **16**, 87-102.

Burns, L.D. (1979). *Transportation, Temporal, and Spatial Components of Accessibility*. Lexington Books, Lexington.

Doherty, S.T. and E.J. Miller (2000). A computerized household activity scheduling survey. *Transpn.*, **27**, 75-97.

Dueker, K.J. and T. Ton (2001). GIS for transportation. In: *Handbook of Transport Modeling* (D.A. Hensher and K.J. Button, eds.), pp.253-270. Elsevier, Oxford.

Ettema, D.F., A.W.J. Borgers and H.J.P. Timmermans (1994). Using interactive computer experiments for identifying activity scheduling heuristics. *Proc. Seventh International Conference on Travel Behaviour*, Valle Nevado, Santiago, Chili.

Golledge, R.G., M.-P. Kwan and T. Gärling (1994). Computational process modeling of household travel decisions using a geographical information system. *Pap. Reg. Sc. Assoc.*, **73**, 99-117.

Hägerstrand, T. (1970). What about people in regional science?. *P. Reg. Sc. Assoc.*, **24**, 7-21.

Joh, C.-H, T.A. Arentze and H.J.P. Timmermans (2002). Modeling individual's activity-travel rescheduling heuristics. *Transpn. Res. Rec.*, **1807**, 16-26.

Joh, C.-H., T. Arentze and H.J.P. Timmermans (2003). Estimating non-linear utility functions of time use in the context of an activity schedule adaptation model. *Paper presented at the 10th International Conference on Travel Behaviour Research*, Lucerne, August 2003.

Jones, P.M. (1982). *'HATS' Educational Manual: Studying Travel in the Context of Household Activity Patterns*. Transport Studies Unit, Oxford University, Ref.193/PR.

Jones, P.M., M.C. Dix, M.I. Clarke and I.G. Heggie (1983). *UnderstandingTravel Behavior*. Gower, Aldershot.

Kemperman, A., C.-H. Joh and H.J.P. Timmermans (2004). Comparing first-time and repeat visitors activity patterns. *Tour. An.*, **8**, 159-164.

Kim, H.-M. and M.-P. Kwan (2003). Space-time accessibility measures: a geocomputational algorithm with a focus on the feasible opportunity set and possible activity duration. *J. Geo. Syst.*, **5**, 71-91.

Kreitz, M. and S.T. Doherty (2002). Spatial behavioural data: collection and use in activity scheduling models. *Transpn. Res. Rec.*, **1804**, 126-133.

Kwan, M.-P. (1997). GISICAS: an activity-based travel decision support system using a GIS-interfaced computational-process model. In: *Activity-Based Approaches to Travel Analysis* (D.F. Ettema and H.J.P. Timmermans, eds.), pp.117-134. Elsevier, New York.

Kwan, M.-P. (1998). Space-time and integral measures of individual accessibility: A comparative analysis using a point-based framework. *Geo. An.*, **30**, 191-216.

Kwan, M.-P. (2003). Geovisualization of activity-travel patterns using 3D geographical information systems. *Paper presented at the 10th International Conference on Travel Behaviour Research*, Lucerne, August 2003.

Lee, M.S. and M.G. McNally (2001). Experiments with a computerized self-administered activity survey. *Transpn. Res. Rec.*, **1752**, 91-99.

Lee-Gosselin, M.E. and T.S. Turrentine (1997). Investigating past and future choice sets of car-users, with special reference to perceptions of safety. *Paper prepared for the 8th Meeting of the International Association of Travel Behaviour Research*, Austin, September 1997.

Lenntorp, B. (1978). A time-geographic simulation model of individual activity programmes. In: *Timing Space and Spacing Time 2, Human Activity and Time Geography*. (T. Carlstein, D. Parks and N. Thrift, eds.), Arnold, London.

Miller, E.J. and Roorda, M.J. (2003) A prototype model of household activity/travel scheduling. *Proc. 83rd Ann. Meet. Transpn. Res. B.,* Washington, D.C. (CD-Rom).

Miller, H.J. (1999). Measuring space-time accessibility benefits within transportation networks: Basic theory and computational procedures. *Geo. An.*, **31**, 187-212.

NTT DoCoMo website, http://www.nttdocomo.co.jp/english/index.shtml

Ohmori, N., N. Harata and K. Ohta (2003). Development and application of GIS-based system for simulating activity-travel patterns under spatio-temporal constraints. *Proc. 8th Int. Conf. Comp. Urb. Plan. & Urb. Mngmnt*, Sendai, Japan (CD-Rom).

Ohmori, N., Y. Muromachi, N. Harata and K. Ohta (1999). A study on accessibility and going-out behaviour of aged people considering daily activity pattern. *J. East. Asia Soc. Transpn. St.*, **3**, 139-153.

Ohmori, N., Y. Muromachi, N. Harata and K. Ohta (2000). Travel behaviour data collected using GPS and PHS. *Proc. ICTTS 2000*, 851-858.

Ohmori, N., Y. Muromachi, N. Harata and K. Ohta (2003). Simulation model for activity planning (SMAP): GIS-based gaming simulation. In: *Selected Proc. 9th WCTR Conf.* (C.-H. Park *et al.*, eds.), Paper 4414. Elsevier, Oxford.

Phifer, S.P., A.J. Neve and D.T. Hartgen (1980). Family reactions to energy constraints. *Transpn. Res. Rec.*, **765**, 12-16.

Rose, G. and E. Ampt (2001). Travel Blending: an Australian travel awareness initiative. *Transpn. Res. D*, **6**, 95-110.

Segawa, S. and Y. Sadahiro (1995). A policy making support system for child care facilities using GIS. *P. & Proc. Geo. Inf. Syst. Assoc.*, 4, 59-64 (in Japanese).

Stopher, P.R., D.T. Hartgen and Y. Li (1996). SMART: simulation model for activities, resources and travel. *Transpn.*, **23**, 293-312.

Taniguchi, A., F. Hara, S. Takano, S. Kagaya and S. Fujii (2003). Psychological and behavioural effects of Travel Feedback Program for travel behavioural modification. *Proc. 82nd Ann. Meet. Transpn. Res. B.,* Washington, D.C. (CD-Rom).

Thill, J.-C. (2000). Geographic information systems for transportation in perspective. *Transpn. Res. C*, **8**, 3-12.

Timmermans, H.J.P., T.A. Arentze and C.-H. Joh (2001). Modelling the effects of anticipated time pressure on the execution of activity programs. *Transpn. Res. Rec.,***1752,** 8-15.

Timmermans, H.J.P., T.A. Arentze and C.-H. Joh (2002). Analyzing space-time behavior: new approaches to old problems. *Progr. Hum. Geo.*, **26**, 175-190.

Weber, J. and M.-P. Kwan (2002). Bringing time back in: A study on the influence of travel time variations and facility opening hours on individual accessibility. *Prof. Geo.*, **54**, 226-240.

Wermuth, M., C. Sommer and M. Kreitz (2003). Impact of new technologies in travel surveys. In: *Transport Survey Quality and Innovation* (P. Stopher and P. Jones, eds.), pp.455-481. Elsevier, Oxford.

Zhou, J. and R.G. Golledge (2004). Real-time tracking of activity scheduling/schedule execution within unified data collection framework. *Proc. 83rd Ann. Meet. Transpn. Res. B.,* Washington, D.C. (CD-Rom).

20

THE RELATION BETWEEN MOTIVES AND FREQUENCY OF TELEWORK: A QUALITATIVE STUDY FROM THE OSLO REGION ON TELEWORK AND TRANSPORT EFFECTS

Randi J. Hjorthol, Institute of Transport Economics, Oslo, Norway

INTRODUCTION

The home as a place of paid work is not new. The production of goods and services at home has occurred throughout history. In the early industrial period, a substantial part of production took place beyond the factory walls. With assembly work and small-scale production taking place in people's homes, the demand for permanent workplaces gradually became an important struggle for the labour unions. The demands for secure and healthy working conditions related to size, lighting, ventilation, and so on, have since been legally constituted. The struggle has also been related to people's right to have a permanent physical place to work. Historically, the home has not been seen as an attractive place to work.

The development of information and communication technology, the flexibilization of working life, and also to some extent environmental problems related to traffic, have made this way of organizing work topical again. Interest in home-based work can be found among both employers and employees, even at the policy level. The motives differ between these groups, however. Employers can save on office or production space; employees might get more freedom in their work situation; and planning authorities foresee traffic and environmental problems being reduced.

Progress in Activity-Based Analysis edited by H. Timmermans

Information and communication technology has made it possible to work independently of place. If suitable equipment is accessible, people can work from home, while travelling, and from almost anywhere. The nomadic worker, the telecommuter or the teleworker are names given to describe people who choose these working arrangements. In the U.S., 'telecommuting' is the term used most, probably because of the effects on transport. Telecommuting has been seen as a means of reducing car traffic and environmental problems. In England and the rest of Europe 'telework' is the more common term. Telework has been seen as a way of work being decentralised to districts, jobs being created for people with physical handicaps and work being moved to places where labour is cheap. This last point is important. Information and communication technology makes it possible to move work tasks not only to countries with low wages, but also to other places at special *times* of the day. Ulrich Beck (2000) gives a good example of how this can be arranged:

It is ten o'clock in the evening. At Berlin Airport a slick-friendly voice informs the weary passengers that their flight to Hamburg is ready for boarding. The voice belongs to Angelika B., who is sitting in front of a console in California – for after 6 p.m. Berlin time, Tegel's announcement service is provided online from California. The reasons are as simple as they are understandable: in California no extra payment has to be made for late working because it is still daytime; and indirect labour costs for the same activity are considerably lower than in Germany. Telecommunications have made this possible by removing what used to seem an inescapable part of the labour system of the industrial society: that is the need for people to work together at a certain place to produce goods and services (Beck 2000, p. 18).

Technology has provided possibilities for work to be organised independently of time and space, but the technology in itself is not always the most important condition. If something sensible is to be said about the diffusion and use of new technology, it is important to examine the social context of its potential use, the institutional framework and the economic and cultural conditions. The technology in itself is only a possibility. The social conditions are essential for its utilization. When people choose teleworking or not, it is important to know which social conditions that motivate people – are they related to work, the household and/or the individual? Statistics in Scandinavia do not indicate a significant increase in teleworking, even when the technological conditions are improving all the time (SIKA, 1998, 2001; Vilhelmson and Thulin, 2001). This fact points toward other than technological explanations, it is the social conditions, which are interesting to know more about.

In this chapter, I present results from a study on the motives behind the decision to telework or telecommute. My hypothesis is that the variations in these motives will have different impacts on the frequency on telework and as such on travel behaviour and transport. The following research questions are addressed:

- What are the motives for teleworking? Which conditions are they related to?
- How do teleworkers organize their work at home, and what is the relation between the motives and how often do they telecommute?
- How does teleworking impact the travel patterns of the teleworker?

The chapter consists of seven parts. Following this introduction, I present some previous work on the relationship between telework and transport. In section three, the different relations between information and communication technology (ICT) and transport are presented and discussed. Data and methods are introduced in section four. Motives for choosing telework are the topic in section five, and section six is about the relation between the motives and transport. Section six also deals with the impacts of telework on other family members. In section seven trends are discussed.

RESEARCH ON TELECOMMUTING/TELEWORK

Ever since the telephone was introduced about 120 years ago, the interaction between travelling and telecommunications technology has been a subject of discussion. Throughout time, the potential saving of both time and money through using the telephone rather than travelling has been emphasised. However, communication has increased exponentially as a result of both transport technology and other communication technologies. Alexander Graham Bell's first telephone call, when he asked his assistant to come to his house, is well known in this respect. The new technology immediately generated a trip. The likelihood of the 'new' ICT replacing travel was believed to be high when it was introduced. Both technology determinism and optimism were striking. It was thought that a substantial part of people's lives would be organised from their homes, and that the environmental problems created by road traffic would be reduced. Nowadays, opinions are more varied.

Within the field of transport, the discussion on the substitution of travel by electronic communication has been going on for more than thirty years. The energy crisis at the beginning of the 1970s was the start of it all (Mokhtarian, 1990). One of the first studies on telecommuting takes this as its point of departure (Nilles *et al.*, 1976). In the debate on how to reduce environmental problems generated by road traffic and to secure more sustainable development, great hope has been placed on stationary means of communication bringing about reduced daily travel by car (Batten, 1989; Capello and Gillespie, 1993; Engström and Johanson, 1996). In debates on planning policy, travelling to work being replaced by telecommuting (e.g., working remotely with the help of a computer either from the home of the employed or from neighbourhood centres) has been emphasised. Most of the debate has been concentrated on the work trip; other types of travel have been discussed but only to a very limited degree. Based on a paradigm of modernity, it is believed that new and modern technology could take over from the old, e.g. that use of information and telecommunication technology to some degree will replace physical travel, the mobile technology.

Research on the substitution of travel has been concentrated on telecommuting and often in pilot and demonstration projects (Nilles, 1991; Hamer *et al.*, 1991; Henderson *et al.*, 1996, Balepur *et al.*, 1998), but the results are not unambiguous. Although some projects have not resulted in reduced travel in total, car-use was reduced on trips to work (Nilles, 1991). In others studies, an increase in car-use was found as a result of more travel outside rush-time hours, but a reduction in the total length of car travel (Balepur *et al.*, 1998). An experiment among 30 employees at the Ministry of Transport in The Netherlands resulted in a reduction in daily trips for telecommuters (Hamer *et al.*, 1991). In a review of eight telecommuting programmes, Mokhtarian *et al.* (1995) claim that the effect of telecommuting has to be analysed in relation to *the total amount of daily travel*, i.e. not just the journey to work. For instance, when travelling to work is eliminated, efficient travel chains (e.g., a combination of different purposes) can be broken and new patterns established that are not so efficient. This might also change travel patterns in the family or the household. What these researchers also found important was that the first telecommuters were different from employees in general. They had to travel further to work than employees on average, and for them the effect of telecommuting was greater than for people with shorter distances to work.

In Norway, no direct studies of the effects of telework on transport have been conducted. Most Norwegian studies on telework have been related to impacts on the working environment, and organisational and/or juridical effects. Connected to the Norwegian national travel behaviour survey in 1997/98 an additional study of the use of ICT in private homes was performed (Hjorthol, 1999, 2002a). The results indicated no significant relation between travel patterns and use of ICT. In Sweden, national communication studies in which telework has been dealt with (SIKA, 1998, 2001) indicate that 5-7 per cent of the workforce telework (whole days, in average two times a week), and that men with higher education and income are representative of the typical teleworker.

Mokhtarian's (1998) conclusion based on the state-of-the-art of the relationship between telecommuting and travel activity is that no significant reduction in travel activity can be expected as a result of any increase in telecommuting. However, she believes that ICT will result in more flexibility in relation to everyday travel. A reduction in time travelling to work could lead for instance to more leisure travelling or shopping trips. Reduced car-use for one member of the family can lead to increased use for another.

In the long term, telecommuting and use of ICT in organising everyday activities can have an impact on land-use. For example, a reduction in the number of trips to work per week can make a long journey to work more acceptable, and people may choose to live in areas further away from city centres (and more attractive), where house prices are lower than in the more central areas of towns and cities. There is little evidence of this, however.

INTERACTION BETWEEN ICT AND TRANSPORT

Mokhtarian (1998) has suggested a classification of the interaction between ICT and transport. She believes that there are four ways in which ICT interacts with transportation:

1. *Substitution or replacement* - new technology (ICT) replace old (transport/travel purpose) without any effects on other parts of the travel pattern or on other household members.
2. *Modification* – new technology is used to conduct or change planned activities.
3. *Generation* – new technology means more information, new acquaintances and possibilities that induce more travel.
4. *Addition* – new technology comes in addition to old, and there is no specific relation between them.

In addition, there can be various combinations of these four "ideal-types". The consequences of telework can be related to these effects. Telework can substitute work trips, modify travel patterns or generate more travel. Since there is no work trip on days the employee works at home, the fourth effect will be of no interest in this context. Working at home for a whole day is work trip substituted by telework. Most research on teleworking and telecommuting stops here. The calculations show that telework reduces transport; transport is reduced in accordance with the number of days the employee works at home; and estimations are often made related to travel distance and mode of transport.

The work trip cannot be regarded separately, but in relation to other activities and tasks. The work trip is often combined with other travel purposes such as shopping, taking children to and from school and kindergarten, and so on. In Norway, nearly 40 per cent have activities in relation to the work trip (Denstadli, 2002). When the work trip ceases, routinized travel patterns are broken and new ways have to be established in reorganising daily activities. A *modification* of travel patterns through teleworking occurs that might also have an impact on the travel patterns of other members of the family, with tasks in the household having to be reorganised. For instance, if the father works at home, previously having taken the child to kindergarten on his way to work, someone else has to do this job or he must take this trip himself even though he is not going to work. The travel pattern can also be modified because the travel destinations change when people telework. People who work at home might choose to shop and have leisure activities in the vicinity on days they are not travelling to work. Teleworking can also have an influence on the working conditions of colleagues at the workplace. Telework requires more planning to coordinate work and meetings for those who do not telework and therefore modifies the travel of those people. When one member of a family teleworks, a car in the household might be free for use by other members of the family for other purposes. This *generates* new trips, which in sum can be longer than the work trip that was substituted. It has also been claimed that people who work at home will need to get out on trips they did not do before (Salomon, 1985). On the aggregate level, telework might *generate* more car traffic because there will be more space on the roads and resulting in people taking their cars into use.

In the longer term these three effects can bring about changes in the localisation of work and housing. Several researchers have foreseen a greater dispersion of urban areas (Nilles, 1991; Helling and Mokhtarian, 2001); some have even claimed the death of the city (Mitchell, 1995, 1999). The argument is that when the work trip becomes less important, people will be free to choose on the housing market and will want to live in places where they find both a good environment and low house prices. The consequences might be greater dependence on the private car because the public transport supply will diminish when work and housing are more dispersed. At the moment there is little empirical evidence of what is actually happening (Lund and Mokhtarian, 1994, Helling and Mokhtarian, 2001).

METHOD AND DATA

The sample of employees in this study was drawn from three companies in Oslo. Knowledge and ICT-based companies are concentrated in the region. Oslo and the county of Akershus between them have half the total employment of the country in cultural service industries, media, data processing, research and development, and other service industries. The composition of the workforce indicates considerable potential for teleworking among employees.

The companies contacted had previously participated in a survey about use of video-conferencing, a study which was part of the same project as this for the Ministry of Transport and Communication. All companies had some sort of telework arrangement for all or some of their employees. The final sample consisted of 16 employees from three different companies. Since I was interested in the impact of telework on transport, I selected only those who worked full days at home. The employees from the selected companies all have higher education and competence. The companies belong to what can be called knowledge and ICT- intensive industries. Since our perspective on the relation between motivation for teleworking and impact on transport has not been focused on before (as far as we know), a qualitative method was chosen to gain insight into, and an adequate understanding of, the phenomenon. Using an interview guide, the in-depth interviews were done by the author. The purpose of choosing a qualitative method was to get a better understanding of what teleworking means when seen in an everyday perspective. When paid work is drawn into the private home, interaction with the organisation of other daily activities and tasks will be different from how it is when work takes place at the usual workplace. There is likely to be an impact on the affairs of other family members.

The sample consists of seven women and nine men, most in their forties. They all have a family; either a partner with or without children or children but no partner. Four of the 16 live within the borders of the city of Oslo, while most of the others live in Akershus (the surrounding county); a few live further away. All but one of the respondents have one or more cars in the household, but most use public transport on their trips to work, mostly because of the central location of their workplaces. The respondents living outside Oslo have on average further to travel to work than workers in the region in average. Those living within the city have on average a short distance to work 3.5 km. For these workers the length of the work trip is obviously not the reason for choosing to telework. Two of the companies have written agreements, which regulate teleworking, while there is an oral agreement in the case of the third company. All respondents have good technical equipment at home with connection to the server at their company. The equipment is either free or greatly subsidised through different arrangements. For all employees the technical standard of the working tools is important. Some had tried to work from home some years previously but had found it frustrating because of "primitive" tools. Most of the respondents had a special room or a physically shielded place to work in at home. Characteristics of the informants and the frequency of teleworking are shown in Table 20.1.

Motives for Teleworking

Time Pressure

Time squeeze for employed parents, and especially the mother, has been put forward as a reason for teleworking (Baily and Kurland, 2002). An American study suggested that women were more likely to mention family benefits as a motivation for telework than men (Mokhtarian *et al.*, 1998). Studies indicate that telework helps in managing family responsibilities (Duxbury *et al.*, 1998). On the other hand, Peters *et al.* (2004) claim that there is little empirical evidence that confirms whether time flexibility improves reconciliation of work and family life.

Table 20.1
Sample Characteristics and Frequency of Teleworking

Gender	Age	Family Situation	Number of Cars in HH	Distance of Work Trip	Transport Mode	Frequency of Teleworking
M	33	Married, children 5, 13 yrs	1	Ca 35 km	Bus, some times car	2-3 times a month
M	44	Married, children 9, 13, 15 yrs	1	Ca 16 km	Bus or train	2-3 times a month
F	48	Unmarried, child 14 yrs	0	Ca 4 km	Bike most of the year	3 times a week during project work, else 1-2 times a month
F	56	Married	1	Ca 5 km	Metro or bike	4-5 times a month
M	48	Unmarried, children 17, 19 yrs	1	Ca 20 km	Bus or bike	Ca 2 times a month
F	43	Married, children 10, 13 yrs	2	Ca 30 km	Train	1- 2 times a month
M	55	Married	1	Ca 3-4 km	Tram or walk, bike	2-3 times a month
F	54	Married	1	Ca 30 km	Bus and train	1-2 times a month
F	37	Unmarried, child 4 yrs	1	2-3 km	Car to the kindergarten, then tram or walk	1-2 times a month
M	53	Married	2	Ca 60 km	Car in the summer, buss in the winter	Summer: 0 Winter: 6-8 times a month
F	36	Married, children 4, 7, 7 yrs (twins)	1	42 km	Bus or car	10-12 times a month
M	43	Married, children 15, 18 yrs	3	Ca 40 km	Express bus or bus+train	1-2 times a month
M	42	Married, children 14, 16 yrs	1	Ca 25 km	Train	1-2 times a month, often parts of a day
M	38	Married, children 4, 13 yrs	1	2-3 km (Oslo)	Metro	8-10 times a month
F	44	Married, children 8,8 yrs (twins)	2	66 km	Bike+train	3-4 times a month
M	40	Married, 3, 5, 7 yrs	1	Ca 125 km	Bike+train+metro	8 times a month

In this study, we argue that in families with small children time pressures and organising a complicated daily life teleworking can help: "… *to make ends meet with small children to and from the kindergarten, to manage the family logistics is most important for me*" said a young mother – a frequent teleworker. Historically, women have combined paid work and looking after children within their homes. Kerstin Hultén (2000) describes the situation of the home sewers at the beginning of the 1900s: ... *"the impossible equation of 10-12 hour workdays and the children having to be looked after: One sews and looks after the children, sews and does the washing, sews and has one eye on the saucepan on the cooker*" (Hultén 2000: 88 (translated from Swedish)). The modern woman in front of her PC is in another situation. The children attend school or kindergarten and housework is different. Time pressures are common to both historical woman and modern woman. It is the mothers who relate about taking children to and from kindergarten, school and other activities as a reason for working at home, and it is still the mothers, more so than the fathers, who are responsible for these tasks (Hjorthol, 2002b). In this study, women more than men mention time pressure and family life as a motive for telework, although men also refer to time pressure and family life.

Distance to Work and Urban Sprawl

Another motive related to time saving is distance to work. However, several studies have reported that travel reduction has not proved to be a very important motive for telework (Mokhtarian and Salomon, 1997; Stanek and Mokhtarian, 1998). Other studies, however, report commuting time to have positive effects on telecommuting adoption (Peters *et al.*, 2004) and several studies show that teleworkers have longer work trips than those who do not (SIKA, 1998, 2001, Olszewski and Mokhtarian, 1994). Distance to work is the most important motive for teleworking in the case of four of the respondents. Loss of job and reorganisation of job in a situation with strong social relations to the residential area made teleworking a good solution for two of them. Based on a preferred residential area, teleworking was the way to overcome distance to the job. In the other cases, families have chosen residential areas which they consider well suited for bringing up children, and long distance to work is the outcome. Teleworking makes it possible to manage the organising of everyday activities. It is a facilitator.

Whether teleworking will lead to more urban sprawl and render small towns in the outer parts of urban areas more attractive to those working in the central areas has been the subject of much discussion. At the moment there is little empirical knowledge (Salomon, 1996; Helling and Mokhtarian, 2001). In the past decade, a tendency towards re-urbanisation has been observed in the larger Norwegian cities (Bjørnskau and Hjorthol, 2003). There has been growth in the population in the bigger cities, while in small towns the population has declined (Juvkvam and Sørli, 2000). Simultaneously, commuting from counties further south of the metropolitan area of Oslo is increasing. There is no information about whether these people have teleworking arrangements. But with a greater flexibility according to such working arrangements, there might be an increase in teleworking among long-distance commuters or the other way around, i.e. an increase in long-distance commuters among those with such contracts of employment.

In Peterborough, 80 miles north of London, both housing prices and commuting have increased, and according to Peter Hall and Stephen Marshall (2002) this trend will continue. The flexible working patterns are demonstrated in the transport system. From Tuesday to Thursday the parking area near the railway station in Peterborough is full, while on Mondays and Fridays it is almost empty (Financial Times, 2003). On those days people work from their homes. For the railway companies this could be a problem if this becomes the travel pattern in the future, providing trains for the demand in the middle of the week having too much capacity on Monday and Friday.

The Double Character of the Working Life

Working life today has been characterized as "greedy" (Brandth and Kvande, 2003). The concept "greedy institutions" was introduced by R. Coser (1974) to characterize institutions based on voluntary but loyal participation and has been used later to characterize modern working life. Time pressure and achievement demands are considerable in modern working life, while at the same time employees have more freedom and flexibility than before. On the one hand, employees have the possibility to develop their own jobs. They are given interesting tasks and the management have confidence in their employees. On the other hand there are expectations of full achievement. Both paid and unpaid overtime are increasing in Norway, especially among people with higher education (Ellingsæter, 2003).

The limits between paid work and leisure time are more diffuse. Some researchers claim that the workplace is a more important arena for social contact and personal acknowledgement than the family (Hochschilde, 1997). Hochschilde states that the workplace is getting more like the home while the home is becoming more like a workplace. For her the workplace is where one's friends are and where people get their daily support. In the home or the family people are experiencing increased time pressures, negotiating about the necessary tasks that have to be done and dealing with ensuing conflicts in relation to these things. As one of our respondents says – *"Everyday life has to be planned, otherwise everything breaks down"* (Woman 44). Several of the respondents claim that the tempo and speed of an ordinary working day, with all the interruptions and enquiries, makes it necessary to work at home once or twice a month. One of the female respondents puts it like this: *"It happens that I work at home a day to recover, or a day when I'm not sick enough to be sick but not well enough to go to work"* (Woman 39). Work is also socially demanding. One has to play the role of a successful and positive colleague and it is not every day that this comes naturally. Teleworking provides the possibility of a legitimate pause. *"I don't have to put on the mask of success. I can be myself without having to perform"* (Woman 43).

Freedom and Control of Work

The freedom to organize work independently of the supervision of an employer and to be in control over the work is an aspect of teleworking that can be seen in relation to the preceding theme. Good reviews of this literature covering these aspects are found in Ellison (1999) and Baily and Kurland (2002). For the majority in this study this is an important dimension of telework. The company has confidence in the employee that he/she does the work without supervision. "*It is a fairly open agreement about what you are going to produce – you do what you have to do, but you decide for yourself when you will do it and how much time you are going to take. If you take six hours rather than eight, that is OK, if you take ten instead of eight that is also OK – it's up to you*" (Woman 48 years). This freedom also has another side, one that shows the double character of working life. The dialogue below indicates that the confidence of the company is rewarded with a significant contribution of work, perhaps more than is "deserved". Both the generosity and greed of working life is demonstrated: *"For one reason or another, the problem is that I do more when I work at home than when I am at the workplace. I get more 'addicted' to work at home. I think this is not what people believe. It is natural to think that it is easier to take breaks when you work at home, but it is not like that"* (Man 33).

Interviewer: Do you think you work too much at home?

"Yes, exactly. I become almost desk-bound. I don't know if it is conscientiousness. I feel I have to prove that I am doing something."

Interviewer: It seems that you are very efficient when you work at home?

"Yes, actually I am. As I said – I think it is my conscience. Nobody can see me at home. Then I have to prove I am here, you see?" (Man 33).

The teleworker becomes his own supervisor, and for some employees this control is stronger than that at the workplace. Self-discipline is severe.

Concentration

An important motive for teleworking in these companies is the possibility to work undisturbed and uninterrupted, especially in periods with tasks that demand concentration. The quality of the technical equipment and the connection between home and work is very good, which means that documents needed for homework are readily accessible. The employees do not have to remember to take documents home for the days they are teleworking. Even if the technology affords good accessibility when working at home, the employees experience fewer interruptions and telephone calls in the home situation.

"When I work at home I get very few calls. The others at the office usually manage the tasks without my help" (Man 33).

Two of the companies had changed their premises during the preceding year from individual (cell) offices to open landscape, and for some employees telework was the means of undisturbed work.

"I think there will be more (telework) after we come down here. Yes, I telework more often. The reason is that I have many tasks that are 'thinking jobs'. When I'm going to make models and read, I feel I can do so better at home than here at the office. I don't get interrupted and it is much quieter. But for many years I have been used to sitting in my own office, so this is very new. I think it is difficult to complete tasks as efficiently as before. When I'm sitting and thinking and somebody bumps into me, the telephone rings or someone laughs, I am disturbed and have to start all over again." (Man 53).

The combination of teleworking arrangements and landscape is one way for companies to save office space.

It is Practical to be Home when the Plumber Comes

For employees who use the teleworking arrangement only sporadically, the days working at home are often related to a practical task or errands that can readily be combined with working at home.

"It (telework) makes it more convenient if a plumber or an electrician is coming, they usually do not arrive at the time we have agreed upon. I don't lose time and the plumber can do his work while I do mine" (Man 44).

Teleworking is often related to one's child being ill or having to be accompanied somewhere.

"It is often related to whether one of the children is going to see the dentist or the doctor, or I have an errand up in Nittedal (the place she lives). Then it takes too long to travel down to town and back by train and coordinate everything" (Woman 43).

RELATION BETWEEN MOTIVES FOR TELEWORKING AND TRANSPORT

These different motives for teleworking can be grouped into five categories: *life cycle, housing-/labour market, the need for concentration, prevention of stress, practical reasons*. These five motives have different impacts on transportation, which I will discuss in relation to the concepts introduced above in section three entitled *substitution, modification* and *generation*. If the motive for working at home is related to life cycle (family obligations), and is connected with the housing market or place of abode, the number of home-based work days will be up to two to three a week. The question is whether this arrangement will last when the children are older and time pressures are less. Will mother or father wish to maintain the frequency of home-based work or will they want a less "home intensive" arrangement?

The adjustment related to the housing and/or labour market, which may result in long work trips, is dependent on preference for housing and opportunities on the labour market, both of which will vary over time. The transportation effect of this adjustment will be about the same per week as the arrangement related to life cycle, and since this may be lasting it will not necessarily change over the lifespan.

The last three motives for selecting telework arrangements are probably fairly common, but the weekly working days will be fewer than for the two previously mentioned adjustments. People using telework as protection or shelter from a demanding job, or for practical reasons, will possibly keep the frequency of working at home on the same level as our respondents, two to three times a month. The need for working undisturbed and concentrated might vary to a much larger degree. In some periods, three times a week will be appropriate; while in others there will be little need to sit at home. Telework related to these three motives takes place independently of travel distance. The impact on travel and transport of these motivated types of arrangement will depend on the willingness of companies to invest in equipment and also in their granting permission for the adjustment, on the one hand, and the employees' wish to work at distance on the other. In our sample, no more than a couple of days a week is wanted.

The majority of people working at home have their own "home-office" or at least a special place in which to work. Most try to maintain a physical division between the private and the working sphere.

Organisation of the work at home is related to the motive for teleworking. Those whose teleworking is life cycle-motivated have a well-planned and structured workday at home. Also others have planned their days, but for the former group planning is a necessary condition for managing the time budget with two employed parents. Almost everyone claims that it is more efficient to work at home than at the workplace, whether they have individual offices or open-plan landscapes. People claim that they are more effective at home and have longer working hours. Some think it is practical to combine different tasks when working at home, and have no problem doing so. For instance, many women report doing domestic work while teleworking at home. Women experience a merging of work-time and family-time. Things are different for men. Most men separate work tasks and household tasks, but very often go back to the PC in the evenings after completing household tasks. The flexibility of the teleworking arrangement is favourable for the employer, according to employees. Employees feel more efficient, they work more concentrated and they do more. But they also experience the work sometimes becoming all-encompassing. It is too easy to log on when the PC is available. To reduce the pressure it is easy to check the e-mail in the evening and prepare for the next day. Even though the work trip is eliminated, the tasks related to it usually have to be done. Children have to be taken to kindergarten or primary school, but the trip might be taken in a different way when mother or father is working at home. Shopping is often done in the local area on days working at home. This is also true for other tasks. The car is used very often for these local trips, even though most of our respondents take public transport on their way to work. Concerning transport, teleworking for one of the members of a household has no unambiguous effect for the other members of the family. In most cases this work arrangement results in more local trips, which also to a certain degree involves others in the family. The main findings of the relation between motives and transport effects are presented in Table 20.2.

Table 20.2
Relation between Motives and Transport Effects

Motives for Teleworking	**Interaction Effects between ICT and Transport**		
	Substitution/the Number of (whole) Days Working at Home	**Generation of Trips**	**Modification of Travel Patterns**
Life cycle	Up to two-three times a week	Few/some	More local trips by car
Housing-/labour market	Up to two-three times a week	Few/none	More local trips by car
Concentration/undisturbed work	Periodic, from very seldom to several times a week	None	Little
Prevention of stress	Two-three times a month	None	Little
Practical reasons	Varying, seldom	Some	More local trips by car

DISCUSSION

From the perspective of employees, the extension of teleworking or telecommuting seems to be limited to 1-2 days a week. No one in our study wanted to telework permanently more than two days (max three for a short period) a week. This is supported in earlier research (e.g., Wells *et al.*, 2001). The reason is, despite the freedom afforded by this way of organizing work, that teleworking also has negative aspects. Social isolation, reduced possibilities for promotion, and reduced impact on one's own work situation are frequently mentioned in our study.

Telework also has negative consequences for employers. The most significant are deficient communication and administrative control problems. The company will have trouble: assessing the capability of their employees to solve problems, motivating the employees, producing relations of loyalty and maintaining a company culture among employees in a more peripheral position as a result of teleworking. Simultaneously, there might be considerable costs related to technical equipment, such as purchase of data programmes, hardware and communication devices.

From the employer's perspective, telework can also have a range of positive effects, e.g. a saving of office space, key personnel can be recruited and encouraged to stay, and productivity can be increased. In our project, the effect on productivity is clear and takes place in two ways. People claim that they work more efficiently at home, and very often work more to prove they are worth the confidence showed in them by the employer. Availability of technical equipment also leads to increased competence of employees, very often in their spare time. As our study revealed, teleworking also contributed to stress reduction, which in the long term will reduce sickness absence.

What makes telework interesting in connection with transport in society today is flexibility in relation to other societal trends of development. Trends in working life, the perception of mobility and development of different lifestyles, a general diffusion/distribution and use of technology in organising everyday life and urban development are all important examples. Teleworking suits the "new" working life characterized by individual rather than collective agreements/contracts, flexibilisation rather than standardisation and a transfer from "lifelong" work relations to short engagements and loose contracts (Sennett, 1998). Individual solutions, work organised as projects, task-oriented engagements and displacement of the limits between private and working life have been emphasised by several researchers (Hochschilde, 1997; Colbjørnsen, 2001; Ellingsæter, 2003).

The framework for the new organizational principles is that it is not enough just to rationalise through specialisation and division of work, companies must also have the ability to readjust quickly in a situation with increased international competition. "Lean production" means reduction of employees, outsourcing and more work tasks on those who remain (Sennett, 1998). Teleworking can be seen within this framework of flexible organisation, individualisation, externalisation of work and displacement of the power relation between employer and employee. The idea that organisations operate in a communication or information room rather than in a material room was launched by Hilz in 1984. Hilz's future vision was that the traditional workplace would eventually give way to a "construction" of relations and network – the intellectual room or space would be the workplace. Even though the network society is postulated by scholars such as Castells (1996), who claims that "the new pattern of sociability is characterized by networked individualism" (Castells, 2003: 129), this and other studies show that the physical meeting, face to face is still important in work relations.

As I have pointed at in this study, the work environment is paramount for the employees in all three companies. First, the workplace is important for social interaction and affirmation of the self (Hochschilde, 1997). Second, visibility in an organisation can be crucial for recognition and career development within the company. The presence of the employees will also be important for organising cooperation at work. Even though ICT is an important tool, and becomes the norm, the need for face-to-face contact will still be significant. Face–to-face interaction was the most important reason for not to participate in a telecommuter program in both a private and a public firm in Minnesota (Wells *et al*., 2001). Meetings are an important everyday part of proceedings in all of the three companies I studied. Some of the workers interviewed expressed their worry that more use of teleworking would complicate the organisation of daily work tasks. Increased teleworking might mean greater pressure on those who do not telework in their taking care of routine activities.

REFERENCES

Baily, D.E. and N.B. Kurland (2002). A review of telework research; findings, new directions, and lessons for the study of modern work. *J. Organ. Beh.,* **23**, 383-400.

Balepur, P.N, K.V. Varma and R.L. Mokhtarian (1998). Transportation impacts of center-based telecommuting: interim findings from the Neighborhood Telecenters Project. *Transpn.*, **25,** 287-306

Batten, D. (1989). The future of transport and interface communication: debating the scope for substitution growth. In: *Transportation for the Future* (D. Batten and T. Roland, eds.), pp. 67-86. Springer Verlag, Berlin.

Beck, U. (2000). *What is Globalization?* Oxford, Blackwell Publishers.

Bjørnskau, T. and R. Hjorthol (2003). Gentrifisering på norsk - urban livsstil eller praktisk organisering av hverdagslivet? *Tidsskrift for samfunnsforskning,* **44**, 169-201.

Brandth, B. and E. Kvande (2003). *Fleksible Fedre. Maskulinitet, Arbeid, Velferdsstat.* Universitetsforlaget, Oslo.

Capello, R. and A. Gillespie (1993). Transport, communications and spatial organisation: conceptual framework and future trends. In: *Europe on the Move* (P. Nijkamp, ed.), pp. 43-66. Avebury, Aldershot.

Castells, M. (1996). *The Information Age: Economy, Society and Culture, Volume I.* Blackwell Publishers, Cornwall.

Castells, M. (2003). *The Internet Galaxy. Reflections on the Internet, Business, and Society.* Oxford University Press, Oxford.

Colbjørnsen, T. (2001). Fjernarbeid og regulering av arbeidsmiljøet. In: *Arbeid på nye Måter. Perspektiver på Fjernarbeid* (J.W. Bakke, E. Fossum, J. Nævdal, H. Vollset and B. Yttri (eds.), pp. 31-44. Fagbokforlaget, Bergen.

Coser, R. (1974). *Greedy Institutions: Patterns of Individual Commitment.* Free Press, New York.

Denstadli, J.M. (2002). RVU 2001. Den Nasjonale Reisevaneundersøkelsen. TØI-Rapport 596, Transportøkonomisk Institutt, Oslo.

Duxbury, L.E., C.A. Higgins and D. Neufeld (1998). Telework and the balance between work and family: is telework part of the problem or part of the solution? In: *The Virtual Workplace* (M. Igbaria and M. Tan, eds.), pp. 218-255. Idea Group Publishing, Hershey, PA.

Ellingsæter, A.L. (2003). Postindustrielle tidskontrakter: fra tidsdisiplin til tidssuverenitet. In: *Den Usynlige Hånd: Kjønnsmakt og Moderne Arbeidsliv* (A.L. Ellingsæter and J. Solheim, eds.), pp.79-109. Gyldendal Akademisk, Oslo.

Ellison, N. (1999). Social impacts: new perspectives on telework. *Soc. Sc. Comp. Rev.,* **17,** 338-356.

Engström, M.G. and R. Johanson (1996). *IT-Utviklingens Effekter på Framtida Res- och Transportstrukturer*. Naturvårdverket Förlag, Stockholm.

Financial Times (2003). Escape from the city. Interview with Peter Hall and Stephen Marshall. October 11-12.

Hall, P. and S. Marshall (2002). *The Land Use Effects of "The 10 Year Plan".* The Independent Transport Commission, London.

Hamer, R., E. Kroes and H. van Oostsroom (1991). Teleworking in the Netherlands, an evaluation of changes in travel behaviour. *Transpn.*, **18,** 365-382.

Helling, A. and P.L. Mokhtarian (2001). Worker telecommunication and mobility in transition: consequences for planning. *J. Plan. Lit.,* **15**, 511-525.

Henderson, D.K., B.E. Koenig and P.L. Mokhtarian (1996). Using travel diary data to estimate the emisions impacts of transportation strategies: the Pudget Sound telecommuting demonstration project. *J. Air & Waste Mngmnt Assoc.,* **46**, 47-57.

Hiltz, S.R. (1984). *Online Communities: A Case Study of the Office of the Future.* Ablex, Norwood, NJ.

Hjorthol, R. (1999). Everyday Travel and Use of Information and Telecommunication Technology at Home. TØI Report 454, Transportøkonomisk Institutt, Oslo.

Hjorthol, R. (2002a). The relation between daily travel and use of the home computer. *Transpn. Res. A*, **36,** 437-452.

Hjorthol, R. (2002b). RVU 2001. Den Nasjonale Reisevaneundersøkelsen: Omsorgsreiser. TØI-Rapport 598, Transportøkonomisk Institutt, Oslo.

Hochschilde, A.R. (1997). *The Time Bind: when Work becomes Home and Home becomes Work.* Metropolitan Books, New York.

Hultén, K. (2000). Datorn på Köksbordet: En Studie av Kvinnor som Distansarbetar i Hemmet. Lund Dissertations in Sociology 33, Sociologiska Institutionen, Lunds Universitet, Lund.

Juvkvam, D. and K. Sørli (2000). Demografiske Hovedtrekk i Fire Storbyregioner. NIBR, Oslo.

Lund, J.R. and P.L. Mokhtarian (1994). Telecommuting and residential location: theory and implication in monocentric metropolis. *Transpn. Res. Rec.*, **1463**, 10 –14.

Mitchell, W. (1995). *City of Bits: Space , Place and Infobahn.* MIT Press, Cambridge.

Mitchell, W. (1999). *E-topia.* MIT Press, Cambridge.

Mokhtarian, P.L. (1990). A typology of relationship between telecommunications and transportation. *Transpn. Res. A,* **24**, 231-242.

Mokhtarian, P.L. (1998). A synthetic approach to estimating the impacts of telecommuting on travel. *Urb. Stud.*, **35,** 214-241.

Mokhtarian, P.L., M.N. Bagley and I. Salomon (1998). The Impact of Gender, Occupation, and Presence of Children on Telecommuting, Motivations and Constraints. University of California Transportation Centre, Berkeley.

Mokhtarian, P.L., S. Handy and I. Salomon (1995). Methodological issues in the estimation of the travel, energy, and air quality impacts of telecommuting. *Transpn. Res. A*, **4**, 283-302.

Mokhtarian, P.L. and I. Salomon (1997). Modeling the desire to telecommute: the importance of attitudinal factors in behavioral models. *Transpn. Res. A*, **31**, 35-50.

Nilles, J.M. (1991). Telecommunting and urban sprawl: mitigator or inciter? *Transpn.*, **18**, 411-432.

Nilles, J.M., F. Carlson, P. Gray and G. Hanneman (1976). *The Communications-Transportation Tradeoff.* John Wiley, New York.

Olszewski, P. and P.L. Mokhtarian (1994). Telecommuting frequency and impacts for State of California employees. *Techn. Forec. & Soc. Change*, **45**, 275-286.

Peters, P., K. Tijdens and C. Wetzels (2004). Employees' opportunities, preferences, and practices in telecommuting adoption. *Inf. & Mngmnt.*, **41**, 469-482.

Salomon, I. (1985). Telecommunication and travel: substitution or modified mobility? *J. Transpn. Econ.& Policy,* **19**, 219-235.

Salomon, I. (1996). Telecommunication, cities and technological opportunism. *Ann. Reg. Sc.*, **30**, 75-90.

Sennett, R. (1998). *The Corrosion of Character. The Personal Consequences of Work in the New Capitalism.* W. W. Norton & Company, New York.

SIKA (1998). IT-Utveklingen och Transporterna 2: Redovisning av en Kommunikations-Undersökning 1997. SIKA, Stockholm.

SIKA (2001). Kommunikastionsmönster hos Befolkningen. Resultat från SIKA's Kommunikationsundersøkningar. SIKA, Stockholm.

Stanek, D.M. and P.L. Mokhtarian (1998). Developing models of preference for home-based and center-based telecommuting: findings and forecasts. *Techn. Forec. & Soc. Change*, **57**, 53-74.

Vilhelmson. B. and E. Thulin (2001). Is regular work at fixed places fading away? the development of ICT-based and travel-based modes of work in Sweden? *Env. & Plan. A,* **33**, 1015-1029.

Wells, K., F. Douma, H. Loimer, L. Olson and C. Pansing (2001). Telecommuting implications for travel behaviour: case studies from Minnesota. *Transpn. Res. Rec.*, **1752**, 148-156.

21

CHANGING TRAVEL CHARACTERISTICS AND ACTIVITY TRAVEL PATTERNS OF HOUSEHOLDS THROUGH TELECOMMUTING?

Andrea F. Glogger, Technische Universität München, Munich, Germany

Thomas W. Zängler, Technische Universität München, Munich, Germany

Georg Karg, Technische Universität München, Munich, Germany

INTRODUCTION

Traffic is increasing steadily. In Germany, from 1976 to 1999, the total distance travelled per person increased by more than 65% (BMVBW, 2001). Taking into account the effect of traffic on the environment and the quality of life, this development is not sustainable. Therefore, a change in travel behaviour is required. The problem here however is that in a free society, such change should ideally take place without limiting the mobility of individuals. Individuals and households need to be mobile because they need to be able to become involved in activities to satisfy their needs and desires. These activities can be in-home or out-of-home, but the latter involve travel (Kitamura, 1988).

In a policy context, it is important to realise that travel behaviour is influenced by various factors. Among others, it depends on the kind of activities generating it, socio-demographics, and temporal and spatial circumstances (Kuppam and Pendyala, 2001; Schwanen and Dijst, 2003). Also, travel behaviour related to frequently conducted activities becomes script-based and habitual. Breaking habits or inertia is difficult (Lotan, 1997; Ouelette and Wood, 1998; Gärling and Golledge, 2000;

Progress in Activity-Based Analysis edited by H. Timmermans

Gärling *et al.*, 2001; Gärling and Axhausen, 2003), as a change calls for the motivation of the individual to learn new routines. At the same time, alternatives to the actual routine have to be available, and the individual has to be aware of them (Gärling and Axhausen, 2003).

Thus, understanding the possible impact of policy decisions on travel behaviour requires a broader perspective. Activity-based models may be relevant in this regard because these models focus on the interactions between travel and activity choice, e.g. the decision on whether, when, and where to work (Golob, 1998; Bhat and Koppelman, 1999), or on comprehensive patterns (Ettema and Timmermans, 1997). The activity-based approach acknowledges the importance of time for understanding travel behaviour. According to Pas (1998), time has many functions. It is, for instance, a resource that has to be allocated to various activities and related travel, a constraint on the allocation of other resources, a dimension for the beginning and the end of travel, a descriptor for the horizon over which travel and its related decisions are made, and an explanatory factor in models of destination, travel mode and route choice.

Recently emerging information and communication technologies (ICT) have been viewed as a potentially important policy to reduce the increase in mobility because it offers, to a certain extent, chances to replace physical by virtual mobility. To date, it is common knowledge that there are various possible interactions between physical and virtual mobility (e.g., Mokhtarian, 2000; Salomon, 2000). The research project "MOBINET- Mobility in Conurbations", sponsored by the German Federal Ministry of Education and Research, examines an option that has been developed as an alternative for work travel: telecommuting. The actual situation in Germany for commuting trips is the following. Over the past 23 years, roughly one fifth of the travelled distance has been allocated to commuter trips (BMVBW, 2001). In this period, commuter trips increased from 124 Mrd km in 1976 to 204 Mrd km in 1999. The share of car use for commuting distances increased from roughly 71% in 1976 to 77% of the travelled distance in 1999, while the time spent on travel remained nearly constant. Telecommuting has different effects on the individuals involved. The aim of the present study is to analyse the effects of telecommuting on travel characteristics of telecommuters and households, especially on time use, and on travel patterns of telecommuters and the other members of their households.

METHOD

In 2000 and 2001, a survey of the travel behaviour among telecommuters and their household members was administered in eight big companies in the Greater Munich area. It involved a before (t_0) - after (t_1) study. Telecommuting was introduced in these companies by the consultancy BPU.

This consultancy was also involved in the development of the survey. The market research institute TNS Infratest (formerly NFO Infratest) carried out the field work.

The telecommuters and their household members of driving age (18 years and older) were asked to fill out a personal and a household questionnaire and a personal travel diary. The observation days in the "before survey" were equally distributed across the 5-day workweek. The data for the "after survey" relate to one telecommuting and one commuting day, the day immediately before or after the telecommuting day. The personal questionnaire consisted of questions about socio-demographics, the expected influence of telecommuting on time use, travel distance and its influence on the telecommuters' professional and private life, the preferred transport mode, habitual travel behaviour and the availability of various means of transport. The household questionnaire was to be completed by a member of the household who was able to provide answers to questions about the availability of public transport and other means of transport, accessibility to daily facilities (e.g., shopping, day care, schools), task allocation in the household, and socio-demographics. Household members and telecommuters were asked to complete a very detailed personal travel diary on the same two adjacent days in t_0 and t_1. For every trip the following characteristics were recorded: day of the week, start and end time, transport mode, type of destination, accompanying person, and the corresponding activity. In addition, questions were asked about subjective reasons for the choice of transport mode (travel by foot, bike, public transport, park and ride (P&R) and car driver or passenger) and the urgency and obligation of the activity. Activities were coded individually and classified into seven activity groups (work, business (business related travel, not the daily commute), education, leisure, accompanying household members, and maintenance).

The micro-analysis of travel behaviour at the individual and household level distinguishes between the following effects of telecommuting on physical travel *P*, represented by the travel indicator trip frequency, travel distance or travel time:

- a substitutional effect if $\Delta P_{(t_1-t_0)} < 0$
- a complementary effect if $\Delta P_{(t_1-t_0)} > 0$
- and a neutral effect if $\Delta P_{(t_1-t_0)} = 0$.

The substitutional and complementary effects can manifest themselves directly or indirectly. We talk of a direct effect if telecommuting influences the commuting trips, of an indirect effect if trips for other activities are affected. According to Mokhtarian (2000) a combination of the mentioned effects is likely.

In the longitudinal analysis a comparison was made between a working week at t_0 (before telecommuting) and a working week at t_1 (after the introduction of telecommuting). At t_0 a working week consisted of 5 working days at the employers' location and at t_1 it consisted of 2.5 commuting and 2.5 telecommuting days, representing the average telecommuting frequency of the sample. It was an appropriate simplification to use the average rather than individual-specific ones since no significant interaction was found between telecommuting frequency and any of the travel indicators analysed in this study (e.g., longer-distance commuters not having higher telecommuting frequencies). That is, Pearson correlations between telecommuting frequency and the transportation measures (trip frequency, travel distance and travel time) were not significant at conventional levels.

RESULTS

Sample Characteristics

A total of 227 households was asked to participate in the study. The response rate was 46% in the first wave and 51% in the second wave. A total of 36 households participated in both waves (16%). 65% of the surveyed households comprised three persons or more, 25% of the households consisted of two persons and only 10% were single households. The average number of persons per household was three. Children were living in over 60% of the sample's households. The average telecommuters' age was 40 years, ranging from 31 to 55. With approximately 40 km, the average daily commuting distance is considerably higher than the average for Munich (12km).

Travel Characteristics

In the following tables the parametric t-test was chosen. Its results were validated by the Wilcoxon-test since the used variables are not normally distributed. Tables 21.1-3 show the effects of telecommuting on the average trip frequency and average travel distance. These effects are shown in Table 21.1 per telecommuter and week, in Table 21.2 per household member and week, and in Table 21.3 per household and week. Table 21.1, which is based on the respondents participating in both waves, shows a 21% decrease in average weekly trip frequency and a 40% decrease in average travel distance. Both findings are significant, the reduction in average travel distance is highly significant. Per household member and week there is a decrease in average trip frequency of 14% and in average travel distance of about 34% (Table 2). The reductions are considerable but neither of them was statistically significant.

Table 21.1
Comparison of Average Trip Frequency and Travel Distance per Telecommuter and Working Week

Travel Characteristics	Mean		Δ Mean	
	t_0	t_1		
	abs.	abs.	abs.	rel.
Trip frequency [trips/pers·week]	18.5	14.5	-4.0 *	-21.4% *
Travel distance [km/pers·week]	483.7	291.4	-192.4 ***	-39.8% ***

Note *** p<0.001 **p<0.01 *p<0.05 (t-test)
Basis: 36 telecommuters

Table 21.2
Comparison of Average Trip Frequency and Travel Distance per Household Member and Working Week

Travel Characteristics	Mean		Δ Mean	
	t_0	t_1		
	abs.	abs.	abs.	rel.
Trip frequency [trips/pers·week]	18.7	16.1	-2.6	-13.9%
Travel distance [km/pers·week]	399.5	263.4	-136.1	-34.1%

Note *** p<0.001 **p<0.01 *p<0.05 (t-test)
Basis: 29 household members of telecommuters

Table 21.3
Comparison of Average Trip Frequency and Travel Distance per Household and Working Week

Travel Characteristics	Mean		Δ Mean	
	t_0	t_1		
	abs.	abs.	abs.	rel.
Trip frequency [trips/hh·week]	33.6	27.5	-6.1 **	-18.2% **
Travel distance [km/hh·week]	805.5	503.6	-301.9 **	-37.5% **

Note *** p<0.001 **p<0.01 *p<0.05 (t-test)
Basis: 36 households

Per household and week there is a decrease in average trip frequency of 18% and in average travel distance of about 37% (Table 21.3). The findings prove significant at the 99% level. Thus, as expected, telecommuting changes the average trip frequency and travel distance of telecommuters, their household members and households. The same phenomenon was observed for travel time and will be discussed in detail below.

Time Use

Table 21.4 and 21.5 show the effect of telecommuting on travel time spent by telecommuters, and households respectively, per week for different activity groups. Per telecommuter and week there is a decrease of approximately 4 hours or 37%. The total decrease is largely caused by a significant decrease in travel time spent for work (approximately 2.5 hours or 39%) and for maintenance (approximately 1.5 hours or 90%). These decreases are counterbalanced to some extent by a non-significant increase in travel time for leisure. Per household and week the total time travelled decreases by approximately 5 hours or 30%, caused mainly by a significant decrease in travel time to work (approximately 3 hours or 31%) and for maintenance (approximately 2.5 hours and 82%). These effects are again diluted to some extent by an increase in travel time for leisure, which is however not significant.

Table 21.4
Comparison of Travel Time of Telecommuters by Activity Group

Activity Group	Travel Time of Telecommuters					
	Mean		Δ Mean			
	t_0	t_1	t_1-t_0			
	abs. [min/tc·week]	abs. [min/tc·week]	abs. [min/tc·week]		rel. [%]	
Work	387.4	238.2	-149.2	***	-38.5	***
Business	44.8	11.3	-33.5	*	-74.8	*
Education	0.0	2.6	2.6		-	
Maintenance	104.7	10.0	-94.6	**	-90.4	**
Accompanying	53.8	49.7	-4.1		-7.6	
Leisure	41.7	85.5	43.8		105.1	
N.a.	0.0	0.0	0.0		-	
Total	632.4	397.4	-235.0	***	-37.2	***

Note *** p<0.001 **p<0.01 *p<0.05 (t-test)
Basis: 36 telecommuters

Table 21.5
Comparison of Travel Time of Households by Activity Group

Activity Group	Travel Time of Telecommuters					
	Mean		Δ Mean			
	t_0	t_1	t_1-t_0			
	abs. [min/hh·week]	abs. [min/hh·week]	abs. [min/hh·week]		rel. [%]	
Work	552.9	379.4	-173.6	***	-31.4	***
Business	133.5	47.6	-85.9	*	-64.3	*
Education	0.0	11.7	11.7		-	
Maintenance	171.0	30.3	-140.7	***	-82.3	***
Accompanying	88.4	88.6	0.2		0.2	
Leisure	86.6	149.2	62.5		72.2	
N.a.	0.0	0.0	0.0		-	
Total	1032.5	706.7	-325.8	***	-31.6	***

Note *** p<0.001 **p<0.01 *p<0.05 (t-test)
Basis: 36 households

Table 21.6
Comparison of Travel Time of Telecommuters for Commuting by Mode of Transport

Mode of Transport	Travel Time of Telecommuters					
	Mean		Δ Mean			
	t_0	t_1	t_1-t_0			
	abs. [min/tc·week]	abs. [min/tc·week]	abs. [min/tc·week]		rel. [%]	
Foot	13.7	3.2	-10.5		-76.7	
Bike	0.0	0.0	0.0		-	
Public Transp.	137.2	55.1	-82.2	*	-59.9	*
P&R	112.1	74.6	-37.5		-33.4	
Car (pass.)	0.0	11.2	11.2		-	
Car (driver)	121.6	94.1	-27.5		-22.6	
Other	0.0	0.0	0.0		-	
N.a.	2.9	0.0	-2.9		-100.0	
Total	387.4	238.2	-149.2	***	-38.5	***

Note *** p<0.001 **p<0.01 *p<0.05 (t-test)
Basis: 36 telecommuters

Tables 21.6 and 21.7 show the effect of telecommuting on travel time for commuting of telecommuters and households, where travel time is differentiated by mode of transport. Per telecommuter and week there is an average decrease of 38.5%. This effect is largely due to over proportional and significant decrease in travel by public transport of 60%, and to a lesser extent to the non-significant under proportional decrease for P&R (-33%) and "car as driver" (-23%). Per household and week there is a decrease in total travel time for commuting of approximately 3 hours or 31%. This decrease in travel time is more than proportion and significant for travel time by public transport (-49%) and less than proportional and not significant for travel time by "car as driver" (-14%). These findings show that decreases in travel time for commuting by public transport and car are largely responsible for the telecommuters' cut in travel time.

Trip Distribution

The temporal daily trip distribution is presented as the weekly average number of trips at t_0 and t_1 (Figure 21.1). Apparently telecommuters tend to avoid peak-hours when travelling at t_1, especially for commuting and maintenance trips. Home trips show a peak in the late afternoon and at night. Significant changes for all activities are observed in the morning hours and late afternoon. Trips for work change significantly during morning hours, maintenance and home trips in the evening. The temporal daily trip distribution for household members does not change significantly.

Table 21.7
Comparison of Travel Time of Households for Commuting by Mode of Transport

Mode of Transport	Travel Time of Telecommuters					
	Mean		Δ Mean			
	t_0 (abs.) [min/hh·week]	t_1 (abs.) [min/hh·week]	t_1-t_0 (abs.) [min/hh·week]		t_1-t_0 (rel.) [%]	
Foot	20.8	3.7	-17.1		-82.2	
Bike	1.2	1.1	-0.1		-8.7	
Public Transp.	158.4	80.1	-78.3	*	-49.4	*
P&R	151.4	107.6	-43.8		-28.9	
Car (pass.)	14.5	11.0	-3.5		-24.1	
Car (driver)	203.7	175.8	-27.9		-13.7	
Other	0.0	0.0	0.0		-	
N.a.	2.9	0.0	-2.9		-100.0	
Total	552.9	379.4	-173.6	***	-31.4	***

Note *** p<0.001 **p<0.01 *p<0.05 (t-test); Basis: 36 households

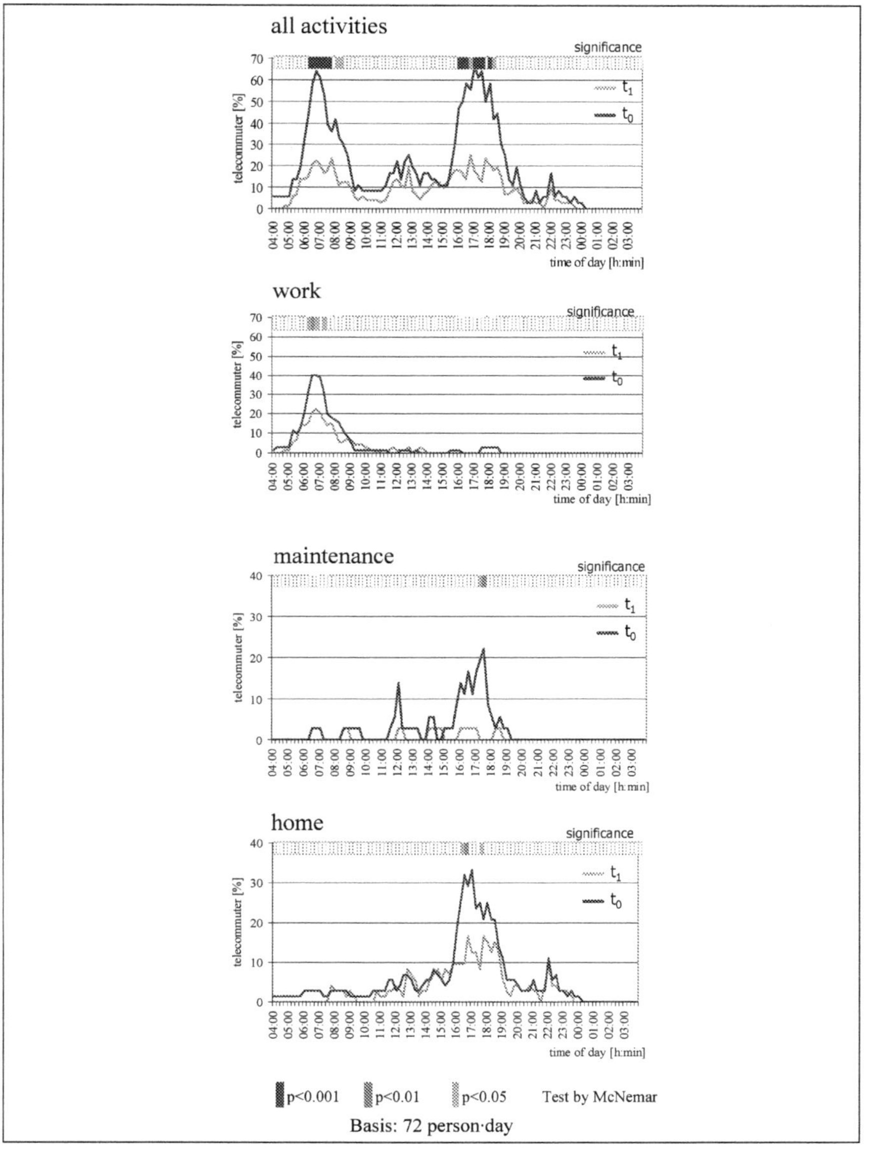

Figure 21.1
Temporal Daily Trip Distribution of Telecommuters by Activity Group

Tours

Tours describe the succession of out-of-home activities. The findings for tours of household members of telecommuters are of special interest (Table 21.8 and 21.9). At t_0 household members undertake on average 1.3 tours per day and person. After telecommuting is introduced there are 1.4 tours per day and person on commuting days and 1.1 tours per day and person on telecommuting days. Household members of telecommuters undertake on average 3.8 trips per day and person in t_0. When telecommuting is introduced trips are reduced to 3.7 trips per day and person (-3%) on commuting days and to 2.7 trips per day and person on telecommuting days (-30%). There seems to be a slight increase in tours on commuting days (7%) and a decrease on telecommuting days (-15%). The trip-tour ratio decreases slightly between t_0 and t_1.

Table 21.8
Comparison of Average Trip and Tour Frequency of Household Members by Survey Day

Travel Characteristics	Survey Day		
	t_0	t_1	t_1
	Commuting Day	Commuting Day	Telecommuting Day
	abs.	abs.	abs.
Trip Frequency [1/d·pers]	3.8	3.7	2.7
Tour Frequency [1/d·pers]	1.3	1.4	1.1
Trips/Tour-Ratio [1]	2.9	2.6	2.5
Basis [person·day]	58	29	29

Table 21.9
Comparison of Tours of Household Members
by Number of Out-of-Home Activities and Survey Day

Number of Out-of-Home Activities	Survey day					
	t_0		t_1		t_1	
	Commuting Day		Commuting Day		Telecommuting Day	
	abs. [1/d·pers]	rel. [%]	abs. [1/d·pers]	rel. [%]	abs. [1/d·pers]	rel. [%]
1	24	65	23	58	25	78
Of which home-work-home	11	30	6	15	14	44
2	6	16	7	18	5	16
3 and more	7	19	10	25	2	6
Total	37	100	40	100	32	100
Basis [person*day]	58		29		29	

Table 21.9 illustrates the findings in tours of the telecommuters' household members. At t_0 65% of the tours comprise only one out-of-home activity, at t_1 it is 58% on commuting days and 78% on telecommuting days. The share of multiple-activity tours (three and more out-of-home activities per tour) is 19% of all tours at t_0, 25% of all tours on commuting days at t_1 and 6% of all tours on telecommuting days at t_1. There is a striking change in the chain home-work-home from 30% on an average working day at t_0 to 15% on commuting and 44% on telecommuting days at t_1. Possibly these findings give some evidence that telecommuting leads to a rescheduling or redistribution of activities within the household. This gives rise to the hypotheses that trip chaining increases with pressure of time. Telecommuting might reduce this pressure of time for certain household activities on telecommuting days and therefore results in fewer chained out-of-home activities for the household members of telecommuters.

DISCUSSION

The discussion is referring to the method used and the results obtained. The method appears to encompass the relevant variables given the objectives of this study. The response rate was quite low and might perhaps be increased by deleting some details, especially in the household and personal questionnaire. It appears that the respondents preferred reporting their mobility behaviour in comparison to giving details on their private environments in their private life. The results may be influenced by the operational decision to assign travel distance to the kind of activity that is conducted at the end of the trip. When for example shopping was conducted at night close to home, but after work involving a long distance from work, this long distance was labelled as travel to shopping respectively maintenance. Thus, there may be a methodological trend of over-emphasising changes in travel time and distance in terms of travel for maintenance purposes.

The extrapolation on an average working week is based on the finding that no correlation was found between the average telecommuting frequency and trip frequency, travel distance and travel time. The procedure has proven useful for the objective of the study. The results of the impact of telecommuting on trip frequency, travel distance and travel time across all activities show a substitution effect. After telecommuting was introduced, time use for physical travel decreases. When distinguishing between activities, this is especially true for work and maintenance travel. In case of maintenance travel, the mentioned distribution problem has to be accounted for. Still, there is a decrease in average trip frequency of maintenance trips. Another possible explanation is a shift in task allocation within the household towards non-surveyed under-aged members or towards non-surveyed days. A large part of the decrease in physical travel is due to the telecommuters' change in travel participation. However, household members cut their travel time and travelled distance by one third as well. When interpreting these findings underreporting, as observed in the State of California telecommuting pilot project (Kitamura *et al.*, 1990), has to be considered. A complementary effect is observed in an increasing travel time for leisure activities. This finding contradicts the findings of Vogt *et al.* (2001). In a Germany-wide study of the impact of telecommuting on travel behaviour they found a decrease in the number of trips and in the distance travelled for leisure purposes. Yet, due to the more than average commuting distance of the sample telecommuters in Munich the findings cannot be generalised to other populations.

The travel mode choice for commuting trips shows an over-proportional decrease in travel time for public transport, both for telecommuters and households. This might be due to the fact that monthly tickets do not pay off when the travel-to-work frequency is reduced as a result of telecommuting. Simultaneously, the travel time by "car as driver" decreases only under-proportionally. These findings agree with the findings of Kitamura *et al.* (1991) in California, Hamer *et al.* in the Netherlands (1991, 1992) and Vogt *et al.* (2001) in Germany. The change in trip distribution by activities agrees with results of Pendyala *et al.* (1991) and Kitamura *et al.* (1991) in California. Telecommuters tend to avoid peak-hour traffic.

The findings about activity chaining in tours by the telecommuters' household members are of special interest. The Dutch study of Hamer *et al.* (1992) reported a constant tour length for household members of telecommuters before and after the introduction of telecommuting. In this study the findings indicate that their tours involve more activities on commuting days and fewer activities on telecommuting days after the introduction of telecommuting. This might reflect a change in activity scheduling or task allocation within the household. A shift towards non-survey days is also possible.

CONCLUSION

Taking into account the limitations of this study with respect to observation area and sample size one can conclude that telecommuting does reduce overall physical travel of telecommuters and their households. Telecommuters reveal however a preference for travel by car over travel by public transport. Trip distribution of telecommuters is more evenly spread and peak-hours are avoided. Thus, introducing telecommuting on a larger scale may results in reduced traffic congestion.

The analysis of tours leads to the conclusion that household members of telecommuters chain fewer activities. This confirms the influence of telecommuting on household members of telecommuters and affirms the need of surveying telecommuters and household members to capture spill-over effects.

Long term effects like residential relocation require a longer observation period. Within the given year, these effects could not be determined (Mokhtarian, 1997; Lyons, 1998). Thus, in future research a larger observation area and a longer observation period should be chosen. The sample size should be increased and questionnaires should be further improved to improve the response rate. It might be worthwhile to observe telecommuters in subsequent waves to gain better insight in the dynamics of the short and long term effects of telecommuting.

ACKNOWLEDGEMENTS

This study was made possible by the support of the German Federal Ministry of Education and Research and the City of Munich.

REFERENCES

Bhat, C.R. and F.S. Koppelman (1999). A retrospective and prospective survey of time-use research. *Transpn.,* **26**, 119-139.

BMVBW (Bundesministerium für Verkehr, Bau- und Wohnungswesen, ed. (2001). Verkehr in Zahlen 2001/2002. Hamburg.

Ettema, D.F and H.J.P. Timmermans (1997). Theories and models of activity patterns. In: *Activity-Based Approaches to Travel Analysis* (D.F. Ettema and H.J.P. Timmermans, eds.), pp. 1-36. Pergamon, Oxford.

Gärling, T. and R.G. Golledge (2000). Cognitive mapping and spatial decision-making. In: *Cognitive Mapping: Past, Present and Future* (R. Kitchin and S. Freundschuh, eds.), pp .44-65. Routledge, London.

Gärling, T. and K.W. Axhausen (2003). Introduction: habitual travel choice. *Transpn.,* **30**, 1-11.

Gärling, T., S. Fujii and O. Boe (2001). Empirical test of a model of determinants of script-based driving choice. *Transpn. Res. F*, **4**, 89-102.

Golob, T.F. (1998). A model of household choice of activity participation and mobility. In: *Theoretical Foundations of Travel Choice Modelling* (T. Gärling, T. Laitila and K. Westin, eds.), pp. 365-398. Elsevier, Oxford.

Hamer, R., E. Kroes and H. van Ooststroom (1991). Teleworking in the Netherlands: an evaluation of changes in travel behaviour. *Transpn.*, **18**, 365-382.

Hamer, R., E. Kroes and H. van Ooststroom (1992). Teleworking in the Netherlands: an evaluation of changes in travel behaviour – further results. *Transpn. Res. Rec.*, **1357**, 82-89.

Kitamura, R. (1988). An evaluation of activity based travel analysis. *Transpn.*, **15**, 9-34.

Kitamura, R., P.L. Mokhtarian, R. Pendyala and K.G. Goulias. (1991). An evaluation of telecommuting as a trip reduction measure. In: *Civilising Transport*, pp. 69-80. PTRC, University of Sussex, Brighton.

Kuppam, A.R. and R.M. Pendyala (2001). A structural equations analysis of commuters' activity and travel patterns. *Transpn.*, **28**, 33-54.

Lotan, T. (1997). Effect of familiarity on route choice behaviour in the presence of information. *Transpn. Res. C*, **5**, 225-243.

Lyons, G.D., A. Kickford and J. Smith (1998). The nature and scale of teleworking's travel demand impact: insights from a U.K. trial. In: *TUCS General Publications, Volume 8, Teleworking Environments* (R. Soumi, P. Jackson., L. Hollmen and M. Aspnäs, eds.), pp. 312-330. TUCS, Turku.

Mokhtarian, P.L. (1997). The transportation impacts of telecommuting: recent empirical findings. In: *Understanding Travel Behaviour in an Era of Change* (P. Stopher and M. Lee-Gosselin, eds.), pp. 91-105. Elsevier, Oxford.

Moktharian, P.L. (2000). Telecommunications and Travel. *Proc. 79th Ann. Meet. Transpn. Res. B.*, Washington, D.C. (CD-Rom).

Ouelette, J.A. and W. Wood (1998). Habit and intention in everyday life: the multi processes by which past behaviour predicts future behaviour. *Psych. Bull.*, **124**, 54-74.

Pas, E.I. (1998). Time in travel choice modeling: from relative obscurity to center stage. In: *Theoretical Foundations of Travel Choice Modelling* (T. Gärling, T. Laitila and K. Westin, eds.), pp. 231-250. Elsevier, Oxford.

Pendyala, R., K. Goulias and R. Kitamura (1991). Impact of telecommuting on spatial and temporal patterns of household travel. *Transpn.*, **18**, 383-409.

Sachs, L. (1992). *Angewandte Statistik*. Springer, Berlin.

Salomon, I. (2000). Can telecommuting help solve transportation problems? In: *Handbook of Transport Modelling* (D.A. Hensher and K.J. Button, eds.), pp. 449-462. Pergamon, Oxford.

Schwanen, T. and M. Dijst (2003). Time window's in workers' activity patterns: empirical evidence from the Netherlands. *Transpn.*, **30**, 261-283.

Vogt, W., S. Denzinger, W. Glaser, M.O. Glaser and T. Kuder (2001). Auswirkungen neuer Arbeitskonzepte und insbesondere von Telearbeit auf das Verkehrsverhalten. Berichte der Bundesanstalt für Strassenwesen M128, Bundesanstalt für Strassenwesen, Bergisch-Gladbach.

Zängler, T.W. (2000). *Mikroanalyse des Mobilitätsverhaltens in Alltag und Freizeit*. Springer, Berlin.

22

COMMUNICATION AND TRAVEL BEHAVIOUR: TWO FACETS OF HUMAN ACTIVITY PATTERNS

Claudia Nobis, German Aerospace Centre, Berlin, Germany

Barbara Lenz, German Aerospace Centre, Berlin, Germany

Colin Vance, German Aerospace Centre, Berlin, Germany

INTRODUCTION

Communication and travel behaviour are two facets of human interaction that lead to particular patterns in the use of time and geographical space. Exploring the question about the relations between communication and travel behaviour has become a prominent topic in transportation and geographical research, as the rate of diffusion of new information and communication technologies (ICT) has been increasing rapidly during the last decade.

To date, research about the effects of ICT use on travel and transport has mostly focussed on observing individual activities with a high affinity for ICT like e-shopping or telework. The primary goal usually is to determine the net effect of each of these activities with respect to substitution or induction of transport and travel. Most of these studies fail to position their findings within the larger context of individual behavioural change and changing societal conditions. A conceptualisation and empirical analysis of how ICT in general influences the way people perform their activities is missing.

The basic idea underlying the present chapter is that the technological developments, both in transport and particularly in communications, increase flexibility and allow a spatial-temporal

Progress in Activity-Based Analysis edited by H. Timmermans

reorganisation of activities. This reorganisation may induce qualitative and quantitative changes in traffic. Against this background, the paper starts with conceptual reflections about the relationship of activities, mobility and the use of ICT. The current empirical deficits are pointed out. Next, the results of empirical analyses based on the data of the first wave of the 'DLR's ICT and Mobility Panel' study are discussed. The data analysis will describe the relationship between communication and mobility patterns and the individual characteristics of people with varying affinities to information and communication technology. Most studies on ICT have shown that heavy ICT users are found particularly among the group of young individuals who are at the same time a group with above average mobility. As there usually is no test whether the relationship between ICT and travel can be observed across all age groups, this relationship will be explored in this study. Finally, by estimating a linear regression model, the factors influencing travel demand are analysed. The specific question of interest is the role of ICT use in this context.

Conceptual Reflections: Activities, Mobility and ICT

A person's daily routine is characterised by activities. Activities are the expression of the human desire (or necessity) to take part and participate in society. Since activities are performed at certain times at various places, each individual needs a spatial-temporal organisation of his or her activities, and some of these activities are associated with travel (Widmer and Axhausen, 2001; Beckmann, 2000). A distinction is made between the 'activity repertoire', the set of potential activities, and the 'activity program', the activities which are actually performed. A very simplified activity program can be, for example, being at home (household activities) – working – shopping - being at home – out-of-home leisure time - being at home. Since activities are performed at different times at different places, the program results in a spatial-temporal activity pattern (Beckmann, 2000). People's out-of-home activity programs show great similarities. Zumkeller (1999) was able to show empirically that a majority of trips undertaken by people in performing their out-of-home activities can be described in terms of just a few basic patterns. However, due to spatial, temporal and individual differences, activity patterns show clear differentiation.

Activity repertoires and programs, as well as the resulting spatial-temporal models, are subject to various factors. On the individual level, socio-demographic features such as age, gender, income etc. play a role, as well as personal preferences, and systems of values and attitudes. In addition to these individual factors, Beckmann (2000) distinguishes the 'supra-individual regime', which includes temporal order, social conditions, and physical configuration. He divides temporal order into natural temporal order (day and night, the seasons) and social temporal order (e.g., work hours, business hours). Both place limits on potential activities. This is also true for 'social conditions', by which he means all economic, legal, and political regulations. 'Physical configuration' refers to the spatial distribution of infrastructure, including locations, the type and quality of opportunity systems, transport networks, and travel options, and communications networks (Beckmann, 2000). The first and basic conceptualisation of these ideas can be found in Hägerstrand's outline of 'time geography' (Hägerstrand, 1960).

Analysing activities and the factors affecting them is much more than just recording trips and their purposes. As is travel, communication is often a precondition for the performance of activities. Beyond that, communication can even replace travel and shift activities to another location. This process has been particularly fostered by the spread of ICT. With the aid of digital information technology, people can participate in economic, social, and cultural activities, at least in principle, from almost anywhere in the world. Thus, accessibility has received a new dimension with various implications concerning the manner in which activities are performed and concerning travel behaviour. The consequence is that activities are no longer tied to specific times and places. This, in turn, has consequences for travel needs and behaviour. For example, work involving the handling and processing of information can be performed at various locations, which may change fundamentally the needs for work-related travel. Activities that were previously restricted to certain times of day, e.g. shopping, can now be performed at any time.

Due to the loss of dependence on time and place, individuals now have the option of dramatically reorganising their activities. This is what Couclelis (2000) calls the 'fragmentation' of activities, a dissolution of activities occurring on several levels: in space, in time and in the manner in which activities are performed. The major outcome of the fragmentation processes will be an increasing demand for contacts, which will be satisfied by the use of ICT. Thereby, the substitution of ICT use for travel is ongoing and will probably continue. However, a higher percentage of the increasing contacts will again be linked to physical travel due to objective or subjective necessities. It is therefore assumed that despite, or, in certain areas, precisely because of, the increasing use of ICT, demand for physical travel will increase, at least under the present conditions that create a relatively low cost of travel. Couclelis therefore raises the hypothesis that "the fragmentation of activity enabled by the spatial technologies of the information age is one of the reasons for the widely observed increases in travel demand in the industrialized world" (Couclelis, 2000, p. 348).

Although the fragmentation concept enriches the theoretical comprehension of the interplay between ICT and travel significantly, it is still weak for its empirical conceptualisation and its empirical verification. Looking at the current state of the art, it can be seen that most studies focus either on 'e-activities', that can be performed either physically or virtually, e.g. telecommuting or e-shopping. Alternatively, the schema developed by Salomon (1986, 1985) describing basic types of interaction between ICT and transport (e.g. Mokhtarian, 1990; Krizek and Johnson, 2003; Marker and Goulias, 1999) is used: substitution (decrease in travel demand through a reduction in total number of trips or in trip duration), complementarity (generation of new trips due to the use of ICT), modification (change of spatial and temporal characteristics of existing travel patterns) and neutrality (no impact of one medium on the other). Looking at the studies conducted to date, two problems are evident: First, although the studies on activities with high ICT affinity allow essential insights into particular fields of e-activities, there is still no complete picture that shows how the different parts of the ICT puzzle fit together. Due to a quite narrow perspective, the observations are not set within a larger context of individual behaviour and changing societal conditions. Second, studies examining whether ICT leads especially to substitution or generation of traffic are, in effect, reversing the logical order of the analysis. Before really understanding the processes of how the interrelation of ICT, activities and mobility is, the impact of the process on travel demand is analysed on aggregate level.

Therefore, this chapter proceeds on the premise that future research should be much more aware of the general change of activities together with ICT use as – apart from the e-activities – also activities like 'going to a restaurant' or 'going to the theatre' may be affected by the use of ICT. The changes are given by the possibility to act spontaneously by using Internet and mobile phone for information about locations and events and for communication to bring together the relevant group of people. While modifying the process of decision-making and the norms of meeting each other, these changes on the micro level can substantially alter activity and mobility patterns. From a research point of view, approaches should move towards a holistic perspective closely linked with efforts to overcome the problem of testing theoretical concepts in empirical work.

USED DATA SET

Many of the salient questions regarding changes in routine behaviour through ICT use can only be answered by a long-term study. This is particularly true for questions pertaining to whether the purchase and increasing use of ICT are associated with changes in activities and travel behaviour. Therefore, the data used here to explore the relation between ICT use and travel on the one hand and activities on the other were drawn from the first wave of a panel survey that will be followed by other surveys in the future.

To generate a statistically measurable relation between activities, travel behaviour, and ICT use that can be differentiated on a group-specific basis, the survey elicited the following information:

- description of activity patterns
- description of activity-specific ICT use
- description of activity-specific travel behaviour, including spatial aspects.

The study was performed as a written survey between the 12th of May and the 3rd of June, 2003. The results are representative for German-speaking residents of the Federal Republic of Germany aged 14 and older. Thus, the results can be projected up to a total of 64.1 million people. The sample size is 3,500 persons. For the present data analysis, however, only data from the first section and data about the respondents' socio-demographic characteristics are used.

RESEARCH QUESTIONS AND HYPOTHESES

The use of ICT has the potential to substantially restructure the organisation of activities concerning the time and the location where they are performed. As communications and travel are both preconditions of the performance of activities, they are closely linked with each other. Therefore, the basic question we address in the empirical part of this paper is: *What is the nature and scope of the relationship between communication and travel?*

As Clouclelis (2000) states, one of the major outcomes of increasing ICT use is an increasing travel demand. Although some trips may be substituted by ICT, the linkage to physical places will remain for many activities. The use of ICT, in general, allows a higher flexibility in the way activities are performed. The hypotheses derived from the question of the general relation between communication and travel therefore are:

Hypothesis 1: A high use of ICT-devices is linked with a high number of trips and a correspondingly high share of out-door-activities. The converse also holds: A low use of ICT-devices correlates with a low number of trips and a low share of out-door-activities.

Hypothesis 2: People with a high use of ICT-devices are more flexible with respect to the time and location of their activities.

Further questions to be answered in this contribution are: (i) Assuming there is a high correlation between communication and mobility, who are the highly mobile and communicative persons? And (ii) which factors play an important role for people having a high number of trips? What is the role of ICT use in this context? To explore these questions, existing studies deliver the basis for the following hypotheses:

Hypothesis 3: Especially young and highly educated people are found within the group of the highly mobile and communicative persons.

Hypothesis 4: Important factors for people having high travel demand are gender, age and employment status. Men, young persons and employed persons are more likely to have a high travel demand. The role of ICT as an influencing factor on travel demand is more important for the younger age groups.

EMPIRICAL RESULTS

What is the Relation Between Communication and Travel Behaviour?

To explore the relation of communication and travel behaviour, three index values were constructed:

- The mobility-index describes the intensity of trip demand.
- The activity-index describes the intensity of out-door-activities.
- The ICT-index describes the intensity of ICT use.

The generation of the index values was carried out by using frequency variables. Based on a seven level scale, the participants could indicate how often they use six different modes of transport and nine different ICT-devices. For the use of ICT, a differentiation between private and job-related purposes was made. An eight level scale was used to identify the frequency of 16 out-of-home activities. These scales were transformed from an ordinal into an interval scale by extrapolating statements like 'once a week' or 'several times daily' up to a frequency number representing the sum over a total of four weeks. This way, the distance between the different groups of people corresponds more closely to the actual frequency.

To build the activity-index, the interval scaled frequencies of all 16 out of home activities were summed. The same was done with the 18 (private and job-related) variables describing the use of ICT and the six variables describing the use of transport modes. The index-values per person cannot be seen as real numbers of trips or ICT use as they are derive from questions concerning general behaviour and not from observed behaviour. But they are suitable measures for describing activity, communication or mobility behaviour with one single value.

The values of these newly built index-variables are either used as mean values to compare different groups or to classify the sample by dividing it into percentiles. The following four percentiles of different intensity of ICT use ('very high ICT use', 'high ICT use', 'low ICT use' and 'very low ICT use') provide the basis for exploring the relationship between communication and mobility behaviour and activity patterns.

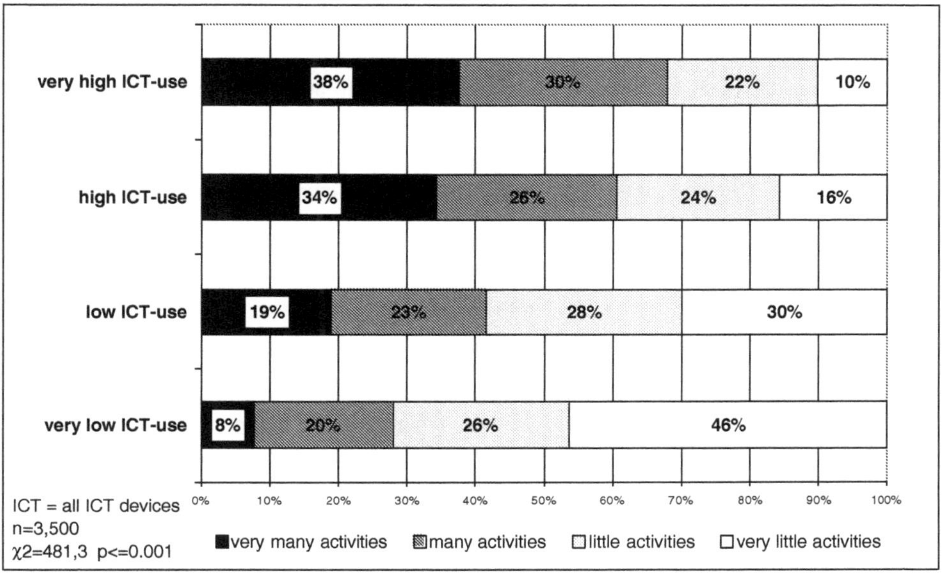

Figure 22.1

Correlation Between Number of Out-Door-Activities and ICT Use

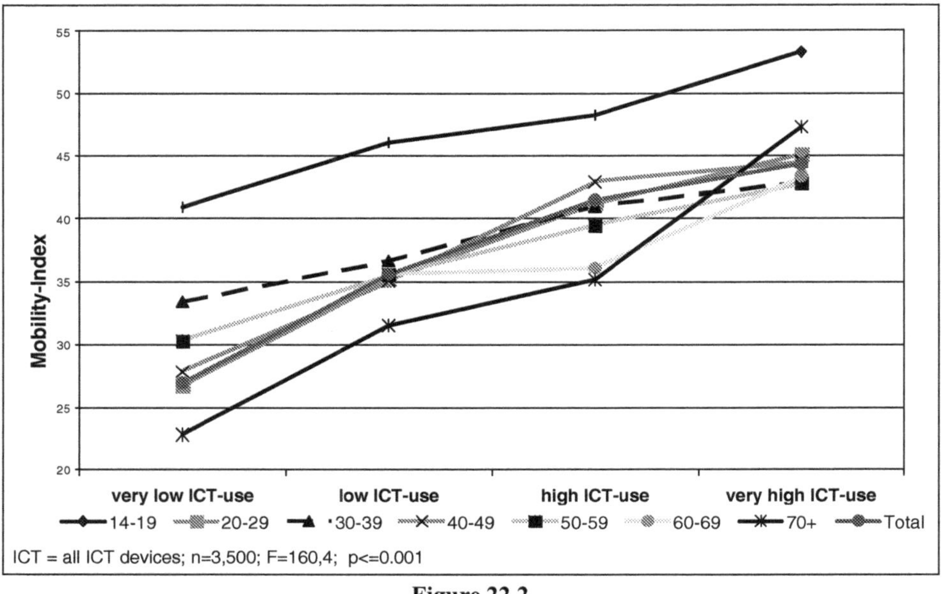

Figure 22.2

Correlation of Communication and Travel Behaviour Depending on Age

The results of the cross table (Figure 22.1) and the comparison of mean values (Figure 22.2) show clearly the strong correlation of the use of ICT and the intensity of trip demand and out-of-home activities. Those individuals performing many activities tend to be characterized by high ICT use (Figure 22.1). Whereas 68% of people with (very) high use of ICT perform (very) many activities, only 28% of the low ICT-users have a (very) high activity-index. Corresponding with this result, the higher the use of ICT devices, the higher the mobility-index (Figure 22.2). Since age has a strong influence on ICT use, a differentiation of age groups was carried out to verify whether the correlation holds within different age brackets. The results of a Chi-square test indicate that the correlation is highly significant.

Figure 22.2 demonstrates that even though the mobility level in general differs, the positive correlation between mobility and use of ICT exists for all age groups. Within each age group people who often use ICT have a much higher trip demand than people who rarely use ICT. As the 14 - 19 age group is characterised by a higher mobility level, it can be seen that even the least mobile in this group still demonstrate a higher mobility than all other age groups over the entire range of ICT-use percentiles. People older than 70 years of age are the counterpart. Within each ICT group, they have the lowest mobility-index value with one notable exception: If people in the 70+ age group exhibit very high ICT use their mobility–index is even higher than that of most other age groups. It has to be taken into account that only very few people aged 70+ are heavy ICT users (3 out of 500) and that only very few of the 14 to 19 year olds rarely use ICT (27 out of 257). Nevertheless, it appears that once belonging to the high ICT users the correlation between high mobility and communication rate exists.

Thus far, the ICT-index considers all devices at once, 'old' ICT devices like the land line telephone as well as 'new' devices like the mobile phone. The two graphs of Figure 22.3 partition the effects of ICT groups by users of the Internet and users of the mobile phone. The results are statistically significant for both devices, with p-values less than 0.001. Whereas the correlation of mobile phone use and the mobility index shows the same definite direction as those obtained for groups derived on the basis of general ICT use (all media), the correlation between Internet use and mobility is less distinct. With the exception of the youngest and the oldest age group, a slight increase in mobility level with increasing use of the Internet is evident. For young people the exact opposite is true. Those 14 to 19 year olds with very low Internet use (n=38) reach the highest mobility level. These findings demonstrate that a distinction of devices is reasonable and that it is not possible to generally talk about the effect of new ICT as their effects can differ substantially by media. One explanation for the observed discrepancies might be the fact that mobile phones can more easily be used on trips than the Internet. Therefore, it fits much better in the daily life of people with many out-of-home activities.

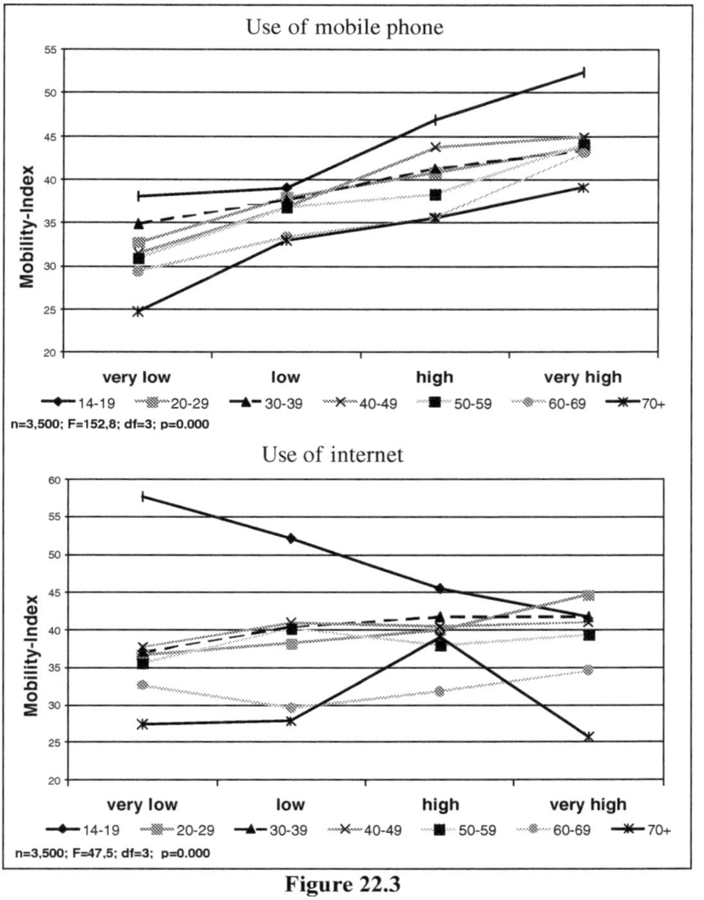

Figure 22.3

Correlation of New Media (Internet and Mobile Phone) Use and Travel Behaviour

There is a mutual relationship: people with a high mobility level possess more often a mobile phone and therefore also use it more often. The converse relation between Internet use and mobility of the youngest age group in comparison to the other age brackets is of special interest. The media often report that young people have very high Internet consumption, which may discourage out-of-home activities for this age group.

In the following, the flexibility in time and location of the performance of activities will be considered. In addition, the distances linked with shopping, work and leisure activities will be analysed as an indicator for the size of the activity space.

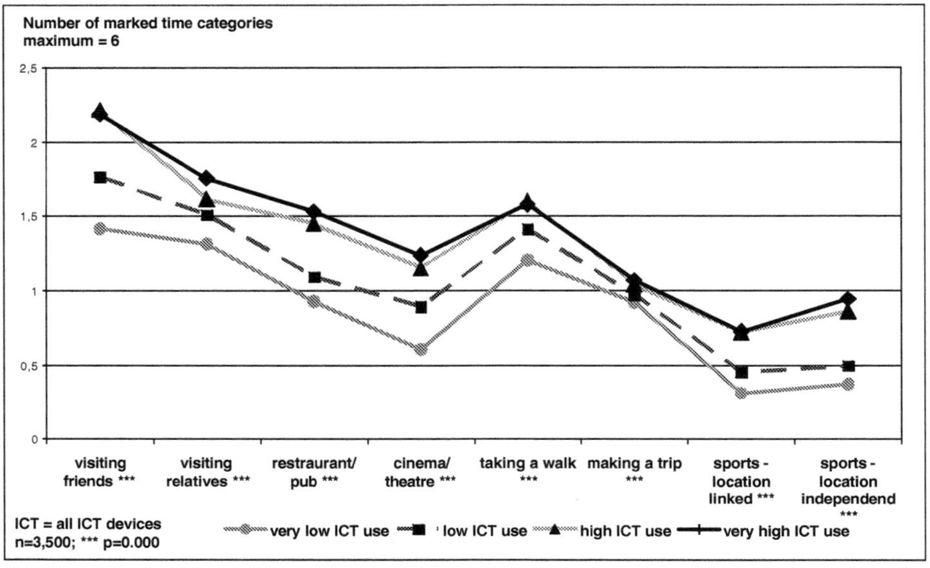

Figure 22.4
Flexibility in Time for Different Leisure Activities

To measure the temporal flexibility of the respondents, they were asked *when* they perform their activities using a multiple response question with the following six categories: on weekends during the day, on weekends at night, on Fridays during the day, on Fridays at night, from Monday to Thursday during the day and from Monday to Thursday at night. As an indicator of temporal variability the marked categories were counted. Therefore, the generated value indicates the number of marked categories of a maximum of six. Figure 22.4 shows that people with (very) high use of ICT have higher scores for all leisure activities higher scores than people with (very) low ICT use.

The higher temporal flexibility of high ICT users could also be verified by analysing self-estimations concerning the use of ICT (Figure 22.5). On a five level scale respondents could indicate whether they agree (+1 or +2), are indifferent (0) or do not agree (-1, -2) with different statements describing the effects of using the mobile phone. Considering only people using a mobile phone, all four ICT groups deny that the mobile phone causes behavioural change as indicated by negative average scores. However, The negation of the statements is less strong for high ICT users than for low ICT users. The differences in mean scores are statistically significant for all statements at the 0.1 % level based on Anova-tests. These findings differ from the findings of Nobis and Lenz (2004), who did identify a group of people with positive scores on the statements.

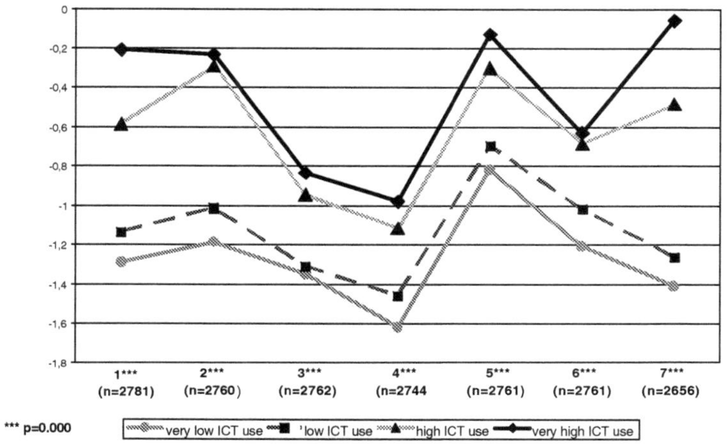

Since I have been owning a mobile phone

1 ...I complete many things while I am on the road.
2 ...I meet friends more spontaneously.
3 ...I am much more on the road because one can reach me anywhere.
4 ...I change appointments more often.
5 ...I am more relaxed concerning appointments because I can let people know when I am late.
6 ...I do not need to fix the time and place for appointments when they are made but decide on that later.
7 The internet allows me to be more flexible so that I can do things at any time of the day.

Figure 22.5
Assessment of ICT Statements Depending on Intensity of ICT Use

By running a cluster analysis on variables describing the frequency of ICT use, the authors show that people with especially high use of new media (mobile phone and Internet) agree at least with some of the statements. A second interesting result of the cluster analysis is that most of the people do not replace the land line telephone by using the mobile phone. Instead they use both media with very high frequency. To explore the variability of locations of the four ICT groups, four variables were used: the number of locations people report to visit for their shopping activities and the average distance of shopping, work and leisure trips. The results are based on general questions and not on a trip diary. The average distance e.g. of leisure trips is calculated on the basis of the respondent's information about the average distance of 16 different leisure activities. High distances are not really an indicator of variability of locations. In the context of the present study, they are still seen as a sign of a generally higher range of physical movements that at least bear a higher potential and likelihood of reaching more locations than persons with generally low average trip distances (Table 22.1). Except for the average distance of leisure trips, the mean values positively correlate with increasing use of ICT and therewith meet the expectations. In comparison with very low ICT users, very high ICT users report a higher number of places to shop. Moreover, their average distance to a shopping place is three kilometres more than that of low ICT users, while their average distance to work is four kilometres more. Only in the case of leisure activities do very low ICT users reach the longest average distance, though the high standard deviation indicates a large spread of the data. Moreover, the differences in both leisure trip distances and work trip distances are not significant. However, if one sees the average distance as a sign of variability of locations, the higher flexibility of high ICT users is at least not true for all activities.

Who are the Highly Mobile and Communicative Persons?

To verify this question, a cross-tab of the four 'communication' and 'mobility types' was produced (Table 22.2). The 16 fields of the cross-tab were collapsed into four groups: people with low ICT use and low mobility, people with low ICT use and high mobility, people with high ICT use and low mobility and people with high ICT use and high mobility. The distribution of these four groups demonstrates again that high ICT use occurs more often in combination with high intensity of mobility (31,3%) and vice versa (33,9%), whereas the two 'mixed' groups are much smaller (low ICT and high mobility: 16.1%; high ICT and low mobility: 18.6%).

Table 22.1
Number of Locations and Average Trip Distances Depending on Intensity of ICT Use

		Intensity of ICT Use				
		Very low	Low	High	Very High	F-value
Number of locations used for shopping	m	3.5	3.7	4.0	4.1	15.9***; df=3
	s	1.8	2.1	2.0	2.1	
Average distance shopping trip	m	10.7	11.0	13.2	13.8	23.8***; df=3
	s	8.5	8.8	10.5	10.2	
Average distance work trips (n=1,833)	m	12.9	16.1	16.3	17.9	1.9 ns; df=3
	s	16.1	24.2	29.2	24.9	
Average distance leisure trip	m	29.3	27.5	25.7	26.4	1.7 ns; df=3
	s	41.4	36.7	34.0	29.4	

n=3,500; m = mean; s = standard deviation; *** p<=0.001; ** p<=0.01; * p<=0.05; ns = not significant

Table 22.2
Relation Between Communication and Mobility Types

	Mobility			
	Very Low	**Low**	**High**	**Very High**
Very low ICT use	389 (11.1%)	260 (7.4%)	97 (2.8%)	101 (2.9%)
Low ICT use	251 (7.2%)	288 (8.2%)	166 (4.7%)	201 (5.7%)
High ICT use	150 (4.3%)	230 (6.6%)	231 (6.6%)	263 (7.5%)
Very high ICT use	78 (2.2%)	193 (5.5%)	321 (9.2%)	281 (8.0%)

The four groups of communication and mobility behaviour have specific socio-demographic characteristics. The differences of all tested variables turned out to be significant (Table 22.3). Briefly summarised, it can be said that the highly mobile and communicative persons tend to be men, of young age, well–educated, and with a high net-income. In addition, they are more likely to live in large communities, and they have the highest values for car ownership, household size, and the number of children below 14 living in the household. The latter observations are quite plausible. As they are the youngest of all groups, their children still live in the household. Therefore, the number of household members is higher and more of them are younger than 14. The assumption that many respondents of this group are themselves the child in the household (lowest age of respondents is 14) can be rejected as this group has the highest share of both full-time and part-time employed persons. These results indicate that the highly mobile and communicative persons are far from being only 'yuppies' living in single or couple households or very young persons still living at home.

The direct counterpart is the group of people with low mobility and low ICT use. As they have the highest share of non-working people, most of them are retired. Looking at Table 22.3, it is striking that the values are either increasing or decreasing from the left to the right except for the index of shopping and leisure activities. This suggests that people with high ICT use and low mobility resemble more the highly communicative and mobile persons, whereas people with low ICT use and high mobility resemble more the low ICT users with low mobility. It becomes obvious that ICT use is the more decisive factor to resemble the highly communicative and mobile persons than intensive mobility behaviour.

Regarding the activities of the four groups, it is plausible that the highly mobile and communicative persons as well as the highly communicative with low mobility have a higher percentage of work trips, as the share of employed persons within these groups is much higher. Consequently, in comparison to the two other groups, their share of shopping and leisure activities is lower. If instead, the absolute values of the indexes are considered, it can be seen that the highly communicative and mobile persons make more trips overall and for specific purposes, except for shopping. Since the mobility and the activity index are closely linked, it was expected that the group with low ICT use and high mobility would have a similar index value for the sum of all activities. Instead, the two 'mixed' groups, even though one having a high and the other having a low mobility index, have more or less the same values. The one of the high ICT users with low mobility is even a bit higher. This might again be a sign that ICT use is the more decisive factor than the mobility index if two groups resemble each other.

Table 22.3
Socio-Demographic Characteristics of Communication and Mobility Groups

		Low ICT Use, Low Mobility	Low ICT Use, High Mobility	High ICT Use, Low Mobility	High ICT Use, High Mobility	
Categorial/ordinal variables		**%**	**%**	**%**	**%**	**Chi-Square**
Gender	Male	40.4	51.0	51.5	54.4	χ^2=50.0***; df=3
	Female	59.6	49.0	48.5	45.6	Cramer-V: 0.12
	Total	100	100	100	100	
Employment Status	Employed full-time	11.9	21.4	52.5	59.2	χ^2=855.8***
	Employed part-time	6.7	8.1	12.9	13.5	df=6
	Not employed	81.3	70.4	34.6	27.3	Cramer-V: 0.35
	Total	100	100	100	100	
Continuous variables		Mean	Mean	Mean	Mean	**F-Value**
Age		58.0	50.3	40.7	37.4	F=369.0***
Education (Index, max.:5)		2.36	2.43	2.85	2.87	F=71.80***
Household income (Index; max.:16)		6.87	7.55	8.90	8.94	F=75.0***
Household size (max.: 4)[1)]		2.28	2.48	2.58	2.65	F=25.6***
Number of children (max.: 5)[1)]		0.20	0.27	0.37	0.41	F=19.9***
Community size of residence (max.: 5)[1)]		3.40	3.46	3.47	3.62	F=4.7**
Number of cars (max.: 3)[1)]		1.14	1.26	1.48	1.57	F=78.1***
Continuous variables		%	%	%	%	**F-Value**
Index work trips		9.0	13.0	25.9	25.1	F=514.0***
Index shopping activities		28.7	23.4	20.5	18.2	F=3.8**
Index leisure trips		62.2	63.6	53.7	56.6	F=61.5***
Index sum of all trips		100.0	100.0	100.0	100.0	F=187.7***

Table 22.4
Parameter Values of Variables Used in the Linear Regression Analysis

Variable	Definition
Gender	0 = female; 1 = male
Age	age of respondent
Education	1 = Primary/secondary school without vocation training
	2 = Primary/secondary school with vocation training
	3 = Further education without high school diploma
	4 = High school diploma
	5 = Advanced education (university college)
Number of cars	0 = no cars, 1 = one car ... 3 = three and more cars
Household size	1 = one person, ..., 4 = four and more persons
Number of children < 14	0 = no children, 1 = one child, ..., 5 = five and more children
Household net-income	1 = < 500 EUR, 2 = 500 to < 750, 3 = 750 to < 1,000, 4 = 1,000 to 1,250
	... 15 = 3,750 to < 4,000, 16 = 4,000 and more
Community size	1 = less than 5,000 inhabitants, 2 = 5,000 to less than 20,000
	3 = 20,000 to less than 100,000, 4 = 100,000 to less than 500,000
	5 = 500,000 inhabitants and more
Employment status	0 = not employed; 1 = part time or full time employed
Use of internet	Index-value; minimum = 0, maximum = 168
Use of mobile phone	Index-value; minimum = 0, maximum = 224

Influencing Factors and Role of ICT

Having described the general correlation between ICT use and mobility, and the socio-demographic characteristics and the activity patterns of the highly communicative and mobile persons, this section will take a closer look at the possible factors that determine a high mobility level. Does the use of ICT have a significant effect on mobility levels when controlling for the effects of socio-demographic variables? To answer this question, a linear regression model was estimated. Table 22.4 defines the variables used in the model, while Table 22.5 presents the standardized coefficient estimates and significance levels. Since age has a strong influence on both communication and mobility patterns, the model was estimated once for the total sample and once for each of three age groups. This differentiation was made to analyse whether the factors affecting the mobility level are distinguished by age bracket. Concerning ICT use, the presented model includes two already outlined index-variables for mobile phone and Internet use to consider their differential influences as evidenced by the foregoing descriptive analyses. The most notable result seen in the table is the positive and statistically significant coefficient of the variable measuring mobile phone use across all four of the models. Use of the Internet, by contrast, has a negative sign in each model and is significant only in the model estimated on the entire sample.

Table 22.5
Results of the Linear Regression Analysis

	14-29		30-49		50-59		Total	
	Beta	**Sig.**	**Beta**	**Sig.**	**Beta**	**Sig.**	**Beta**	**Sig.**
Gender	0,04		0,05		0,05		0,09	***
Age	-0,19	***	0,00		0,01		-0,18	***
Education	0,03		0,04		-0,03		0,01	
Number of cars	-0,09		-0,01		0,10	**	0,03	
Household size	0,13		0,04		-0,02		0,03	
Number of children < 14	-0,05		-0,01		0,00		-0,02	
Household net-income	0,00		0,01		0,06		0,02	
Community size	0,02		0,06	*	0,05		0,04	*
Employment status	0,10		0,13	***	0,10	**	0,08	***
Use of internet	-0,01		-0,03		-0,02		-0,04	*
Use of mobile phone	0,19	***	0,18	***	0,18	***	0,17	***
Constant	48,35	***	29,07	***	23,87	**	35,99	***
N	591		1,154		990		3,154	

*** p<=0.001; ** p<=0.01; * p<=0.05; ns = not significant

This suggests that the small but positive correlation between Internet use and mobility evident in Figure 22.3 for age brackets over 29 fades away when controlling for the effects of other factors. Gender, age, the number of cars, community size and employment are the other variables found to be significant in at least one of the four models estimated. With reference to the model of the total sample, the indicator for males, community size, and employed status all have positive effects on mobility, while age has a negative effect. The number of cars has a positive effect, but only for individuals in the 50 to 59 age bracket. With respect to the magnitude of the coefficients, those for mobile phone use are by far the largest. Although possible multi-collinearity complicates the direct comparison of the coefficients, the big difference between the mobile phone coefficient and all other coefficients in all models argues at least for the tendency of a high importance of mobile phone use.

SUMMARY AND CONCLUSIONS

Summarising, it can be stated that hypothesis one, two and three could be confirmed. Within all age groups a high number of trips correlates with a high ICT use. However, a distinction should be made in terms of which specific ICT device is used, since the correlation between the different devices and mobility – as presented for mobile phones and Internet – is not the same. Moreover, taking all ICT devices into account it was shown that high ICT users are more flexible at least concerning the timing of leisure activities. There are also signs of a greater variability in the location of activities, as evidenced by the higher number of shopping locations visited by high ICT-users. Finally, as stated in hypothesis three, the share of young and well-educated persons is above average within the group of highly communicative and mobile persons. However, the highly mobile and communicative tend to be employed persons living in large households with children below 14. Therefore, this group does not appear to fit the 'yuppie' class of individuals living in single or couple households or very young persons still living at home.

Hypothesis four was partially verified. The expected influence of gender, age and employment status was found to be significant. Specifically, young and employed men are more likely to have a high travel demand. However, the influence is considerably less strong than assumed. Instead, the use of Internet and the mobile phone turned out to be the strongest factors influencing the magnitude of the coefficients, not only for young persons – as expected – but also for the higher age groups.

A tentative conclusion to draw from the analyses is that high use of ICT, and the mobile phone, in particular, positively impacts mobility. Nevertheless, it should be noted that people using especially new ICT may have already been highly mobile before. Thus, the use of ICT may simply fit better their lifestyle and travel behaviour. Further investigation of the extent to which ICT use affects mobility will be possible pending the collection of panel data. With data about the same individuals over several dates, it will be possible to conduct inter-temporal and intra-personal comparisons of respondent's behaviour, allowing one to better isolate causal relationships.

Despite these limitations of this study the results allow the conclusion that there is in fact a relation between communication and mobility behaviour, but that we have to look for causalities that lie beyond this direct relation – in the person who is acting, his or her functions, his or her context. The identification of the causal relationship will be one major objective of future research.

REFERENCES

Beckmann, K. (2000). Umweltgerechtes verkehrsverhalten beginnt in den köpfen. In: *Mobilitätsforschung für das 21. Jahrhundert: Verkehrsprobleme und Lösungsansätze*, pp. 213-238. TÜV Energie und Umwelt GmbH, Köln.

Couclelis, H. (2000). From sustainable transportation to sustainable accessibility: can we avoid a new tragedy of the commons? In: *Information, Place and Cyberspace* (D.G. Janelle and D.C. Hodge, eds.), pp. 341-356. Springer Verlag, Berlin.

Hägerstrand, T. (1970). What about people in regional science? *Pap. Reg. Sc. Assoc.*, **24**, 7-21.

Krizek, K. and A. Johnson (2003). Mapping the terrain of information and communications technology (ICT) and household travel. *Proc. 82nd Ann. Meet. Transpn. Res. B.*, Washington, D.C. (CD-Rom).

Marker, J.T. and K. Goulias (1999). A framework for the analysis of grocery teleshopping. *Proc. 79th Ann. Meet. Transpn. Res. B.*, Washington, D.C. (CD-Rom).

Mokhtarian, P. (1990). A typology of relationships between telecommunications and transportation. *Transpn. Res. A*, **24**, 231-242.

Nobis, C. and B. Lenz (2004). Results of the DLR ICT and Mobility Panel. Unpublished Manuscript, German Aerospace Centre (DLR), Institute of Transportation Research, Berlin.

Salomon, I. (1985). Telecommunications and travel. *J. Transp. Econ. & Pol.*, **19**, 219-235.

Salomon, I. (1986). Telecommunications and travel relationships: a review. *Transpn. Res. A*, **20**, 223-238.

Widmer, P. and K.W. Axhausen (2001). *Aktivitäten-orientierte Personenverkehrsmodelle: Vorstudie*. Frauenfeld, Zürich.

Zumkeller, D. (1999). Verkehr und Telekommunikation – Grundlagen und Simulationsansätze. IfV – Report, 3, Arbeitsberichte des Instituts für Verkehrswesen, Universität Karlsruhe, Karlsruhe.

23

CHOICES OF ACTIVITY- AND TRAVEL-CHANGE OPTIONS FOR REDUCED CAR USE

Peter Loukopoulos, Göteborg University, Göteborg, Sweden

Tommy Gärling, Göteborg University, Göteborg, Sweden

Cecilia Jakobsson, Göteborg University, Göteborg, Sweden

Solveig Meland, SINTEF, Trondheim, Norway

Satoshi Fujii, Tokyo Institute of Technology, Tokyo, Japan

INTRODUCTION

Travel including private car use is primarily dependent on needs, desires, or obligations to participate in out-of-home activities (e.g., Root and Recker, 1983; Jones *et al.*, 1983; Axhausen and Gärling, 1992; Gärling *et al.*, 1994; Ettema and Timmermans, 1997; Vilhelmson, 1999). For this reason, any reduction or change in car use needs to be conceptualized as an adaptation by car users with potential consequences for their engagement in out-of-home activities and the satisfaction they derive from this. In order to understand such adaptations, we propose that research focuses on an in-depth understanding of the *process of change* (Pendyala *et al.*, 1997, 1998; Kitamura and Fujii, 1998; Gärling *et al.*, 2000, 2002; Arentze *et al.*, 2004; Cao and Moktharian, 2005a, 2005b). The purpose of the present chapter is to illustrate this approach by investigating actual observed changes by car users after the introduction of a new road pricing system in Trondheim, Norway. More specifically, the aim was to see if any differences could be observed in how frequently different adaptation options are chosen and whether such choices differ with increases in the size of the car-

Progress in Activity-Based Analysis edited by H. Timmermans

use change goal that has arisen as a result of the introduction of the toll ring. In the next section, we introduce a theoretical framework. We then define the purpose of the data analysis followed by a description of the data. Finally, the results are presented and discussed. The main contribution of this research project is the tentative suggestion of a cost-minimization principle that drives the process of change that is generalizable and not linked to any specific policy measure for the reduction of private car use.

THEORETICAL FRAMEWORK

Gärling *et al.* (2002) proposed a theoretical framework with the aim of analyzing the multi-facetted nature of adaptations of private car use in response to the implementation of policies designed to reduce car use (frequently referred to as Travel Demand Management (TDM) measures). In this theoretical framework, travel choices are assumed to be made of trip chains defined as bundles of attributes (purposes or activities, destinations, modes, departure and arrival times, travel times, and monetary costs). Travel choices are assumed to be influenced also by the goals individuals or households are trying to achieve. In self-regulation theory in social psychology (Carver and Scheier, 1998), such goals form a hierarchy from concrete (action programs) to abstract levels (principles) that function as reference values in nested negative feedback loops regulating ongoing behaviour or changes in behaviour. After having set car-use reduction goals in response to TDM measures, individuals or households are assumed to form plans for how to achieve the set goals and to make commitments to execute the plan. In social psychological research, this process is referred to as the formation of implementation intentions (Gärling and Fujii, 2002; Gollwitzer, 1993). The plan that is formed consists of predetermined choices contingent on specified conditions (Hayes-Roth and Hayes-Roth, 1979). In making plans for achieving the reduction goal, households consider a range of change options. It is hypothesized that individuals or households seek and select change options or adaptation alternatives that lead to the goal they have set. We do not assume, however, that this process necessarily entails a simultaneous utility-maximizing choice among all available change options. Consistent with the notion of satisficing (Simon, 1990; Gigerenzer *et al.*, 1999), it is instead claimed that options are chosen and evaluated sequentially. The results of experimental laboratory-based research (Payne *et al.*, 1993) imply that people make tradeoffs between accuracy and effort, and that such tradeoffs are frequently optimal. Yet, in contrast to microeconomic utility-maximization theories (e.g., Zwick *et al.*, 1999; McFadden, 2001), it is not assumed that people invariably invest the required degree of effort. It is assumed that both the degree in reduction or change of car use (effectiveness) and the costs of the chosen adaptations are evaluated. Costs are defined broadly, in a psychological sense, and may be expressed not only as a function of money or time but also as a function of convenience, planning, comfort, stress and so on. If the achieved degree of reduction or change differs from the set goal, more costly adaptations are chosen. Costs are assumed to be taken into account first because these are in general immediately felt, whereas effectiveness is evaluated over time on the basis of negative feedback. Thus, we propose that a sequential cost-minimizing principle dictates the initial choices of adaptation alternatives. Even though people make optimal accuracy-effort tradeoffs in laboratory experiments, we do not know whether they do so in real life when making complex travel choices. As has been noted (Gärling and Axhausen, 2003), habitual car use and related habitual and routine activities cause inertia. Research has also demonstrated that a bias exists such that the current state is overvalued (e.g., Samuelson and Zeckhausen, 1988), thus making changes less attractive. In particular, if the car-use reduction goal is vague, evaluating whether or not an adaptation alternative is effective may possibly be biased towards confirming the expectation that it is (e.g., Einhorn and Hogarth, 1978; Klayman and Ha, 1987). Furthermore, previous research has demonstrated that immediate clear feedback is essential (e.g., Brehmer, 1995). On the basis of our theoretical framework, we posit the existence of a hierarchy of adaptation alternatives that vary from less to more costly. In Table 23.1 we operationalize this hypothesis by specifying three different classes of potential adaptations together with their associated costs. An additional assumption is that adding the adaptations to each other implies increased costs. Furthermore, each adaptation may be performed to varying degrees (frequency over time), thereby also increasing its costs. We further assume that a direct relationship may frequently exist between effectiveness and costs for these adaptation alternatives. Thus, more costly adaptations are expected to be chosen when the size of the reduction or change goal increases. In addition, the number of chosen adaptations as well as their frequency should increase.

As indicated in Table 23.1, the first stage involves making car use more efficient by chaining trips, car pooling, or choosing closer destinations. The cost is an increased need to plan ahead (e.g., trip chaining may require rescheduling activities in time; car pooling may require organizing meeting times with fellow travelers; choosing closer destinations may require taking steps to find destinations closer to one's home that fulfil the functions served by the original further destination). The resulting change in car use may, however, not be sufficient to achieve a set car-use reduction goal. In a second stage, trips may also be suppressed in order to achieve greater car-use reduction; that is, the second stage involves both more efficient car use and trip suppression. In addition to increased planning, trip suppression implies changes in activities. Although in extreme cases this may necessitate lifestyle changes, the required changes are generally likely to be minor, perhaps solely involving the suppression of isolated shopping trips. Leisure activities are next most likely to be removed from the activity agenda or substituted by in-home activities. Least likely are more consequential changes in work hours. The car-use reduction goal may still not be attained unless other modes are chosen. For instance, since work cannot easily be suppressed, public transport may be chosen for such trips. Additional planning, increased time pressure, and inconveniences are possible costs associated with switching mode. That is, in the third stage, mode switching is assumed to be added to the already-performed adaptations of using the car more efficiently and suppressing certain trips. It should be noted that other possible operationalizations of the principle of sequential cost-minimizing of choices of adaptation alternatives may be proposed. In fact, survey results (Gärling *et al.,* 2000; Loukopoulos *et al*., 2004) suggest that the hypothesized hierarchy may vary with trip purpose. Furthermore, effectiveness may not increase with costs.

PURPOSE OF DATA ANALYSIS

A survey including retrospective reports of actual car-use changes or reductions in response to the introduction of a toll ring in Trondheim, Norway, provided the data utilized in the present study. A primary aim was to examine differences in how frequently different adaptation options are chosen, and whether these choices differ with increases in the size of the car-use change goal that has arisen as a result of the introduction of the toll ring. It is expected that more costly adaptation options are added to those already chosen. Ideally, a direct measure of cost of adaptation alternatives should be examined. Although such data were not collected, this may not be so problematic. The present study is still able to examine whether the number of chosen adaptations increases as the size of the reported change goal increases and whether a particular adaptation option is performed more frequently as the size of the change goal increases. In both cases, costs are assumed to increase. Failure to find such patterns would challenge the proposed change hierarchy outlined in Table 23.1. A secondary aim was to determine whether these differences generalize across trip purpose.

Table 23.1
Adaptation Alternatives and their Hypothesized Associated Costs

Adaptation	Possible Cost
More efficient car use:	
Trip chaining	Additional planning
Car pooling	
Choosing closer destinations	
More efficient car use	Additional planning
Trip suppression	Activity suppression
More efficient car use	Additional planning
Trip suppression	Activity suppression
Mode switching	Increased time pressure
	Inconveniences

DESCRIPTION OF DATA SET

The data are taken from the Trondheim Panel Travel Surveys of 1990 (pre Toll Ring) and 1992 (post Toll Ring) (Meland, 1994). The initial survey gathered information about individual and household socioeconomic characteristics, car, motorcycle and cycle ownership, public transport usage and a host of other factors, as well as trip-based information in the form of a trip diary for a given day. The second survey gathered similar information to the first, also including a travel diary. Of particular interest here, additional questions were asked specifically about the extent to which the toll ring had led to a change being made in trips to work, for shopping and for other purposes. For all three purposes, respondents indicated whether they had made "no change", "little or some change", or "substantial change" by ticking the box that best described their situation. These last two categories are henceforth labelled small and large change, respectively. Responses were interpreted as retrospective reports of the size of the change goal that had been set. It was possible for respondents to indicate different change goals for the different trip purposes. Furthermore, the questionnaire also examined which changes had been made to respondents' trips as a result of the toll ring. The examined adaptations were: change in timing, change in mode, change of destination, fewer trips, more trips with others, and change of route. Respondents indicated for each trip purpose by ticking a box whether or not they had adopted a particular adaptation alternative.

The Trondheim Toll Ring began operation in October, 1991. There were 11 toll stations forming a cordon around the city (population 138 000, 40 percent of the population living within the tolled area). The entry charge for a small vehicle was NOK 10 (NOK 1.00 ≈ EUR 0.12), and the charging period was Monday to Friday 6 a.m. to 5 p.m., with reduced charges being levied between 10 a.m. and 5 p.m. Only inbound crossings were charged. There were no charges for evening and weekend crossings. Initial contact was made via telephone at the household level, with questionnaires being distributed to all individuals in the household (although only those members over the age of 13 years received the travel diary). A total of 3087 (76.0 percent response rate) households returned completed questionnaires in 1990 and, of these households, 2780 were successfully contacted in 1992 with 1869 (67.2 percent) households returning completed questionnaires and travel diaries. Sample descriptives at the individual level are provided in Table 23.2. The analyses and results are based only on those respondents with a driving license and access to a car. Prior to discussing choice of adaptation strategies, it is important to rule out the possibility of other factors that may have occurred between the two data collection waves and that may have led to changes in observed driving patterns. Meland (1994) reports no evidence of a change in car ownership at the aggregate level (8 percent of households increased the number of cars they owned and 8 percent had decreased the number of cars), no evidence of a change in the percentage of participants listing paid work as their main occupation and a general shift towards higher income level.

Table 23.2
Sample Descriptives

Characteristics (n = 3490)	N	Descriptive
Gender (% men)	3402	48.0
Age (years) (M/SD)	3403	41.7/16.3
Tertiary education (%)	3336	76.5
Gross income, in '000 NOK (M/SD)	2658	182.0/139.9
In paid employment (excluding full-time study) (%)	3373	62.3
In full-time employment (%)	2193	78.4
Fixed work hours (%)	2140	65.3
Possession of driving license and access to a car (%)	3283	78.2
Place of work inside tolled area (%)	2030	67.5
Residence inside tolled area (%)	3339	34.1

There was no evidence that people had bought two-wheelers, which were toll exempt; in 1990, 2 percent of the respondents had access to a motorcycle or moped, with the corresponding figure being 3 percent in 1992. The average size of households participating in the surveys did not change from 1990 to 1992 (2.7 members) but 24 percent of households had either lost or gained members. While it is likely that some of those households that could not be contacted in 1992 (9 percent) had moved, it is not possible to say with certainty how many households had moved and if the move affected all members or only some. Finally, it should be mentioned that the questionnaire asked respondents to explicitly state the extent to which the toll ring had brought about any changes in their travel. Taken together, the above observations and the narrow two-year time window suggest that other factors (e.g., birth of a child, acceptance of a new job, residential move) would play a small role in explaining the changes in car-driving behaviour between 1990 and 1992.

RESULTS

Table 23.3 presents the relative frequency of the different types of adaptation for each level of degree of change (no change, small, substantial change) and for each trip purpose (work, shopping, other). The most striking feature is that most respondents did not change their travel as a result of the introduction of the toll ring. As expected, the percentage of respondents indicating that they had made no change and that they had adopted a particular adaptation alternative is very small for all trip purposes. Also as expected, this number increases from no to small change for all adaptation alternatives and trip purposes. For shopping and other trips there are also increases from small to large change. On the other hand, for work trips some changes decrease in frequency from small to large change. Change was more frequent among men and those with lower income.

Table 23.3
Degree of Change by Percentage of Respondents Choosing Each Change Option, by Gender, Mean Age, and Mean Income

	Adaptation	No Change (n = 1807)	Small Change (n = 247)	Large Change (n = 101)
Work (n = 2155)	Change in timing	0.4	13.8	13.9
	Change in mode	0.6	35.2	43.6
	Change of destination	0.1	1.6	2.0
	Fewer trips	0.4	11.7	2.0
	More trips with others	0.4	16.6	6.9
	Change of route	0.2	20.6	18.8
	Sex (% male)	52.0	58.0	66.0
	Age (years)	40.1	39.5	38.7
	Income ('000 NOK)	204.6	222.3	185.1
	Adaptation	**No Change (n = 1390)**	**Small Change (n = 673)**	**Large Change (n = 236)**
Shopping (n = 2299)	Change in timing	0.7	50.1	66.9
	Change in mode	0.1	4.5	8.5
	Change of destination	0.4	30.3	36.4
	Fewer trips	0.7	26.9	30.1
	More trips with others	0.2	2.4	1.7
	Change of route	0.5	11.9	19.9
	Sex (% male)	52.0	54.0	56.0
	Age (years)	41.5	40.8	41.2
	Income ('000 NOK)	208.2	190.8	173.1
	Adaptation	**No Change (n = 1238)**	**Small Change (n = 515)**	**Large Change (n = 149)**
Other (n = 1902)	Change in timing	0.6	49.5	66.4
	Change in mode	0.1	6.0	6.7
	Change of destination	0.1	9.7	11.4
	Fewer trips	1.0	28.0	28.2
	More trips with others	0.3	2.7	7.4
	Change of route	0.3	15.3	22.1
	Sex (% male)	52.0	56.0	55.0
	Age (years)	41.7	39.7	41.1
	Income ('000 NOK)	210.8	189.7	159.8

Table 23.4
Results of Multinomial Logit Analyses

	Adaptation	Small Change		Large Change		Difference	
		B	**SD**	**B**	**SD**	**B**	**SD**
Work (n = 2155)	Intercept	-4.21‡	.52	-4.83‡	.63	-.62	.81
	Change in timing	4.96‡	.46	4.76‡	.53	-.20	.70
	Change in mode	6.14‡	.46	6.08‡	.50	-.07	.68
	Change of destination	4.66‡	1.20	4.24‡	1.46	-.43	1.89
	Fewer trips	4.64‡	.52	2.98†	.86	-1.66*	1.00
	More trips with others	4.87‡	.45	3.77‡	.59	-1.10	.74
	Change of route	6.66‡	.75	6.60‡	.78	-.05	1.08
	Gender	.187	.26	.70**	.33	.51	.42
	Age	.01	.01	.02	.01	.01	.02
	Income	.00	.00	-.01**	.00	-.01**	.00
	Model χ^2 = 1028.23 (df = 18); Nagelkerke Pseudo R^2 = .65						
	Adaptation	**Small Change**		**Large Change**		**Difference**	
		B	**SD**	**B**	**SD**	**B**	**SD**
Shopping (n = 2299)	Intercept	-3.45‡	0.40	-5.47‡	0.52	-2.02‡	0.66
	Change in timing	5.86‡	0.37	6.79‡	0.40	0.93†	0.54
	Change in mode	22.13‡	0.00	22.70‡	0.35	0.56	0.35
	Change of destination	6.17‡	0.60	6.60‡	0.62	0.43	0.86
	Fewer trips	4.73‡	0.36	5.11‡	0.40	0.37	0.54
	More trips with others	1.92	1.22	2.10	1.36	0.18	1.83
	Change of route	3.75‡	0.49	4.45‡	0.52	0.70	0.71
	Gender	-.06	.21	.09	.27	.14	.34
	Age	.02**	.01	.03**	.01	.01	.01
	Income	.00	.00	-.01**	.00	.00	.00
	Model χ^2 = 1894.94 (df = 18) Nagelkerke Pseudo R^2 = .75						
	Adaptation	**Small Change**		**Large Change**		**Difference**	
		B	**SD**	**B**	**SD**	**B**	**SD**
Other (n = 1902)	Intercept	-2.46‡	0.37	-4.34‡	0.54	-1.89‡	0.65
	Change in timing	5.60‡	0.43	6.54‡	0.47	0.94	0.64
	Change in mode	4.89‡	1.05	4.88‡	1.13	-0.01	1.54
	Change of destination	4.99‡	1.05	4.69‡	1.12	-0.30	1.53
	Fewer trips	4.31‡	0.34	4.47‡	0.40	0.16	0.53
	More trips with others	2.53‡	0.69	3.05‡	0.81	0.52	1.06
	Change of route	4.57‡	0.54	5.07‡	0.59	0.50	0.81
	Gender	.47**	.21	.54	.29	.07	.36
	Age	.00	.01	.02	.01	.01	.01
	Income	-.01**	.00	-.01‡	.00	-.01**	.00
	Model χ^2 = 1285.57 (df = 18); Nagelkerke Pseudo R^2 = .68						

*Note: *, **, †, ‡ denote significance at the .10, .05, .01 and .001 level of significance, respectively.*

Three separate multinomial logit analyses were performed, one for each of work, shopping and other trips. Since Chi-squared values for each of the three estimated models are significant at the $p = .001$ significance level and the pseudo R^2 measures (Nagelkerke R^2) are acceptable, the null models can be rejected. Parameter estimates provided in Table 23.4 for the independent variables represent the difference between the reference category (no change) and either small change or large change. The independent variables in each of the estimated models were the six adaptation strategies (0 = not adopted, 1 = adopted), gender (0 = female, 1 = male), age, and income. In addition, differences between the obtained parameter estimates for degree of change were tested for significance, for each independent variable and each trip purpose. The objective of the multinomial logit analyses is not to develop statistical models to explain or predict the dependent variable (i.e., degree of change), but to analyze the relationship between degree of change and performance of an adaptation strategy while controlling for personal attributes. This is achieved through the examination of the coefficient estimates for each adaptation alternative, which reflect the partial correlation between an adaptation alternative and degree of change. As an example, if a significant positive coefficient were to be obtained for a given adaptation for a small change, the coefficient would indicate the existence of a positive correlation between the tendency to choose the adaptation and the tendency to change to a small extent. Note, however, that causality is not implied.

As the significance tests indicated, all adaptation alternatives are significantly related to both small and large change with the single exception of "more trips with others" for shopping trips. However, there were only a few significant differences between small and large change: for work trips "fewer trips" was significantly less frequently chosen for large than for small change; for shopping trips "change in timing" was significantly more frequently chosen for a large than for a small change. For both shopping and other trips, the increase in the number of chosen change options was significant. Of the socio-demographic factors, gender was significantly related to large change in work trips and to a small change in other trip, indicating that more men than women changed. Increases in age were also related to small and large changes in shopping trips. Income was significantly related to a large change in work trips and shopping trips, and both a small and large change in other trips. Those with higher income had changed their travel behaviour less as a result of the introduction of the toll ring.

DISCUSSION

The data analyses provide partial support for the proposed sequential cost-minimization principle. Specifically, the results for shopping trips and other trips showed that the frequency of and the number of chosen adaptations increase with the size of the change goal. Yet, only the parameter estimates for the total number and, for shopping trips, change in timing significantly increased with degree of change. The most frequently chosen adaptations were, as expected, some of those associated with more efficient car use (change in timing, change of destination), next most frequently fewer trips (trip suppression), and less frequently change of mode. However, conducting more trips with others (car pooling) was less frequently chosen. Change of route was also frequently chosen. The existence of a toll ring might have motivated some to drive around it.

Work trips differed from the remaining trip purposes in that the frequency of some adaptations decreased, most clearly "fewer trips" and "more trips with others". This resulted in an absence of increase in the number of chosen adaptations with the size of the change. Conducting "more trips with others" for work trips may be costly in terms of planning and coordination, and conducting "fewer trips" may also be costly if it implies a change in work hours or scheduling. Additionally, a change in mode or a change of route may be relatively less costly for work trips (presumably because most public transport is geared to supporting the work commute). Such changes naturally reduce the need for or even prevent some other adaptations. Thus, substitutions of adaptation alternatives are also a possibility that should be taken into account.

In sum, the data suggest that the proposed sequential cost-minimization principle is applied differently for different trip purposes. As has been indicated, the reason may be that the costs of alternative adaptation options differ, depending on trip purpose. The benefits of the present research project and the utilized framework are that generalisable principles of change are identified that are not tied to the specifics of a TDM measure, adaptation, or reduction goal. Therefore, a practical implication would be to design policies that assist the adaptation process in response to implemented TDM measures. Examples may include improvements in public transport networks, local urban planning emphasizing mixed zoning thereby making walking or cycling viable alternatives for a range of trip purposes, or the greater market penetration of technological improvements (Gärling *et al.*, 2004). Of course, the assumption behind these suggestions is that each policy assists in reducing the costs, and in increasing the effectiveness, of adaptation alternatives. However, firm conclusions cannot be drawn unless data are obtained on costs. Furthermore, it would be desirable to examine the assumption that effectiveness increases with costs.

In Loukopoulos *et al.* (2004) hypotheses derived from the sequential cost-minimizing principle were tested for data that were collected to assess stated adaptation intentions. The virtue of the present study is that it analyzes what changes people actually made. A possible limitation is, however, that information about the change goals was retrospective and, as a result, respondents may have adapted their change goals to knowledge of the actual changes and adaptations that they had made. Furthermore, respondents were self-selected to the change goals causing potential confounding with socio-demographic differences. Another limitation is that road tolls imposed the changes in car use. It is possible then that monetary costs play a more important role than would otherwise be the case. Still another limitation is that both change goals and adaptations were measured coarsely (on binary or ternary scales) which may have contributed to the lack of significant differences being detected between large and small change. Ideally, a more rigorous test would entail utilizing a scale with a finer resolution for both the dependent variable and the change options.

ACKNOWLEDGEMENT

This research was financially supported by grant #2002-00434 from the Swedish Agency for Innovation Systems and grant #25.9/2001-1763 from the Swedish Research Council for Environment, Agricultural Sciences and Spatial Planning.

REFERENCES

Arentze, T.A., F. Hofman and H.J.P. Timmermans (2004). Predicting multi-faceted activity-travel adjustment strategies in response to possible pricing scenarios using an Internet-based stated adaptation experiment. *Transp. Pol.*, **11**, 31-41.

Axhausen, K. and T. Gärling (1992). Activity-based approaches to travel analysis: Conceptual frameworks, models, and research problems. *Transp. Rev.*, **12**, 323-341.

Brehmer, B. (1995). Feedback delays in complex dynamic decision making. In: *Complex Decision Making: The European Perspective* (P. A. Frensch and J. Funke, eds.), pp. 103-130. Erlbaum, Hillsdale, NJ.

Cao, X. and P. L. Mokhtarian (2005a). How do individuals adapt their personal travel? a conceptual exploration of the consideration of travel-related strategies. *Transp. Pol.,* in press.

Cao, X. and P.L. Mokhtarian (2005b). How do individuals adapt their personal travel? objective and subjective influences on the consideration of travel-related strategies. *Transp. Pol.,* in press.

Carver, C.S. and M.F. Scheier (1998). *On the Self-Regulation of Behaviour*. Cambridge University Press, Cambridge.

Einhorn, H.J. and R.M. Hogarth (1978). Confidence in judgment: persistence of the illusion of validity. *Psych.. Rev.*, **85**, 396-416.

Ettema, D.F. and H.J.P. Timmermans (1997). Theories and models of activity patterns. In: *Activity-Based Approaches to Travel Analysis* (D. Ettema and H.J.P. Timmermans, eds.), pp. 1-36. Pergamon, Oxford.

Gärling, T. and K. Axhausen (2003). Habitual travel choice (introduction to special issue). *Transpn.*, **30**, 1-11.

Gärling, T., D. Eek, P. Loukopoulos, S. Fujii, O. Johansson-Stenman, R. Kitamura, R. Pendyala and B. Vilhelmson (2002). A conceptual analysis of the impact of travel demand management on private car use. *Transp. Pol.*, **9**, 59-70.

Gärling, T. and S. Fujii (2002). Structural equation modeling of determinants of implementation intentions. *Scand. J. Psych.*, **43**, 1-8.

Gärling, T., A. Gärling and A. Johansson (2000). Household choices of car-use reduction measures. *Transpn. Res. A*, **34**, 309 – 320.

Gärling, T., A. Gärling and P. Loukopoulos (2002). Forecasting psychological consequences of car-use reduction: achallenge to an environmental psychology of transportation. *Appl. Psych. - Int. Rev.*, **51**, 90-106.

Gärling, T., C. Jakobsson, P. Loukopoulos and S. Fujii (2004). Adaptation of private car use in response to travel demand management measures: potential roles of intelligent transportation systems. *ITS J.,* **8**, 189-194.

Gärling, T., M.-P. Kwan and R.G. Golledge (1994). Computational-process modeling of household activity scheduling. *Transpn.. Res. B*, **25**, 355-364.

Gigerenzer, G., P.M. Todd and the ABC Research Group, eds. (1999). *Simple Heuristics that Make Us Smart.* Oxford University Press, Oxford.

Gollwitzer, P.M. (1993). Goal achievement: The role of intentions. *Eur. Rev. Soc. Psych.*, **4**, 141–185.

Hayes-Roth, B. and F. Hayes-Roth (1979). A cognitive model of planning. *Cogn. Sc.*, **3**, 275–310.

Jones, P., M.C. Dix, M.I. Clarke and I.G. Heggie (1983). *Understanding Travel Behaviour*. Gower, Aldershot, UK.

Kitamura, R. and S. Fujii (1998). Two computational process models of activity-travel choice. In: *Theoretical Foundations of Travel Choice Modeling* (T. Gärling, T. Laitila and K. Westin, eds.), pp. 251-279. Elsevier, Amsterdam.

Klayman, J. and Y.-W. Ha (1987). Confirmation, disconfirmation, and information in hypothesis testing. *Psych. Rev.*, **94**, 211-228.

Loukopoulos, P., C. Jakobsson, T. Gärling, C.M. Schneider and S. Fujii (2004). Car user responses to travel demand management measures: goal setting and choice of adaptation alternatives. *Transpn. Res. D*, **9**, 263-280.

McFadden, D. (2001). Disaggregate behavioural travel demand's RUM side – A 30 years retrospective. In: *Travel Behaviour Research* (D.A. Hensher, ed.), pp. 17-63. Elsevier, Oxford.

Meland, S. (1994). Road Pricing in Urban Areas. The Trondheim Toll Ring - Results from Panel Travel Surveys, Report STF63 S94006, SINTEF, Trondheim, Norway.

Payne, J.W., J.R. Bettman and E.J. Johnson (1993). *The Adaptive Decision Maker*. Cambridge University Press, Cambridge.

Pendyala, R.M., R. Kitamura, C. Chen and E.I. Pas (1997). An activity-based micro-simulation analysis of transportation control measures. *Transp. Pol.*, **4**, 183-192.

Pendyala, R.M., R. Kitamura and D.V.G.P. Reddy (1998). Application of an activity-based travel demand model incorporating a rule-based algorithm. *Env. & Plan. B*, **25**, 753-772.

Root, G.S. and W.W. Recker (1983). Towards a dynamic model of individual activity pattern formulation. In: *Recent Advances in Travel Demand Analysis* (S. Carpenter and P. Jones, eds.), pp. 371-382. Gower, Aldershot, UK.

Samuelson, W. and R. Zeckhausen (1988). Status quo bias in decision making. *J. Risk Uncertainty*, **1**, 7-59.

Simon, H.A. (1990). Invariants of human behaviour. *Ann. Rev. Psych.*, **41**, 1-19.

Vilhelmson, B. (1999). Daily mobility and the use of time for different activities: the case of Sweden. *GeoJournal*, **48**, 177-185.

Zwick, R., I. Erev and D. Budescu (1999). The psychological and economic perspectives on human decisions in social and interactive contexts. In: *Games and Human Behaviour* (D.V. Budescu, I. Erev and R. Zwick, eds.), pp. 3-20. Erlbaum, Mahwah, NJ.

www.ingramcontent.com/pod-product-compliance
Ingram Content Group UK Ltd.
Pitfield, Milton Keynes, MK11 3LW, UK
UKHW020846290926